Nanoelectronics

Principles and Devices

For a list of recent related titles from Artech House,
turn to the back of this book.

Nanoelectronics

Principles and Devices

Mircea Dragoman
Daniela Dragoman

BOSTON | LONDON
artechhouse.com

Library of Congress Cataloging-in-Publication Data
A catalog record for this book is available from the U.S. Library of Congress.

British Library Cataloguing in Publication Data
Dragoman, Mircea, 1955–.
Nanoelectronics: Principles and Devices.
1. Molecular electronics 2. Nanotechnology
I. Title
621.3'81

ISBN-10 1-58053-694-8

Cover design by Igor Valdman

685 Canton Street
Norwood, MA 02062

International Standard Book Number: 1-58053-694-8

10 9 8 7 6 5 4 3 2 1

Contents

Preface

Nanoelectronics will be the main research area of electronics, at least in the near future. Nanoelectronics is the successor of today's microelectronics, which has produced an unprecedented revolution in communications and computing during the last 20 years. Microelectronics is based on silicon technologies developed for microscale features of electrical devices, attaining an amazing degree of efficiency, and thus producing very low-cost devices with very high performance. For example, the number of transistors contained in all DRAMs fabricated in 2002 was greater than the number of grains of rice produced in the same year, but the price of a grain of rice is one hundred times greater than that of a transistor. A complex electronic device such as a microprocessor contains 10^8 transistors that play the role of logical switches, connected by various electrical paths, which have a total length of over 20 km in a single chip, but are packed in an area of a few cm^2. Basically, in the last 25 years, the transistor dimensions were reduced by 30% every 2 to 3 years, doubling the functions of a single chip. The gate length of today's transistors is 60 nm, a value that is expected to reach 18 nm in the next 5 years and only 9 nm in the next 10 years.

However, the jump from the submicron scale to the nanoscale (1 nm = 10^{-9} m) cannot simply follow the same path; that is, the downscaling of geometrical dimensions of transistors using improved technological processes and machines. The reason is that new physical phenomena appear as soon as at least one geometrical dimension of a device is in the nanoscale range. The electrons and holes behave completely different compared to other scales since their motion is confined in tiny regions of space. As a major consequence, the diffusive transport is replaced by the ballistic transport, in which electrons travel unscattered over large distances, their behavior being rather wave-like than particle-like. The striking outcome of this manifestation is the quantum mechanical behavior of any electronic or molecular device at the nanoscale. As a result, Ohm's law and Kirchoff's law, which are the fundamentals of electronics, are no longer valid and must be replaced by the quantum mechanical Landauer formula, which links via an integral a macroscopic observable – the current – with quantum mechanical parameters, such as the transmission probability of the electron wavefunction and the Fermi distribution of carriers. Many other phenomena appear at the nanoscale, in particular tunneling in electronic devices and the action of the van der Waals and Casimir forces in mechanical systems, which require that a large part of electronics must be re-written and that the nanoscale electrical and optoelectronic devices must be based on different principles. The ballistic transport and the possibility of nanomanipulation with nanoelectromechanical systems can be fruitfully exploited to build electronic and optoelectronic devices that have no analog at larger scales. This book deals mainly with these devices.

The technologies that realize such nanoscale devices are, at present, more sophisticated and more expensive than those used for larger scales. The lithography used to process Si requires the replacement of the optical lithography currently used with particle lithography based on electrons or ions, or with UV or even X-ray lithography. All these lithographical methods are expensive and necessitate different materials for mask fabrication than optical lithography. However, complex geometries of particles or wire assemblies can be also built at the nanoscale by using simple and natural chemical or biological reactions often assisted by external electric, magnetic or electromagnetic fields. Unfortunately, the inexpensive, complex, self-assembly techniques cannot always be applied, and therefore the lithography accompanied by different deposition techniques of thin layers of semiconductors, dielectric, and metals still preserves, at the nanoscale the power gained at the microscale in the fabrication of very-large scale integrated devices, such as microprocessors and memories. But, at the nanoscale, more flexibility must be added to all these technological methods, termed *nanotechnologies*, because Si is no longer the predominant material. Silicon will still play a major role at the nanoscale, but other materials, such as carbon, AIII-BV semiconductors, and biological materials, such as DNA, play the leading role at this scale. The occurrence of biological materials in connection with semiconductor materials is a novelty encountered for the first time at the nanoscale. This happens because many biological life constituents have dimensions in the nanoscale range. For example, the dimensions of the proteins are 10^{-8}–10^{-9} m, a water molecule has a length of 1 nm, while the dimensions of cells are in the order of 10 μm.

The instrumentation for nanoscale is also specific. New tools like AFM (atomic force microscope), STM (scanning tunneling microscope), or SNOM (scanning near-field optical microscope) are used to observe nanosized devices, to fabricate and manipulate them, or even to measure their electrical or optical characteristics. The instrumentation at the nanoscale displays unprecedented sensitivity and flexibility, being able to sense a single atom and to build strange new forms of matter by manipulating single atoms and organizing them in prescribed structures. The reason is that very often MEMS (microelectro-mechanical systems), which integrate tiny mechanical movable parts, such as cantilevers terminated with nanosized tips, are combined with electrical devices and circuits, forming sensing elements.

We believe that this brief account describing the significant changes in the physical phenomena, materials, technologies, and instrumentation is convincing in our demonstration that a large part of electronics, encompassing electrical devices and circuit theory, technology, and instrumentation, must be completely rewritten when going to the nanoscale in the context of ubiquitous presence of quantum mechanical phenomena. The result is a new science termed nanoelectronics, which is the subject of this book. This science has its roots in the mesoscopic devices and their theory developed in the last years, in the AFM, STM, and SNOM techniques, and mainly in the microelectronics technology, which is further developed and enriched, having as its end result new electrical and optoelectronic devices with astounding properties and performances. Moreover, new devices, such as spin devices or biological and molecular devices, never encountered at larger scale are the best example that nanoelectronics is exploring the ultimate properties of matter for increased performances of communication, computing and life understanding.

We are now witnessing the period when various sciences are exploring and exploiting their capabilities at the nanoscale. Indeed, the physical, chemical, and biological phenomena at the nanoscale can be fully understood only by the thorough cooperation of several sciences, such as solid-state physics, quantum mechanics, chemistry, biology, and electronics. Often, this assembly of sciences is referred to as nanosciences. The prefix "nano" is often abusively used today as a prefix to many scientific words due to the billions of dollars and euros involved yearly in nanosciences. We hope that the reader will be convinced that nanoelectronics is not another nano-hoax but the leading discipline of nanosciences.

The book begins with the basic physical effects, nanotechnologies, and nanomaterials encountered in nanoelectronics. The references listed at the end of Chapter 1 will help the reader to understand in-depth the theoretical and technological facts, which are at the foundation of nanoelectronics. Chapter 2 is dedicated to the specific instrumentation used in nanoelectronics such as AFM, STM, and SNOM. Chapter 3 is focused on carbon nanotubes and the electrical devices based on them, while the optoelectronic devices based on carbon nanotubes are studied in Chapter 6. The carbon nanotubes are actually the most studied materials at the nanoscale and thus the electrical and optoelectronic devices based on them are extensively presented in the book. Chapter 4 deals with spintronics, which deals with devices that exploit a quantum mechanical property of electrons: the spin. The spin injection and manipulation are used to implement innovative nanoscale devices and possible quantum computers. Chapters 5 and 6 present the main electronic and optoelectronic devices designed for the nanoscale. Here Si still plays an important role when it is structured in the shape of nanosized dots and wires. Chapter 7 is dedicated to molecular and biological nanodevices, which begin to have an important role in nanoelectronics.

Nanoelectronics is a very young discipline since 80% of the references used in this book are works published in the last three years. Therefore, this book is only an account of the nanoelectronics today, and it is likely that in a few years we will be able to add another volume to the present one containing new and astonishing facts about the adventure of electronics at the nanoscale or even at lower geometrical scales.

Mircea Dragoman
Daniela Dragoman

Bucharest, 2005

Chapter 1

Physical Principles of Nanostructures and Nanomaterials

Nanomaterials are metallic, semiconductor, or isolator materials that have at least a nanosize dimension and, hence, produce quantum confinement of electron motion and render size-dependence of physical properties. Nanomaterials display unusual properties, not encountered in bulk materials, which will be presented in this chapter, together with the technologies needed for the fabrication of nanomaterials, and will be exploited in the remainder of the book for the implementation of nanoelectronic devices. Examples of invaluable properties of nanomaterials include the superparamagnetic behavior of nanomagnetic materials, the ability to generate a reversible metal-semiconductor transition in carbon nanotubes by adjusting their bandgap, and the bandgap tunability of nanoparticle assemblies, which is enhanced as soon as the dimensions are reduced, causing a blue shift of optical spectra.

1.1 PHYSICAL PROPERTIES OF NANOSCALE STRUCTURES

All physical properties of nanoscale structures, including the transport and heat exchange mechanisms, differ dramatically from the properties of bulk materials because quantum effects become significant. The confinement of carrier wave-functions in spatial regions with nanoscale dimensions induces a discretization of the energy spectrum of carriers and a corresponding discontinuity in the density of states. The influence of these quantum effects on the physical properties of nanoscale structures depends on the number of dimensions along which the carrier motion is confined.

In bulk materials with macroscopic scale dimensions (i.e., with dimensions larger than 1 mm) conduction electrons are scattered by impurities and phonons and thus have a random movement. The charge transport in the presence of an applied electric field has, in this case, a diffusive character, which is described by the stochastic Boltzmann equation. This equation is no longer valid when the dimensions of the sample are scaled down to nanometer sizes; the charge transport at the nanoscale is characterized by distinct features that are due to the manifestation of the coherent, wave-like behavior of charged particles. The nanoscale is sometimes called mesoscale since it refers to an intermediate scale between the macroscopic scale of bulk materials and the microscale of atoms and molecules with dimensions of the order of 1 Å = 10^{-10} m, at which the laws of atomic and molecular physics are valid.

Electron transport at the nanoscale depends on the relation between the sample dimensions and three important characteristic parameters: 1) the mean free path L_{fp}, which represents the distance that an electron travels before it suffers collisions with impurities or phonons that destroy its initial momentum, 2) the phase relaxation length L_{ph}, which is the distance after which the phase memory of electrons (the electron coherence) is lost due to time-reversal breaking processes such as dynamic scattering processes, electron-electron collisions, or impurity scattering if an internal degree of freedom of the impurity (the spin, for example) varies during the process, and 3) the electron Fermi wavelength λ_F [1].

While the number of scattering events decreases with the dimensions of the sample, and so the random character of electron motion becomes less evident, a limiting electron behavior is attained when the dimensions of the sample are smaller than both the mean free path and the phase relaxation length (sometimes called phase coherence length). In this case, the electron motion becomes collisionless and the transport regime is called ballistic; the electron wavefunction in the ballistic regime is coherent. Because of the impressive progress in semiconductor technology, and mainly in the lithography and epitaxial growth techniques (see Section 1.2), ballistic transport can be encountered in high-mobility semiconductors at low temperatures ($T < 4$ K), for which L_{fp} and L_{ph} reach tens of μm (a value of 64 μm is reported in [2] at 0.3 K). On the contrary, in polycrystalline metallic films L_{fp} is of the order of tens of nm. The coherence of electron wavefunction and the associated phenomena, such as resonant electron tunneling and Coulomb blockade, have wide applications in the design of new electronic and optoelectronic devices [3], the ballistic transport regime being essential for high-frequency operation.

1.1.1 Energy Subbands and Density of States in Nanoscale Structures

In the ballistic transport regime, in which at most a few elastic scattering processes occur, the motion of electrons with constant energy E is described by the time-independent Schrödinger equation

$$-\frac{\hbar^2}{2}\{m^{\alpha}\nabla[m^{\beta}\nabla(m^{\alpha}\Psi)]\}+V\Psi=E\Psi \tag{1.1}$$

when coupling between different electron bands is negligible (see [3] and the references therein). In (1.1) Ψ is the envelope electron wavefunction, which varies slowly over the unit cell, m is the electron effective mass, and V is the potential energy, which includes the discontinuities of the conduction band in heterojunctions, the electrostatic potential due to ionized acceptors and donors, and the self-consistent Hartree and exchange potentials caused by free carriers; the first contribution to V is predominant in the low doping case. The material-related parameters α and β satisfy $2\alpha+\beta=-1$ and take the values $\alpha=0$, $\beta=-1$ in AlGaAs semiconductor compounds, in which the ballistic transport has been primarily evidenced.

The constraints on electron motion appear in the boundary conditions associated with the Schrödinger equation. A nanoscale structure in which electrons are confined by potential barriers along the z direction in a region of width L_z but are free to move along the transverse x and y directions is called quantum well. In a quantum well with abrupt infinite-height barriers the solution of the Schrödinger equation satisfies the boundary conditions $\Psi(x,y,0)$ $\Psi(x,y,0)=\Psi(x,y,L_z)=0$; for $V=0$ this solution is $\Psi(x,y,z)=(2/L_zL_xL_y)^{1/2}$ $\times\sin(k_zz)\exp(ik_xx)\exp(ik_yy)$, with L_x and L_y the dimensions of the sample along x and y. In ballistic structures L_z is comparable to the electron Fermi wavelength and $L_z<L_x,L_y<<L_{fp},L_{ph}$. The boundary conditions induce a discrete spectrum $k_z=p\pi/L_z$ for the momentum component along z and hence discrete energy levels along the direction of constraint; the integer p labels the discrete subbands (also called transverse modes). The energy dispersion relation in the quantum well measured from the bottom of the conduction band E_c is

$$E(k_x,k_y,k_z)=E_c+\frac{\hbar^2}{2m}\left(\frac{p\pi}{L_z}\right)^2+\frac{\hbar^2}{2m}(k_x^2+k_y^2)=E_{s,p}+\frac{\hbar^2}{2m}(k_x^2+k_y^2), \tag{1.2}$$

where $E_{s,p}$ is the cut-off energy of the subband p. The energy spacing between subbands increases as the electrons become more confined, i.e., as L_z decreases.

The spin-degenerate density of states, defined by

$$\rho(E)=(2\pi)^{-3}\int_{\Sigma}\frac{dS}{|\nabla_{\boldsymbol{k}}E|_{E=const.}} \tag{1.3}$$

for any energy distribution in the $\boldsymbol{k}$ space, with Σ the surface in the $\boldsymbol{k}$ space on which $E(\boldsymbol{k})$ is constant, takes for the quantum well case the expression

$$\rho_{\mathrm{QW}}(E) = \frac{m}{\pi\hbar^2 L_z} \sum_p \vartheta(E - E_{\mathrm{s},p}), \tag{1.4}$$

where ϑ is the unit step function. Unlike in bulk semiconductors where no spatial constraint is imposed upon electron motion, the density of states in the quantum well is discontinuous. ρ_{QW} is represented in Figure 1.1.

The discrete energy levels of the quantum well in equilibrium at temperature T are occupied by electrons according to the Fermi-Dirac distribution function

$$f(E) = 1/\{1 + \exp[(E - E_{\mathrm{F}})/k_{\mathrm{B}}T]\}, \tag{1.5}$$

where E_{F} is the Fermi energy level and k_{B} the Boltzmann constant, the equilibrium electron density per unit area of the quantum well being given by [4]

$$n = L_z \int_0^{\infty} \rho_{\mathrm{QW}}(E) f(E) dE = k_{\mathrm{B}}T \frac{m}{\pi\hbar^2} \sum_p \ln[1 + e^{(E_{\mathrm{F}} - E_{\mathrm{s},p})/k_{\mathrm{B}}T}]. \tag{1.6}$$

The Fermi-Dirac distribution function becomes proportional to $\vartheta(E_{\mathrm{F}} - E)$ at low temperatures or in the degenerate limit, when $k_{\mathrm{B}}T \ll E_{\mathrm{F}}$, so that in these conditions all electron subbands above the Fermi energy are empty and all subbands below it are filled up. Electrons with energy E occupy at low temperatures a number of subbands $M(E)$ obtained by counting the number of transverse modes with cut-off energies smaller than E.

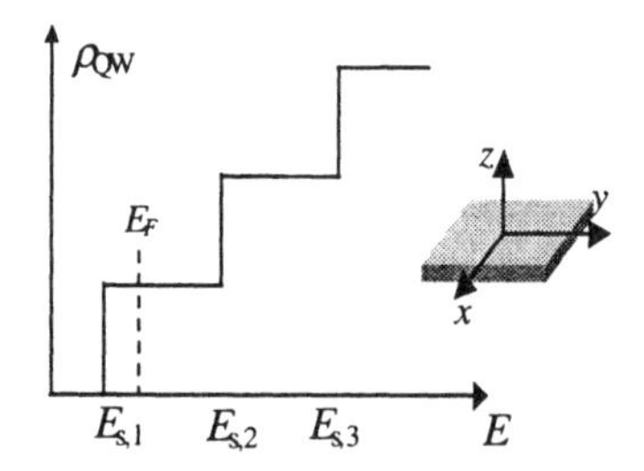

Figure 1.1 Density of states in a quantum well. The position of the Fermi energy corresponds to two-dimensional electron gas systems.

When the Fermi level in a quantum well is located between the first and the second energy subband, as shown in Figure 1.1, the Fermi wavenumber k_{F} is defined by the kinetic energy of electrons as $E_{\text{kin}} = E_F - E_{s,1} = \hbar^2 k_F^2 / 2m$, and becomes related to the electron density per unit area $n = (m / \pi\hbar^2)(E_F - E_{s,1})$ through [4]

$$k_F = (2\pi n)^{1/2} . \tag{1.7}$$

The Fermi wavelength is related to the electron kinetic energy via $\lambda_F = 2\pi / k_F$. This case corresponds to a two-dimensional electron gas (2DEG), which behaves like a metal since E_F is inside the conduction band.

A quantum wire is a nanoscale structure in which the electron motion is constrained by energy potentials along two directions: y and z, but is free along x. For infinite-height potentials the electron wavefunction is given by $\Psi(x, y, z) = [2/(L_y L_z L_x)^{1/2}] \sin(k_y L_y) \sin(k_z L_z) \exp(ik_x x)$ and similar boundary conditions as in the quantum well case require that $k_y = p\pi / L_y$, $k_z = q\pi / L_z$, with p, q integers. The energy dispersion relation becomes now

$$E(k_x, k_y, k_z) = E_c + \frac{\hbar^2}{2m}\left(\frac{p\pi}{L_y}\right)^2 + \frac{\hbar^2}{2m}\left(\frac{q\pi}{L_z}\right)^2 + \frac{\hbar^2 k_x^2}{2m} = E_{s,pq} + \frac{\hbar^2 k_x^2}{2m} \tag{1.8}$$

and the corresponding density of states is given by

$$\rho_{\text{QWR}}(E) = \frac{(2m)^{1/2}}{\pi\hbar L_y L_z} \Sigma_{p,q} (E - E_{s,pq})^{-1/2} . \tag{1.9}$$

This energy-dependent density of states is represented in Figure 1.2(a).

In analogy to optical waveguides, the quantum wires for which $L_y, L_z \cong \lambda_F$ and $L_y, L_z < L_x << L_{\text{fp}}, L_{\text{ph}}$ are also called electron waveguides when the energy separation between different subbands is larger than both the thermal energy $k_B T$ and the eventual potential drop eV along the waveguide, with V the applied bias [5]. In modulation-doped AlGaAs/GaAs heterostructures at temperatures lower than 4 K this condition is satisfied for $V < 1$ mV and $L_y, L_z \cong 0.1$–0.5 μm. Natural electron waveguides are carbon nanotubes, which will be treated in Chapter 3.

In a quantum dot, the electron motion is spatially confined along all directions in regions with dimensions much smaller than the mean free path and the phase relaxation length, the discrete energy spectrum

$$E(k_x, k_y, k_z) = E_c + \frac{\hbar^2}{2m}\left(\frac{p\pi}{L_x}\right)^2 + \frac{\hbar^2}{2m}\left(\frac{q\pi}{L_y}\right)^2 + \frac{\hbar^2}{2m}\left(\frac{r\pi}{L_z}\right)^2 = E_{s,pqr} , \tag{1.10}$$

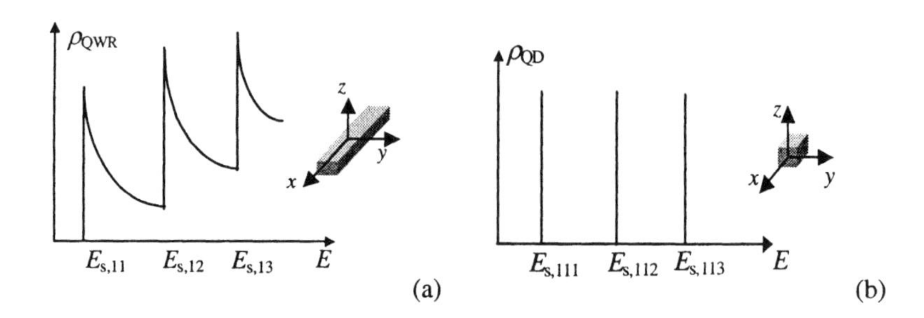

Figure 1.2 Density of states in: (a) a quantum wire, and (b) a quantum dot.

imposing a density of states proportional to the Dirac function:

$$\rho_{\mathrm{QD}} \propto \delta(E - E_{\mathrm{s},pqr}) . \tag{1.11}$$

ρ_{QD} is shown in Figure 1.2(b). Because the discrete energy spectrum of quantum dots resembles that of atoms or molecules, quantum dots are regarded as artificial atoms.

1.1.2 Electron Transport in a Two-Dimensional Electron Gas

2DEG systems arise, for example, in modulation-doped GaAs/AlGaAs heterojunctions, at the interface between intrinsic GaAs and *n*-doped AlGaAs layers. The extremely low electron scattering rate in a 2DEG is assured by the spatial separation of the free electrons that form an inversion layer on the GaAs side of the interface from dopant atoms in AlGaAs. According to (1.7) the electron density in the 2DEG can be modified by varying the electron kinetic energy through the application of negative electrostatic voltages on Schottky surface gates located in close proximity to the 2DEG. The negative bias depletes the 2DEG and modifies the electron wavenumber beneath the gate and laterally from its geometric edge. For a certain depletion bias, no free electrons are left, and the gate becomes a barrier for electrons.

In a 2DEG at low temperatures, the transport of free electrons is ballistic, which means that the electron wavefunction is coherent and given by (1.1), and all electrons have the same wavenumber k_{F}. The coherence of the electron wavefunction is a unique feature of nanoscale devices that can be used, in connection with the possibility of manipulating the electron wavenumber in 2DEGs via gate voltages, to implement electron counterparts of geometrical optical systems or interference devices. Electrons in ballistic systems propagate in a similar way to rays in classical optics [1, 6].

The analogy between electromagnetic wave propagation and electron propagation in ballistic conductors is based on the formal similarity between the Helmholtz equation $\nabla^2 F + k^2 F = 0$ satisfied by any electromagnetic field component F and the time-independent Schrödinger equation for the electron wavefunction. In particular, at an interface between two 2DEG systems with electron densities n_1 and n_2 ballistic electrons refract according to Snell's law: $\sin\theta_1 / \sin\theta_2 = (n_2 / n_1)^{1/2}$, where θ_i is the angle between the electron trajectory in region i and the normal to the interface. Similar to classical optics, the wavefunction of ballistic electrons is partly transmitted and partly reflected at the interface, and coherent parts of the same electron beam can interfere after being reunited. Moreover, by exploiting the similarity between the equation satisfied by the scalar electron wavefunction $m^{\alpha}\Psi$ and different components F of the vectorial electromagnetic field if $[2m(E-V)]^{1/2}/\hbar$ is replaced by the wave-number k, as well as the analogies between the respective boundary conditions, different sets of quantitative analogies between electron and light parameters (energy, potential energy, effective mass, and frequency, electric permittivity, magnetic permeability, respectively) can be derived. For details, see [6].

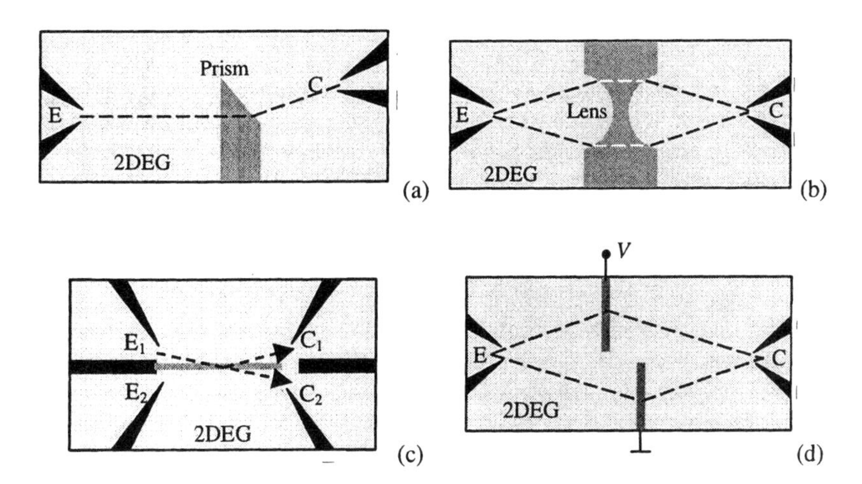

Figure 1.3 Schematic representation of (a) an electron prism, (b) a refractive electron lens, (c) a beam splitter, and (d) an interferometer. The electrons are emitted from the narrow orifice E and are collected by the collector C. Dashed lines represent electron trajectories (*After*: [6]).

As an example of ballistic electron propagation, we have represented in Figures 1.3(a)-(d) a prism, a refractive lens, a beam splitter, and an interferometer for ballistic electrons, respectively; their operating principles are similar to that of the corresponding optical devices and can be followed by tracing the electron trajectories. The electron sources with wide angular spectrum, denoted by E in Figure 1.3, are narrow constrictions in the 2DEG delimited by totally depleting gates, depicted as black lines, while partially depleting gates that refract electron waves are represented in Figure 1.3 by gray electrodes. The narrow constrictions C act as electron collectors. In Figure 1.3(d) the applied voltage V changes the relative phase between the electron beams that interfere at the collector.

1.1.3 Resistance of a Ballistic Conductor

Unlike in the diffusive transport regime, which involves electrons with a wide-spread energy distribution, in the ballistic regime only electrons with energies close to E_F participate at transport. However, when a one-dimensional (1D) ballistic conductor is placed between two contacts that act as electron reservoirs, an external bias V, besides inducing net electron transfer, drives the electrons in the ballistic conductor away from equilibrium. In this case, there is no common Fermi energy level, but, instead, a spatially varying local quasi-Fermi level can be defined, which equals E_{FL} and E_{FR} in the left and right contacts, respectively. A 1D ballistic conductor is a quantum wire. For reflectionless contacts and for a small bias $eV = E_{FL} - E_{FR}$, where it is assumed that $E_{FL} > E_{FR}$, there is current flow at zero temperature only in the electron energy range $E_{FR} < E < E_{FL}$. Moreover, for a ballistic conductor with a constant cross-section, in which there is no scattering of electrons from one subband to another, each occupied subband contributes to the net current $I = ev\delta n$. Here $\delta n = (dn/dE)eV$ is the extra density of electrons in the left contact and $v = \hbar^{-1}(dE/dk)$ is the electron velocity along the current flow direction. Then $I = (2e^2/h)MV$ if the number of subbands $M(E)$ is constant over $E_{FR} < E < E_{FL}$, the corresponding conductance [4]

$$G = I/V = 2e^2M/h \tag{1.12}$$

is an integer multiple of the so-called conductance quantum $G_0 = 2e^2/h$, and the resistance is given by $R = 1/G = 1/(MG_0) \cong 12.9\,\text{k}\Omega/M$. In the absence of collisions, the resistance can only originate in the mismatch at the conductor/contact interface between the infinite number of transverse modes carried by contacts and the finite number of transverse modes in the ballistic conductor. For this reason, R is called contact resistance and its value decreases from the quantum value $R_0 = 12.9\,\text{k}\Omega$ when the number of energy subbands in the ballistic conductor increases. Note that, unlike in bulk materials, where Ohm's law implies that the conductance is inversely proportional to the length of the sample, the conductance of ballistic structures is independent of the length of the conductor. G depends, however, on the width W of the 1D conductor because the number of subbands occupied by electrons propagating with the Fermi wave-number is $M \cong \text{Int}[k_F W / \pi]$, where Int symbolizes the integer value.

This expression has been demonstrated experimentally, the measured conductance of a ballistic conductor with a variable width depending on the number of occupied subbands as shown in Figure 1.4(a). The conductance at low temperatures increases in steps of G_0 each time M increases with one unity, this behavior being "smoothed" by thermal motion as the temperature increases. A 1D ballistic quantum wire can be delimited from a 2DEG using a split-gate geometry, represented in Figure 1.4(b) [5]. In this geometry a narrow slit, with a width W comparable to λ_F, is cut in a depleting metallic gate that is patterned on top of the 2DEG. As the effective width W of the conductor decreases continuously by applying an increasingly negative gate voltage V, the number of modes M varies in steps. The split-gate structure is commonly referred to as quantum point contact if the length of the constriction is comparable to its width, and is called electron waveguide if its length is much larger than W.

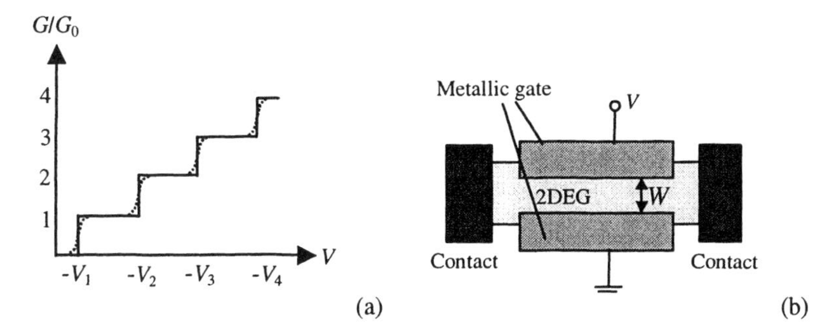

Figure 1.4 (a) The dependence on the gate voltage at zero temperature (solid line) and finite temperatures (dotted line) of the conductance of (b) a 1D ballistic conductor with variable width.

1.1.4 Landauer Formula

The implicit assumption made in the derivation of (1.12), namely that all electrons injected by the left contact arrive to the right contact, is not always satisfied. In particular, if the ballistic conductor consists of several parts with different widths or potential energies, the electrons from one contact are only partially transmitted to the other. If T denotes the transmission probability of the ballistic conductor connected to reflectionless contacts by ballistic leads, as represented in Figure 1.5, then the conductance that would be measured between the contacts at zero-temperature is given by the so-called Landauer formula [1]

$$G = \frac{2e^2}{h} MT , \tag{1.13}$$

where M is the number of transverse modes in the leads. The current between the contacts is $I = (2e^2 / h)MTV$.

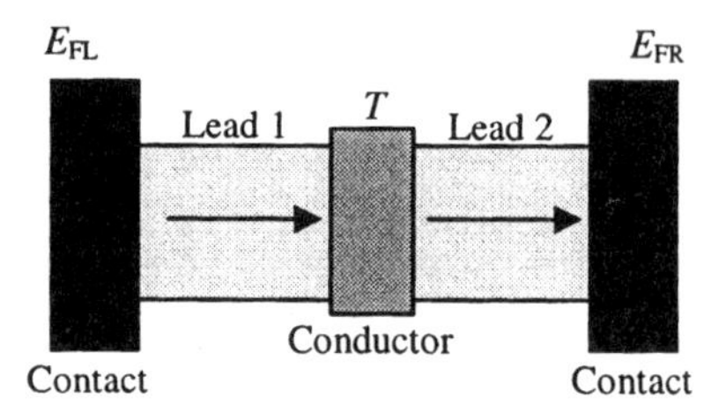

Figure 1.5 Model of a ballistic conductor.

Then, the total resistance between the contacts, $R = h/(2e^2MT)$, can be expressed as a sum between the contact resistance $h/(2e^2M)$ and the resistance of a scatterer with transmission T, $R_s = h(1-T)/(2e^2MT)$. Analogously, the resistance of several scatterers with transmissions T_i connected in series is given by $R_s = \sum_i R_{s,i}$, where $R_{s,i} = h(1-T_i)/(2e^2MT_i)$. This formula is equivalent to writing the total transmission probability as $(1-T)/T = \sum_i (1-T_i)/T_i$, expression that can be alternatively found by adding the successive partial transmitted waves.

At higher temperatures, when the Fermi-Dirac distribution $f(E)$ can no longer be approximated with a step function, the electrons that participate at transport have energies in the range $E_{FR} - \Delta E < E < E_{FL} + \Delta E$, where ΔE is of the order of a few $k_B T$. Then, the net current between the left and right contacts, characterized by Fermi-Dirac quasi-distribution functions $f_L(E)$ and $f_R(E)$, respectively, is [1]

$$I = \frac{2e}{h} \int M(E)T(E)[f_L(E) - f_R(E)]dE . \tag{1.14}$$

This expression can be generalized to account for the existence of several contacts or terminals by modeling the zero-external current terminals, such as those used to measure the voltage drop along a conductor (see Figure 1.6), as scatterers characterized by transmission probabilities. In this case, the net current flow through the pth terminal at finite temperatures is given by [1]

$$I_p = \frac{2e}{h}\int \sum_q [\overline{T}_{qp}(E) f_p(E) - \overline{T}_{pq}(E) f_q(E)] dE \,, \tag{1.15}$$

where $G_{pq} = (2e^2/h)\overline{T}_{pq}$ is the conductance associated to current flow from terminal q with the quasi-Fermi energy level $E_{Fq} = eV_q$ to terminal p with the quasi-Fermi level $E_{Fp} = eV_p$, $\overline{T}_{pq}$ being a product between the number of modes M and the transmission probability per mode T_{pq}. Because at equilibrium there is no net current flow, the conductances satisfy the relation $\sum_q G_{qp} = \sum_q G_{pq}$. The zero-temperature counterpart of (1.15) is $I_p = (2e^2/h)\sum_q [\overline{T}_{qp}V_p - \overline{T}_{pq}V_q] = \sum_q [G_{qp}V_p - G_{pq}V_q]$, known as the Büttiker formula.

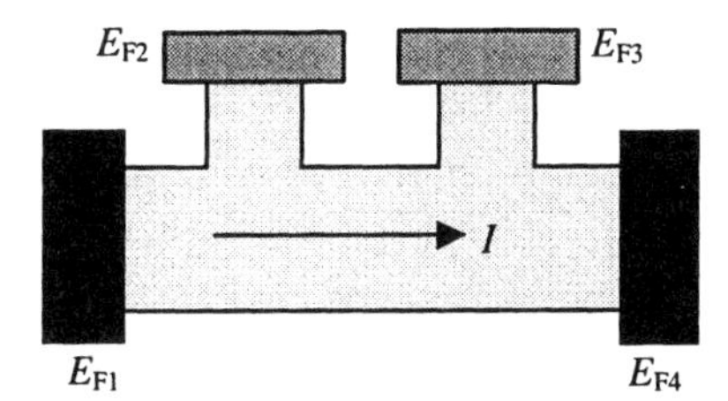

Figure 1.6 Schematic representation of a four-terminal configuration in which the current flows between terminals 1 and 4, terminals 2 and 3 being used to measure the voltage drop along the conductor.

1.1.5 Transmission Probability Calculation

In the previous section, we have seen that both the conductance and the net current through a 1D ballistic conductor connected to reflectionless contacts by ballistic leads is determined by the transmission probability between the leads. Although there are many methods to calculate the transmission probability [1, 4], which include the Green's function approach, the transfer Hamiltonian formalism or the Kubo formalism, the easiest method is the matrix formalism.

In the simplest situation, when the 1D conductor consists of (or can be approximated as a succession of) several regions with different but constant potentials and electron effective masses, the solution of the Schrödinger equation (1.1) in the ith region, $\Psi_i(x) = A_i \exp(ik_i x) + B_i \exp(-ik_i x)$, is a superposition of forward- and backward-propagating waves with wavenumber $k_i = \hbar^{-1}\sqrt{2m_i(E-V_i)}$ [4]; the x direction is the direction of electron propagation. The wavefunction and its x-derivative are continuous inside each layer, while at each interface $x = x_i$ between layers i and $i+1$, represented in Figure 1.7, the wavefunction and $\nabla\Psi \cdot \hat{x} / m^{\alpha+1}$, with $\hat{x}$ the unit vector along x, must be constant.

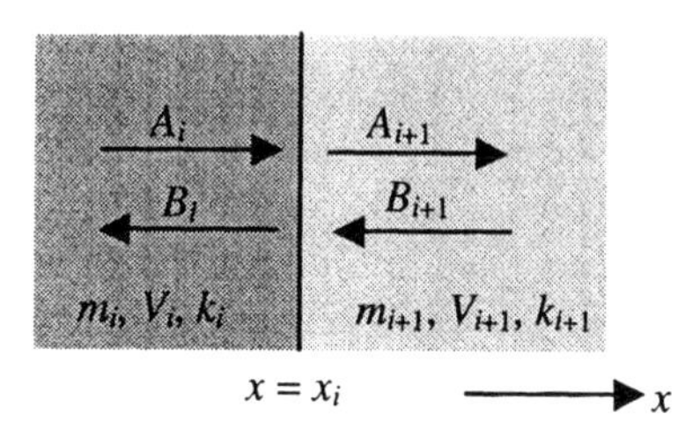

Figure 1.7 The forward- and backward-propagating components of the electron wavefunction at the interface between layers i and i + 1.

For $\alpha = 0$ these continuity conditions relate the wavefunction components on each side of the interface through a transfer matrix

$$\begin{pmatrix} A_i \exp(ik_i x_i) \\ B_i \exp(-ik_i x_i) \end{pmatrix} = \frac{1}{2}\begin{pmatrix} (1+v_{i+1}/v_i) & (1-v_{i+1}/v_i) \\ (1-v_{i+1}/v_i) & (1+v_{i+1}/v_i) \end{pmatrix}\begin{pmatrix} A_{i+1}\exp(ik_{i+1}x_i) \\ B_{i+1}\exp(-ik_{i+1}x_i) \end{pmatrix}, \qquad (1.16)$$

with $v_i = \hbar k_i / m_i$ the electron velocity in the ith layer, the corresponding transfer matrix for free propagation across the ith layer, between planes $x = x_{i-1}$ and $x = x_i$ being diagonal, with elements $\exp[ik_i(x_i - x_{i-1})]$ and $\exp[-ik_i(x_i - x_{i-1})]$. Then, the transmission probability for a structure composed of N layers, $T = v_N |A_N|^2 / (v_1 |A_1|^2) = v_N / (v_1 |M_{11}|^2)$, is simply related to the element M_{11} of the total transfer matrix with elements M_{pq}, p,q = 1,2, which is obtained by multiplying the matrices for each interface and across each layer.

Besides the transfer matrix in (1.16), another matrix can be defined that relates the outgoing amplitudes $B_i \exp(-ik_i x_i)$, $A_{i+1}\exp(ik_{i+1}x_i)$ to the incoming amplitudes $A_i \exp(ik_i x_i)$, $B_{i+1}\exp(-ik_{i+1}x_i)$ on either side of the interface. This matrix is known as the scattering matrix, because its elements for a succession of N layers are related to the reflection and transmission amplitudes of the structure from left to right and right to left, respectively, r, t, and r', t', as $S_{11} = r$, $S_{12} = t'$, $S_{21} = t$, $S_{22} = r'$. The scattering matrix is antidiagonal for free propagation across the ith layer, with elements $\exp[-ik_i(x_i - x_{i-1})]$ and

$\exp[ik_i(x_i - x_{i-1})]$. Unlike the transfer matrix in (1.16) the scattering matrix is unitary [1], property that reflects current conservation across the structure. The S matrix can be used as an alternative to (1.16) to calculate the transmission probability through the layered medium as $T = v_N |S_{21}|^2 / v_1$.

A more complicated expression of the transmission probability occurs in ballistic conductors with variable cross-sections (see [4], [6], and the references therein), since in this case the transverse mode matching condition for the electron wavefunction at different interfaces between regions with different wavenumbers k_i and different number of transverse modes M_i generates scattering between transverse modes. However, the procedure for determining the transmission probability is the same as before: it follows from the continuity condition imposed on the electron wavefunction, which in the ith layer is $\Psi_i(x, y, z) = \sum_{k=1}^{M_i} [A_{ik} \exp(ik_i x) + B_{ik} \exp(-ik_i x)] \phi_{ik}(y, z)$, with $\phi_{ik}(y, z)$ the transverse part of the kth mode in the ith layer.

1.1.6 Electron Tunneling

The transmission probability calculation method described in the previous section is valid for either real or imaginary values of the wavenumbers k_i. The wavenumber is imaginary if the electron energy is smaller than the potential energy value, the electron propagation being forbidden from a classical point of view. The layer in which the wavenumber is imaginary acts as a barrier for electron propagation, the electron wavefunction decaying exponentially inside this region, similar to the evanescent propagation of electromagnetic waves. The transmission probability across such a layer vanishes unless its width is small enough. The quantum phenomenon of electron propagation with constant energy E through thin potential barriers (in general through a succession of barriers separated by quantum wells, that is, by regions with real wavenumbers) is called tunneling. Electron tunneling, which necessitates low temperatures that prevent thermal excitation and ballistic structures (no inelastic scatterings), is pivotal in modern electronic and optoelectronic devices [3].

Classical transport across a barrier is only allowed if extra energy is provided to the electron, the process in which this extra energy is of thermal nature being called thermionic emission; thermionic emission accompanies tunneling at finite temperatures and becomes the predominant transport mechanism across barriers at high temperatures [7]. The thermionic emission contribution of the net current across a rectangular barrier of height ϕ depends on temperature as $I \propto T^2 \exp(-\phi / k_B T)$.

In tunneling devices the direction x of electron propagation is that of an applied electric field, the mathematical treatment in Section 1.1.5 being valid only if Ψ is the x-dependent part of the envelope electron wavefunction and if the electron motion in the transverse direction (along y and z directions) can be separated from that along the longitudinal direction (along x). Then, the transmission probability through the simplest tunneling structure shown in Figure 1.8 and composed of a single barrier region for electron propagation, in which $k_2 = -i\gamma_2$ is purely imaginary, surrounded by regions labeled by 1 and 3 with real propagation constants is given by

$$T = \frac{4v_1 v_3}{(v_1 + v_3)^2 + [(v_1^2 + v_2^2)(v_2^2 + v_3^2)/v_2^2]\sinh^2(\gamma_2 L)}, \tag{1.17}$$

where $v_2 = \hbar\gamma_2 / m_2$. The transmission probability is generally defined as the ratio between the incident and transmitted electron probability currents, defined in any j layer as $J_j = (\hbar / 2m_j i)(\Psi_j^* \partial_x \Psi_j - \Psi_j \partial_x \Psi_j^*)$, where * stands for complex conjugation and ∂_x is a shorthand notation for partial derivation with respect to x.

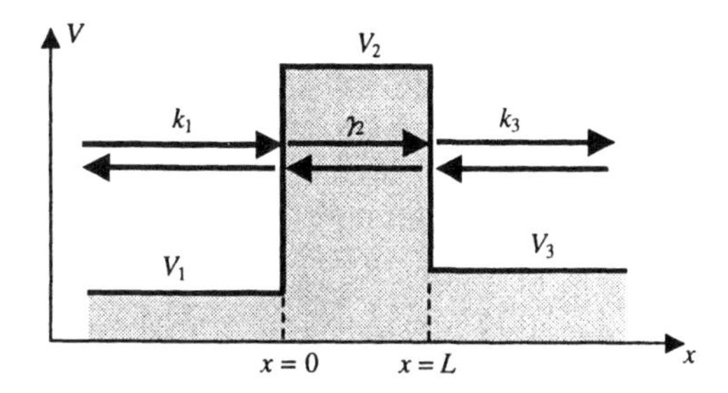

Figure 1.8 Geometry of a single-barrier tunneling structure.

Expression (1.17) shows that for $\gamma_2 L >> 1$ the transmission probability depends exponentially on the width of the barrier L: $T \propto \exp(-2\gamma_2 L)$. This formula can be replaced by $T \propto \exp[-2L\sqrt{2m_2(V_2 - E_F)} / \hbar]$ at low temperatures, when all electrons that take part in transport have energies close to E_F [8]. At low temperatures, the current density through the structure has the same exponential dependence on L as the transmission probability, since it is proportional to T, and the conductance is much smaller than G_0.

The study of electron transmission through a single barrier is significant for scanning tunneling microscopy [9], in which electrons tunnel (usually through the vacuum) from the tip of the scanning probe to the surface under study. Scanning tunneling microscopy, which will be detailed in Chapter 2, is a technique that can be used to map solid surfaces with atomic resolution, to manipulate the surface structure, or to analyze single molecules on surfaces via spectroscopy methods.

Tunneling through a single barrier is also relevant to vacuum microelectronic devices, which include electron sources for microscopes and flat panel field emission displays, in which electrons tunnel from a solid surface into vacuum under the influence of a high electric field [8]. A nanoscale device based on a combination of electron selective tunneling and thermionic emission into vacuum under a small external voltage (of a few V) can also be used in certain conditions as a nano-refrigerator, for cooling applications [10].

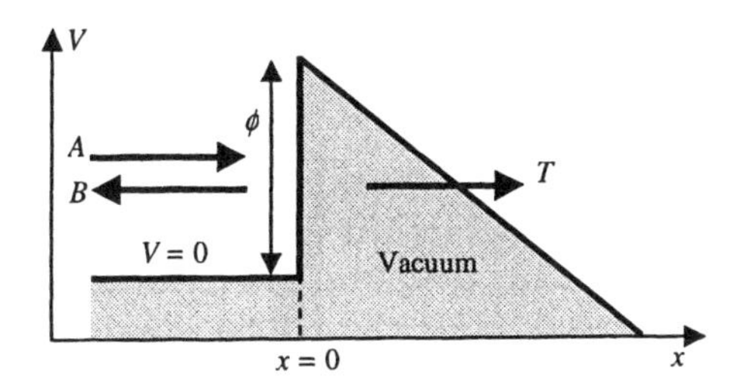

Figure 1.9 Geometry of electron emission in vacuum via tunneling through a triangular barrier.

However, (1.17) is not appropriate to describe the transmission coefficient and the low-temperature current through field emission devices since the potential energy at high electric fields can be better approximated by a triangular rather than a rectangular shape; a different dependence of T on the height and width of the potential barrier are expected in this case. In the vacuum region the barrier determined by the electron affinity ϕ decreases with the applied electric field F as $\phi - eFx$ (see Figure 1.9). The electron wavefunction in vacuum is a solution of the Schrödinger equation (1.1) with a triangular barrier, and for $\alpha = 0$ it is a combination of Ai and Bi Airy functions: $\Psi_{vac}(\xi) = A_{\mathrm{vac}}\mathrm{Ai}(\xi) + B_{\mathrm{vac}}\mathrm{Bi}(\xi)$, with $\xi = [2m_0/(eF\hbar)^2]^{1/3}(\phi - eFx - E)$ and m_0 the electron mass in vacuum. The condition imposed on the electron wavefunction, that of current flow along the positive x direction, imposes that $A_{\mathrm{vac}} = 1$, $B_{\mathrm{vac}} = i$, the transmission coefficient T across the barrier being again given by the ratio between the incident and transmitted electron probability currents. The result is

$$T = \left(\frac{2eF}{\hbar^2 m_0^2}\right)^{1/3} \frac{m}{\pi} \frac{|A|^2}{k}, \tag{1.18}$$

if the electrons with mass m and energy E incident on the barrier have a wavefunction $\Psi(x) = A\exp(ikx) + B\exp(-ikx)$ with $k = (2mE/\hbar^2)^{1/2}$. The A and B

coefficients are found from the boundary conditions for the wavefunction and its derivative at $x = 0$.

In the limit of high electric fields, low temperatures and $m = m_0$ the expression (1.18) for the transmission coefficient leads to a current density of the form

$$J = \frac{em_0 k_B T}{\pi^2 \hbar^3} \int T(E) \ln\left[1 + \exp\left(\frac{E_F - E}{k_B T}\right)\right] dE \propto F^2 \exp\left(-\frac{4}{3\hbar F}\sqrt{2m_0 \phi^3}\right), \quad (1.19)$$

where the logarithmic term accounts for the contribution to the Fermi-Dirac distribution function of the transverse degrees of freedom. Formula (1.19) is known as the Fowler-Nordheim equation, and describes the current-voltage characteristic of field-emission devices, although image charge corrections and many-body effects need to be taken into account in some situations [11].

1.1.7 Resonant Tunneling Devices

When electrons tunnel through a structure that contains two (or more) barriers separated by quantum well regions, the coherent nature of ballistic electron propagation generates constructive or destructive interference between the waves that are partially reflected and transmitted at the interfaces. Therefore, in a geometry such as that represented in Figure 1.10, in which a quantum well with width L in which electrons propagate freely with a wavevector k is surrounded by two thin barriers, high and low values of the transmission probability are expected, similar to the appearance of high and low intensity values in interference between coherent light beams. Resonant tunneling refers to the case of a large transmission probability through a structure containing two (or more) barriers.

The transfer matrix theory in Section 1.1.5 can be conveniently applied also to this case, the transmission probability T of the whole structure being expressed in terms of the M_{11} element of the total transmission matrix M that is obtained by the multiplication of the transmission matrix of the left barrier, M_L, with the diagonal matrix of free propagation across the quantum well with elements $\exp(-ikL)$ and $\exp(ikL)$, and finally with the transmission matrix of the right barrier, M_R. The result is $M_{11} = M_{L,11} M_{R,11} \exp(-ikL) + M_{L,12} M_{R,21} \exp(ikL)$, with $M_{L,11}$ designating the 11-element of the M_l matrix, and thus

$$T = \frac{T_L T_R}{(1 - \sqrt{R_L R_R})^2 + 4\sqrt{R_L R_R} \cos^2 \theta} \quad (1.20)$$

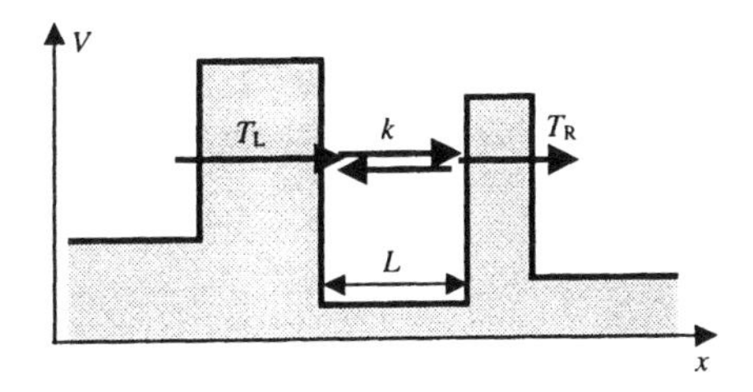

Figure 1.10 Geometry of the double-barrier tunneling structure.

where, as shown in Figure 1.10, T_L and T_R are the transmission probabilities through the left and right barrier, respectively, R_L and R_R are the corresponding reflection probabilities, and the angle $\theta = kL + (\arg M_{L,12} + \arg M_{R,21} - \arg M_{L,11} - \arg M_{R,11})/2$.

From (1.20) it follows that T is significant even if T_L and T_R are small, so that $1-\sqrt{R_L R_R} \cong (T_L + T_R)/2$, if the energy resonance condition $\theta = (2n+1)\pi/2$ with n integer is satisfied. At resonance $T_{res} = 4T_L T_R /(T_L + T_R)^2$ is unity if $T_L = T_R$, or is approximately $4\min(T_L, T_R)/\max(T_L, T_R)$ if T_L and T_R are very dissimilar. When the incident electron energy matches a resonant energy value E_{res}, the electron is not only transmitted with high probability, but also in a shorter time compared to off-resonance conditions; resonant tunneling devices are generally ultrafast devices.

Near resonance

$$T(E) \cong \frac{\Gamma_L \Gamma_R}{(\Gamma_L + \Gamma_R)^2/4 + (E - E_{res})^2} \tag{1.21}$$

with $\Gamma_L = (dE/d\theta)T_L/2$ and $\Gamma_R = (dE/d\theta)T_R/2$ (divided by $\hbar$) the rates at which an electron in the well leaks out through the left and right barrier, respectively [1]. The transmission probability is, in this case, very sensitive to the position of the resonant energy level or to the electron energy. The first parameter can be easily controlled by a bias applied across the double-barrier structure.

It must be emphasized that, although in deriving (1.20) it was assumed that the electron transport is coherent, that is, the electrons are transmitted from left to right in a single quantum process, this expression remains valid even if sequential tunneling takes place [1]. In a sequential tunneling process the electron tunnels first into the quantum well, loses its phase memory through a scattering process, and then tunnels out of the well. Unlike in coherent tunneling, in the sequential process the time spent by the electron in the resonant state (i.e., the eigenstate lifetime) is much longer than the scattering time. Electron tunneling is coherent when the barriers are thin enough so that $\Gamma_L + \Gamma_R >> \hbar / \tau_{ph}$, while sequential tunneling is important in devices with thicker barriers, for which $\Gamma_L + \Gamma_R \leq \hbar / \tau_{ph}$, where $\tau_{ph} = L_{ph} / v_F$ is the phase relaxation time of electrons with Fermi velocity $v_F = \hbar k_F / m$.

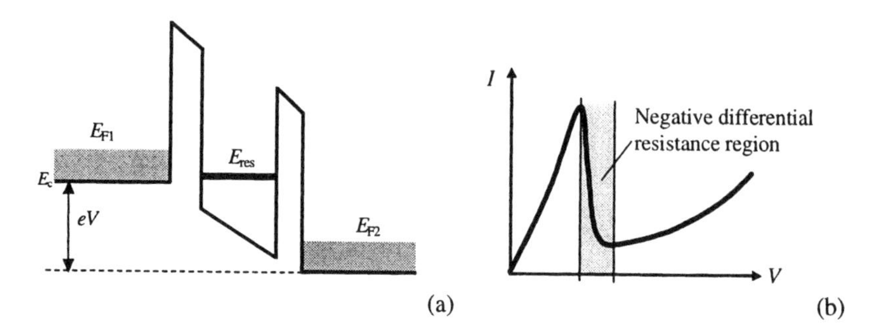

Figure 1.11 Schematic representation of: (a) a resonant tunneling diode, and (b) its typical current-voltage characteristic.

One of the first and best-known devices based on resonant tunneling is the resonant tunneling diode (RTD), schematically represented in Figure 1.11. It consists of a biased double-barrier structure with a single resonant energy level E_{res} that acts as an energy filter for the electrons in the left contact (emitter). That is, from all electrons in the emitter with energies between the bottom of the conduction band and the Fermi energy E_{F1} only those with energy E_{res} can tunnel toward the right contact (collector), which is characterized by the Fermi energy E_{F2}. In consequence the $I-V$ characteristic of the RTD exhibits a negative differential resistance region because at a sufficiently high bias E_{res} drops below the conduction band edge E_c in the emitter, and so no electrons can tunnel into the collector. As a result, the current decreases significantly. The negative differential resistance region of the RTD is maintained even at room temperature. Its existence assures RTD applications such as oscillators, bistable elements, or components in logic circuits (see also Chapter 5). The optoelectronic applications of resonant tunneling structures are described in [3].

1.1.8 Coupled Nanoscale Structures and Superlattices

If a quantum well (or another confined structure such as a quantum wire or a quantum dot) is placed in close proximity to another quantum well (quantum wire or quantum dot), the electrons in one confined structure can interact or not with the electrons in the other depending on the height and width of the barrier that separates the two structures. In the first case, the quantum wells (quantum wires or quantum dots) are called coupled and the electron wavefunction extends throughout the entire structure, while in the second case we deal with a succession of noninteracting quantum wells (quantum wires or quantum dots), for which the electrons localized in one confined structure can be transferred in the other only through sequential tunneling. Coupling of nanoscale structures is possible since, unlike in wells surrounded by infinite barriers, in finite barrier regions the electron wavefunction decays exponentially, and hence the envelope electron wavefunctions in adjacent wells can overlap if the barrier that separates them is sufficiently thin.

When two identical quantum wells, QW_1 and QW_2, are coupled, the degeneracy of the resonant energy levels in the individual quantum wells is removed and the electron wavefunction splits into a symmetric part, Ψ_{sym}, and an antisymmetric part, $\Psi_{antisym}$, that extend over the whole structure, as shown in Figure 1.12. When two nonidentical quantum wells are coupled the electron wavefunction remains still confined in one well or the other unless an applied bias aligns the different energy levels in the individual wells and brings the structure to resonance. Only in this last case the electron wavefunction becomes delocalized and splits into a symmetric and an antisymmetric part. Coupled quantum wells, quantum wires or quantum dots have countless applications in modern electronics, optoelectronics, and even quantum computing, as will become evident throughout the book.

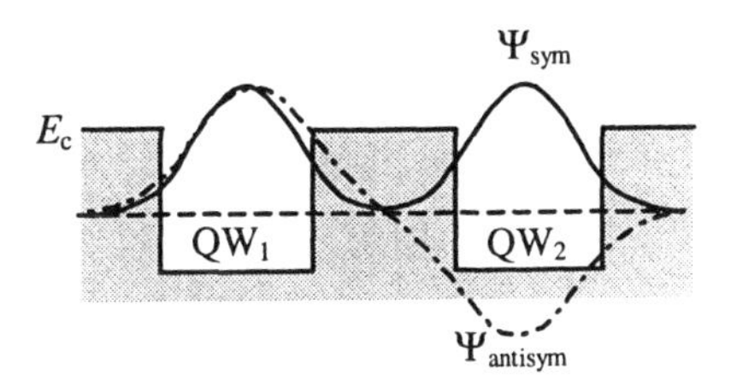

Figure 1.12 Electron wavefunction splitting into a symmetric and an antisymmetric part when two identical quantum wells are brought in close proximity.

When several identical quantum wells are grown in a periodic structure, with a period Λ, and are sufficiently close to one another so that they become coupled, the delocalized electron wavefunction feels a periodical energy potential with a period Λ. In a structure with N periods, called a multiple quantum well, the N

degeneracy of each level is split, the result being a band of discrete levels, such that each well contributes one state in each band. However, for structures with many periods and sufficiently thin barriers (almost) continuous permitted and forbidden electron energy bands develop (see Figure 1.13), similar to bulk materials, in which the periodicity is that of the crystalline lattice. This artificial periodic structure, which is an artificial lattice with a controllable cell, is called a superlattice. The positions and widths of the electron energy bands are determined by the form of the periodic potential and therefore can be engineered using advanced semiconductor growth techniques.

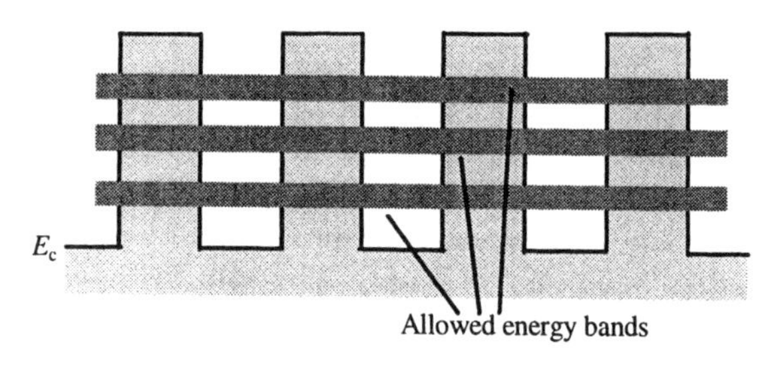

Figure 1.13 Allowed and forbidden energy bands for conduction electrons in the periodic potential of a superlattice.

The possibility to control the energy dispersion of electrons in a superlattice is particularly important in optoelectronic applications, especially when the superlattice period is comparable to the electromagnetic field wavelength. In this case, allowed and forbidden frequency bands appear also for electromagnetic fields, similar to those for electrons in bulk materials or superlattices; the width and position of these energy bands can be engineered at will. A theory of quantum transport in superlattices based on nonequilibrium Green functions is developed in [12].

1.1.9 Coulomb Blockade

In tightly confined nanostructures, particularly in quantum dots, the Coulomb interaction becomes significant and leads to the dependence of electronic states on the discrete number of particles (electric charges) in the dot. A phenomenon that illustrates this dependence is the Coulomb blockade, which consists of the appearance of a gap at the Fermi level in the energy spectrum of electrons confined in semiconductor quantum dots or small metallic clusters (generically, islands) that are coupled to metallic leads through tunneling barriers; Figure

1.14(a) is a schematic representation of the Coulomb blockade device [4]. This energy gap, similar to the energy gap in semiconductors, can be viewed as the extra energy needed, due to Coulomb interaction between electrons in the island, for an electron to tunnel in or out of the island.

This extra energy, equal to $e^2/2C$ in metallic islands where C is the capacitance between the island and the environment, corresponds to an energy gap of e^2/C in the electron spectrum at the Fermi level because not only electrons need an extra energy $e^2/2C$ to tunnel into the island but so do the holes. The Coulomb blockade is observed at low temperatures, when $e^2/C >> k_B T$, and if the number of electrons on the island is fixed, which implies that the Coulomb charging energy e^2/C should be much larger than the lifetime broadening $\hbar/\tau$, with τ the electronic lifetime. In terms of an effective RC-time this condition becomes $R >> h/e^2$, a condition that imposes a decoupling of the island from the reservoirs via tunneling barriers with resistances much larger than the quantum resistance.

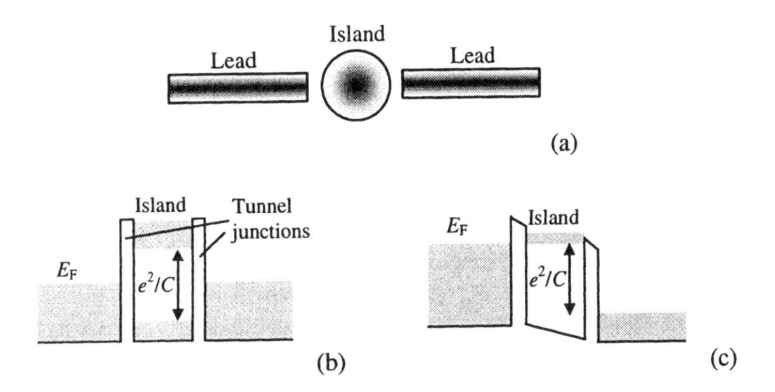

Figure 1.14 (a) Island connected to external leads through tunneling junctions. (b) The opening of a gap in the energy spectrum at the Fermi level due to the additional Coulomb energy needed by an electron to tunnel in or out of the island can be overcome (c) by applying a bias $V = e/C$.

As shown in Figures 1.14(b)-(c), electron tunneling into a metallic island in the presence of Coulomb blockade is only possible when a sufficiently large applied bias, $V > e/C$, compensates this charging energy; the $I-V$ characteristic of the island around zero bias is thus characterized by a very low current value. At $V = e/C$ an electron can tunnel in the island from one lead, the Fermi energy in the island raises again by e^2/C and further tunneling is forbidden by the occurrence of a new energy gap, unless the bias increases to $V > 3e/C$ or the extra electron in the island tunnels out, into the other lead. In general, the average number of electrons on the island increases by one each time the voltage increases in steps of $2e/C$. These correlated tunneling processes into and out of the island induce a net current, the $I-V$ characteristic having a staircase shape if the two junctions are very dissimilar [4].

In semiconductor ballistic islands the Coulomb blockade can be similarly treated, but the effect of energy quantization must be accounted for (in metallic clusters the size quantization effects do not exist since the ballistic transport conditions are generally not met). More precisely, the extra energy needed to add a charge to the island is in this case $e^2/C+\Delta E$, where ΔE is the energy difference between adjacent quantum states.

The Coulomb blockade is a single-electron phenomenon, which originates in the discrete nature of electric charge that can be transferred from a conducting island connected to electron reservoirs through thin barriers; in contrast, resonant tunneling devices rely on the discrete spectrum of resonant energy levels in a quantum well coupled through thin barriers to electron reservoirs. The Coulomb blockade allows a precise control of a small number of electrons, with important application in switching devices with low power dissipation and a corresponding increased level of circuit integration. Single-electron devices based on the Coulomb blockade have generally an extra control of the charge in the island through an additional gate electrode, which induces periodic oscillations of the current through the leads as a function of the gate voltage, and some of them even consist of an array of islands [13]. A more detailed account of the Coulomb blockade and the single-electron device based on it will be given in Chapter 5. We only mention here that single-electron devices, in particular the single-electron transistor, are extremely sensitive electrometers that can detect sub-single-electron variations in the external charge.

Another example of the influence of Coulomb interactions in nanoscale structures is the Coulomb drag phenomenon, which consists of the drag of carriers in a quantum well (or quantum wire) due to the Coulomb interaction with carriers that move under the influence of an electric field in another quantum well (or quantum wire) located in close proximity. If the first, or drag, layer, is electrically isolated so that no current can flow through it, the positive and negative charges in the drag layer accumulate at opposite ends until the drag force is compensated by a drag voltage V_{d}. The drag voltage has a negative sign compared to the voltage drop in the second, drive layer, if the carriers in both layers are of the same type: electrons or holes. A review of Coulomb drag between ballistic nanowires, which includes also references regarding the drag phenomenon between ballistic 2DEGs, can be found in [14].

1.1.10 Quantization of Thermal Conductance in Ballistic Nanostructures

Thermal conductivity in nanoscale devices is also important for the fabrication of high-density integrated circuits, since power dissipation considerations often limit the integration density. In bulk materials the thermal conductivity is determined by the phonon Boltzmann transport equation. In nanoscale devices the interfaces between different layers modify the predictions of the bulk phonon modes, while in superlattices the phonon spectrum differs from that of the constituent materials due to the new imposed periodicity and the consequent interference that appears between transmitted and reflected phonon waves at the interfaces if the sample is smaller than the phonon mean free path (see [15] and the references therein). Thus, both in-plane and cross-plane thermal conductivities of superlattices can significantly differ from the bulk properties and can, as well, differ notably one from the other; they can be engineered in a properly designed superlattice. The phonon Boltzmann transport equation is no longer valid when the sample size is smaller than both the phonon mean free path and phase relaxation length, which have the same significance as the corresponding parameters for electrons, and must be replaced by a counterpart of the Landauer formula for electric transport.

Amazingly, although electrons and phonons have completely different properties (the first are fermions, while the latter are bosons), the phonon thermal conductance is quantized in ballistic phonon structures, similar to the quantization of conductance in units of $G_0 = 2e^2/h$ [16]. To find the thermal conductance quantum, let us consider a 1D bridge in which transverse and/or longitudinal massless phonon modes propagate ballistically between two right and left reservoirs with temperatures T_R and T_L, respectively. For electrically insulating reservoirs the thermal energy is carried only by phonons. In this structure the spatial confinement induces the appearance of finite gaps in the phonon frequency dispersion relation, and the quantization of thermal conductance at low temperatures follows from the definition of the reservoir-to-reservoir thermal dielectric wire conductance $\kappa = \dot{Q}/(T_R - T_L)$. Here

$$\dot{Q} = (2\pi)^{-1} \sum_{\alpha} \int_{\omega_\alpha(0)}^{\infty} \hbar\omega T_\alpha(\omega)[\eta_R(\omega) - \eta_L(\omega)]d\omega \tag{1.22}$$

is the Landauer energy flux, where the sum must be performed over the massless phonon modes α with cut-off frequency $\omega_\alpha(0)$, phonon transmission probability through the wire $T_\alpha(\omega)$, and thermal distribution $\eta_i(\omega) = [\exp(\hbar\omega / k_B T_i) - 1]^{-1}$ of phonons with frequency ω in the reservoir with temperature T_i, i = R, L. (In (1.22) it is assumed that phonon modes propagate to arbitrary large frequencies above the cut-off frequency; if this assumption is not valid the upper limit of integration must be replaced by a finite value.) The phonon transmission probability through the dielectric wire can be calculated as in Section 1.1.5, by matching the wave equation at interfaces and using the appropriate boundary conditions [16]. As shown in [17], the cut-off frequency of vibrational modes that propagate along x through narrow bridges with transverse dimensions L_y and L_z, described by a scalar model of elasticity, are

$$\omega_{\alpha,mn}(0) = c_\alpha \pi[(m/L_y)^2 + (n/L_z)^2]^{1/2} \tag{1.23}$$

with m, n integers. The phonon dispersion is $\omega_{\alpha,mn}(k) = [\omega^2_{\alpha,mn}(0) + c^2_\alpha k^2]^{1/2}$. Note the similarity between (1.23) and (1.8), and between the definition of κ and G, which is equal to the ratio between the current (the rate of charge flow) and the voltage difference between the electron reservoirs.

For the case of perfect adiabatic contact between the ballistic phonon wire and the thermal reservoirs, for which $T_\alpha(\omega) = 1$, the integration of (1.22) gives

$$\kappa = \frac{k_B^2 \pi^2}{3h} T N_\alpha + \frac{k_B^2}{h} T \sum_{\alpha'}^{N_{\alpha'}} \left(\frac{\pi^2}{3} + 2\text{dilog}[\exp(x_0)] + \frac{x_0^2 \exp(x_0)}{\exp(x_0) - 1} \right), \tag{1.24}$$

in the limit $T_R - T_L \to 0$, where $T = T_R = T_L$, N_α is the number of phonon modes with zero cut-off frequency, $N_{\alpha'}$ is the number of higher-energy phonon modes with finite cut-off frequencies $\omega_{\alpha'}(0) \neq 0$, and $x_0 = \hbar\omega_{\alpha'}(0)/k_B T$. The dilogarithm function is defined as $\text{dilog}(x) = \int_x^0 dt \log(1-t)/t$. So, the contribution of phonon modes with zero cut-off frequency at κ can be expressed in terms of a universal thermal conductance quantum $\kappa_0 = k_B^2 \pi^2 T / 3h$; in general, in the expression of the quantum conductance $T = (T_R + T_L)/2$. The contribution of higher-order phonon modes to (1.24) is independent of the phonon dispersion curve and depends only on the respective cut-off frequencies, but the higher-order phonon modes have an exponentially small contribution at low temperatures. Similar to electron ballistic transport through an electron waveguide with a variable number of transverse modes, the thermal conductance is expected to increase in steps of κ_0 each time a new phonon mode contributes to the heat transfer; the steps are smoothed with increasing temperature or time since phonon branches have finite relaxation times. The discreteness of the thermal conductance can be observed for a nonthermal narrow-frequency-band phonon distribution, for which the heat transport rate increases in steps as the central frequency of the distribution passes through a phonon mode cut-off frequency, the step sharpness increasing with frequency [17].

Thermal transport through a mesoscopic weak mechanical link, in which the reservoirs are mechanically connected through one or a few atomic bonds or very narrow dielectric bridges, is studied in [18]. In this regime of thermal flow, which is very similar to that of electron tunneling through a barrier, the thermal conductance is much smaller than κ_0 and is determined by the product of the local vibrational density of states of the reservoirs at the point of connection. However, "true" phonon tunneling does not exist since there are no "classically forbidden" regions for thermal vibrations. Therefore, the length dependence of the transmission probability of tunneling electrons and of the thermal conductance of phonons propagating through thin bridges are different: exponential decay with L (see Section 1.1.6) and, respectively, proportional to L^{-2}.

In conducting ballistic wires there is an additional electronic contribution to the thermal conductivity. Remarkably, if Joule heating is neglected, which implies a negligible current flow, the same quantum of thermal conductance, κ_0, is found to characterize the electronic heat transfer. The quantum of electronic thermal conductance is a consequence of Heisenberg's uncertainty principle [19], and can be estimated from its definition $\kappa_e = J_e / \Delta T$, where $\Delta T = T_R - T_L$ is the temperature difference between the electron reservoirs connected by the ballistic conductor and $J_e = dE / dt$ is the net electronic heat current. Under the assumption that there is no net electrical current and that the width of the ballistic conductor is on the scale of λ_F, while its length is smaller than the electron mean free path, $J_e = k_B \Delta T / \Delta t$, where Δt is the transit time limited by the Heisenberg's uncertainty principle and $dE = k_B (T_R - T_L)$ is the net energy carried across the conductor in the extreme quantum limit, in which a single carrier moving from left to right carries an energy $k_B T_L$ while a carrier moving in the opposite sense carries an energy $k_B T_R$. Because only electrons with energy spread $\Delta E \cong k_B T$ around the Fermi level propagate across the ballistic conductor, where $T = (T_R + T_L)/2$, the thermal conductance becomes $\kappa_e \cong k_B^2 T / (\Delta E \Delta t)$, or $\kappa_e \cong 2k_B^2 T / h$ if the spin degeneracy and the Heisenberg's uncertainty relation $\Delta E \Delta t \cong h$ are taken into account. More detailed calculations confirm that the quantum of thermal conductance is $\kappa_0 = \pi^2 k_B^2 T / 3h$ (see the reference in [19]).

1.1.11 Nonballistic Electron Propagation

The transition between the diffuse charge transport regime, characterized by the Boltzmann equation, and the ballistic regime is not a sudden one; there is an intermediate regime of electron propagation characterized by the preservation of coherence of the electron wavefunction when static and elastic scattering processes occur. Specific phenomena such as universal conductance fluctuations, coherent electron backscattering and electron localization occur in this intermediate regime, in which the dimensions of the sample are larger than the mean free path but smaller than the phase coherence length (see [1], [6], and the references therein). These phenomena cannot be treated by the Boltzmann equation, which is valid for uncorrelated scattering events or, equivalently, when the electron visits every scatterer at most once so that no loops are allowed.

In media with random scattering, the electronic wavefunctions are no longer extended throughout the medium, as in perfect crystals, but are solutions of the Schrödinger equation (1.1) with a random potential energy and thus become localized if the medium is sufficiently disordered. Electron localization due to quantum interference in random conducting media should be distinguished from localized electronic states in insulators or undoped semiconductors, in which electron transport occurs through hopping.

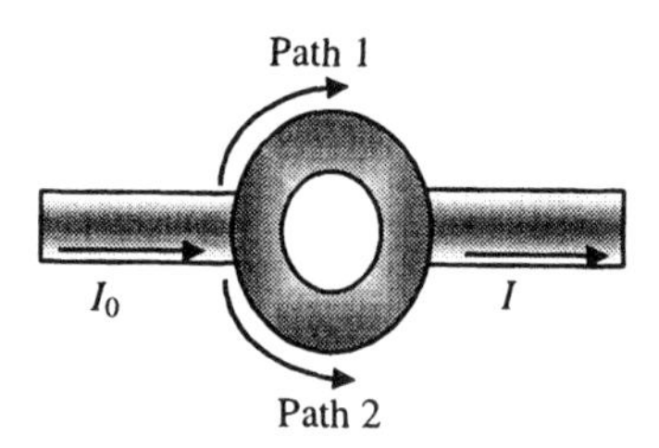

Figure 1.15 Interference experiment with coherent electrons.

Electron coherence in nonballistic nanoscale structures is evidenced in electron interference experiments. As long as the device length is smaller than L_{ph} interference between two parts of an electron beam that are reunited occurs, even if random elastic scattering processes occur along the electron paths. The current at the output of a symmetric ring-shaped conductor as that in Figure 1.15 is [1]

$$I = I_0(1 + \cos\varphi), \tag{1.25}$$

where φ is the difference between the phases of the electron beams that propagate along paths 1 and 2, similar to the intensity in an optical interference experiment. Current oscillations with φ can be observed, for example, if magnetic fields are applied, the phase difference in this case being proportional to the magnetic flux enclosed by the ring. The output current at constructive interference, for which $\cos\varphi = 1$, is double the value obtained when no interference occur, in case which $I = I_0$ since the cos term disappears through averaging. On the contrary, the output current at destructive interference, for which $\cos\varphi = -1$, vanishes.

Interference effects between coherent electron beams can be observed also in nanoscale structures that have no ring shape. They manifest themselves in the enhancement of backscattered electrons and universal conductance fluctuations, which are characteristic features of the weak localization regime of electrons in random media that appears when the sample length is much smaller than the localization length of electrons. Enhanced electron backscattering is a result of interference between time-reversed paths that connect spatially identical initial and final states. An incident electron that enters a random scattering conductor and is reflected at a certain angle θ arrives at its final state after propagation along numerous possible paths, each of them being a succession of elastic scattering events that do not generally interfere because the phases of various paths are random. The only exception is scattering in the opposite direction, a case in which the initial and final state can be connected by different time-reversed paths (paths that can be followed in the opposite direction by simply reversing the order of the scattering events), which interfere constructively and hence double the reflection probability (and electron current) in the backscattering direction with respect to the reflection probability in any other direction. A backscattering cone in electron current, as shown in Figure 1.16, would be observed if one would be able to collect electrons at different scattering angles; in solid-state measurements the backscattering cone is evidenced through resistance measurements along the incident electron direction by applying a magnetic field that destroys the enhanced backscattering since it breaks the time-reversal symmetry of the paths that interfere constructively in its absence. The backscattering cone corresponds to a negative magnetoresistance, although a dip (a positive magnetoresistance) can also be observed in some solid-state systems due to spin-orbit scattering.

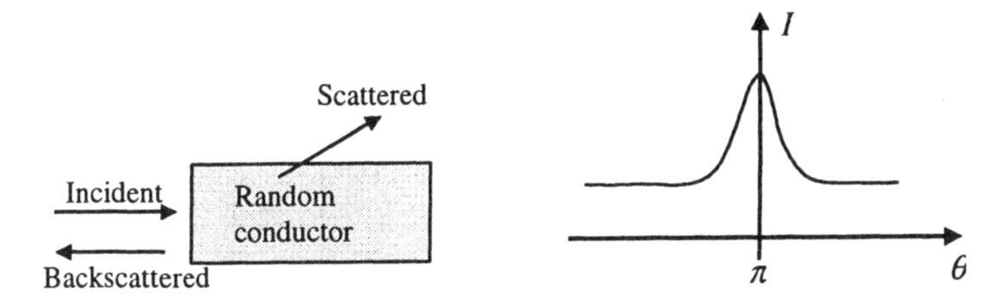

Figure 1.16 Enhanced backscattering of electrons from a random scattering medium.

Universal conductance fluctuations, of the order of e^2/h, which are independent of the background conductance, appear if interference occurs between electron trajectories that encircle an impurity. The quantum conductance depends in this case on the position of the impurity in the sample, unlike the classical conductance, which depends only on the number of impurities but not on their positions. The average quantum conductance is smaller than the classical conductance, calculated assuming incoherent electron wavefunctions, by approximately e^2/h, a value independent of the size of the system [1]. The fluctuation of the quantum conductance, defined as $(\langle G^2\rangle - \langle G\rangle^2)^{1/2}$, where the average is taken over an ensemble of similar nanoscale structures with different positions of the scattering centers, becomes comparable to the conductance itself; in the classical theory of conductance the fluctuations are expected to decrease with the size of the system. Time-independent fluctuations in the conductance of the order e^2/h are observed as a function of the impurity arrangement in the sample, magnetic field or electron density, which influence the phase of the electron wavefunction. Universal conductance fluctuations can be seen as quantum corrections to the classical conductance, and disappear at high temperatures when the discrete electron level spacings in nanoscale structures become smaller than $k_B T$.

In the strong localization regime the resistance of the sample increases exponentially with its length L over a characteristic length scale called the localization length L_{loc}. An exponential increase of the resistance can be predicted starting from the formula $R_s = h(1-T)/(2e^2T)$ that describes the resistance of a scatterer for a single transverse electron mode (see Section 1.1.4) and (1.20), which expresses the composition law of transmission coefficients through subsequent barriers (or scatterers). The ensemble-averaged normalized resistance $\rho = R_s(2e^2/h)$ of two scatterers in series is then $\rho = \langle (1-T)/T\rangle$, or

$$\rho = \rho_1 + \rho_2 + 2\rho_1\rho_2, \tag{1.26}$$

where $\rho_1 = (1-T_1)/T_1$, $\rho_2 = (1-T_2)/T_2$ are the normalized resistances of the individual scatterers and $T = T_1T_2/[1-2(R_1R_2)^{1/2}\cos\theta + R_1R_2]$ with θ the phase shift in one round-trip between the scatterers. Expression (1.26) is different from the formula $\rho = \rho_1 + \rho_2$ valid in bulk materials, which would have been obtained if the ensemble average had been performed as $\rho = (1-\langle T\rangle)/\langle T\rangle$, and leads to an exponential dependence of ρ with length if the second scatterer region has an incremental length dL. Then, the incremental increase $d\rho = \rho - \rho_1$ of the normalized resistance due to the addition of the second scatterer with $\rho_2 = \alpha dL$ is $d\rho \cong \alpha(2\rho+1)dL$, which leads to $\rho(L) = (1/2)[\exp(2\alpha L) - 1]$. Although there are different procedures for ensemble averaging depending on which quantity is averaged first (see [6] and the references therein), the exponential dependence of resistance with L characterizes the strong localization regime, in which electrons are not free to move across the whole sample, as in the ballistic regime, but are confined due to quantum interference in a region of dimensions of the order of $1/\alpha$.

It is interesting to mention that the similarity between electron wavefunction propagation and electromagnetic radiation can be extended from the ballistic regime to the case in which random elastic scatterings occur. Enhanced back-scattering, light intensity correlations, universal light conductance fluctuations, and localization of light can be treated in a similar way to the corresponding phenomena for electrons due to their common nature: the wave character of electrons and phonons, and have been already evidenced experimentally [6].

1.2 NANOTECHNOLOGIES

The nanotechnologies originate from the technologies used in electronics to produce micron and submicron-sized integrated circuits based on Si and complex circuits based on AIII-BV semiconductors. Silicon, which is the key material at the microscale, is still the key material at the nanoscale, but carbon, AIII-BV semiconductors, and assemblies of various molecules or biomolecules will doubtless surpass its supremacy.

There are two major steps to be followed by any nanotechnology method: 1) deposition of one or multiple layers on a substrate and 2) the transfer of desired patterns onto different layers grown in the first step, a process which requires intermediate steps to remove the unnecessary parts of various materials that are not included in the imposed pattern. The first step is termed deposition technique, while the second is termed lithography. At nanoscale we are dealing with nanolithography, which simply means the capability of a lithographic technique to create patterns on a surface with nanosized features.

There is also a large category of nanotechnology techniques, which have no analog in the technology used in electronics. This self-assembly technology does not use masks, but specific chemical reaction to create prescribed three-dimensional (3D) patterns composed of nanosized materials, which could be either metals or semiconductors. These maskless techniques are very important because they are less expensive than masks with nanosized features. However, very often nanoelectronic devices are implemented using a certain combination of standard technologies and self-assembly techniques.

1.2.1 Deposition Techniques for Nanoscale Devices

In this section, we review briefly the main deposition techniques applicable at the microscale and particularly at the nanoscale. An updated review and extensive references can be found in [20].

The oxidation of silicon to obtain SiO_2 is a key process at the microscale and is equally important at the nanoscale. SiO_2 is a dielectric, which can be deposited on a silicon substrate with a thickness ranging from a few nm up to 2 μm, and is a key element in the isolation of various metallic electrodes from other conductive substrates. SiO_2 can be found in any transistor or in many MEMS (micro-electro-mechanical systems), such as switches. It is grown at temperatures of about 1000–1200 °C in the presence of oxygen or water into a furnace consisting of a quartz tube, an electrical resistance heater, and a quartz wafer holder. The desired SiO_2 thickness is obtained by monitoring the temperature and the gas flow.

Semiconductor doping is a key technological process in microelectronics and accompanies the deposition processes. It is well known that *p*- or *n*-doping of a semiconductor changes significantly the electrical, chemical, and mechanical characteristics of semiconductors. These changes are often referred to as the functionalization or engineering of the material. The electrical changes are used to produce almost any active electronic component, such as transistors and diodes, while the mechanical changes create MEMS devices with prescribed mechanical characteristics [21]. Impurities of *p*- or *n*-type are controllable introduced in intrinsic silicon via diffusion into furnaces at high temperatures from liquid or solid sources or via ion implantation techniques, which are more accurate with respect to the amount of impurities/area and the prescribed impurity profile. The variety of functionalization techniques at nanoscale is much richer compared to that of microscale transistors, diodes, and integrated circuits. The doping is still used, for example, for nanotube transistors, but the functionalization of nanowires or dots can be done via oxygenation, hydrogenation, adsorbing of molecules, biomolecules, and in many other ways, which are presented throughout the book.

Deposition techniques for nanoscale devices also include CVD (chemical vapor deposition), which implies the deposition on a substrate of a thin film material, with a thickness of up to a few angstroms, via a chemical reaction of some gaseous components. The chemical reaction in the CVD requires the consumption of a high quantity of energy, which is provided by several means: 1) by heating the substrate at very high temperatures, 2) by plasma excitation, or 3) by optical excitation. The latter two procedures require lower temperatures than the first case.

There are several types of CVD techniques, the two most basic ones being low pressure CVD (LPCVD) and plasma enhanced CVD (PECVD). The LPCVD takes place in electrical heated furnaces where the pressure is kept very low (at 0.1–0.7 torr) with the help of a pumping system. The material is deposited on both sides of a wafer placed in a holder inside the furnace. With the help of LPCVD one can deposit SiO_2 (using as gaseous components $SiCl_2H_2$ and N_2O at 900 °C), Si_3N_4 (from gaseous components SiH_4 and NH_3 at 800 °C), or polysilicon (from gaseous SiH_4 at 600 °C). A large category of metals such as Ti, Cu, Mo, and Ta can be deposited using LPCVD.

The PECVD is realized in the plasma created by a powerful RF source. The advantage of this method is the lower temperature (100–300 °C) at which the substrate is heated. The PECVD is produced in a plasma reactor equipped with an RF source, two parallel plates, and a system of pumps, which introduces the gases inside the reactor and creates a vacuum in the reactor chamber. The RF signal is applied on one plate and the RF generator is grounded on the other plate, which contains the wafer and which is placed above an electrical heater. Typical materials, which can be deposited via PECVD, are Si_xN_y (the nonstoichiometric version of the silicon nitride), SiO_2, amorphous silicon, and carbon nanotubes produced from a gas mixture consisting of NH_3 and C_2H_2.

The epitaxial growth techniques using CVD methods have revolutionized the semiconductor industry, allowing the growth of monolayers of various AIII-BV materials, such as GaAs/AlAs or InP/GaInAs. These heterostructures are being used in the design and fabrication of new devices based on quantum wells, wires, and dots, which are the building blocks of nanoelectronics. In the CVD-based epitaxial growth a single crystal is grown from the substrate material. A different single crystal can be further grown if it is lattice matched with the crystal beneath it, or even when the two layers are not matched, a case in which a slight strain is introduced between them. This strain is extremely useful in some applications, as, for example, to tune the bandgap of the heterostructure to a prescribed value necessary to produce a certain emission wavelength in a quantum wire laser.

There are two main CVD-based growth techniques able to grow heterostructures: metal organic chemical vapor deposition (MOCVD) and molecular beam epitaxy (MBE). MOCVD uses vapors of organic compounds with group-III atoms and group-V hydrides, which are introduced in a CVD chamber characterized by a fast gas switch capability. The MBE requires a high-vacuum CVD chamber in which the substrates are placed. Inside this chamber, molecular beams originating from thermally evaporated basic sources are used, which are switched on or off depending on the heterostructure to be grown.

1.2.2 Nanolithography

Nanoscale lithography is of paramount importance in nanotechnologies and a basic review can be found in [22]. Presently, the patterning of various surfaces with the existing nanolithographical methods attains a resolution (i.e., the smallest possible feature) of a few nm. The hope of nanoelectronics resides in the capabilities to increase this resolution and thus to reduce the smallest possible features.

In lithography, a desired geometrical pattern is transferred onto a substrate; the geometrical pattern is a mask containing the paths to be imprinted on the substrate. The most encountered method to perform this transfer is to cover the substrate with a resist material, with a thickness ranging in the 0.2–2 μm interval, using a spinner. The deposited resist is sensitive to illumination with light or energetic particles, such as electrons or ions, the desired pattern being imprinted in the resist material after illumination and the subsequent developing process. The illumination of the resist is made through a mask, which transmits the excitation (optical, or particle fluxes) in the desired regions and absorbs the excitation in the undesired regions.

1.2.2.1 Optical Nanolithography

In optical lithography the mask is a glass or fused silica plate, which is selectively covered with a Cr layer that absorbs the light and has a thickness of 80–100 nm. The fabrication of the mask is a difficult step when nm resolutions and low defect rates are required. In this case, a pattern generator is needed, which writes the desired pattern into a resist deposited over the Cr layer via an electron or a laser beam, a process followed by resist developing and etching of the Cr layer. Generally, the mask is a reduced version of an initial pattern designed at a much larger scale. This procedure has the advantage that during the reduction procedure all errors and defects are reduced in size with the same scale. However, there are also masks directly made at the 1:1 scale, as in the case of X-ray lithography. The resist materials experience selective chemical reactions due to illumination. In the resist developing stage the illuminated areas will be dissolved and removed in the case of positive resists, while the exposed areas are kept intact in the case of negative resists (see Figure 1.17). The result is that the desired pattern is imprinted in the resist deposited on the substrate.

The lithographical process can be repeated a number of times with various resists and masks, which are subsequently aligned, the procedure being accompanied by intermediate steps of etching and deposition processes in order to achieve the final architecture of a certain device or integrated circuit. Mask alignment (inter-level alignment values) and resolution are the key parameters of the lithographical system. Although the above lithography method is widespread in present-day technology for the realization of electron devices and integrated circuits at the microscale, the lithography based on masks and resist dominates also at the nanoscale, but as we will see below, high-accuracy lithographical methods without masks and resists are equally applicable at nanoscale. The resists are in general polymers, such as PMMA (polymethylmethacrylate) or DNQ (diazonaph-thoquinone) in the case of optical lithography, the type of resist depending on the illumination type (optical, particle beams, X ray). Resist exposure is realized either in a direct write mode, when the resist is exposed point-by-point, or as in Figure 1.17, when the mask is illuminated and the desired pattern is transferred in the resist either by contact or proximity alignment of the mask to the resist surface, or by projection of the mask image. In the contact exposure method the mask is in direct contact with the resist, conferring a good resolution, but the mask is often damaged in the process. In the proximity method, the mask is placed at a certain distance, $d_{\mathrm{m-r}}$, from the resist, strongly limiting the resolution by diffraction. The minimum feature size in this case is given by

$$w_{\min} \approx (\lambda d_{\mathrm{m-r}})^{1/2}, \tag{1.27}$$

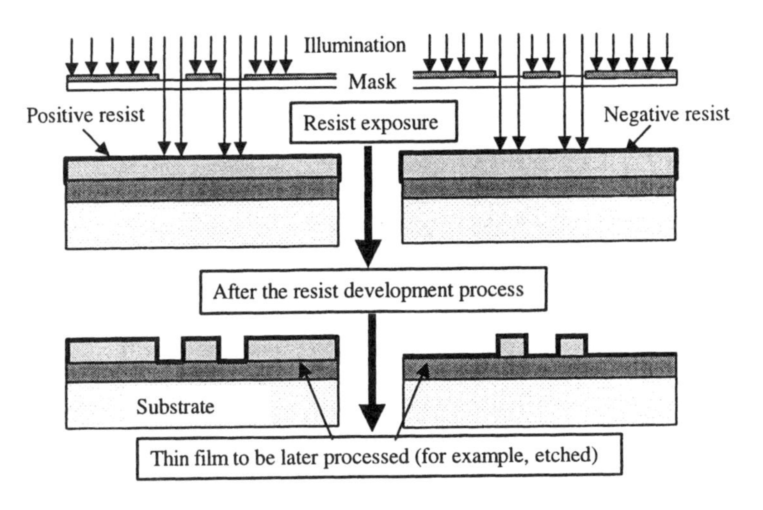

Figure 1.17 Principle of the lithography process.

where λ is the wavelength of the illumination source. The distance d_{m-r} is very small and difficult to keep fixed in order to obtain reproducible results. Another drawback is that the printed feature size is the same as that on the mask and the corresponding image.

In the projection exposure method an optical system projects the mask image at a certain distance on the resist, offering the best resolution, which in this case is

$$R = \beta\lambda / \text{NA}, \tag{1.28}$$

where $\beta \cong 0.8$ is a constant and NA is the numerical aperture of the imaging lens. The ratio β / NA is almost 1 in the optical projection printing method, and thus the resolution is directly proportional with the wavelength of excitation. An improved resolution means a lower value of R, that is, lower wavelengths and higher numerical apertures. The latter solution can be implemented via sophisticated lens systems, which compensate the aberrations at certain wavelengths, but a dramatic improvement of the resolution is realized by decreasing the excitation wavelength. Thus, a nanolithographical system requires excitations with wavelengths much shorter than in the optical spectrum. In an optical lithographic system mercury arc lamps, which have pronounced emission peaks at 435 nm and 365 nm, are used for geometrical line widths exceeding 0.25 μm. Thinner line widths, attaining 0.13 μm, are obtained with excimer lasers such as KrF ($\lambda = 248$ nm), ArF ($\lambda = 193$ nm), and F_2 ($\lambda = 157$ nm).

The enhancement of the resolution of optical lithography up to 130 nm implies a complete change in the mask design and fabrication; the type of resist and the accompanying chemistry are also changed, including the developing solutions. As soon as the wavelength decreases the issues of practical realizations of masks and resists become very difficult; the inter-level alignment reaches 30 nm.

Higher resolutions, attaining 30 nm, and inter-level alignments of only 10 nm are obtained with extreme ultraviolet lithography (EUV), which works at wavelengths of 10–14 nm, placed in the soft X-ray spectrum. The EUV is based on the combination between small-NA reflective optical systems and small wavelengths, much lower than the dimensions of the device to be realized. The reflective elements are multilayers that form Bragg mirrors, the mask in EUV being constituted from reflective portions of the same type of multilayer deposited on Si. A plasma or syncrotron source illuminates the mask, its patterns being imaged on the resist deposited on a substrate. The imaging system for the projection exposure system must have tolerances of only a few angstroms. Despite the difficulties to realize masks and thick resists, the EUV attains very high performances, including a high speed of patterning features, of about 10^{11} features/s, but the price of such a nanolithographical instrument is startling.

Lithography with X-rays, which are the extreme electromagnetic field wavelengths, is realized using high-energy X-ray sources that provide a few keV, such as electron syncrotrons or Cu target systems that emit X-rays. X-ray lithography is working in the proximity-printing mode and is able to attain a resolution of 50 nm, its main drawback being the significant difficulties to fabricate masks at the 1:1 scale.

1.2.2.2 Particle Nanolithography

The particles used in nanolithography are electrons or ions, the lithographical process showing impressive performances, with resolutions of 50 nm in the case of electrons and 10 nm in the case of ions. Low dimension features, of even 5 nm, can be aligned in these advanced nanolithographical methods.

The electron beam (EB) lithography has two configurations. The direct writing EB lithography consists of an electron source, which is focalized and directed to a substrate or to a substrate covered with a resist material, the latter method being more efficient. The electron beam is scanned via electrical or magnetic deflection systems to imprint the desired pattern in the resist-coated substrate. However, the writing process is very slow and takes hours to write a pattern of high resolution. Therefore, modern techniques with short writing times use electron apertures with shapes similar to the repeating areas of a certain pattern, which are rapidly replicated. The second EB method is electron beam projection lithography, which is the electron analog of the corresponding optical lithographical technique. The electron beam is directed to a mask, which projects an image of the desired pattern to a resist-coated substrate via an imaging system. The mask is a membrane with holes that constitute the desired pattern. An updated review of EB lithography and its recent developments can be found in [23,24].

The focused ion beam lithography (FIB) is based on an energetic ion beam scanned and focused directly onto a substrate. FIB does not use masks or resists and the lithographical process is a point-by-point technique, which consists either of subtracting surface atoms or of decomposition of an organic vapor over the substrate. In the former case, the sputtering of atoms forms the desired pattern imprinted directly on the substrate via scanning. In the latter case, the deposited material on the substrate is the desired pattern. The emitted ions and electrons resulted due to the interaction between the ion beam and substrate are used to monitor and image in real time the FIB lithographical process. The number of features patterned in one second, called the throughput, is low (10–10^2 features/s) for direct writing EB and FIB methods, and very high (10^{10} features/s) when the EB projection method is used, being similar with that obtained with optical nanolithographical techniques.

1.2.2.3 Nanoimprint

The nanoimprint combines the high resolution of particle nanolithography with specific techniques that are able to provide very high reproducibility in a short time. The throughput of this method attains the highest value of any nanolithographical technique: 10^{12} features/s, and the resolution is of 10 nm.

In principle, a nanoimprint technique uses a master or mold made from Si or SiO_2 that contains the desired pattern, which is realized via an EB or FIB nanolithographical technique. This mold represents a rigid stamp, which is used to imprint a resist. The resist is a polymer, which can be thermoplastic (must be heated beyond its glass temperature), like PMMA, or UV-curable. After the pattern is stamped in the polymer, the resist residues are eliminated by an etching technique, such as the reactive ion etching (RIE) represented in Figure 1.18. The final step is the processing of the substrate, a process that depends on the pattern imprinted on it, and the removing of the resist. Reviews on nanoimprint techniques are found in [25,26]. The fastest nanoimprint technique is the laser-assisted direct imprint (LADI) where the stamping process is realized in 250 ns by illuminating with a XeCl excimer laser a quartz mask in direct contact with a Si substrate. A molten Si forms at the interface between the quartz mask and the Si substrate, which is subsequently stamped by the quartz mask and cooled, the pattern being transferred from the mask onto the silicon substrate [27].

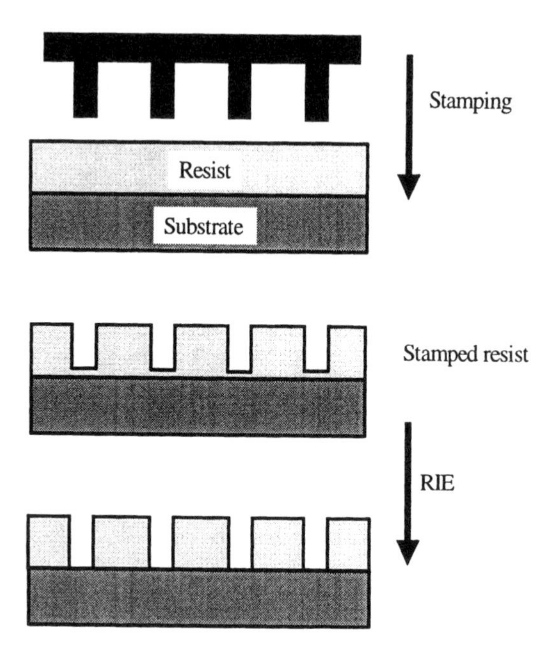

Figure 1.18 The nanoimprint technique.

1.2.2.4 Scanning Probe Nanofabrication

The scanning probe consists of a very sharp tip, which is scanned over a sample. The nature of the interaction between the tip and the sample is different for the main scanning probe techniques. In the case of scanning tunneling microscope (STM), the tunneling current between the tip and the sample is monitorized. On the contrary, the atomic forces acting at a surface of a sample, such as van der Waals forces, are evidenced via atomic force microscopy (AFM). The same features can be observed using the near optical fields produced by a nanosized aperture located at a few nm over the substrate in the scanning near-field optical Microscope (SNOM) technique. These important devices for nanoelectronics are explained in detail in Chapter 2, which is fully dedicated to scanning probe techniques. Here, we emphasize only that STM, AFM, or SNOM is able to manipulate single atoms, by placing them at desired positions, in order to create nanostructures in the form of clusters, corrals, lines, and so on. This manipulation is possible due to various forces that act between the tip and the displaced atom:

van der Waals forces in the AFM techniques, adhesion forces in STM, or optical forces in SNOM. An introductory review of nanomanipulation can be found in [28]. The tip is used either to remove locally the substrate or to add atoms in a prescribed manner and thus to pattern the substrates.

An example of resist removing via an STM technique is displayed in Figure 1.19. The tip of the STM emits electrons, which remove the resist in a controllable manner. This is an EB-like lithography, but much slower and at much lower energies.

An example of scanning probe deposition technique is the dip-pen lithography, where the tip of an AFM is coated with a certain chemical substance that plays the role of the "ink." The chemical substance flows from the tip on the substrate, "writing" on it a certain pattern when the tip is scanned over the substrate. The writing process is enhanced by the nanolayer of water that is always present between the tip and the layer when the deposition is made in air. The AFM "fountain pen" linewidth is 12 nm wide and has a resolution of 5 nm.

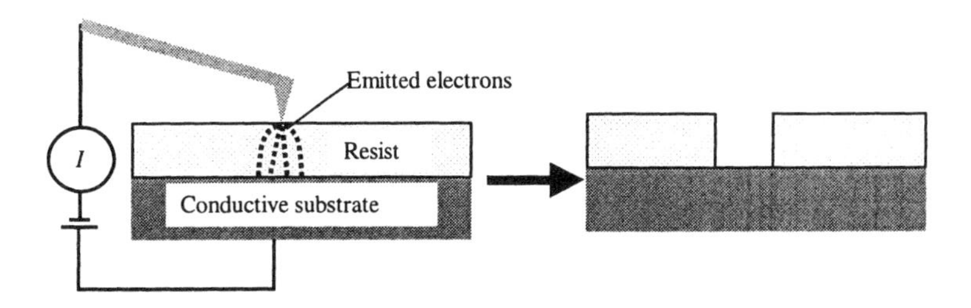

Figure 1.19 STM technique of resist removing.

The scanning probe nanofabrication procedures are no longer pure lithographical techniques, the nanostructures being made either by placing atom-by-atom or by sputtering atom-by-atom from a substrate. Therefore, we refer to these techniques as one-shot nanofabrication of a desired nanostructure. The resolution is very high, attaining 1 nm, but the throughput is low. A review about the AFM and STM nanofabrication is found in [29], while [30] is a review about nanofabrication and atom manipulation using the SNOM.

1.2.2.5 Atom Lithography

The ultimate nanolithography technique uses atomic beams, which are collimated and focused by atomic optical devices, such as atomic lenses, apertures, and so on. The atom beam interacts often with laser beams and the optical-atomic interaction forces are used either to deposit atoms in order to built nanostructures or to pattern desired shapes via a lithographical technique. A comprehensive review on atom lithography is [31]. An atomic deposition method consists of an incoming atomic beam, which is directed to a standing wave produced by a laser and a mirror system, as displayed in Figure 1.20. The standing wave acts as a light mask for atom beams; it is equivalent to an array of lenses that directs the atoms in the optical nodes or antinodes, depending on the sign of the difference between the resonant frequency of atoms and the laser frequency. So, in the mask light of atom lithography the desired pattern is created via nodes and antinodes of a light beam. The incoming atom beam passes through the mask and the atoms are deposited over a resist, which is later developed.

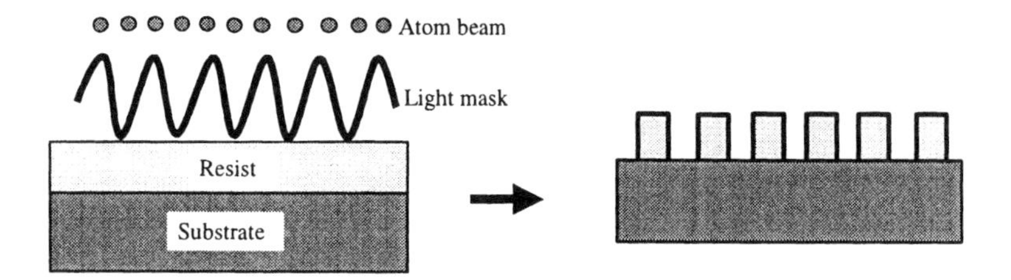

Figure 1.20 Atomic beam lithography.

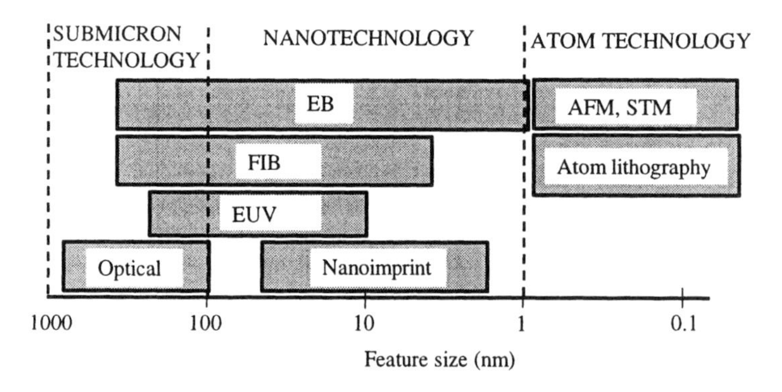

Figure 1.21 The performance of the main lithographical techniques.

In conclusion, in the nanotechnology several lithographical techniques are available and the choice of one of them depends on the required resolution and reproducibility of a certain nanoelectronic circuit. Figure 1.21 displays the main lithographical techniques and their applicability.

1.2.3 Self-Assembly Techniques

The self-assembly process refers to the spontaneous organization of various components (molecules or various nanosize objects such as nanoparticles) into a single, ordered aggregate. The organization process is made into a desired structure via physical, chemical, or biochemical interactive processes involving, for example, electrostatic and surface forces, hydrophobic and hydrophilic chemical interactions. All these processes, irrespective of their origin, are very selective and reject defects so that the resulting desired structure is characterized by a high degree of perfection [32]. Several self-assembly techniques are briefly described below.

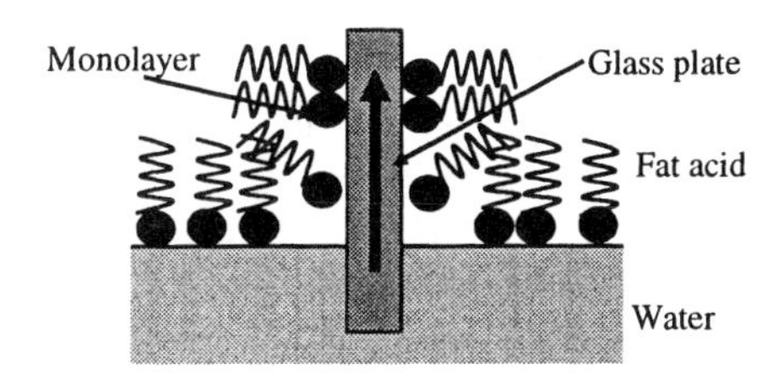

Figure 1.22 The principle of Langmuir-Blodgett technique.

The Langmuir-Blodgett (LB) technique dedicated to thin film realization is a well-spread self-assembly technique in which the organized aggregate is built by growing one monolayer at a time. A monolayer of a desired material, which is initially adsorbed at a gas-liquid interface, is transported to a substrate on which the self-assembly structure will be grown. For example, a monolayer of some molecular species, such as a fatty acid, is spread over the surface of the water. In the water there is a glass microscope slide, which plays the role of the self-assembly substrate. If we pull out the glass slide the monolayer will be attached to it. If we then pass the glass slide several times through the water surface we will deposit on it a monolayer at each passing (see Figure 1.22). The LB technique is implemented with specialized instruments comprising a Langmuir trough, a dipping device for the substrate that will be raised or lowered, and a movable barrier controlled by a pressure sensor, which slides on the gas-liquid interface to maintain a certain surface pressure. A comprehensive review about LB films is found in [33]. Although 2D gold nanoparticle arrays, semiconducting quantum dots and polymeric films were realized using the LB technique, the method is quite difficult and requires expensive instrumentation.

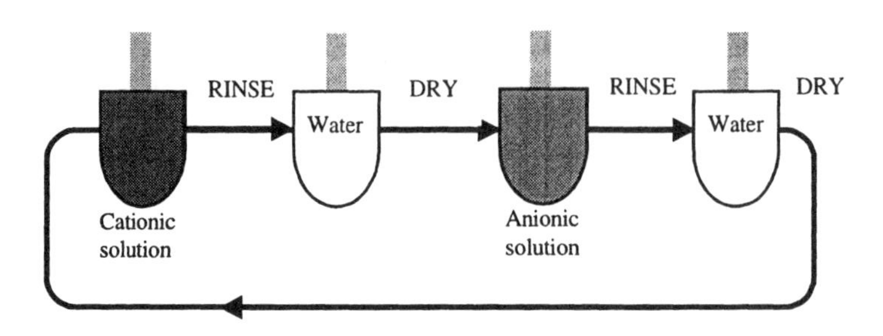

Figure 1.23 The electrostatic self-assembly process.

The electrostatic self-assembly is realized through the electrostatic interaction between molecules or nanoparticles. This self-assembly technique originates from the fabrication of multilayer films where each layer is composed of positive- and negative-charged colloid particles such as Si and Al. The method is widespread in the implementation of nanostructured films containing metals, semiconductors, magnetic materials, polymers, or organic molecules. The resulting film is uniform and stable due to strong ionic bonds between negative- and positive-charged particles, and the defects are minimized due to the repulsive force between the particles involved in the process. The utilization of polymers in combination with layers of charged nanoparticles is the best way to minimize the defects. The electrostatic self-assembly process starts with the immersion of a clean substrate into a cationic solution, followed by a dip of the substrate coated with cations into an anionic solution where the adsorption of anionic molecules takes place at a molecular level. The substrate is rinsed and dried after the cationic immersion and after dipping into the anionic solution. If the above procedure is repeated cyclically the nanostructured film is built up one layer/cycle, as shown in Figure 1.23. For example, positively charged gold nanoparticles are self-assembled on a negative glass substrate by immersion into 4-ATP(aminotiophenol)-capped gold solution with pH = 4. A layer of negatively charged Ag nanoparticles can be further deposited by immersion of the glass coated with Au particles into a 4-CTP(carboxythiophenol)-capped silver solution with pH = 8.5 [32]. Thus, it is possible to obtain a heterostructure consisting of successive layers of Au and Ag nanoparticles. Another example of a self-assembled structure is a multilayer film formed from nanosized iron oxide and conducting polymer chains, which benefits from both the mechanical properties of polymers (flexibility and strength) and the magnetic properties of iron. In both examples the final aggregate properties are a combination of the properties of its constituents, which is a common characteristic of self-assembled nanomaterials.

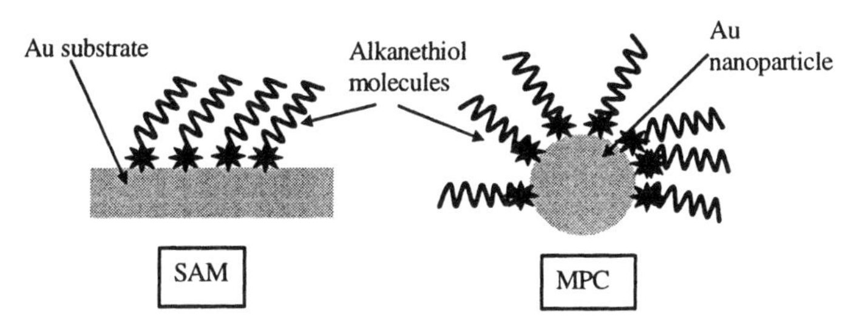

Figure 1.24 Schematic representations of the SAM and MPC methods.

The chemical self-assembly methods are the most commonly used techniques. Among them, SAM (self-assembled monolayers) and MPC (monolayer-protected clusters) are the most prominent. SAM deals with the spontaneous formation of monolayers via immersion of a suitable substrate into a solution, while MPC refers to nanoclusters whose surface is derivatized by ligand molecules with the help of chemisorption [32], as shown in Figure 1.24.

Biomolecular self-assembly uses DNA or proteins as basic constituents to realize: 1) biomolecular-metal complexes such as DNA-Au complexes, 2) self-assembly of semiconducting nanoparticles such as CdSe quantum dot ensembles, or 3) functionalized metallic nanoparticles.

There are many other methods for self-assembly of a final aggregate. The use of surface forces, as the control of the solvent evaporation that can produce an ordered array of colloids, is an alternative method of self-assembly. Yet another self-assembly technique is the utilization of self-organizing block copolymers. In summary, the self-assembly strategy is one of the main promoters of the realization of new materials or devices. The review paper [34] can be used for a more detailed analysis of the powerful self-assembly and self-organization methods.

Here, we will provide an example to give the reader a clear image as to the astonishing results that can be obtained by following the self-assembly strategy. Metal nanowires with a width in the range of 2–50 nm and lengths between 20 and 50 μm can be implemented via self-assembly techniques in a prescribed 3D geometry, without using lithographical methods [35]. This could be a breakthrough in the realization of interconnections between nanoscale circuits.

The method for producing these ultrathin and long metal nanowires consists of PECVD coating a Si substrate by SiO_2 that contains OH impurities, which induce compressive stress in the film. The Si/SiO_2 film is then annealed for a long time at 600 °C in order to remove the OH impurities; during this process tensile stress is produced in the film. The film ultimately cracks, and the nanoscale cracks can eventually reach the Si substrate. A metal is then deposited in these cracks, and finally the oxide is removed via wet etching, the network formed from metal nanowires being patterned on the substrate. The crack geometry is dependent on the type of the induced stress. For example, a uniaxial stress produces a parallel array of cracks and hence a parallel array of metal nanowires, as shown in Figure 1.25. Other examples regarding the self-assembly techniques will be found in the forthcoming sections of this chapter.

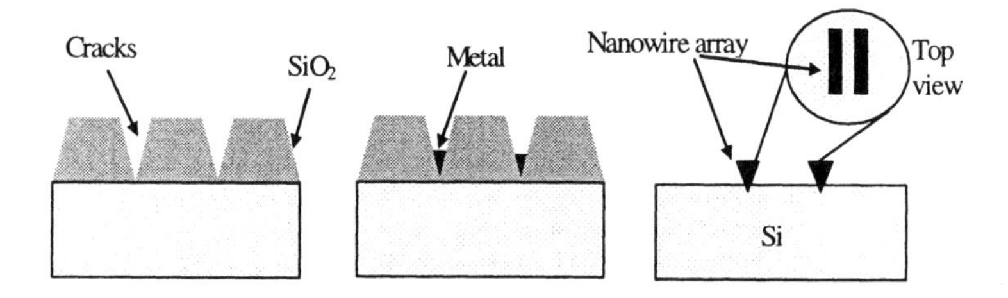

Figure 1.25 Self-assembly of a metal nanowire array (*After*: [35]).

1.3 NANOMATERIALS

The nanomaterials, which are the building blocks of the nanoelectronic devices, are realized with the nanotechnologies described in the previous section. The nanomaterials are very different in size, shape, and properties compared to the constituent bulk materials. All nanomaterials exhibits strong quantum behavior, which is utilized to implement various functionalities of nanoelectronic devices.

1.3.1 Nanoparticles

The physical properties, and especially the electronic structure of metal nanoparticles, are very different from those of bulk metals and are strongly dependent on their size. For example, the conduction band of bulk metals is replaced by discrete energy states in a metal nanoparticle. Thus, the metal particle behaves like a quantum dot, in which electrons are confined in all directions, in contrast with the free electrons of bulk metals. Metal particles covered with organic molecules such as thiols are able to self-organize in 1D, 2D and 3D arrays [36].

The discreteness of electronic states in a metal nanoparticle is characterized by the Kubo gap (i.e., the average spacing between successive quantum levels)

$$\delta = 4E_F / 3n , \tag{1.29}$$

where E_F is the Fermi energy, and n is the number of electrons in the nanoparticle. The nanoparticle is metallic if

$$k_B T > \delta . \tag{1.30}$$

For example, a silver particle with a diameter of 3 nm, for which $\delta = 10$ meV, is metallic at room temperature since $k_B T \cong 25$ meV for $T = 300$ K.

Many metal nanoparticles, such as Au, Ag, Pd, Ni, and Cu, are obtained by the evaporation of the corresponding bulk metal in vacuum. This method is suitable for materials with a low core-binding energy. The core-binding energy ΔE_b increases significantly when the diameter of the nanoparticle decreases, due to the weak screening of the core-hole and due to the occurrence of the metal-insulator transition, which is size dependent in the 1–2 nm range (i.e., for nanoparticles containing 200–400 atoms). The core-binding energy can attain values as large as 1.3 eV in Pd nanoparticles.

Another category of nanoparticles includes bimetallic or alloy colloid nanoparticles, which are obtained by chemical reduction of salt mixtures. Typical examples of bimetallic nanoparticles are Ag-Pd, Cu-Pd, Ni-Cu, Ag-Au, and Ni-Pd. The core-binding energy increases also in this case with the nanoparticle size but has now two components: one originating from alloying and the other from the size effect. At relative large diameters of bimetallic nanoparticles the alloying component is dominant, while the size component becomes more important and eventually prevails over the alloying component for nanoparticles with small size.

2D nanoparticle arrays can be realized when a hydrosol containing metal nanoparticles (Au, Pd) with a certain diameter is mixed with a toluene solution of alkali thiol. The metal nanoparticles mixed with thiol are then deposited on a solid surface and form a large regular 2D array in which the 3–4 nm nanoparticles are equally spaced by about 5–6 nm in the x and y directions [36]. This 2D array is basically an array of quantum dots in which the Coulomb blockade $I-V$ staircase behavior can be recognized in measurements. The charging energy increases when the diameter of the metal nanoparticles decreases.

3D assemblies can be implemented via adsorption of dithiol molecules and metal particles by plummeting the substrate into the respective solutions, washing with toluene solution, and drying. In this way, metal nanoparticle-metal nanoparticle or metal nanoparticle-semiconductor heterostructures can be realized, as can be seen from Figure 1.26.

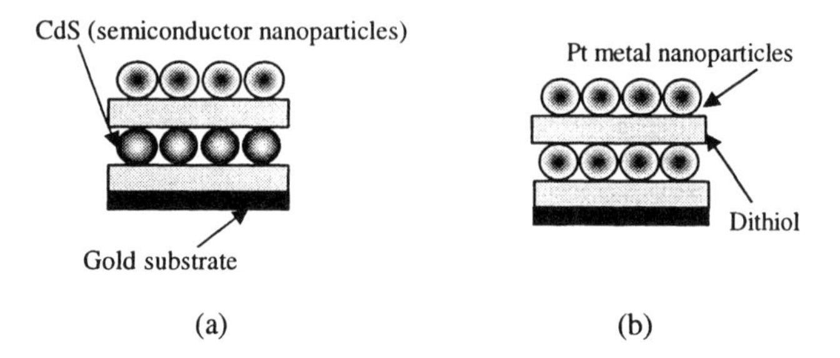

Figure 1.26 3D assemblies of nanoparticles: (a) semiconductor-metal and (b) metal-metal heterostructures.

There are not only metal nanoparticles, but also semiconductor nanoparticles such as silicon nanoparticles, III-V nanoclusters, elemental II-VI semiconductor quantum dots in solution or gaseous phases. The variety of methods for obtaining either metallic or semiconductor nanoparticles is presented in detail in [37]. Nanoparticles are hosted in either organic (see the example above) or dielectric media such as SiO_2 and are used especially in optoelectronic applications due to their enhanced optical nonlinearities.

1.3.2 Nanowires

The nanowires, called also quantum wires, form a large category of nanomaterials with many applications. The physical properties of nanowires differ significantly from those of the corresponding bulk materials. For example, nanowires show a quantized conductance behavior with a staircase shape, display Luttinger liquid signature, and support charge density waves.

A nanowire is a metallic, semiconducting, superconducting, or magnetic physical system, which is confined in two dimensions. The nanowire has a transverse area of a few nm and can attain 100–300 nm in length, although longer nanowires, up to 1 μm, have also been fabricated. Nanowires can be realized via many methods. A review of the bottom-up approaches to obtain atom scale nanowires can be found in [38]. Atom-scale contacts representing rudimentary nanowires originate from a break junction realized using a nanotip attached to an STM, which is pressed over a metal surface and then withdrawn;

nanomanipulation with the STM is treated in the paragraph following and in Chapter 2. By repeating this operation the metal is thinned until the contacts are a single-atom wide. Another way to obtain nanowires uses Si(001) since Si is more reactive in the regions where it is not hydrogenated. Using hydrogen as a nanoscale mask, materials can be adsorbed on the nonhydrogenated regions and thus form a nanowire. Fe, Ga, Al, Co, or Ag nanowires can form in the depassivated regions but are qualitatively poor, with defects and variable widths.

Self-assembled nanowires are made in various ways, especially using surface forces and specific chemical reactions. For example, rare-earth nanowires are obtained by depositing the rare-earth material on a Si(001) substrate, followed by annealing. The resulting material is an XSi_2 compound, which matches the Si lattice only in one direction. Due to the inherent strain, long and straight rare-earth metallic nanowires are finally obtained. Another example is the Bi nanowires, which form spontaneously when Si(001) covered with Bi is annealed near 600 °C. Bi nanowires are 1.5 nm wide and have a length in the range of 200–600 nm.

The most encountered method for nanowire growth is based on the template synthesis. The template consists of nanosized pores or voids that are arranged in a prescribed and ordered manner in a host material. The pores of the template are then filled with the material to be grown as a nanowire, called source material. Two very recent reviews dedicated to nanowire growth are [39] and [40].

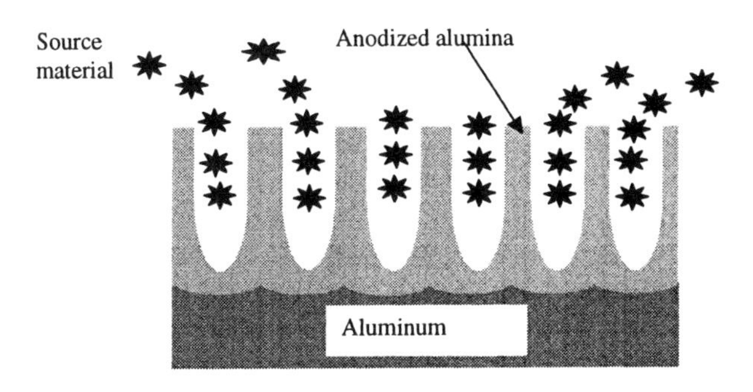

Figure 1.27 Nanowire growth in an alumina template.

The most commonly used template for nanowire growth is anodic alumina, although nanochannel glasses and etched polymers are also widely used. The alumina template is obtained by anodizing Al films in certain acids. During the anodization process an electrical current flows between the cathode and an Al thin film, which plays the role of the anode, the result of this etching process being an alumina film (membrane) that contains a regular hexagonal array of parallel and almost identical cylindrical channels, as shown in Figure 1.27. The alumina pores have a diameter in the range of 20–200 nm, are separated by 50–400 nm, and have a density of 10^9–10^{11} cm^{-3} depending on the etching conditions. A two-step anodization process is used to obtain very regular pore arrangements. In the first step, highly ordered imprints of alumina are produced, which are retained and filled with the source material in the second step.

Chemical etching of particle tracks due to ion bombardment is another template technique. Even DNA molecules can be used as templates. Almost all nanowire types can be grown using the template technique, which is accompanied by a deposition method that creates the nanowires, followed by the extraction of the source material after the nanowire is formed in the nanopores of the template.

Electrochemical deposition employed for thin films is also frequently used for the growth of nanowires. The deposition process is confined to a template of ordered nanopores, which is coated on one side with a thin metallic film that acts as a cathode. The electrochemical deposition can be used to create metallic nanowires of Fe, Co, Cu, Au, Ag, Ni, Pb, semiconductor nanowires of CdS, or superlattices A/B formed from two constituents A and B, for example, Cu and Co.

Nanowires with ultra-small diameters are grown using CVD or MOCVD. In this method, the nanowire precursor material is heated to produce vapors that penetrate the nanopores of the template, which is then cooled to get the solidified nanowires. Nearly single-crystal nanowires are obtained with the CVD method, while in the rest of the above-mentioned methods mainly polycrystalline nanowires are fabricated. Single-crystal nanowires of Bi, GaN, GaAs, and InAs, with diameters less than 10 nm, can be grown with the CVD techniques. Carbon nanotubes (CNTs), which have tremendous applications in nanoelectronics (see Chapter 3), are also grown in alumina templates via CVD techniques. CNTs with very small diameters, of a few nm, can be grown by replacing the alumina template with a zeolite template [40]. A comprehensive and updated review about CNT growth methods can be found in [41].

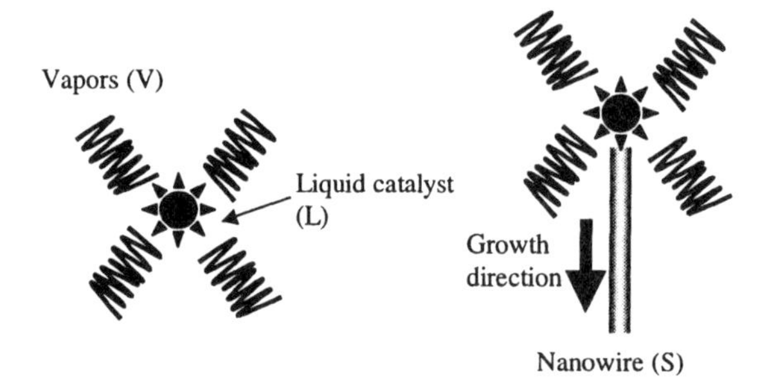

Figure 1.28 The VLS method of nanowire growth.

The vapor-liquid-solid (VLS) nanowire growth method is based on the fact that the vapors (V) of the source material can be absorbed into a liquid (L) droplet of a catalyst. The nanowire is obtained due to the solidification (S) of the source material as a result of the saturation of the liquid alloy followed by a nucleation process, which creates a preferential site for further deposition at the liquid boundary. In this way, other nucleation processes are disabled while the growth in one direction is enabled, as shown in Figure 1.28. Nanowires of Si, Ge, and ZnO are grown using the VLS method.

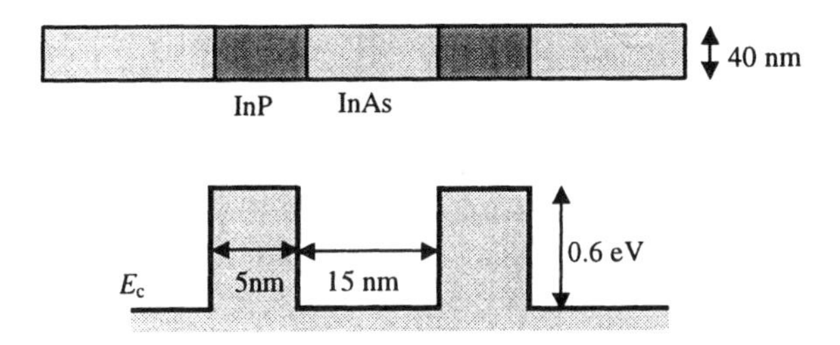

Figure 1.29 Circular nanowire InP/InAs superlattice and its conduction band diagram.

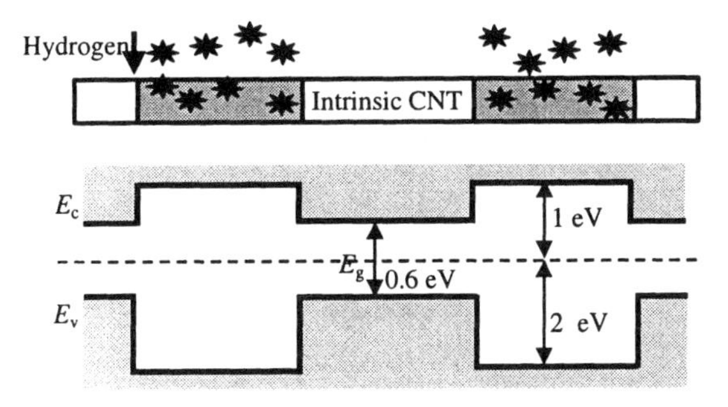

Figure 1.30 Carbon nanotube superlattice and its energy band diagram.

The arrangement of nanowires in a certain 1D, 2D, or 3D structure is a central issue of nanomaterial science. A typical example of a 1D nanowire arrangement is the nanometric superlattice formed from two types of nanowires, A and B, the resulting structure being of the form ABABAB…[40]. Hence, a nanometric superlattice is a nanowire whose material properties are periodically changed. Figure 1.29 displays an InP/InAs superlattice formed from AIII-BV semiconductor materials [42], while Figure 1.30 shows the energy band of a superlattice formed from an alternation of hydrogen functionalized semiconducting CNT and an intrinsic CNT; each material has a different energy band gap E_g [43]. Both nanowire superlattices show a pronounced negative differential resistance and can be used as resonant tunneling diodes (see Chapter 3 and 5), for generation of very high frequency electromagnetic radiation or as logic elements. In the former case, InP/InAs quantum dots are fabricated, which are geometrically confined in all three dimensions, the transport through the structure occurring via tunneling between consecutive quantum dots.

Nanowire superlattices are fabricated using VLS, CVD, or electrochemical methods. The examples presented up to now refer to an axial modulation of the nanowire composition. However, it is possible to create nanowire superlattices with a radial periodicity by growing multiple nanowire shells with different properties. This was demonstrated by the realization of coaxial Si/Ge and Ge/Si nanowires, fabricated by growing one nanowire via the VLS method followed by a CVD deposition of the second nanowire on the surface of the first one [44].

A simple way to obtain an array of nanowire superlattices, in fact an array of vertically ordered quantum dots, is to etch a 3D wafer on which two alternative materials are grown. In order to form a chain of coupled quantum dots the etching of the wafer is done using a template containing nanopores (an alumina template, for example) [40]. This is a direct method to obtain a variety of quantum dot arrays enclosed in nanowires, as illustrated in Figure 1.31.

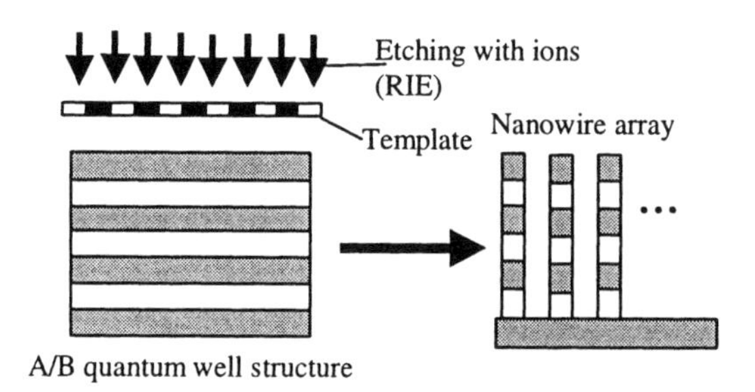

Figure 1.31 Fabrication method of nanometric superlattice arrays.

There are many other methods to fabricate nanowire arrays and to align them using self-assembly techniques, such as LB nanowire assemblies or electric-field-assisted assemblies. In this latter method, the nanowires are aligned between two electrodes, which are capacitively coupled with two buried electrodes excited by an ac voltage. When a metallic nanowire produces a shortcut between the two electrodes, the electric field is interrupted and no more nanowires are deposited in that place [45].

Millions of CNTs can be aligned and integrated using a large-scale assembly technique inspired from biomolecular self-assembly processes [46]. In this method, chemically functionalized patterns on a surface are first realized, on which millions of CNTs spread in a solution are then aligned. Two distinct regions coated via direct deposition with polar groups and nonpolar groups, respectively, create the functionalization of the surface. By placing the functionalized surface into a liquid suspension containing millions of CNTs the nanotubes are attracted by the polar regions and millions of CNTs are aligned in less than 10 s, as shown in Figure 1.32. The electrostatic attraction force rotates the CNTs towards the polar region and confines them only in this region. The efficiency of alignment is very high, about 90%.

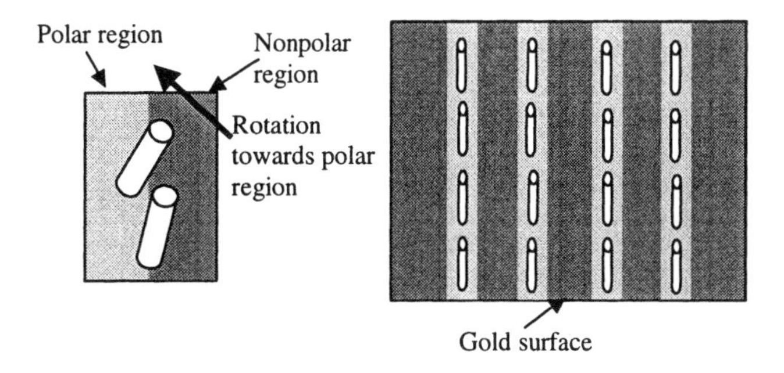

Figure 1.32 The self-assembly principle for aligning millions of carbon nanotubes.

1.3.3 Nanomagnetic Materials

Nanomagnetic materials have a huge variety of shapes and are formed from different material combinations. They have major applications in future memory systems, sensors or spintronic devices (see Chapter 4). The main categories of nanomagnetic materials are nanoparticles, nanowires, dots, and antidots, which are grown using the methods described above, in which the source material is magnetic. Examples of different types of nanomagnetic materials are shown in Figure 1.33. A comprehensive review on nanomagnetic materials can be found in [47].

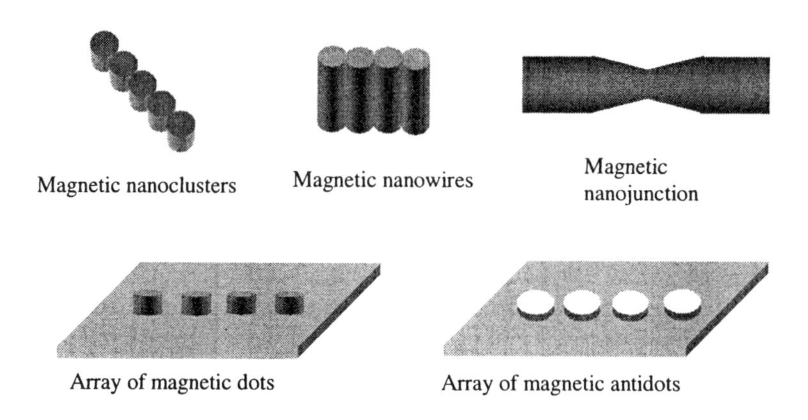

Figure 1.33 Different types of nanomagnetic materials (*After*: [47]).

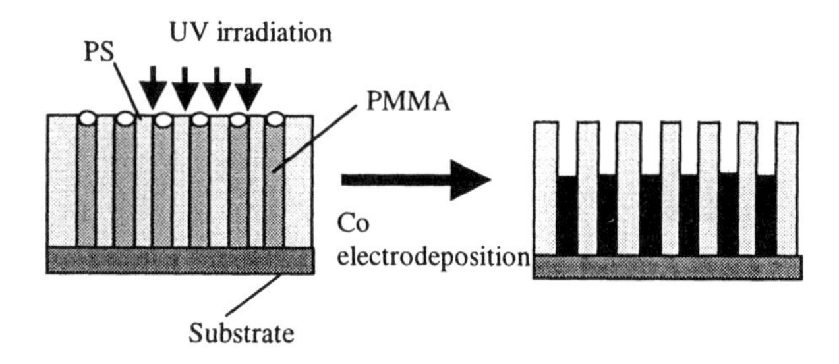

Figure 1.34 Implementation of a dense magnetic nanowire array via a copolymer template.

Arrays of metallic nanowires with magnetic properties, such as Co nanowires, are very important for the realization of ultra-high-density data storage, which requires a very dense and large array of nanowires with a high aspect ratio. This complicated array is grown through a template method that uses a PS-b-PMMA copolymer template, formed from PS (polystyrene) and PMMA monomers. The copolymer is sandwiched between metal electrodes and a dc voltage is applied beyond its glass transition temperature to anneal it. The result is a self-organization of the copolymer into a bi-dimensional array of cylinders of PMMA, with diameters of the order of nm, embedded into a matrix of PS, as shown in Figure 1.34. A subsequent UV irradiation degrades only the PMMA, creating a porous PS substrate that contains a 2D array of cylinders filled mainly with air. The Co nanowires, which are 500 nm long, are grown inside these cylinders by electrodeposition [48].

1.3.4 Nanostructured Surfaces

Nanostructured surfaces have geometrical characteristics that do not exceed 100 nm; a detailed review about nanostructured surfaces can be found in [49]. Nanostructured surfaces show different and very often surprising electrical, magnetic, or mechanical characteristics in comparison with the 3D solid from which they originate. One of the most notorious nanostructured surfaces is porous Si. Porous silicon is formed by electrochemical etching of Si in an HF solution. At the interface between Si and the electrolyte a random arrangement of nanopores appears, with diameters ranging between 10 nm and 100 nm. Very regular arrays of pores can also be obtained with much larger features. These macropores have a diameter of 1 μm, an array pitch larger than 2 μm, and a depth of 100 μm. Porous silicon is used for visible light emitting diodes and various sensors, including biosensors. The nanosize features of the silicon surface induce the appearance of new physical effects such as photoluminescence quenching due to long-lived electron-hole pairs, linear polarization memory, and strong optical nonlinearities. For a review of the main optical properties of porous silicon see [50].

REFERENCES

[1] Datta, S., *Electronic Transport in Mesoscopic Systems*, Cambridge: Cambridge University Press, 1997.

[2] Spector, J., et al., "Refractive switch for two-dimensional electrons," *Appl. Phys. Lett.*, Vol. 56, No. 24, 1990, pp. 2433-2435.

[3] Dragoman, D. and M. Dragoman, *Advanced Optoelectronic Devices*, Berlin: Springer, 1999.

[4] Ferry, D.K. and S.M. Goodnick, *Transport in Nanostructures*, Cambridge: Cambridge University Press, 1997.

[5] del Alamo, J.A., et al., "Electron waveguide devices," *Superlattices and Microstructures*, Vol. 23, No. 1, 1998, pp. 121-137.

[6] Dragoman, D. and M. Dragoman, *Quantum-Classical Analogies*, Berlin: Springer, 2004.

[7] Appenzeller, J., et al., "Tunneling versus thermionic emission in one-dimensional semiconductors," *Phys. Rev. Lett.*, Vol. 92, No. 4, 2004, pp. 048301/1-4.

[8] Zhu, W. (ed.), *Vacuum Microelectronics*, New York: John Wiley & Sons, 2001.

[9] Meyer, E., H.J. Hug, and R. Bennewitz, *Scanning Probe Microscopy*, Berlin: Springer, 2004.

[10] Hishinuma, Y., et al., "Refrigeration by combined tunneling and thermionic emission in vacuum: use of nanometer scale design," *Appl. Phys. Lett.*, Vol. 78, No. 17, 2001, pp. 2572-2574.

[11] Jensen, K.L., "Theory of field emission," in *Vacuum Microelectronics*, W. Zhu (ed.), New York: Wiley, 2001, pp. 33-104.

[12] Wacker, A. and A.-P. Jauho, "Quantum transport: the link between standard approaches in superlattices," *Phys. Rev. Lett.*, Vol. 80, No. 2, 1998, pp. 369-372.

[13] Likharev, K.K., "Single-electron devices and their application," *Proc. IEEE*, Vol. 87, No. 4, 1999, pp. 606-632.

[14] Debray, P., et al., "Coulomb drag between ballistic one-dimensional electron systems," *Semicond. Sci. Technol.*, Vol. 17, No. 11, 2002, pp. R21-R34.

[15] Chen, G., "Thermal conductivity and ballistic-phonon transport in the cross-plane direction of superlattices," *Phys. Rev. B*, Vol. 57, No. 23, 1998, pp. 14958-14973.

[16] Rego, L.G.C. and G. Kirczenow, "Quantized thermal conductance of dielectric quantum wires," *Phys. Rev. Lett.*, Vol. 81, No. 1, 1998, pp. 231-234.

[17] Angelescu, D.E., M.C. Cross, and M.L. Roukes, "Heat transport in mesoscopic systems," *Superlattices and Microstructures*, Vol. 23, No. 3/4, 1998, pp. 673-689.

[18] Patton, K.R. and M.R. Geller, "Thermal transport through a mesoscopic weak link," *Phys. Rev. B*, Vol. 64, No. 15, 2001, pp. 155320/1-7.

[19] Ciraci, S., A. Buldum, and I.P. Batra, "Quantum effects in electrical and thermal transport through nanowires," *J. Phys.: Condens. Matter*, Vol. 13, No. 29, 2001, pp. R537-R568.

[20] Ziaie, B., A. Baldi, and M.Z. Atashbar, "Introduction to Micro/Nanofabrication," in *Springer Handbook of Nanotechnology*, B. Bhushan (ed.), Berlin: Springer, 2004, pp. 147-184.

[21] Dragoman, D. and M. Dragoman, "Micro/Nano-Optoeletromechanical systems," *Prog. Quantum Electron.*, Vol. 25, No. 6, 2001, pp. 229-290.

[22] Harriott, L.R. and R. Hull, "Nanolithography," in *Introduction to Nanoscale Science and Technology*, M. Di Ventra, S. Evoy, and J.R. Heflin Jr. (eds.), Dordrecht: Kluwer Academic Publishers, 2004, pp. 7-40.

[23] Matsui, S., "Nanostructure fabrication using electron beam and its application to nanometer devices," *Proc. IEEE*, Vol. 85, No. 4, 1997, pp. 629-643.

[24] Tseng, A.A., K. Chen, and K.J. Ma, "Electron beam lithography in nanoscale fabrication: recent developments," *IEEE Trans. on Electronic Packaging Manufacturing*, Vol. 26, No. 2, 2003, pp. 141-1949.

[25] Zankovych, S., et al., "Nanoimprint lithography," *Nanotechnology*, Vol. 12, No. 1, 2001, pp. 91-95.

[26] Guo, L.J., "Recent progress in nanoimprint technology and its applications," *J. Phys. D*, Vol. 37, No. 11, 2004, pp. R123-R141.

[27] Chou, S.Y., C. Keimel, and J. Gu, "Ultrafast and direct imprint of nanostructures in silicon," *Nature*, Vol. 417, No. 6891, 2002, pp. 835-837.

[28] Colton, R.J., "Nanoscale measurements and manipulation," *J. Vac. Sci. Technol. B*, Vol. 22, No. 4, 2004, pp. 1609-1635.

[29] Snow, E.S., P.M. Campbell, and F.K. Perkins, "Nanofabrication with proximal probes," *Proc. IEEE*, Vol. 85, No. 4, 1997, pp. 601-611.

[30] Ohtsu, M., et al., "Nanofabrication and atom manipulation by optical near-field and relevant quantum optical theory," *Proc. IEEE*, Vol. 88, No. 9, 2000, pp. 1499-1518.

[31] Meschede, D. and H. Metcalf, "Atomic nanofabrication: atomic deposition and lithography by laser and magnetic forces," *J. Phys. D*, Vol. 36, No. 3, 2003, pp. R17-R38.

[32] Huie, J.C., "Guided molecular self-assembly: a review of recent efforts," *Smart Mater. Struct.*, Vol. 12, No. 2, 2003, pp. 264-271.

[33] Rietman, E.A., *Molecular Engineering of Nanosystems*, New York: Springer, 2001, pp. 158-185.

[34] Shenhar, R., T.B. Norsten, and V.M. Rotello, "Self-assembly and self-organization," In *Introduction to Nanoscale Science and Technology*, M. Di Ventra, S. Evoy, and J.R. Heflin, Jr. (eds.), Dordrecht: Kluwer Academic Publishers, 2004, pp. 41-74.

[35] Saif, T., E. Alaca, and H. Sehitoglu, "Nano wires by self assembly," 16th IEEE Annual Conference on Micro Electro Mechanical Systems MEMS 03, Kyoto, Japan, 19-23 January 2003, pp. 45-47.

[36] Rao, C.N.R., et al., "Metal nanoparticles, nanowires and carbon nanotubes," *Pure Appl. Chem.*, Vol. 72, No. 1-2, 2000, pp. 21-33.

[37] Adair, J.H., et al., "Recent developments in the preparation and properties of nanometer-size spherical and platelet-shaped particles and composite particles," *Materials Science and Engnieering*, Vol. R 23, No. 2, 1998, pp. 139-242.

[38] Bowler, D.R., "Atomic-scale nanowires: physical and electronic structure," *J. Phys.: Condens. Matter*, Vol. 16, No. 24, 2004, pp. R721-R754.

[39] Dresselhaus, M.S., et al., "Nanowires," in *Springer Handbook of Nanotechnology*, B. Bhushan (ed.), Berlin: Springer, 2004, pp. 99-145.

[40] Chik, H. and J.M. Xu, "Nanometric superlattices: non-lithographic fabrication, materials, and properties," *Materials Science and Engineering*, Vol. R43, No.1, 2004, pp. 103-138.

[41] Monthioux, M., et al., "Introduction to carbon nanotubes," in *Springer Handbook of Nanotechnology*, B. Bhushan (ed.), Berlin: Springer, 2004, pp. 99-145.

[42] Björk, M.T., et al., "Nanowire resonant tunneling diode," *Appl. Phys. Lett.*, Vol. 81, No. 23, 2002, pp. 4458-4460.

[43] Gülseren, O., T. Yildirim, and S. Ciraci, "Formation of quantum structures on a single nanotube by modulating hydrogen adsorption," *Phys. Rev. B*, Vol. 68, No. 11, 2003, pp. 115419/1-6.

[44] Lauhon, L.J., et al., "Epitaxial core-shell and core-multishell nanowire heterostructures," *Nature*, Vol. 420, No. 6911, 2002, pp. 57-61.

[45] Smith, P.A., et al., "Electric field assisted-assembly and alignment of metallic nanowires," *Appl. Phys. Lett.*, Vol. 77, No. 9, 2000, pp. 1399-1401.

[46] Rao, S.G., et al., "Large-scale assembly of carbon nanotubes," *Nature*, Vol. 425, No. 6953, 2003, p. 36-37.

[47] Skomski, R., "Nanomagnetics," *J. Phys. C*, Vol. 15, No. 20, 2003, pp. R841-R896.

[48] Thurn-Albrecht, T., et al., "Ultra-high density nanowires arrays grown in self-assembled diblock copolymer templates," *Science*, Vol. 290, No. 5499, 2000, pp. 2126-2129.

[49] Rosei, F., "Nanostructured surfaces: challenges and frontiers in nanotechnology," *J. Phys.: Condens. Matter*, Vol. 16, No. 17, 2004, pp. S1373-S1436.

[50] Dragoman, D. and M. Dragoman, *Optical Characterization of Solids*, Berlin: Springer, 2002, pp. 362-371.

Chapter 2

Instrumentation for Nanoscale Electronics

This chapter presents the main instrumentation techniques utilized at nanoscale to observe, manipulate, and measure nanosized components and devices. Unlike instrumentation techniques at other scales, which are extremely specialized, the nanoscale instrumentation has multiple purposes. For example, using an atomic force microscope (AFM) it is possible to manipulate nano-objects, to measure the distribution of the electrical resistivity or the electromagnetic field over a certain surface, to perform surface topography and lithographical processes, both with Å resolution. Of course, a lot of instrumentation techniques, such as Raman or luminescence test equipments, is common for different device scales; these must however be properly changed in order to work with material samples, components, or devices that have at least one dimension of less than a few nm.

This chapter presents the instrumentation specific for nanoelectronics, which is mainly based on the principles of micro-electro-mechanical systems (MEMS) and nano-electro-mechanical systems (NEMS). These systems are the first to be treated in this chapter, but not with the aim to provide a full description of MEMS/NEMS and their applications, which could be the subject of a distinct book. In what follows we describe only those MEMS and NEMS devices with direct relevance to nanoelectronics and in particular, to the instrumentation used at the nanoscale. The specific instrumentation of nanoelectronics based on MEMS and NEMS, such as the above-mentioned AFM, will be subsequently described, with an emphasis on its capabilities and performances.

2.1 MEMS AND NEMS

2.1.1 Micro and Nanocantilevers

MEMS and NEMS combine mechanical and electrical devices that have, respectively, micronic (10^{-6} m) and nanometer (10^{-9} m) dimensions. The resulting system is called very often "smart" because it can sense the environment changes and optimally reconfigure its functions to respond to new external conditions.

A key component of MEMS or NEMS is the cantilever, which is also a basic device for many instruments used in nanoelectronics. The cantilever, represented schematically in Figure 2.1, can be monolithically integrated with electronic and/or optoelectronic components for various purposes. The basic function of the cantilever is its controlled bending due to an applied actuation force, which can have an electric, magnetic, thermal, or optical origin. The electrical actuation is the most encountered, in this case the cantilever being very often terminated with a nanometer size metallized tip. Electrical forces are generated by applying a voltage V between the cantilever and the metal deposited on the substrate (the substrate electrode), the cantilever being thus attracted towards the substrate.

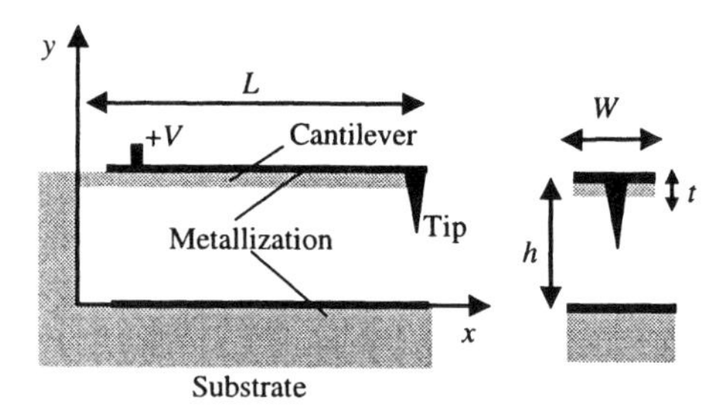

Figure 2.1 Schematic representation of a cantilever with rectangular cross-section.

The electrostatic force $F_{\text{e-s}}(y)$ applied along the y direction on a cantilever with width W, length L, and thickness t can be calculated in terms of the energy

$$U = CV^2/2 \tag{2.1}$$

stored in the capacitor C formed between the cantilever and the substrate as

$$F_{\text{e-s}}(y) = dU/dy = (V^2/2)dC/dy\,. \tag{2.2}$$

Introducing in (2.2) the expression of the capacitance C we obtain

$$F_{\text{e-s}}(y) = \frac{V^2}{2}\frac{\varepsilon_0 WL}{[y+(t/\varepsilon)]^2}\,, \tag{2.3}$$

where ε_0 is the free-space electrical permittivity and ε is the permittivity of the cantilever material.

The cantilevers are commonly fabricated from semiconductor materials, especially Si [1], but very thin GaAs cantilevers were also produced using micromachining technologies [2]. Metallic cantilevers are also frequently encountered [3], and in many cases semiconductor cantilevers are coated with various materials able to adsorb chemical or biological substances that provoke an additional bending of the cantilever, which is essential for sensing applications. Micronic cantilevers have lengths of tens of μm and widths and thicknesses of a few microns, the much smaller nanocantilevers having widths and thicknesses of a few nm and lengths of the order of a few microns.

The bending of the cantilever due to an applied force is expressed by the equation

$$d^2y/dx^2 = M(x)/EI\,, \tag{2.4}$$

where E is the Young modulus of elasticity,

$$I = \int_{-t/2}^{t/2} y^2 dy \int_{-W/2}^{W/2} dz = Wt^3/12 \tag{2.5}$$

is the moment of inertia and

$$M(x) = (1/L)\int_{x'=x}^{L} F_{e\text{-}s}(y(x'))(x-x')dx' = -\varepsilon_0 W \int_{x'=x}^{L} (x-x')/[y(x')+(t/\varepsilon)]^2 dx' \tag{2.6}$$

is the total moment at position x, completely determined by the electric force. Thus, the bending of the cantilever is described by the integro-differential equation

$$d^2y/dx^2 = -V^2(6\varepsilon_0/Et^3)\int_{x'=x}^{L} (x-x')/[y(x')+(t/\varepsilon)]^2 dx', \tag{2.7}$$

which cannot in general be solved analytically. However, if $y(x') \approx h$, which implies that $y(x)$ is not very different from the initial air gap value h in the absence of an applied force, an approximate solution of (2.7) is

$$y(x) = h - V^2(6\varepsilon_0/Et^3)[L^2x^2/4 - Lx^3/6 + x^4/24]/[h+(t/\varepsilon)]^2\,. \tag{2.8}$$

The maximum deflection obtained for $x = L$ is then

$$y(L) = h - V^2(3\varepsilon_0 / 4Et^3)L^4 / [h + (t / \varepsilon)]^2 . \qquad (2.9)$$

The cantilever deflection is usually detected by an optical system as that represented in Figure 2.2. A laser beam is focused on a cantilever or an array of cantilevers, and the reflected light, which contains information about the deflection, is sensed by an optical array detector. More sophisticated detection schemes for deflection measurement exist, as discussed in Section 2.2.1, but the large majority of them are based on the simple principle described in Figure 2.2, that is, the detection of a reflected light beam.

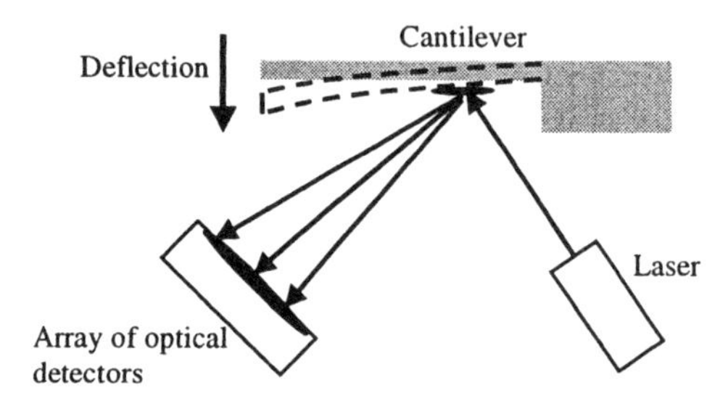

Figure 2.2 Detection principle of cantilever deflection.

Another way to calculate the deflection parameters of the cantilever is to consider that the deflection caused by the force $F(y(x)) = f(x)$, which is concentrated in a single point x, is given by [4]

$$y_{\max} = x^2(3L - x)Wf(x)dx / 6EI . \qquad (2.10)$$

Then, the total deflection due to the electrostatic force that is uniformly distributed along the cantilever is

$$y_{\max} = W\int_0^L (3L - x)x^2 f_{\text{e-s}}(x)(6EI)^{-1} dx , \qquad (2.11)$$

where

$$f_{e\text{-}s}(x) = -(\varepsilon_0 / 2)V^2 / [t - y(x)]^2 \tag{2.12}$$

is the electrostatic force per unit area.

In contrast with equation (2.7), the integral in (2.11) can be exactly solved for a square-law behavior $y(x) = (x/L)^2 y_{max}$ of the beam deflection [4]. The deflection can be determined numerically in this case from the equation

$$\begin{aligned}\varepsilon_0 WL^4V^2 / 2EIh^3 = (4\bar{y}_{max})[(2/3)(1-\bar{y}_{max}) - \tanh^{-1}(\bar{y}_{max})^{1/2} / (\bar{y}_{max})^{1/2} \\ - \ln(1-\bar{y}_{max}) / 3(\bar{y}_{max})]^{-1}.\end{aligned} \tag{2.13}$$

Figure 2.3 represents the dependence of the normalized deflection $\bar{y}_{max} = y_{max} / h$ on the right-hand side of equation (2.13), which stands for a normalized load dependent on the voltage V.

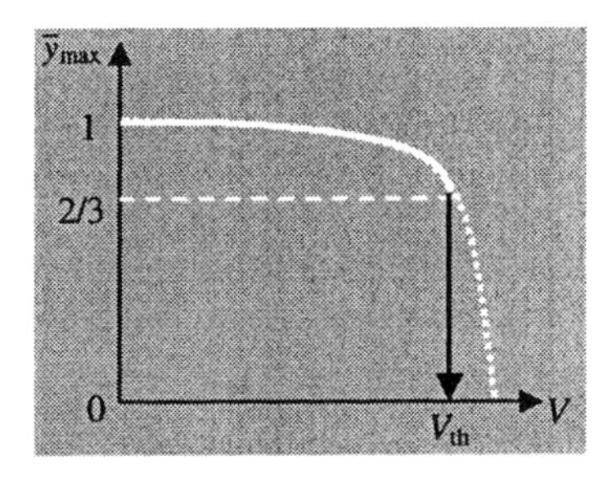

Figure 2.3 Normalized deflection as a function of the applied voltage.

The cantilever is in a dynamical equilibrium defined by the condition that the sum between the electrostatic force and the elastic force $F_{el} = -Ky$ vanish:

$$F_{e\text{-}s}(y) - Ky = 0, \tag{2.14}$$

where K is the spring constant given by $K = 3EI / L^3 = Et^3W / 4L^3$ for a cantilever beam. The equilibrium is maintained up to one-third of the cantilever substrate height h, after which the balance between the electrostatic and elastic force is broken and the cantilever collapses on the substrate electrode. The threshold voltage for which the collapse occurs is given by

$$V_{th} = \sqrt{8Kh^3 / 27WL\varepsilon_0}, \tag{2.15}$$

or [4].

$$V_{\text{th}} = \sqrt{8EIh^3 / 9\varepsilon_0 L^4 W}\ . \tag{2.15.a}$$

These formulas are very important for many MEMS applications that employ cantilevers [5]. However the above formulas are no longer valid if the cantilever, as well as h, is of nanometric size. Then, one has to take into account the forces that act on the cantilever at the nanoscale only.

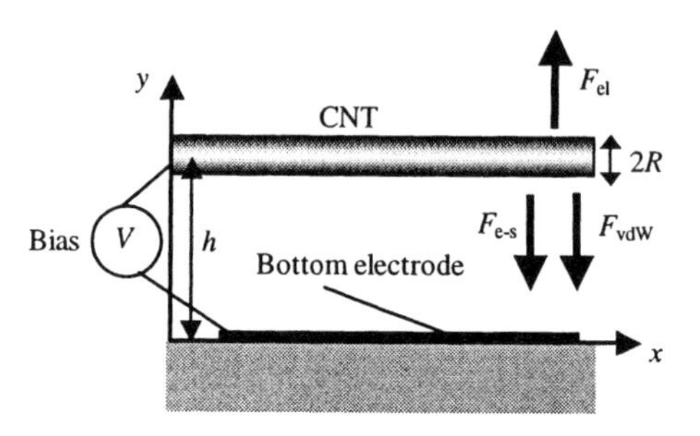

Figure 2.4 Metallic carbon nanotube cantilever and the forces acting on it at the nanoscale.

For example, let us consider a nanocantilever made from a multiwalled carbon nanotube (CNT), as that displayed in Figure 2.4. The integro-differential equation (2.7) remains valid for the CNT cantilever if it encompasses the van der Waals force F_{vdW}, which is an attractive force that acts only at the nanoscale. (The van der Waals force originates from intermolecular attraction and repulsion, and can be either repulsive or attractive, but for nanocantilevers only the attractive component is taken into account, the repulsive part becoming important after the cantilever comes in contact with the substrate electrode.) In this case [6]

$$EId^4 y / dx^4 = d[F_{\text{e-s}} + F_{\text{vdW}}] / dy = f_{\text{e-s}} + f_{\text{vdW}}\ , \tag{2.16}$$

where the electrostatic force per unit length is

$$f_{\text{e-s}} = \frac{\pi\varepsilon_0 V^2}{R[y(y+2R)/R^2]\log^2\{1 + y/R + [y(y+2R)/R^2]^{1/2}\}} \tag{2.17}$$

with R the CNT radius, and

$$f_{vdW} \propto (1/h^7)[1/(y+2R)^2 - 1/y^2]. \quad (2.18)$$

The van der Waals force has a significant contribution to the actuation process of nanocantilevers. Because the van der Waals force is attractive in nature, it can pull down the cantilever even when no voltage is applied on it. The occurrence of specific physical phenomena at nanoscale (the van der Waals force, for example, is negligible at larger scales) implies that nanoscale devices are not simply downscales of bigger, micron-scale devices. This important conclusion should be kept in mind throughout the book.

From the above discussion it is apparent that the threshold voltage for nanocantilevers is different than that for cantilevers with micronic dimensions. If we approximate our nanotube with a filled narrow rectangle with a thickness $t = 2R$, and an effective width W obtained from matching the moment of inertia with that of the circular geometry [6], the threshold voltage is obtained from the condition $F_{el} + F_{e\text{-}s} + F_{vdW} = 0$ and is given by

$$\begin{aligned} V_{th} &= \{K(h - y_{th}) - C_1[1/y_{th}^3 - 1/(y_{th} + t)^3]\}^{1/2} \\ &\times \{2y_{th}^2 / [(1 + y_{th} / \pi W)WL\varepsilon_0]\}^{1/2} \end{aligned} \quad (2.19)$$

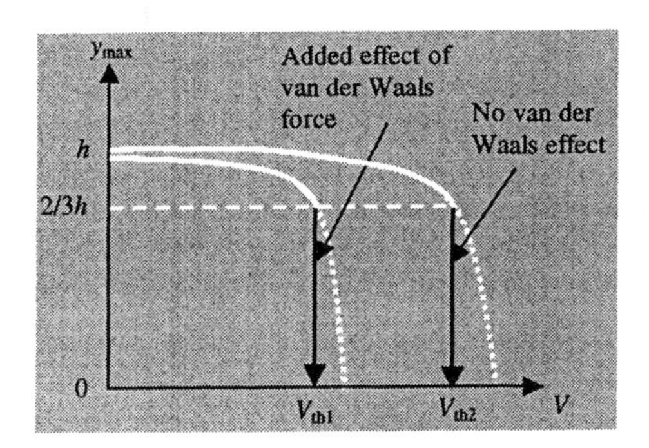

Figure 2.5 Deflection of a nanocantilever as a function of voltage in the presence and absence of van der Waals force.

where $y_{th} \cong 2/3h$ and C_1 is a constant. The first term in (2.19) includes the effects of the elastic and electrostatic forces while the second term originates from van der Waals forces. If the van der Waals contribution is neglected, as is the case at larger scales, then the threshold voltage becomes similar to that given in (2.15). However, the van der Waals force has a significant effect on the nanocantilever deflection, as can be seen from Figure 2.5, which displays this effect for a 50 nm long CNT that has a diameter of 2 nm and is positioned at 4 nm above the substrate electrode.

The influence of the van der Waals force at the nanoscale is strongly dependent on the geometry of the cantilever. Unlike the case of a free-standing cantilever, represented in Figure 2.5, the van der Waals force has no significant effect when the CNT is double-clamped. A double-clamped CNT cantilever can work either as a switch, when the NEMS has a substrate electrode, as in Figure 2.6(a), or as a mechanical resonator, when the substrate electrode is removed. In the latter case, represented schematically in Figure 2.6(b), the resonator has an oscillation frequency $f_{osc} \cong 1.03(E/\rho)^{1/2} \times (2R/L^2)$, where ρ is the density of the material.

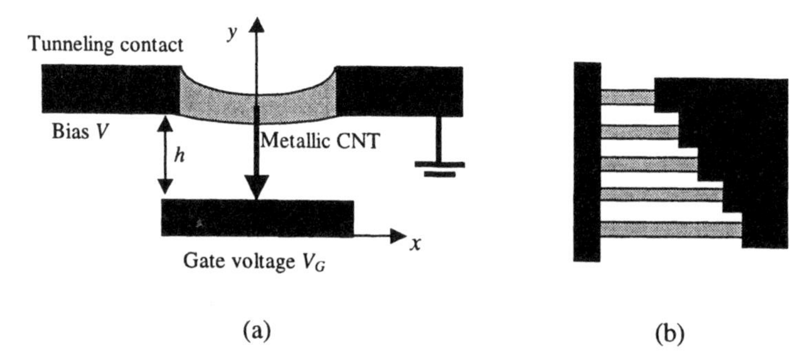

Figure 2.6 The double-clamped CNT NEMS working as: (a) a nanoswitch, and (b) an array of resonators.

Even double-clamped NEMS, for which van der Waals forces have lesser effects, are not replicas of identical devices at larger scales. The reason is that in this case the deflection of the nanostructure represented in Figure 2.6(a) changes in discrete steps each time an electron injected from the bias contact tunnels through the nanotube [7]. The CNT is actuated by the gate voltage V_G but, when the bias V is raised, the tunnel contact injects a discrete number of charges ne, which represents an additional electrostatic contribution to the nanotube bending. The maximum displacement is then

$$y_{max} = 0.013(ne)^2 L^2 / ER^4 h \tag{2.20}$$

when the total stress $T = (\pi ER^2 / 2L)\int_0^L [h - y(x)]^2 dx$ satisfies $T << EI / L^2$, and takes the form

$$y_{max} = 0.24(ne)^{2/3} L^{2/3} / (E^{1/3} R^2 h^{1/3}) \tag{2.21}$$

when $T << EI / L^2$.

The inequalities dealing with the total stress can be rewritten as a function of the discrete n value as $n << ER^5 h / (e^2 L^2)$ and $n >> ER^5 h / (e^2 L^2)$, respectively, where

$$n = \text{Int}\left[\frac{V_G L}{2R\ln(2h/R) + 1/2}\right]. \tag{2.22}$$

The above relation reveals an astounding characteristic of the nanocantilever, which has a quantized mechanical movement since its deflection changes in discrete steps every time an electron tunnels through the tunneling contact. In this case, the dependence of the maximum nanocantilever deflection on the actuation voltage is represented in Figure 2.7.

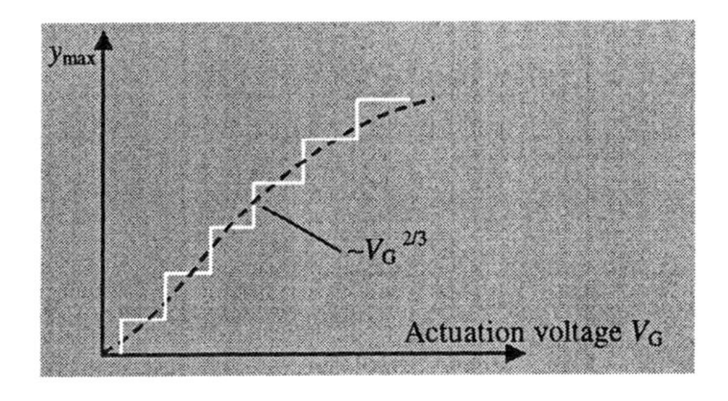

Figure 2.7 The quantized motion of a nanocantilever (*After*: [7]).

Other interesting effects besides the van der Waals force, which actuate supplementary the nanocantilevers, are characteristic for nanostructures for which the distance between the actuated element and the substrate electrode is in the nanometer range [8]. In this regime quantum forces fully manifest their strength.

An example in this sense is the Casimir force, which can be understood only in the realm of quantum field theory. According to this theory the vacuum is not empty but filled with virtual particles, which are continuously created and annihilated. The effect of this vacuum fluctuation is manifested in the Casimir force between two arbitrary bodies separated by a distance less than 1 μm. The strangest aspect of Casimir force is its dependence on geometry and the boundary conditions. For example, the Casimir force between two conducting plates in a vacuum is attractive due to the fact that the reflective surfaces of the metal plates cancel the virtual photons with wavelengths longer than the separation distance. On the other hand, if the plates are replaced by hemispherical shells the Casimir force becomes repulsive.

The Casimir force between two metallic plates of area S and separated by a distance d is given by

$$F_{\text{Cas,plate}} = -\pi^2 \hbar c S / 240 d^4 . \tag{2.23}$$

For an area of 1 cm^2 and $d = 500$ nm the Casimir force is about 3 μN (i.e., 3×10^{-6} N), which is easily detected by cantilevers as will be discussed in the next section; under certain conditions the cantilevers are able to detect forces as small as 1 aN (i.e., 10^{-18} N).

The Casimir force between a sphere of radius R and a plate is given by

$$F_{\text{Cas,sphere}} = -\pi^3 \hbar c R / 360 d^3 , \tag{2.24}$$

where d is now the separation between the plate and the sphere; additional expressions of Casimir forces between different bodies and their behavior as a function of temperature can be found in [8].

The Casimir force is a mesoscopic force, the spatial dependence and sign of which can be engineered via the shapes of the interacting bodies. The engineering of a force action can be used to implement nonlinear mechanical oscillators and switches at the nanoscale [9–11]. A Casimir NEMS device consists, for example, of a metallic plate, which is able to move under the action of an elastic force, and a sphere placed over the plate, as indicated in Figure 2.8. The distance between the sphere and the plate was chosen 40 nm in [9]. The nonlinear dependence of the energy of the Casimir oscillator on the position, beyond a certain critical distance, described by the formula [11]

$$U = Kx_0^2(\bar{x}^2 / 2) - (\hbar c \pi^2) / [720(1-\bar{x})^3 x_0^3] , \tag{2.25}$$

where $\bar{x} = x / x_0$, with x_0 the equilibrium position, is displayed also in Figure 2.8 in terms of the unit of elastic energy $U_0 = Kx_0^2$. From this figure it follows that the NEMS oscillates around the local minima and can switch between its minimum and maximum positions.

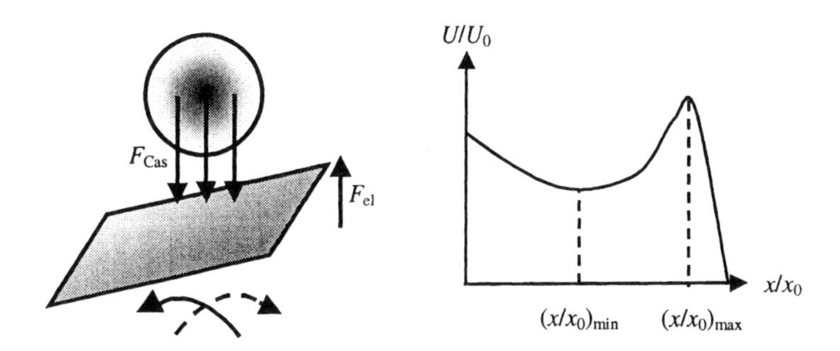

Figure 2.8 The Casimir switch (*After*: [11]).

2.1.2 Frequency Analysis of Micro and Nanocantilevers

For a complete analysis of the cantilever, the basic equation (2.4) must be replaced by a generalized equation in which $y(x)$ is replaced by the deflection $y(x,\tau)$ that is a function of the space coordinate x and the temporal coordinate τ, and (2.4) is replaced by [12]

$$EI(d^4y/dx^4)+m_B(d^2y/d\tau^2)=0. \tag{2.26}$$

Here $m_B = \rho A$ is the cantilever mass per unit length with ρ the density of the cantilever material and $A = Wt$ the cross-sectional area of the cantilever. Assuming that the cantilever vibrates only in one of its natural resonance frequencies ω_n, so that it has a harmonic behavior, the solution of (2.26) has the form

$$y(x,\tau) = X(x)\cos(\omega_n\tau+\theta). \tag{2.27}$$

By introducing the notation

$$k^4 = \omega_n^2 \rho Wt / EI \tag{2.28}$$

equation (2.26) can be rewritten as

$$d^4X/dx^4 = k^4X(x), \tag{2.29}$$

its general solution having the form [12]

$$X = A[\cos(kx + \cosh(kx)] + B[\cos(kx - \cosh(kx)] + C[\sin(kx) + \sinh(kx)] \\ + D[\sin(kx) + \sinh(kx)] \tag{2.30}$$

where the constants A, B, C, and D are found by imposing appropriate boundary conditions. More precisely, it is assumed that the deflection, which is proportional with X, and its slope, which is proportional with $X' = dX / dx$, vanish at the bukit-in end of the cantilever. We assume also that the moment, which is proportional with X'', and the shear, which is proportional with X''', are zero at the free end of the cantilever. So, the boundary conditions in (2.30) are

$$\begin{cases} y = 0 \\ \partial y / \partial x = 0 \end{cases} \text{at } x = 0, \qquad \begin{cases} \partial^2 y / \partial x^2 = 0 \\ \partial^3 y / \partial x^3 = 0 \end{cases} \text{at } x = L \tag{2.31}$$

from which we get $A = C = 0$,

$$D / B = [\cos(kL) + \cosh(kL)]/[\sin(kL) + \sinh(kL)] \\ = [\sin(kL) - \sinh(kL)]/[\cos(kL) + \cosh(kL)] \tag{2.32}$$

and the natural frequencies of the cantilever are given by

$$\cos(k_n L)\cosh(k_n L) = -1 . \tag{2.33}$$

The first four natural resonance frequencies of the cantilever satisfy the relations

$$k_0 L = 1.875 , \tag{2.34a}$$

$$k_1 L = 4.694 , \tag{2.34b}$$

$$k_2 L = 7.855 , \tag{2.34c}$$

$$k_3 L = 10.996 , \tag{2.34d}$$

the first resonance frequency determined from (2.28) being

$$\omega_0 = 2\pi f_0 = 1.015(t / L^2)(E / \rho)^{1/2} . \tag{2.35}$$

In general, the natural frequencies of the cantilever can be put in the form

$$\omega_n = (c_n / L)^2 (EI / m_B)^{1/2} , \tag{2.36}$$

where the constants c_n are determined from (2.28) and (2.34a)–(2.34d); for example $c_1 = 1.9$.

Formula (2.36) is valid however only if the cantilever moves in vacuum. But how do the resonant frequencies change when the cantilever moves without friction in an incompressible medium, which could be air, water, or various liquids? This situation is mostly encountered in the nanoscale instrumentation that will be described in the next sections. Equation (2.26) remains valid even in this case, but the mass per unit length differs. It consists now, besides the mass per unit length of the cantilever, from an additional contribution originating from the moved medium (M) due to the movement of the cantilever. Thus we have [13]

$$m_{\mathrm{B}} \mapsto m_{\mathrm{B}} + m_{\mathrm{M},n} = (1 + \mu_{\mathrm{M},n}) m_{\mathrm{B}} , \tag{2.37}$$

where the dimensionless quantity

$$\mu_{\mathrm{M},n} = WL\rho_{\mathrm{M}}\pi[(n-1)+1/4]/3A\rho(2n-1)^2 = (L/t)(\rho_{\mathrm{M}}/\rho)g_n \tag{2.38}$$

describes the mass to be moved by the cantilever when it is excited by the nth natural frequency, with ρ_{M} the medium density, ρ the cantilever density, and g_n the nth eigenmode. The co-moved mass of the medium is modeled as a cylinder that has in its center a certain node of the cantilever vibration. In consequence, the natural vibrating frequencies of the cantilever (2.36) are modified as

$$\omega_n^{\mathrm{M}} = \omega_n / [(1 + (L/t)(\rho_{\mathrm{M}}/\rho)g_n]^{1/2} , \tag{2.39}$$

the cantilever resonant frequencies depending significantly on the medium where the cantilever is used.

Equation (2.26) is a valid irrespective of the cross-section of the cantilever. We have discussed above the case of a cantilever with a rectangular cross-section. If the cantilever with length L has a cylindrical cross-section of radius R, as in the case of the optical fiber cantilevers [14] or CNT cantilevers [15], the momentum of inertia becomes $I = \pi R^4 / 4$, the area is $A = \pi R^2$ and the spring constant $K = 3\pi E R^4 / 4L^3$. Because the equation satisfied by the cantilever and the boundary conditions are the same as for a rectangular cantilever, the resonant frequencies follow also from (2.36), in which the proper momentum of inertia and area must be used. The resonant frequencies of a cylindrical cantilever are thus

$$f_n = (k_n L)^2 R (E/\rho)^{1/2} / (4\pi L^2) . \tag{2.40}$$

When the cylindrical cantilever consists of an external cylinder of length L and radius R that encompasses an internal cylinder of the same length and internal radius r the natural frequencies are

$$f_n = (\beta_n^2 / 8\pi L^2)(4R^2 + 4r^2)^{1/2}(E/\rho)^{1/2} \quad (2.41)$$

where $\beta_1 = 1.875$, $\beta_2 = 4.694$, and so on. This last formula is valid, for example, for cantilevers made from multiwalled CNTs [15].

Equation (2.26) is valid only for cantilevers that oscillate at their natural frequencies. The solution of the cantilever deflection when an external force $F_{\text{appl}}(y,t)$ is added is very difficult to obtain and can be found only using specialized numerical routines. However, in most encountered practical situations the cantilever oscillates at the first natural frequency, this approximation greatly simplifying the solution because the cantilever motion can now be modeled as a damped harmonic oscillator that oscillates on the first frequency of the cantilever. The corresponding equation is thus

$$m_B d^2y/dt^2 + \gamma dy/dt + Ky = F_{\text{appl}}(y,t) \quad (2.42)$$

where $y(t)$ is the displacement of the cantilever, γ is the damping coefficient, K the linear spring constant, and $F_{\text{appl}}(x,t)$ is the force applied on the cantilever. This equation describes the majority of the instrumentation used at the nanoscale. Its solutions and significance will be described in the next section.

Table 2.1

Cantilever Resonant Frequencies (*From*: [16])

Resonant frequency	*SiC*	*Si*	*GaAs*
L = 100 μm, W = 3 μm, t = 0.1 μm	19 KHz	12 KHz	6.5 kHz
L = 10 μm, W = 0.2 μm, t = 0.1μm	1.9 MHz	1.2 MHz	0.65 MHz
L = 1 μm, W = 50 nm, t = 50 nm	93 MHz	60 MHz	32 MHz
L = 100 nm, W = 10 nm, t = 10 nm	1.9 GHz	1.2 GHz	0.65 GHz

Table 2.1 shows the first resonant frequency of cantilevers with various dimensions that are fabricated from three basic semiconductors: SiC, Si, and GaAs [16]. It follows that, depending on the dimensions of the cantilevers, the resonant frequencies can vary from kHz up to GHz. The cantilevers with micronic dimensions oscillate at MHz while those with nanometric sizes oscillate at GHz frequencies as many other nanomechanical systems. It is interesting to point out that at nanoscale the nanomechanical oscillators attain mechanical frequencies in

the microwave range, which is amazing because at the macroscopic scale we assign low frequencies to mechanical oscillations. If the dimensions of the nanomechanical oscillators are only a few nm the mechanical oscillations are in the THz range, being similar to the vibrations of molecules or the quantized mechanical vibrations of crystals named phonons.

2.1.3 Quality Factor and Noise of Cantilevers

The quality factor is a physical parameter that describes any resonant system irrespective of its physical origin (mechanical, electromagnetic, electric, or magnetic), characterizing the degree of confinement of the free energy of the physical system due to resonance, or, in other words, telling us how sharp the resonance is. In the case of the cantilever the quality factor is defined as

$$Q = 2\pi E_0 / \Delta E\,, \tag{2.43}$$

where E_0 is the stored vibrational energy and ΔE is the total energy lost per cycle of vibration.

It is rather difficult to calculate directly the quality factor using the above definition; therefore, the quality factor is expressed as a sum of effects that could cause losses of a cantilever oscillating at its fundamental (first order) resonance. Using this approach the cantilever quality factor is given by [17]

$$1/Q = 1/Q_c + 1/Q_{th} + 1/Q_v + 1/Q_s\,, \tag{2.44}$$

where $Q_c = 2.2(L/t)^3$ is the clamping loss, which is of the order $10^3 - 10^5$.

Q_{th} characterizes thermoelastic losses, which are negligible for submicron-thick cantilevers operating at kHz or MHz frequencies, but represent the dominant source of energy dissipation for thicker cantilevers. Thermoelastic losses originate from the heat flow during the harmonic movement of the cantilever and depend on the temperature and the material from which the cantilever is made. As soon as a cantilever is bent, a temperature gradient (i.e., a heat flow) appears inside it. The compressed part of the cantilever warms up, while its expanded part cools down, the heat flow being irreversibly directed from the warmer part towards the cooler part. The thermoelastic losses are expressed through

$$Q_{th} = 1/2\Gamma(T)S_{th}(f)\,, \tag{2.45}$$

where

$$\Gamma(T) = \alpha^2 TE / 4\rho C_P\,, \tag{2.46a}$$

$$S_{th}(f) = (2f / f_{th}) / [1 + (f / f_{th})^2], \tag{2.46b}$$

α is the thermal expansion coefficient, T is the cantilever temperature, C_p is the specific heat capacity, and $f_{th} = \pi\sigma_{th} / 2\rho C_p t^2$ is a thermal characteristic frequency of the cantilever due to heat flow, with σ_{th} the thermal conductivity. The thermal characteristic frequency is 40 GHz for a 60 nm thick Si cantilever and only 390 kHz for a Si cantilever with a thickness of 2 μm. In the last case, the thermoelastic losses are important.

The losses due to internal frictions in the cantilever (volume contribution) are modeled via a complex Young modulus $E = E_1 + iE_2$, with E_1 the conventional Young modulus and E_2 a dissipation part due to friction, and are defined as

$$Q_v = E_1 / E_2 . \tag{2.47}$$

The surface contribution to the quality factor, that is, the losses of a layer with thickness δ that models a disruption of the atomic lattice at the cantilever surface, or a surface contamination, is given by

$$Q_s = WtE_1 / 2\delta(3W + t)E_2 . \tag{2.48}$$

A general loss parameter, which accounts better for the cantilever performances and which is directly connected to the minimum detectable force F_{min}, can be defined as $\gamma = K / \omega_0 Q$. For a rectangular cantilever this parameter becomes

$$\gamma = 0.246Wt^2(E\rho)^{1/2} / LQ . \tag{2.49}$$

Because the cantilever is bent under the action of an applied external force, it is of paramount importance for many applications to define a minimum detectable force F_{min} that is able to deflect the cantilever. F_{min} is related to the thermo-mechanical noise, which has as its source the Brownian motion of the cantilever around its equilibrium position resulting from the immersion of the cantilever in a thermal bath with temperature T. This vibrational noise, which is the mechanical analog of the Johnson noise (white noise), is produced by an equivalent noise force with a constant power spectrum, which is localized at the free end of the cantilever and which has as its main effect the generation of an rms amplitude y_{rms}. This amplitude is given by [1]

$$y_{rms} = \int_0^\infty S_F |G(f)|^2 df , \tag{2.50}$$

where $G(f) = (f_0^2 / K)/[(f_0^2 - f^2) + i(ff_0 / Q)]$ is the frequency-dependent cantilever transfer function and S_F is the flat power spectrum. The solution of (2.50) for $Q >> 1$ yields

$$S_F^{1/2} = (2 / \pi Q f_0)^{1/2} K y_{rms} . \qquad (2.51)$$

On the other hand, if we consider that the cantilever is in thermal equilibrium, the equipartition theorem imposes that

$$y_{rms} = (k_B T / K)^{1/2} , \qquad (2.52)$$

where $k_B T$ is the thermal energy with k_B the Boltzmann constant. From the last two equations it follows that

$$S_F^{1/2} = (2 K k_B T / \pi Q f_0)^{1/2} . \qquad (2.53)$$

The minimum detectable force (simply called force noise) in a bandwidth B is given, for a rectangular cantilever, by

$$F_{min} = S_F^{1/2} B^{1/2} = (Wt^2 / LQ)^{1/2} (E\rho)^{1/4} (k_B T B)^{1/2} \qquad (2.54)$$

or, using the loss definition in (2.49),

$$F_{min} = (\gamma k_B T B / 0.246)^{1/2} . \qquad (2.55)$$

Thus, the minimum detectable force is directly related to the cantilever losses and to the temperature.

Relation (2.55) is illustrated in Figure 2.9 for two distinct cantilevers. It shows that ultrasensitive cantilevers for force detection should be long, narrow, and thin, as the cantilevers represented in Figure 2.10 that can be made from silicon [1] or can be metallic [18]. One of the most sensitive cantilevers reported in literature is 220 μm long, 5 μm wide, and 60 nm thick and has a shape similar to that presented in Figure 2.10(a); it has a quality factor of 6700, being able to detect forces in the range of attonewtons (5.6×10^{-18} N) at 5 K in a bandwidth of 1 Hz [1]. Also, a tungsten nanocantilever with a mass of only 10^{-14} kg, having a similar form as that indicated in Figure 2.10(b) (neck form) with a diameter of $D =$ 50 nm, was used to detect individual molecules [18].

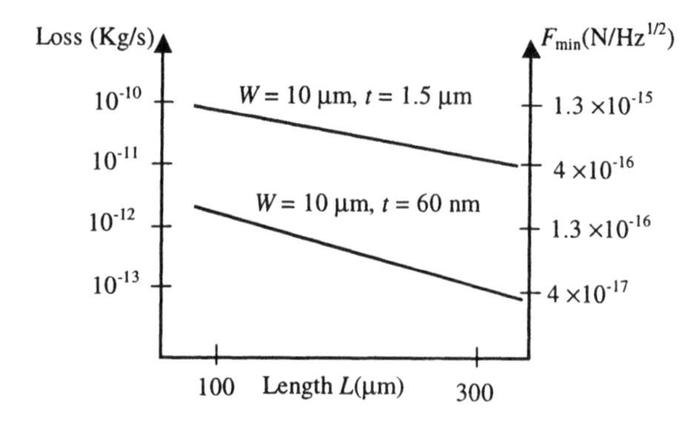

Figure 2.9 The dependence of loss and force noise on the cantilever parameters (*After*: [17]).

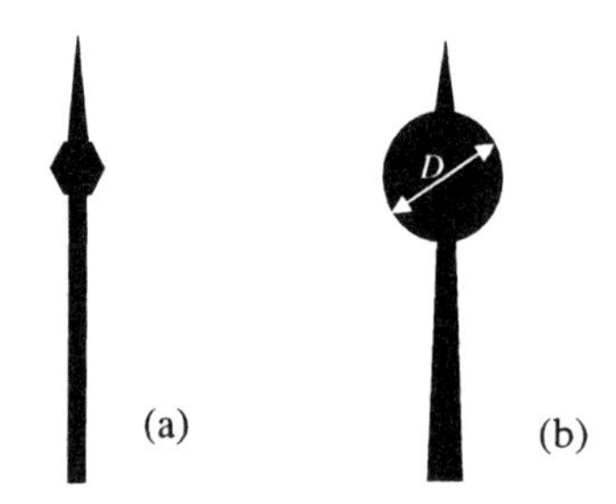

Figure 2.10 Basic shapes of cantilevers for ultrasensitive force detection.

Table 2.2

Cantilever Mode-Dependent Spring Constants and Minimum Detectable Forces (*From*: [19])

Mode-dependent spring constant K_n^m	$F_n^{\min} = (2k_B TBK_n^m / Q\pi\omega_n)$
$K_0^m = 3EI / L^3$	$F_0^{\min} = (2k_B TBK_0^m / Q\pi\omega_0)^{1/2}$
$K_1^m = 121.3\ K_0^m$	$2.54\ F_0^{\min}$
$K_2^m = 951.6\ K_0^m$	$4.25\ F_0^{\min}$
$K_3^m = 3654.3\ K_0^m$	$5.951\ F_0^{\min}$

A more detailed analysis of the noise of cantilevers, based on a modal description of the cantilever motion, is presented in [19]. This approach leads to the calculation of mode-dependent spring constants K_n^m, which in their turn define mode-dependent minimum detectable forces. Their expressions and values are summarized in Table 2.2.

It is worth noting that the quality factor decreases when the cantilever operates in higher modes, its highest value being attained for the fundamental mode of oscillation [19]. Although the sensitivity of cantilevers cannot be increased working in higher modes, it is possible to tune the cantilever frequency over a range of 300%, without a serious degradation of the quality factor [12], by scanning the edge of a cantilever with a tip excited by an ac voltage. This tip transfers its vibrations to the cantilever, playing the role of a point-like actuator that induces a driving force on the cantilever, as illustrated in Figure 2.11.

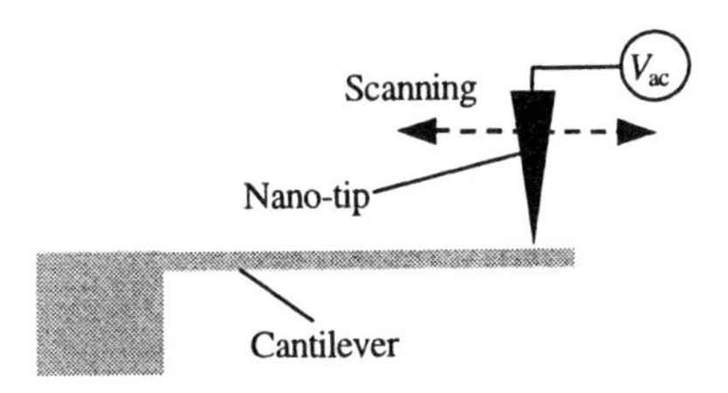

Figure 2.11 Tuning of the cantilever frequency.

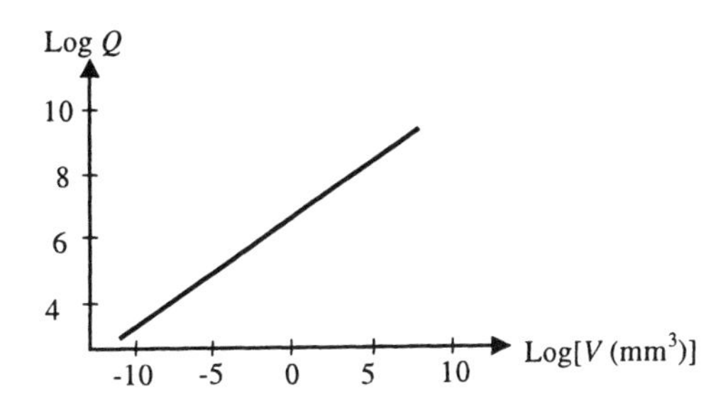

Figure 2.12 Size dependence of the quality factor (*After*: [16]).

Another quantity that depends on the quality factor is the minimum operating power [16], defined as

$$P_{\min} = k_B T \omega_0 / Q \,. \tag{2.56}$$

The minimum detected power for nanocantilevers with resonant frequencies beyond 100 MHz and with quality factors of $10^4 - 10^5$ is of the order of 10 aW at low temperatures, and becomes about 1 μW at room temperature. This power is very low compared with the dissipation power, which is its analog in electrical devices. A high Q factor is again the key parameter for very low minimum powers. The Q factor dependence on the volume V of the mechanical resonator is displayed in Figure 2.12.

Although the sensitivity of the cantilever is limited by ambient thermal fluctuations, which are the origin of the thermomechanical noise, it can multiply by increasing the quality factor. However, Q cannot increase indefinitely by modifying only the geometrical dimensions of the cantilever. A three-order-of-magnitude increase of Q has been achieved, for example, employing a positive feedback of the cantilever system, which consists of an amplifier and a phase shifter [20]. The sensitivity of the cantilever can also increase by coupling it with a single-electron transistor [21].

We have discussed up to now only the influence of internal losses (thermoelastic or surface losses) on the quality factor of cantilevers. However, Q is also influenced by external losses such as losses in airflow or losses due to the radiation of elastic waves at the cantilever support [22].

By modeling the air damping as the effect of independent collisions of non-interacting air molecules with the surface of the cantilever, the quality factor due to the cantilever-air interaction can be expressed as

$$Q_{\text{air}} = 2\pi f_n h \rho / \beta_{\text{gas}} P \,, \tag{2.57}$$

where P is the air pressure and $\beta_{\text{gas}} = (32 m_{\text{air}} / 9\pi RT)^{1/2}$, with R = 8.31×10^3 J/K the universal gas constant and T = 300 K. Air damping becomes important and thus damages the quality factor of the cantilever when the surface-to-volume ratio of the cantilever is large. Equation (2.57) tells us that, since the quality factor is inversely proportional to the air pressure, a stable operation of nanocantilevers requires their introduction in a high-vacuum environment.

The contribution to the quality factor of the external loss due to the radiation of elastic waves at the cantilever support is given by

$$Q_{\text{supp}} = 0.34 (L/h)^3 \,. \tag{2.58}$$

This support loss becomes important for ultrathin cantilevers.

Table 2.3

Main Material Parameters for MEMS/NEMS Cantilevers

Material	Young modulus($\times 10^{11}$ Pa)	Density (g/cm^3)	Thermal expansion ($\times 10^{-6}/C$)
Silicon	1.5	2.4	2.5
Carbon nanotube	10	1.4	10
Steel	2.1	8	12

Cantilevers can be realized from many materials but frequently they are fabricated from Si or GaAs semiconductors, metals, or CNTs. Table 2.3 summarizes the main material parameters of Si and CNT MEMS/NEMS cantilevers and compares them with steel, known for its elasticity and hardness. This table indicates that Si and CNT cantilevers have parameters comparable or even better than steel. Therefore, although MEMS/NEMS are tiny, they are also very robust.

Table 2.4

Bulk and Micro/Nanoscale Material Parameters

Parameter	*Bulk*	*MEMS/NEMS*
Young modulus (GPa)	E	$0.8E$–$0.9E$
Density (kg/m^3)	ρ	ρ
Thermal linear expansion coefficient (K^{-1})	α	α
Poisson ratio	0.25	0.25

Care must be taken, however, when reading Table 2.3, since the various material parameters presented in it, which are essential for MEMS/NEMS devices, are extracted from bibliographical sources that refer to bulk materials. It is only natural to ask if these parameters remain the same at the micro- or nanoscale. In Table 2.4 we indicate some basic bulk material parameters, which are specific for MEMS/NEMS, and the way in which they must be corrected in order to be usable for the micro- and nanoscale [23]. This table indicates that, while some parameters are the same for bulk and MEMS/NEMS, others, like the Young modulus, change at the micro- and nanoscale with up to 10%–20% from the bulk value.

2.1.4 Magnetic and Optical Actuation of Cantilevers

Although the electrostatic actuation of cantilevers, which was discussed up to now, is the most common form of actuation, it is not the only way in which cantilevers, and in general MEMS and NEMS, can be actuated. For example, for large deflections over several hundreds of micrometers, bi-directional actuation between a permanent magnet and an electromagnet is used. Changing the direction of the current through the electromagnet can generate attractive or repulsive actuation force between the magnets. The main problem is to fabricate micro- or nanomagnets on the surface of the cantilever. But, as reported in [24], it is possible to realize an array of CoNiMnP micromagnets on the Si surface of a cantilever; the magnetic bi-directional cantilever is illustrated in Figure 2.13.

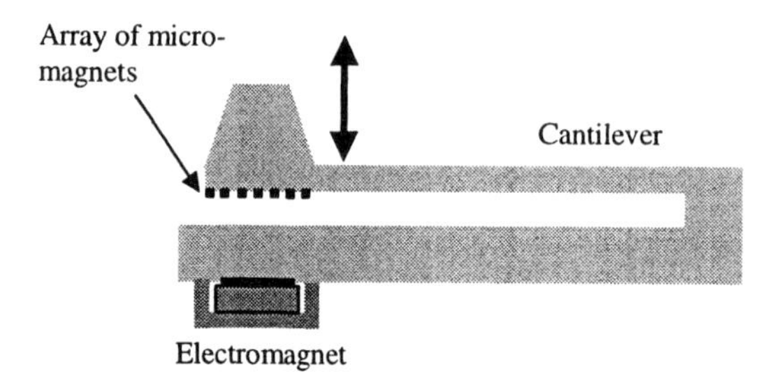

Figure 2.13 A magnetically actuated cantilever (*After*: [24]).

The magnetic deflection of such a cantilever of length L can be expressed as

$$y_{\max} = 2VM_z(\partial B_z / \partial z)a^2(3L - a) / EWt^3, \qquad (2.59)$$

where V is the magnet volume, M_z its magnetization, B_z is the flux density of the magnet, and a is the distance from the fixed end of the cantilever to the electro-magnet. The magnetic actuation of cantilevers is used in many optical applications such as switches and scanners.

In optical actuation a laser excites a cantilever or an array of cantilevers and provokes their bending due to the applied optical force [25] (see Figure 2.14). The pressure produced by an optical field on nonabsorbing objects is given by

$$p_{\text{opt}} = F_{\text{opt}} / A = 2RS / c, \qquad (2.60)$$

where A is the area upon which the optical force F_{opt} acts, S (W/m^2) is the modulus of the Poynting vector, R is the reflection coefficient, and c the speed of light. F_{opt} is not sufficient to produce observable movements of massive objects since it does not exceed tens of nanonewtons (nN) for moderate optical powers of less than 1 W, even for reflection coefficients near unity. However, it produces observable effects for MEMS with a typical mass of 10^{-20} kg, in particular for micromechanical cantilevers. The advantages of optical actuation include remote actuation, and a direct use of optical power without the need of transducers; the optical actuation of cantilevers has been demonstrated experimentally [26].

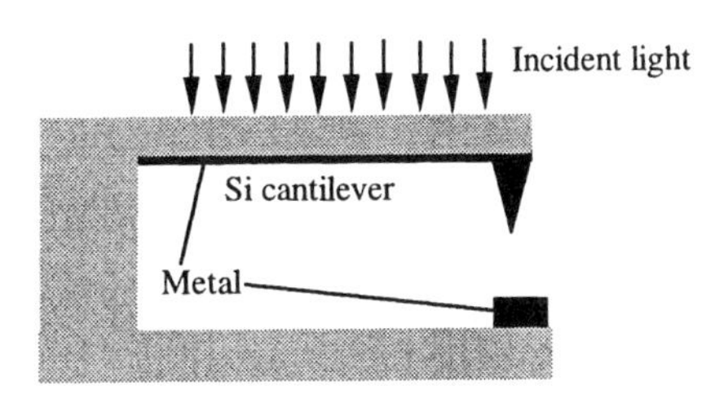

Figure 2.14 An optically actuated cantilever.

The deflection of the cantilever due to an optical field is given by [25]

$$y_{\max} = 2RP_{\text{opt}}L^3 / cWEt^3, \tag{2.61}$$

where $P_{\text{opt}} = SWL$ is the optical power applied on the cantilever. Optical actuation has important applications in signal processing, optical memories, and optical measurement techniques.

More details about MEMS/NEMS, especially about fabrication techniques and applications can be found in [27], which is a recent and comprehensive review of the field. In what follows, we apply the knowledge gained in this section to a series of instruments used at the nanoscale. The large majority is based on the MEMS/NEMS devices described above.

2.2 SCANNING PROBE INSTRUMENTATION FOR NANOELECTRONICS

Scanning probe techniques, also known as scanning probe microscopy (SPM) techniques, are the basic instrumentation that is able to reveal various physical properties (mechanical, electromagnetic, thermal) at the nanoscale. Some of these instruments, which will be described below, are also invaluable tools for the manipulation of nano-objects and the fabrication of nanodevices.

The common working principle of all SPM techniques resides in a three-dimensional scanning of a nanoprobe across a surface, which provides an SPM image of the surface with nanometric or even atomic resolution. The nanoprobe is located at nanometric distances over the probe. Different SPM techniques exist that use different physical effects to probe the surface. For example, the (AFM), uses as probe a cantilever ended with a nano-sized tip, which is deflected by the force between the tip and the surface of the material that can be a conductor, semiconductor or an isolator. A tip scanned over the surface to be tested is also used in the scanning tunneling microscope (STM), but here the tip-sample tunneling current is detected. Similarly, in the scanning near-field optical microscope (SNOM) an optical probe (a nanometer-sized aperture) is scanned over an illuminated object, the intensity in the near field being monitored during scanning. In all cases the data collected as a result of the interaction between the scanned probe and the surface offer invaluable information about the atomic structure and the properties of the various investigated media (solids, liquids, biological cells, gases, vacuum) at low, ambient, or high temperatures, with an unprecedented resolution (of about fractions of Å). In Table 2.5 we have summarized the physical principles of various SPM techniques.

Table 2.5

Physical Principle of Various STM Techniques

STM technique	*Physical principle*
AFM	Force detection between a cantilever terminated with a tip and a surface
STM	Tunneling current detection of electrons that tunnel from a nanosized tip into a surface
SNOM	Optical intensity measurement due the near field interactions between a nano-sized aperture and an illuminated surface

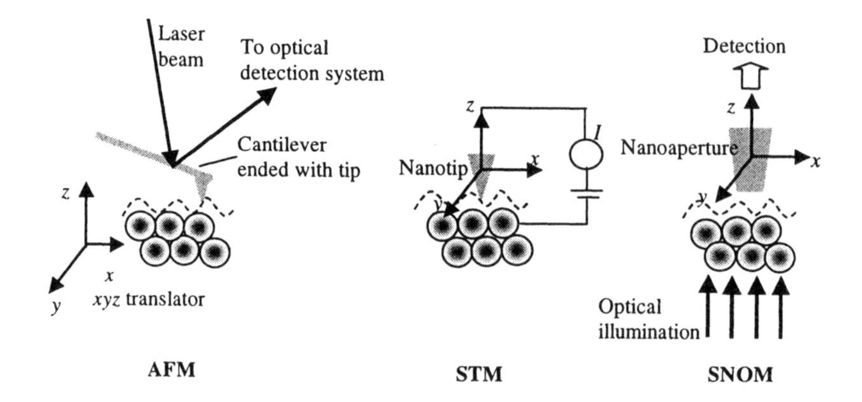

Figure 2.15 Basic SPM techniques.

The basic configurations of the instruments used in nanoelectronics are represented in Figure 2.15. As can be observed from this figure, STM requires an electric conductive sample, while in the case of SNOM the sample must be optically transparent. These restrictions do not appear in the case of the AFM, which is therefore the most spread instrumentation for nanoscale characterization. In what follows, we analyze the main SPM techniques and indicate their most important applications.

2.2.1 The Atomic Force Microscope (AFM)

The AFM measures very weak forces, up to fractions of attonewtons (1 aN = 10^{-18} N) [1], which are exerted between the tip and a surface under test. These forces are measured via the deflection of a cantilever terminated with a tip that has a very small mass. The deflection of the cantilever is sensed and measured using different techniques, the most common being the optical interferometry, which allows the detection of displacements of less than 10^{-4} Å [28].

The AFM senses different types of forces that determine the tip-sample interaction, depending on the distance between the tip and the sample. In this respect, at small tip-sample distances the interatomic forces are dominant. They consist from a short-range ($\leq 0.1\,\text{nm}$) Born repulsive force component and a long-range (up to 10–15 nm) van der Waals force component. If the distance between the tip and the sample increases the van der Waals forces decay exponentially and long-range forces, such as electrostatic, magnetic, electromagnetic, or capillarity, dominate the tip-sample interaction.

At tip-sample distances of a few nm the van der Waals forces, which are usually attractive, are sufficiently strong to produce a detectable deflection of the AFM cantilever. For two interacting simple molecules, such as gas molecules, the van der Waals potential is given by

$$U_{\text{vdW}} \approx -C_1 / z^6, \tag{2.62}$$

where the constant C_1 is known as the London coefficient, and z is the distance between the molecules. At shorter distances, when the wavefunction of electrons located at the end of the tip overlaps the wavefunction of electrons located on the surface sample, strong repulsive forces called Born forces are produced. The potential of these forces is modeled as

$$U_{\text{Born}} = C_2 / z^{12}, \tag{2.63}$$

where C_2 is another constant. Thus, the total intermolecular potential is given by

$$U = C_2 / z^{12} - C_1 / z^6 \tag{2.64}$$

and is known as the Lenard-Jones potential. The force detected with the AFM is determined by the sum of all repulsive and attractive potentials encountered during the tip-sample interaction.

The attractive van der Waals force exerted between two macroscopic bodies, such as two spheres of radius R, is given by

$$U_{\text{vdW}} = -(AR/6z)\,, \tag{2.65}$$

the van der Waals force being

$$F_{\text{vdW}} = -dU_{\text{vdW}}/dz = -AR/6z^2\,, \tag{2.66}$$

where $A \propto \pi^2 C_1 \rho_1 \rho_2$ is the Hamaker constant and ρ_1 and ρ_2 are the densities of the two bodies.

The repulsive force between macroscopic objects is modeled as an indentation force that acts on a sphere pressed on a flat surface. This force, known as the Hertz force, is expressed through

$$F_{\text{Hertz}} = (K^2 R h_{\text{ind}}^3)^{1/2}\,, \tag{2.67}$$

where R is the radius of the sphere and h_{ind} is the depth of indentation.

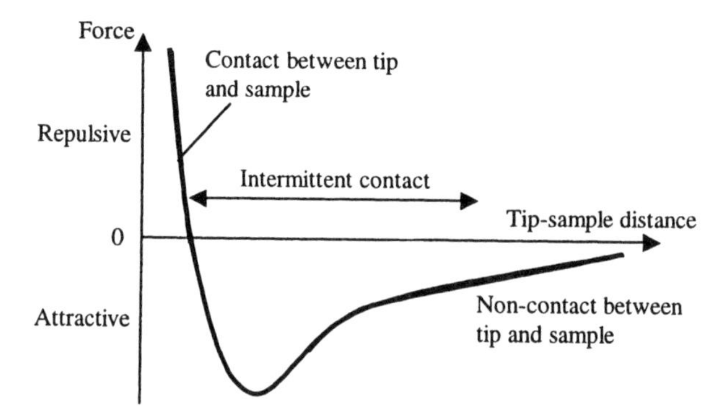

Figure 2.16 AFM modes of operation as a function of the tip-sample separation.

The distance between the tip and the sample and thus the dominant forces that occur at specific distances between them completely determine the different types of AFM operations. The main three AFM modes of operation are schematically indicated in Figure 2.16. These are the dynamical, static, and tapping modes of AFM operation.

In the *dynamical mode*, also called attractive force imaging or noncontact imaging mode, the cantilever tip is not in direct contact with the surface under test but is placed well above it. The cantilever is driven to vibrate mechanically at a high frequency f using a piezoelectric actuator placed over it. There are two possibilities of modulating and hence working with the AFM in the noncontact mode: frequency modulation (FM) and amplitude modulation (AM). In the FM case the resonant frequency f_0 of the cantilever is slightly shifted due to the interaction force between the tip and the sample, the shift Δf being directly proportional with the gradient of the interaction force between the tip and the sample. The frequency shift can be measured and from it the interacting force can be determined by integration. If the cantilever is scanned over the entire surface of the sample we can obtain the two-dimensional distribution of the interacting force. In the AM case, the gradient force between the tip and the sample is directly proportional with the amplitude of the cantilever and the phase of the deflected signal, which is monitored with the help of a lock-in amplifier. In both cases, a feedback signal is generated to keep constant either the frequency or the amplitude of the vibrating cantilever. The feedback signal produces a repositioning of the cantilever tip over the surface using a (x, y, z) piezo-translator, so that what is actually measured is the cantilever response (frequency shift or phase) over the (x, y) plane. This mode of operation is useful when we want to map the distribution of electrostatic, magnetic forces or the topography of the sample without touching it.

In the *static mode*, also called repulsive mode or contact mode, the cantilever is in contact with the sample. A weak repulsive force acts on the tip atoms, which are in contact with the atoms of the surface, producing a cantilever deflection $z(x, y)$ that is monitored and measured by several detection techniques; these techniques will be described below. In the contact mode of operation, a feedback system insures that a constant force is always exerted on the contact surface such that the cantilever deflection is maintained at a constant level, which is usually less than 0.1 nm. The measured cantilever deflection via optical means is used to determine the surface topography $z(x, y)$. In this mode of operation amazing horizontal and vertical resolutions, of 0.1 nm, were obtained. Because the displacement sensitivities are about 0.01 nm, it is possible to measure forces in the range of 10 nN – 10 pN, situated between the chemical ionic bonding forces, of about 100 nN, and the hydrogen bond, which corresponds to 10 pN [29]. The surface topography obtained in the contact mode is thus revealed at an atomic scale, individual atoms, or clusters of a few atoms being imaged with this technique. This atomic resolution is possible if special V-shaped cantilevers (as indicated in Figure 2.17) are used in order to minimize the lateral (torsion) forces that could damage them. These cantilevers also have a low spring constant, of about 0.04–1 N/m, which is below the value of an equivalent atomic spring constant. The equivalent atomic spring constant can be estimated as $\omega^2 m \approx 10$ N/m if the atoms bound by molecules or solid crystals, with a mass of 10^{-25} Kg, are assumed to vibrate at 10 THz. The cantilevers used in the AFM contact mode are terminated with very sharp pyramidal tips that must have a few nm at their end. These cantilevers are usually fabricated by etching silicon nitride.

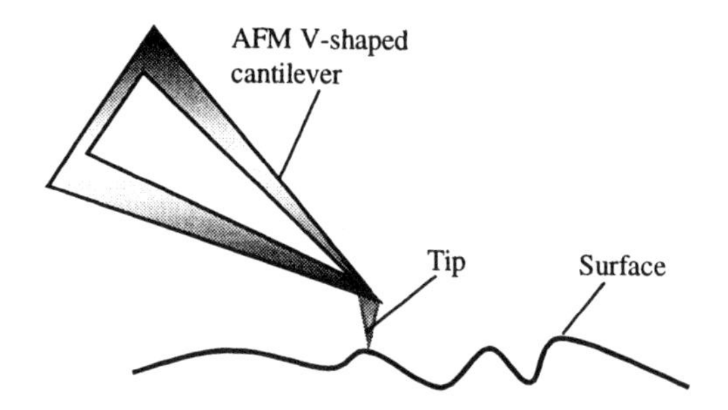

Figure 2.17 The static AFM mode.

Although the AFM contact mode has impressive performances, there are also some disadvantages: some certain surfaces can be damaged due to the direct contact with the tip, a typical example being biological surfaces, and there are increased normal and lateral forces due to the capillarity action originating from the thin water layer adsorbed by any surface at ambient temperature. These disadvantages are overcome by the third AFM mode of operation.

The *tapping mode* of operation is characterized by the fact that the cantilever touches the sample periodically. The cantilever is driven again into oscillation at its mechanical frequency by a driver (a piezoelectric crystal), and its changes in amplitude or in phase due to the periodic tip-surface interaction are monitored. In contrast to the static mode of operation, the cantilever is, in this case, more rigid: it has a spring constant in the range of 20–70 N/m and oscillates at much higher amplitudes (50–120 nm). One of the parameters of the vibrating cantilever—frequency, phase, or amplitude—can be measured via a feedback regulation system included in the AFM system. Most frequently, the amplitude of the tapping cantilever is recorded during the surface scanning operation, as in the case of the static mode. The feedback maintains a constant deflection for a set of desired points on the surface by adjusting the distance between the cantilever and the surface. The feedback signal, which regulates the vertical movement (along the z direction) of the cantilever during scanning in the (x,y) plane, is recorded, and thus provides a topographical image of the surface.

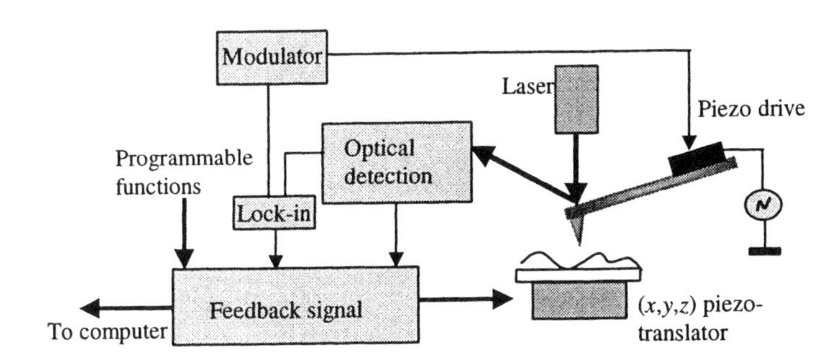

Figure 2.18 The block diagram of the AFM.

A block diagram of the AFM, valid for all three operating modes, is given in Figure 2.18. The reader can browse several Web sites containing AFM image galleries for an in-depth knowledge of the AFM capabilities and the wealth of atomic details of solid-state matter, gases, liquids, biological cells, and genetic components, such as DNA, that can be imaged with the AFM. Among the many Web sites containing thousands of AFM images, two sites are really impressive:

http://www.jpk.com/spm/gallery1.htm,

http://www.quesant.com/Gallery/gallery_contents.htm.

The key component of an AFM is the sensor, which is able to sense the weak mechanical deflections of the cantilever that can be less than 0.1 nm in the static mode. There are two main types of detection systems: electronic and optical. The most encountered detection systems are based on optical techniques.

Optical interferometry, homodyne, or heterodyne interferometric techniques are very sensitive; a detailed analysis of these techniques can be found in [30]. The same reference can be perused by readers interested in electronic detection techniques, which can be either piezoresistive, when the cantilever is made from piezoelectric materials, or capacitive, when the cantilever is placed within the plates of a capacitor. We only present here, in Figure 2.19, the scheme of the optical lever, which is the most commonly used detection method. The optical lever detection scheme measures the detection of the light deflected by the cantilever vibrations using a photodetector array containing two or four photodetectors. The signal difference between the various photodetectors is monitored, amplified, and transmitted to the AFM feedback system.

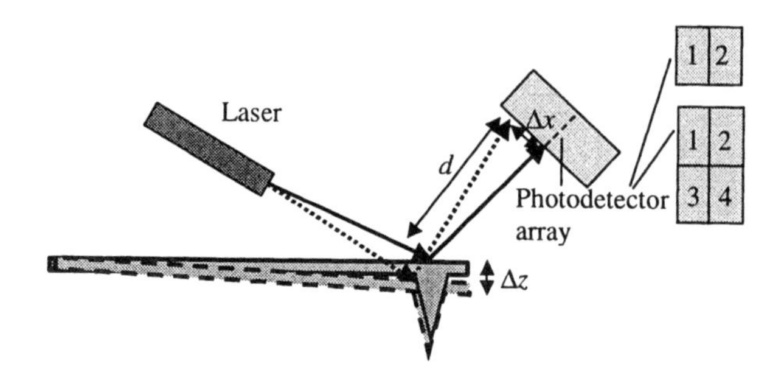

Figure 2.19 The principle of optical lever detection.

In the case of two photodetectors the difference between the photodetector currents is given by

$$\Delta I = I_2 - I_1 \propto L\Delta x \times N_0 \propto d\Delta z \times N_0 , \tag{2.68}$$

where N_0 is the number of photons in the incoming laser beam, L is the length of the cantilever, and the other parameters are described in Figure 2.19. Four photodetector elements are used for monitoring the lateral forces with the aim of preventing them to twist the cantilever or in order to determine the friction and adhesion effects. The computer attached to the AFM calculates in this case the following signals:

$$\Delta_{\text{normal force}} \propto [(I_1 + I_2) - (I_3 + I_4)] , \tag{2.69a}$$

$$\Delta_{\text{lateral force}} \propto [(I_2 + I_4) - (I_1 + I_3)] . \tag{2.69b}$$

The optimal sensitivity is $2 \times W \times I_{\text{laser}} / \lambda$ (mW/rad), where W is the cantilever width, I_{laser} is the laser intensity, and λ its wavelength.

An interdigital cantilever represents an alternative detection solution, which attains the high sensitivity of optical interferometer detection and preserves the simplicity of the optical lever [31]. In this case the cantilever is micromachined in the form of an interdigitated finger structure, which forms a diffraction grating.

Optical interference occurs between alternative fingers, which are either deflected or fixed, so that the deflection can be measured via a laser intensity measurement. As in the case of the optical lever detection, this method requires only a laser source and a photodetector.

The ultimate AFM resolution is dictated by the tip geometry and its characteristics. Usually, AFM tips are fabricated by micromachining silicon or silicon nitride, the resulting pyramidal tips having a high aspect ratio. The latest solution for very sharp AFM tips are CNTs with a diameter of a few nanometers, which are grown on a Si cantilever or are even attached on its pyramidal tip [32]. The resolution of the AFM depends also on its operating mode. In the contact mode two spikes are resolved at a minimum separation distance $2(D\Delta z)$, where D is the diameter of the tip and Δz is the minimal detectable depression of the AFM. D is typically around 10–20 nm for Si tips and about 5–6 nm for CNTs, while Δz is typically a fraction of a nanometer. Thus, the resolution is in the range of 2–5 nm. In the noncontact mode, the resolution is $0.8h$, where h is the height of the tip above the surface. Atomic resolution can be reached if the tip is placed in high vacuum, the resolution decreasing when the tip works at ambient temperature because the tip must be positioned well above the surface in order to avoid attraction by the superficial water layer that exits on any material.

The silicon, silicon nitride, and CNTs are not the only materials for AFM cantilevers and tips. Monomode optical fiber cantilevers for AFM applications are fabricated using etching techniques [14]. GaAs/AlGaAs self-sensing cantilevers were also produced, having a FET integrated into the cantilever in order to sense the strain due to the piezoelectric effect. In this way, low noise and high sensitivity, of about 0.002 Å/(Hz)$^{1/2}$ at 4.2 K, are achieved [33]. GaAs submicronic cantilevers with a thickness of 100 nm were fabricated using GaAs grown on an AlAs layer that acts as an etch stop layer [2]. This 100 μm long cantilever has a huge aspect ratio and thus a very low mechanical spring constant of 10^{-4} N/m, allowing the detection of forces as small as 100 fN. Micromachined metal cantilevers with silicon tips are also used for high sensitive AFM [3].

One million single-crystal cantilevers per square centimeter, terminated with a tip, were realized using micromachining techniques [34]. Cantilevers with the length and the thickness in the range of 0.5–200 μm and 30–100 nm, respectively, were tailored to form such huge arrays. Because AFM mapping of a surface is time consuming due to the mechanical scanning of the tip over the surface, an AFM based on millions of cantilevers acting in parallel and scanning different portions of the same surface has tremendous applications, since it decreases dramatically the time needed to obtain AFM images of the surface. Another way to increase the speed of an AFM is to use nanocantilevers with resonant frequencies beyond 100 MHz and up to 1 GHz [35]. Such nanocantilevers with a length of a few microns and a thickness of about 100 nm were fabricated using Si micromachining techniques; they attain resonant frequencies of about 100 MHz.

2.2.1.1 The AFM Noncontact Mode

Although the AFM static mode has demonstrated atomic resolutions in some cases, it has severe limitations, some of them being mentioned above. Others are related to a strong $1/f$ noise, which could be reduced only by working at low temperatures, and to the presence of long-range attractive forces, which can be cancelled only by imposing restrictions on the cantilever motion, such as immersion in a liquid or the application of an additional electromagnetic force. Therefore, the most encountered AFM operation mode is the noncontact mode, which has some unique characteristics with regard to other AFM modes:

1) it is nondestructive;
2) it has atomic resolution;
3) it is able to measure atomic forces, such as van der Waals forces;
4) it can measure electric, magnetic, electromagnetic, and elastic fields.

As we have already pointed out, the AFM in the noncontact mode can be implemented in the frequency modulation, FM, or amplitude modulation, AM, modes. In the FM mode the signal resulted from the cantilever deflection is shifted in phase with $\Delta\varphi$ and is used, via an automatic gain device, as an excitation signal for the piezoelectric actuator of the cantilever. The mechanical movement of the cantilever is frequency modulated, the entire AFM loop system oscillating at the frequency f, which becomes the natural frequency of oscillation f_0 when $\Delta\varphi = \pi/2$. The AFM system preserves in this case a constant amplitude A of the cantilever by using the automatic gain, which in the same time records the changes in frequency Δf and/or an average of the tip-sample dissipation.

The link between the tip-sample force gradient $F_{\text{t-s}}$ and the frequency shift is given by the following relation:

$$\Delta f = f_0(-\partial F_{\text{t-s}}/\partial z)/2K\,. \tag{2.70}$$

The FM method was applied successfully to an impressive range of materials that includes semiconductors, insulators, thin films, organic films, or even individual molecules. Atomic resolution surface imaging was obtained in all these cases.

In the AM mode, the cantilever is excited at its natural resonance frequency or at a certain frequency f, which is close to f_0, while a feedback loop keeps the excitation amplitude at a constant value. The detected signal due to the cantilever deflection excites a lock-in amplifier, which uses the excitation signal of the cantilever piezoelectric actuator as reference signal. The output amplitude signal of the lock-in amplifier is the feedback signal for the tip-sample distance control, while the output phase of the lock-in amplifier is the recorded parameter. The tip-sample force gradient is given in this case by

$$\partial F_{\text{t-s}} / \partial z = K(1 - A_0 / A \cos\varphi)\,, \tag{2.71}$$

where A_0 is the free oscillation amplitude and A is the cantilever amplitude at frequency f.

In the noncontact mode, the mechanical motion of the cantilever is modeled as a damped harmonic oscillator and is described by the differential equation

$$m_B d^2 z / d\tau^2 + \gamma dz / d\tau + Kz = F_{\text{appl}}\,, \tag{2.72}$$

where m_B is the cantilever mass and γ is the damping coefficient. If the applied force that excites the cantilever through the piezoelectric element has a harmonic time dependence of the form $F_{\text{appl}} = F \exp(\mathrm{i}\omega\tau)$, the displacement should have the same temporal dependence; that is, $z = A\ \exp(i\omega\tau)$, so that the frequency response of the cantilever is described by

$$A(\omega) = |F| / [\omega^4 + \omega^2(\gamma^2 / m_B^2 - 2K / m_B) + K^2 / m_B^2]^{1/2}\,. \tag{2.73}$$

Then, using (2.72) and taking into account that $\gamma = m_B \omega_0 / Q$, the modulus and the argument of the amplitude are found to be

$$|A(\omega)| = FQ\omega_0^2 / [\omega^2 \omega_0^2 + Q(\omega_0^2 - \omega^2)]^{1/2}\,, \tag{2.74}$$

$$\varphi(\omega) = \arctan[\omega\omega_0 / Q(\omega_0^2 - \omega^2)]\,. \tag{2.75}$$

The first natural resonance frequency of (2.73) is

$$\omega_0 = 2\pi f_0 = (1/2\pi)(K / m_B)^{1/2}\,, \tag{2.76}$$

the minimization of the cantilever response taking place at the frequency

$$\omega_{\min} = (\omega_0^2 - \gamma^2 / 2m_B^2)^{1/2}. \tag{2.77}$$

Relation (2.77) shows that in the damped case the frequency response of the cantilever is shifted from the natural frequency, this shift being zero in the undamped case. By decreasing the cantilever dimensions up to the order of microns, much higher resonant frequencies are obtained compared with larger cantilevers (>500 kHz in air), without significantly modifying the spring constants (<100 mN/m).

The equation of motion of a harmonic oscillator can also take a different form than in (2.72):

$$d^2z/d\tau^2 + \Gamma dz/d\tau + \omega_0^2 z = 0\,, \tag{2.78}$$

where

$$\Gamma = 2(\partial F/\partial z)^{1/2}\,, \tag{2.79}$$

the resonance frequency shifting from ω_0 when no external force is applied to

$$\omega_1 = [(K - \partial F/\partial z)/m_{\mathrm{B}}]^{1/2} \tag{2.80}$$

in the presence of an applied force. The solution of (2.78) for $\tau > 0$ is

$$z(\tau) = Y_{\mathrm{s}}\exp(-\tau\Gamma/2)\sin(\omega_1 t) + Y_{\mathrm{c}}\exp(-\tau\Gamma/2)\cos(\omega_1 t)\,. \tag{2.81}$$

For example, in the case of an electrostatic external force produced by biasing the cantilever with the voltage V, we have $\Gamma = 2(C_{\text{t-s}}^3/m_{\mathrm{B}}\varepsilon A)^{1/2}V$, where $C_{\text{t-s}}$ is the capacitance between the tip and the sample.

Equation (2.78) can be also expressed as a set of two first-order differential equations in the phase space $(z, dz/d\tau = p)$:

$$\begin{cases} dz/d\tau = p \\ dp/d\tau = -\Gamma p - \omega_0 z \end{cases} \tag{2.82}$$

from which it is possible to find the cantilever response in the joint time-frequency representation. This constitutes a significant step forward in the analysis of cantilever dynamics with respect to either temporal- or frequency-domain approaches [36]. The solutions of (2.82) for the amplitude of the cantilever and its time derivative can be put in a matrix form

$$\begin{pmatrix} z \\ dz/d\tau \end{pmatrix} = \begin{pmatrix} A & B \\ C & D \end{pmatrix}\begin{pmatrix} z(0) \\ dz/d\tau(0) \end{pmatrix}, \tag{2.83}$$

where the values $z(0)$, $dz/d\tau(0)$ of the cantilever amplitude and its derivative, respectively, correspond to $\tau = 0$. The matrix elements in (2.83) are given by

$$A = \exp(-\tau\Gamma/2)[\cos(\omega_1 t) + \Gamma\sin(\omega_1 t)/2\omega_1]\,, \tag{2.84a}$$

$$B = \exp(-\tau\Gamma/2)\sin(\omega_1\tau)/\omega_1\,, \tag{2.84b}$$

$$C = -\exp(-\tau\Gamma/2)(\omega_1 + \Gamma^2/4\omega_1)\sin(\omega_1 t)\,, \tag{2.84c}$$

$$D = \exp(-\tau\Gamma/2)[\cos(\omega_1 t) - \Gamma\sin(\omega_1 t)/2\omega_1]. \tag{2.84d}$$

The matrix in (2.83) has a determinant equal to $\exp(-\Gamma t)$, which approaches the value 1 when $\Gamma \to 0$. Because unit-determinant matrices characterize both temporal lenses and free space sections, it is possible to find an optical analog of the AFM in the noncontact mode; that is, it is possible to find an optical system that mimics the cantilever motion. To this end, (2.83) should be scaled in order that the new unit-determinant matrix that expresses the cantilever motion becomes decomposable in a sequence of matrices that describe temporal lenses and free spaces. More precisely, by writing (2.83) as

$$\begin{pmatrix} z\exp(\tau\Gamma/2) \\ (dz/d\tau)\exp(\tau\Gamma/2)/\omega_1 \end{pmatrix} = \begin{pmatrix} A' & B' \\ C' & D' \end{pmatrix} \begin{pmatrix} y(0) \\ (dy/d\tau(0))/\omega_1 \end{pmatrix}, \tag{2.85}$$

the new matrix

$$\begin{pmatrix} A' & B' \\ C' & D' \end{pmatrix} = \begin{pmatrix} A\exp(\tau\Gamma/2) & B\exp(\tau\Gamma/2)\omega_1 \\ \exp(\tau\Gamma/2)/\omega_1 & D\exp(\tau\Gamma/2) \end{pmatrix} \tag{2.86}$$

decomposes as follows:

$$\begin{pmatrix} A' & B' \\ C' & D' \end{pmatrix} = \underbrace{\begin{pmatrix} 1 & 0 \\ -\Gamma/2\omega_1 & 1 \end{pmatrix}}_{\substack{\text{divergent temporal lens} \\ \text{with } f_T = 2\omega_1/\Gamma}} \underbrace{\begin{pmatrix} \cos\alpha & \sin\alpha \\ -\sin\alpha & \cos\alpha \end{pmatrix}}_{\substack{\text{fractional Fourier} \\ \text{transformer of order} \\ \alpha = \omega_1 t}} \underbrace{\begin{pmatrix} 1 & 0 \\ \Gamma/2\omega_1 & 1 \end{pmatrix}}_{\substack{\text{convergent temporal lens} \\ \text{with } f_T = 2\omega_1/\Gamma}} . \tag{2.87}$$

The first and the last matrices in the right hand side of (2.87) are matrices that characterize temporal lenses with focal lengths $f_T = 2\omega_1/\Gamma$ and $f_T = -2\omega_1/\Gamma$, respectively. It is quite remarkable that an external force acting on a cantilever has the same effect, in the time-frequency domain, as a pair of temporal lenses, one divergent and the other convergent. The second matrix in the right hand side of (2.87), which characterizes a temporal fractional Fourier transform of order $\alpha = \omega\tau$, illustrates the outcome of the undamped component of the cantilever motion.

The temporal Wigner transform of an arbitrary function $z(\tau)$ is defined as

$$W(\tau,\omega) = \int z(\tau + \tau'/2)z(\tau - \tau'/2)\exp(i\tau'\omega)d\tau' \tag{2.88}$$

and has, for the solution of the cantilever vibration amplitude

$$z(\tau) = \begin{cases} 0, & \tau < 0 \\ \exp(-\tau\Gamma/2)\sin(\omega_1\tau + \theta), & \tau \geq 0 \end{cases} \tag{2.89}$$

an analytic expression of the form

$$W(\tau,\omega) = \begin{cases} 0, & \tau < 0 \\ \exp(-\tau\Gamma)\left(\dfrac{\sin[2\tau(\omega+\omega_1)]}{2(\omega+\omega_1)} + \dfrac{\sin[2\tau(\omega-\omega_1)]}{2(\omega-\omega_1)} - \dfrac{\sin(2\omega\tau)}{\omega}\cos[2(\omega_1\tau+\theta)]\right), & \tau \geq 0 \end{cases} \tag{2.90}$$

Figures 2.20 and 2.21 represent the Wigner transform (2.90) and its contour plot, respectively, in normalized coordinates $T = \tau\omega_0$ and $w = \omega/\omega_0$, for two applied bias values: V = 0.1 V and V = 0.5 V, respectively [36]. These simulations were performed for the following values of cantilever parameters: A = 0.016 μm^2, K = 0.05 N/m, ω_0 = 2.7 10^5 Hz, m_B = 1.16 10^{-12} Kg, and $C_{t\text{-}s}$ = 1.41 10^{-17} F. As can be seen from Figures 2.20 and 2.21, the shape of the Wigner transform is quite different for these two cases. For a low bias V (and low Γ) the oscillatory nature of $y(\tau)$ is apparent from the closed curves in the phase-space (w,T). As the bias increases the oscillations are quickly damped, allowing a good response of the cantilever at its fundamental resonance frequency.

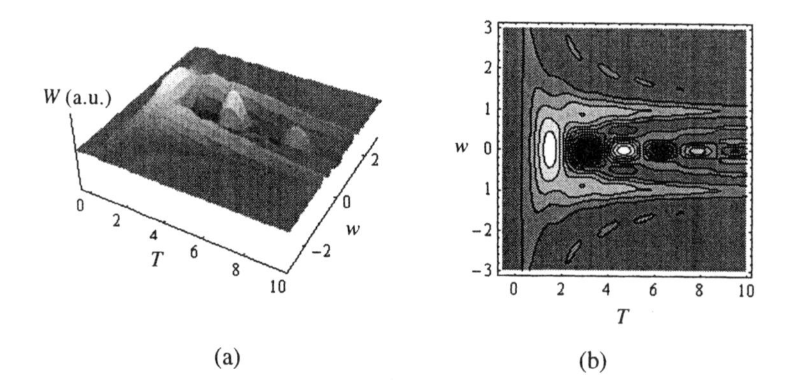

Figure 2.20 (a) Wigner transform of the cantilever amplitude and (b) its contour plot for an applied bias of 0.1 V.

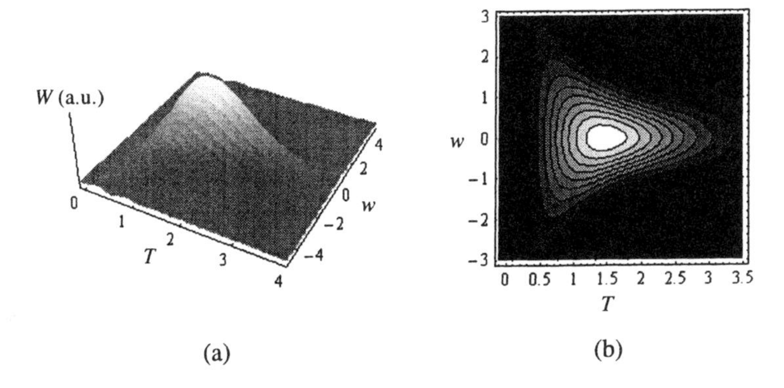

Figure 2.21 (a) Wigner transform of the cantilever amplitude and (b) its contour plot for an applied bias of 0.5 V.

The noncontact AFM is not only the mode in which the AFM attains an atomic resolution when measuring the surface sample via the distribution $z(x, y)$, but the mode in which electric and magnetic long-range forces distributed on a surface sample can be measured. In this last case, the terminology used is electrostatic force microscopy and magnetic force microscopy, respectively.

In the electrostatic force microscopy the cantilever is biased by both a dc and an ac voltage [37]. The aim of this method is to determine the distribution of voltages of either low or high frequency, and even to determine the charge distribution on certain samples, such as semiconductor devices. Because the cantilever is biased by $V = V_{dc} + V_{ac} \sin \omega t$,an electrostatic force is exerted between the tip and the sample, and the tip-sample interaction can be modeled (see Figure 2.22) as a capacitor $C_{t\text{-}s}$ with an effective area A [38].

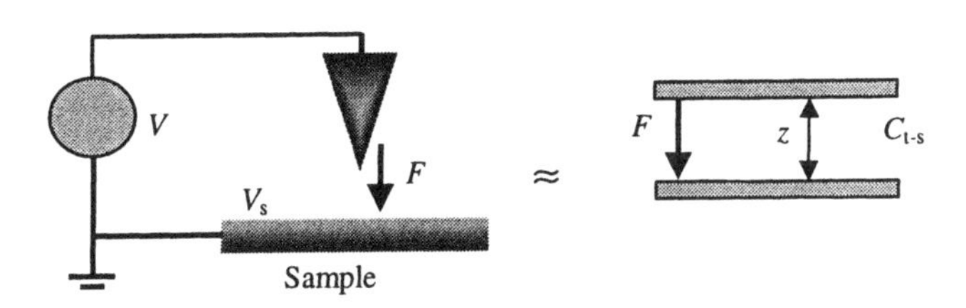

Figure 2.22 The electrostatic force microscopy principle.

The electrostatic force between the tip and the sample can then be written as

$$F = (1/2)(dC_{\text{t-s}}/dz)V_{\text{tot}}^2, \tag{2.91}$$

where the total voltage V_{tot} is a sum between $V = V_{\text{dc}} + V_{\text{ac}} \sin \omega t$ and the voltage to be measured V_{s}. An explicit calculation of (2.91) shows that the electrostatic force has three frequency components: a dc term, a component of frequency ω, and one of frequency 2ω. The dc component, which continuously bends the cantilever, cannot be easily detected. The component of the electrostatic force of frequency ω,

$$F_{\omega} = (1/2)(dC_{\text{t-s}}/dz)(V_{\text{dc}} + V_{\text{s}})V_{\text{ac}} \sin \omega t, \tag{2.92}$$

is proportional to the voltage distribution on the sample, and so this voltage distribution can be easily determined with good contrast over the entire sample (x, y). When distinct charges are involved instead of voltages, their distribution on the sample q_{s} and sign can be obtained using a modified version of (2.92):

$$F_{\omega} = (1/2)(dC_{\text{t-s}}/dz)[V_{\text{dc}} - q_{\text{s}}/(4\pi\varepsilon_0 z^2)]V_{\text{ac}} \sin \omega t. \tag{2.93}$$

The component of frequency 2ω of the electrostatic force is given by

$$F_{2\omega} = -(1/4)(dC_{\text{t-s}}/dz)V_{\text{ac}}^2 \cos(2\omega t). \tag{2.94}$$

If the ac signals for the tip and sample have different frequencies, namely if $V_{\text{ac}} = V_{\text{p}} \cos \omega_{\text{p}} t - V_{\text{s}} \cos \omega_{\text{s}} t$, where V_{p} and ω_{p} are the amplitude and frequency of the ac signal that pump the cantilever, and V_{s} and ω_{s} are the amplitude and the frequency of the signal that excite the sample, and if $|\omega_{\text{p}} - \omega_{\text{s}}| = \Delta\omega \le \omega_0$, it follows that $F_{2\omega} \to F_{\Delta\omega}$. The electrostatic force is given in this case by

$$F_{\Delta\omega} \cong (1/2)(dC_{\text{t-s}}/dz)V_{\text{p}}V_{\text{s}} \cos \Delta\omega t. \tag{2.95}$$

Equation (2.95) implies that the cantilever responds only to the beating components of the cantilever pump signal and the sample signal, although both can be much larger than the cantilever resonant frequency. In this case, the AFM acts as a high frequency mixer that works at the frequency $\Delta\omega$, and the sample potential distribution $V_{\text{s}}(x, y)$ can be determined from (2.95) as discussed above.

Time-domain or frequency-domain signals can be thus measured with the AFM in time intervals or at frequencies that exceed the range of usual instrumentation. In principle, there is no upper limit for the signal frequencies that are measurable with an AFM. For example, the amplitude and phase of 1 ps transient signals [39] or microwave signals exceeding 100 GHz were measured in integrated circuits using this method, which is specific for nanoscale devices [40]. This shows the impressive range of applications of AFM techniques. An AFM system for the characterization of high frequency circuits is schematically displayed in Figure 2.23. The minimum detectable voltage is in the range of 2–20 mV depending mainly on the cantilever Q factor and on the sharpness of the tip.

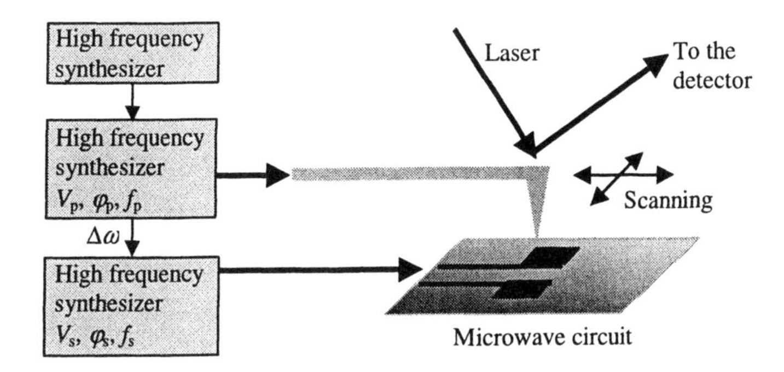

Figure 2.23 AFM system for high frequency measurements.

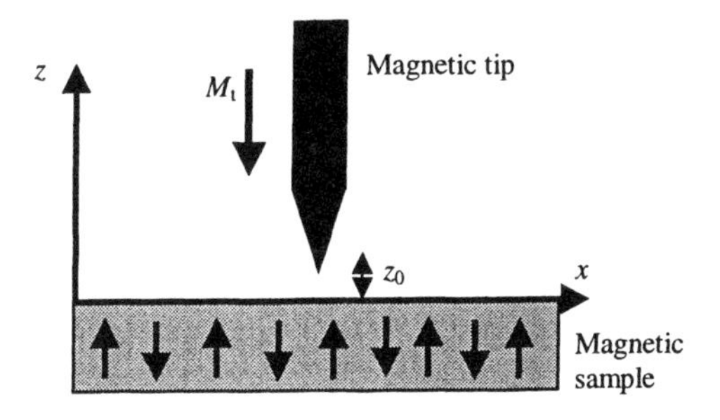

Figure 2.24 Schematic representation of the MFM technique.

The magnetic fields that exist at the surface of a magnetic material, the dynamics of magnetic domains and their structure can be determined and visualized using magnetic force microscopy (MFM). This is a noncontact AFM technique for which the tip is made either from a ferromagnetic material or coated with a thin magnetic film. For a comprehensive review of MFM, see [41]. In the MFM method the magnetic tip is scanned at a distance of 10–20 nm over the sample, the gradient force being detected and recorded. From these measurements one can then subtract information about the magnetic field or the structure of the magnetic walls. The MFM system is presented in Figure 2.24.

Considering that the tip with an area A and a length L_t is magnetized perpendicular to the sample, so that $M_z = M_t$, and that the sample magnetization is $M_z = M_s \cos k_x x$, the force gradient between the tip and the sample is [42]

$$\partial F / \partial z = -\partial^2 / \partial z^2 \int_{\text{tip}} M_t M_s \cos(k_x x)[1-\exp(-kxt)]\exp(-kxz)dV_{\text{tip}} . \tag{2.96}$$

Various calculations of (2.96) for different tip geometries of a Co tip with M_t = 1422 kA/m and with geometrical dimensions A = 20 nm×100 nm, L_t = 1 μm, placed at 10 nm over a sample with a thickness of 10 nm, and M_s = 295 kA/m, have shown that the optimal tip shape is elliptical; in this case it is possible to obtain a resolution of 10 mN/m.

The magnetic domains can be imaged and visualized even when the sample is very thin, as in the case of magnetic cobalt nanowires, for example [43]. A gallery of MFM images can be found in [44]. In particular, MFM shows high contrast for lithographically patterned particles. Single domain walls can be detected in the case of Ni particles (size 200 nm×70 nm×15 nm), as shown in Figure 2.25(a), the black spots indicating a repulsive force between the tip and the sample. In contrast, the white spot in Figure 2.25(a) indicates the attractive nature of the tip-sample force. Elliptical or disk-shaped particles, such as the Permalloy disk with a diameter 700 nm represented in Figure 2.25(b), show a vortex magnetic state. MFM is able to visualize also the magnetic moments of magnetic multilayers. These magnetic moments can be either parallel or antiparallel, a mixture of them being found in the most common situations, as illustrated in Figure 2.25(c).

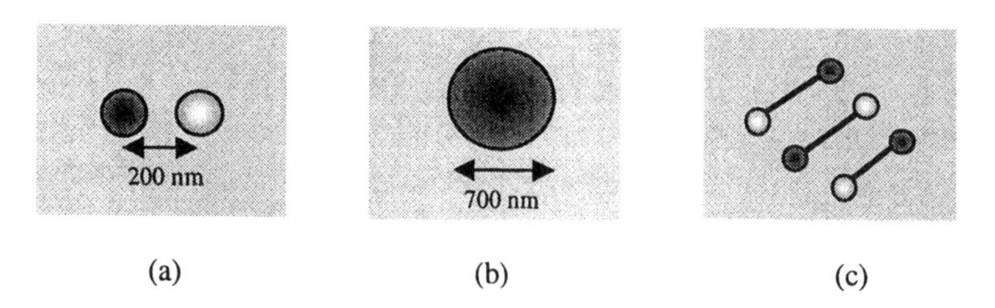

Figure 2.25 Schematic representation of MFM images: (a) magnetic single-domain, (b) vortex, and (c) parallel and antiparallel magnetic moments.

2.2.2 Scanning Tunneling Microscopy

Basically, STM provides images of surfaces $z(x, y)$ with atomic resolutions and even images of individual surface atoms. The STM is based on monitoring the tunneling current between a sharp metallic tip located at nanometric distances over a conductive sample. When a voltage V is applied between the tip and the sample, the electrons of energy E tunnel from the tip to the sample through a vacuum barrier of height ϕ, where $\phi > E$, as shown in Figure 2.26. The tunneling current density at low voltages is given by [45,46]

$$i = V\rho_{\text{surf}}(E_{\text{F}})\exp\{[-2m(\phi - E)]z/\hbar\} \propto V\rho_{\text{surf}}(E_{\text{F}})\exp(-1.025\phi^{1/2}z), \qquad (2.97)$$

where $\rho_{\text{surf}}(E_{\text{F}})$ is the density of states of the sample at the Fermi edge, z is the distance between the tip and the sample (the barrier width), and m is the electron mass. For a barrier height of 5 eV, the current decreases with one order of magnitude when the barrier width z is modified with only 0.1 nm. The tunneling current is usually of tens or hundreds of pA, an amplifier being needed to evidence it. In typical STM mode working conditions the current variation is about 2%–3%, which means, using (2.97), that the gap varies with about 0.001 nm. Practically, in this case the tunneling current is simply the density of states at the surface, and thus direct topography of the surface with atomic resolution becomes achievable.

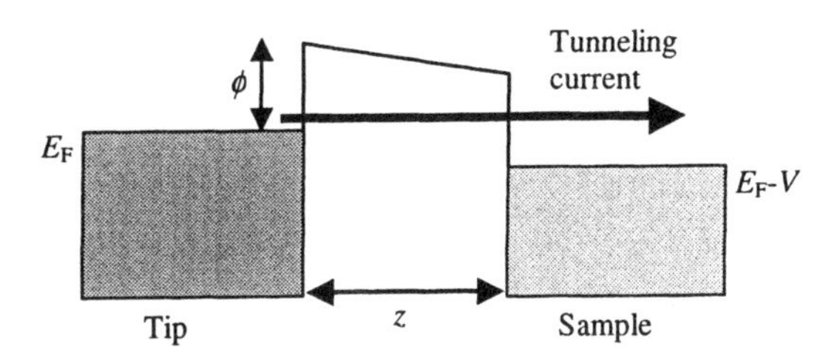

Figure 2.26 The physical principle of the STM.

The above considerations are only intuitive and valid in one dimension; in reality the mapping of a surface via the tip-surface tunneling current is a more complicated problem, which must be treated in three dimensions. A complete theory of STM imaging, the so-called Tersoff-Hamann model, can be found in [45]. The main result of this theory can be summarized as:

$$I(V) \propto \int_{E_F}^{E_F+eV} \rho(r,E)dE\ , \qquad (2.98)$$

where

$$\rho(r,E) = \sum_i |\Psi_i(\boldsymbol{r})|^2\ \delta(E - E_i) \qquad (2.99)$$

is the local density of states (LDOS) of the sample, with $\Psi_i(\boldsymbol{r})$ the 3D wave-function of the sample at energy E_i. The normalized conductance determined from the $I-V$ curve, $(dI/dV)/(I/V)$, is proportional to the LDOS of the sample. Thus, the Tersoff-Hamann model shows that the tunneling current is fully determined by the LDOS of the sample at the Fermi edge. This conclusion is valid only if the tip is very sharp (has a radius of a few Å) and is located at a few nm from the sample. Only under these conditions is the tunneling current confined to a thin filament between the end of the tip and the surface.

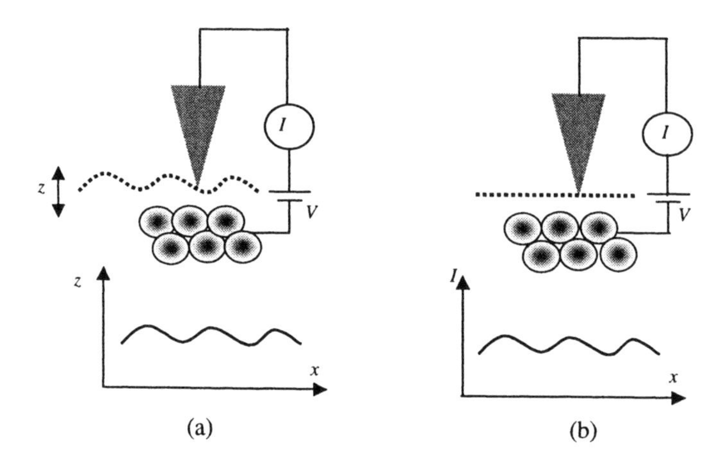

Figure 2.27 STM operation modes: (a) constant current mode, (b) constant height mode.

There are two basic modes of STM operation, described schematically in Figures 2.27(a) and 2.27(b). In the constant current mode the tunneling current is measured when the tip is positioned near the surface and is suitable based in the 5 mV – 2 V range. The tip is scanned over the surface and any change in the tunneling current is sensed by a feedback loop, which then changes, if necessary, the height z such that the current remains constant. Finally, a map $z(x, y)$ is obtained. In the constant height mode, the tip is scanning the surface at a constant height and the current is directly measured obtaining finally the map $I(x, y)$.

Some STM systems operate nowadays at room temperature and in normal atmospheric pressure, but are accompanied by sophisticated signal processing units able to calibrate very precisely the entire instrument and to remove the thermal noise and the hysteresis [47]. Other STM systems work in extreme conditions: low temperatures (350 mK), high magnetic fields (11 T), and ultrahigh vacuum (UHV) (10^{10} Pa) [48]. In particular the low-temperature and high-vacuum extreme conditions offer an exceptionally high resolution and a clean environment for studying nanomaterials and quantum processes. Single molecules on surfaces, adsorbates, liquid crystals, DNA sequencing, local density of states of metals, semiconductors, superconductors, and even of thin insulating films grown on conductive substrates (for example, metal oxides Al_2O_3, NiO, FeO) were imaged with the STM at low temperatures and UHV. STM can also be used to investigate the magnetic properties of materials with the help of magnetic tips, such as Fe tips. In this case, spin-polarized tunneling is employed in UHV conditions, the tunneling current being dependent on the magnetic moment orientation of the tip. There are also many STM image galleries, one of the richest being that provided by the company who invented the STM:

http://www.almaden.ibm.com/vis/stm/catalogue.html.

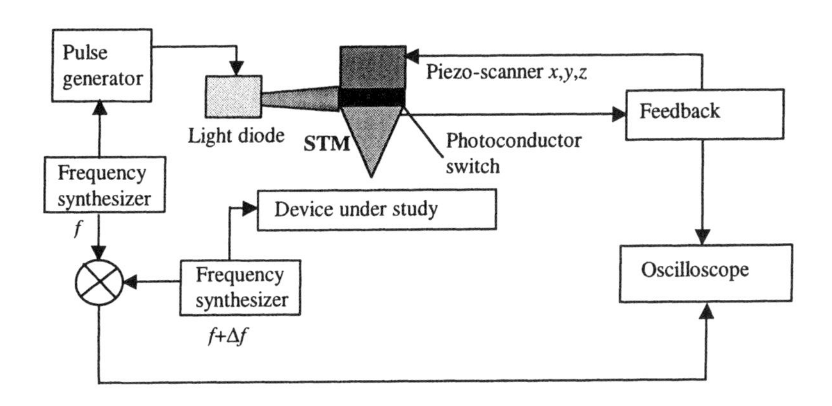

Figure 2.28 STM scheme for ultrafast signal detection (*After*: [49]).

As in the case of AFM, the STM can be used to measure ultrafast electrical signals. In this case, the tip is connected to an optical switch, which is driven in the ON or OFF state by an optical pulse with a repetition rate $1/f$. If the device under test is excited by a slightly different frequency, $f + \Delta f$, the waveform of the tunneling current is the replica of the response of the device under test at f, but extended with $f/\Delta f$ along the time axis. If $\Delta f > f_{\text{fb}}$, where f_{fb} is the frequency of the feedback system, the control of the tip position is not affected by the waveform measurement [49]. Pulses of ps duration and tens of mV in amplitude can be measured using this technique, schematically represented in Figure 2.28.

STM has also important applications in the study of electrodynamics at the nanoscale [50]. In this respect, STM can induce photon emission from molecules and thus can be used as a spectroscopic tool. The photons are emitted from a guest molecule placed in the gap between the tip and the sample, as illustrated in Figure 2.29. Several mechanisms of STM-induced photon emission exist, depending on the characteristics of the surface sample: 1) excitation of localized surface plasmons on metallic surfaces due to inelastic tunneling, 2) radiative decay due to inelastic tunneling into surface states on metal and semiconductor states, and 3) luminescence due to electron-hole recombination on semiconductor surfaces.

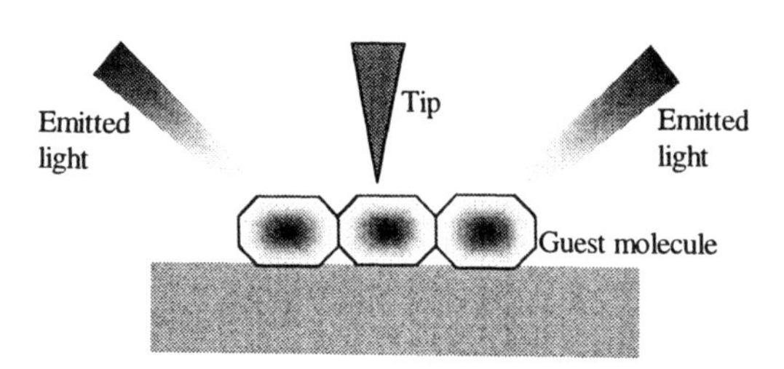

Figure 2.29 STM-induced photon emission.

A simple physical model, which explains the photon emission during the tunneling process, is based on the experimental observation that the conductance (which is proportional with the single electron tunneling rate in the barrier) shows certain jumps when the bias voltage is varied. Increasing the bias the current increases due to the elastic tunneling, up to a certain threshold given by

$$eV_{\text{th}} = n\hbar\omega . \tag{2.100}$$

where n is an integer. Beyond this threshold a new current channel is opened, which is assigned to inelastic tunneling through the guest molecule that excites a molecular vibration at the frequency ω, with the net result that a photon with energy $\hbar\omega$ is emitted. This process corresponds to a kink in the dI/dV curve and a peak in the d^2I/dV^2 curve. In this way, the vibrational spectra of the molecule can be determined. The re-radiated light due to photon-electron interaction in the tunneling junction is optically detected and important chemical information about molecules, such as their vibrational spectra, can be collected from it. A spatial map of the collected optical signals can thus be translated into a chemical map of the sample, which can be a molecule or an assembly of molecules, adding in this way important information to STM images.

2.2.3 Scanning Near-Field Optical Microscopy

The SNOM is based on the induced short range near field of a sample, which is excited by an electromagnetic source that radiates an electromagnetic wave with wavelength λ. A probe is scanned over the sample surface, and, depending on the particular SNOM configuration, it can have one of the following roles: 1) radiates the electromagnetic field in the near-field region of the sample, which is then detected with a photodetector in the far field, 2) detects the near field of the sample when the surface is excited by an external electromagnetic field placed far away from the sample, or 3) emits and detects the reflected wave from a surface when it is placed in the near-field radiation region of the sample.

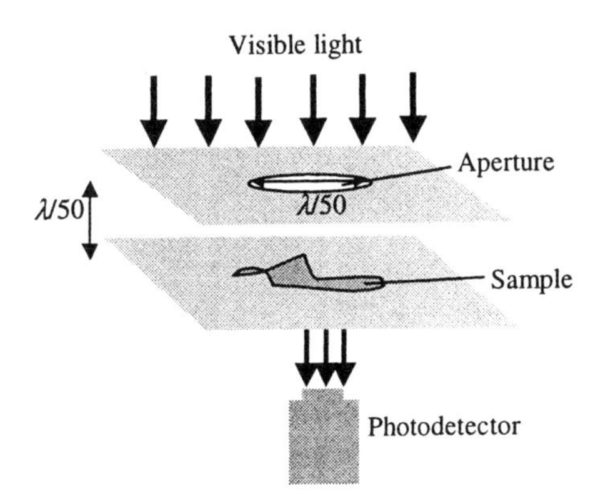

Figure 2.30 The SNOM principle.

The distance between the probe and the sample, the size of the probe as well as the physical features of the sample must all be much smaller than the wavelength. The diffraction limit in the far field depends on the wavelength and is equal to $\lambda / 2$. However, in the near field the diffraction limit does not depend on the wavelength and so the diffraction limit can be overcome. In the near field the SNOM can attain a super-resolution of a few nm in visible, in sharp contrast with a normal microscope, which cannot have a resolution better than 250 nm in the visible spectral region. The SNOM physical principle is displayed in Figure 2.30.

A simple physical example can help us understand better the SNOM principle [51]. Let us consider two point sources separated by a distance d, which oscillate with the same frequency and the same phase. The problem is to determine d from measurements of the field emitted by the sources at a certain distance $\boldsymbol{r}$. The amplitude at $\boldsymbol{r}$ is a sum of two waves:

$$A(\boldsymbol{r}) \propto \exp(ik\,|\boldsymbol{r}+\boldsymbol{d}/2|)/|\boldsymbol{r}+\boldsymbol{d}/2| + \exp(ik\,|\boldsymbol{r}-\boldsymbol{d}/2|)/|\boldsymbol{r}-\boldsymbol{d}/2|. \quad (2.101)$$

with k the wavenumber of the emitted radiation. In the far field, where $kr >> kd >> 1$, the amplitude in (2.101) becomes

$$A(\boldsymbol{r}) \propto [\exp(ikr)/r]\{2\cos[kd\cos(\theta)/2]\}, \quad (2.102)$$

where θ is the angle between $\boldsymbol{r}$ and $\boldsymbol{d}$. It is clear from (2.102) that the amplitude depends on the wavenumber k, and thus on the wavelength. To determine the distance d the amplitude A must be measured at two observation points for which θ is zero and $\pi / 2$, respectively.

If we analyze the expression of the amplitude (2.101) in the near-field range, in which $kd \leq kr << 1$, we get

$$A(\boldsymbol{r}) \propto [|\boldsymbol{r}+\boldsymbol{d}/2|^{-1} + |\boldsymbol{r}-\boldsymbol{d}/2|^{-1}]. \quad (2.103)$$

This expression is independent of the wavelength! The distance d is computed only from near-field measurements of A taken, for example, for $6\,|_{r=r_0} = 0$, case in which $d = 2(4 - r_0^2\,|A(r_0)|^2)^{1/2}/|A(r_0)|$.

The fabrication of near-field apertures with diameters of tens of nm, as well as the ability to position them at a height of 10–20 nm above the sample, was only possible after the development of AFM and STM techniques, and only after the MEMS technology was able to create such small apertures. SNOM benefits from all the equipment included in an AFM, such as the feedback system, the piezo-scanners, and so on.

SNOM probes are fabricated via various techniques and physical principles. The most encountered aperture is made from an optical fiber, which is sharpened through chemical etching and then covered with an opaque metal (for a review, see [52]). The apex of the probe is then cut to realize a nanosized aperture, as shown in Figure 2.31. A large category of SNOM probes is realized by integrating the nanosized apertures with AFM cantilevers (see Figure 2.32); this can be achieved using MEMS techniques (for a recent review, see [53]). Moreover, VCSEL lasers can be integrated with the cantilever and the nanoaperture [54], offering a high-resolution SNOM probe as that in Figure 2.33.

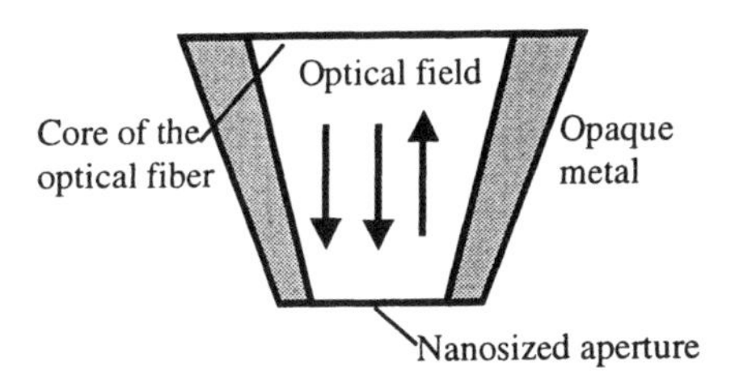

Figure 2.31 SNOM nanosized aperture made from an optical fiber.

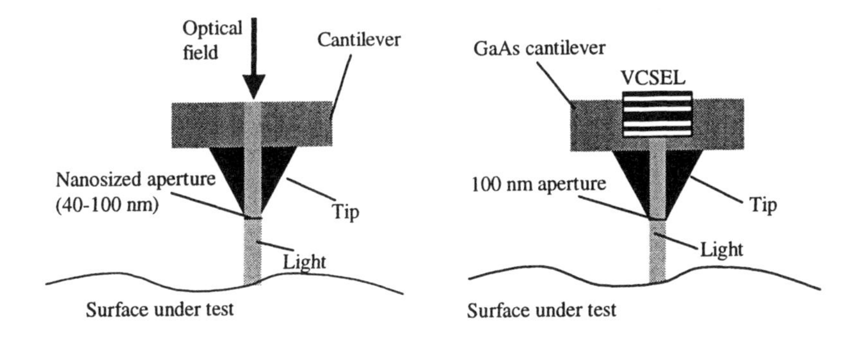

Figure 2.32 A cantilever integrated with a tip ended with a nanosized aperture.

Figure 2.33 A cantilever integrated with a VCSEL laser and a tip ended with a nanosized aperture (*After*: [27]).

SNOM can also be realized with apertureless probes placed in the near field of the sample. In this case, the probe is either a metal nanoparticle attached to a cantilever or a metallic tip positioned in the near field. An optical field excites the apertureless probes, and the scattered field in the near-field range is then detected and processed. The resolution of these probes can be higher than in the case of nanosized optical apertures.

SNOM is of paramount importance in some applications related to the study of single molecules and thin films, and in general in the study of nanomaterials whose properties are dictated by the structure and size of the constituents (for a review, see [55]). Biomolecular systems are also analyzed with the SNOM (see [56] for a comprehensive review). In this respect, single DNA molecules and human chromosomes, molecular motors, single chemical reactions, and the cell dynamics were imaged using SNOM. Quantum devices and nanophotonic structures can be also imaged with help of SNOM [57]. Recently, the SNOM was used to detect the amplitude and phase distributions of optical fields inside a photonic band gap structure composed of nanosized periodic materials [58].

It is worth mentioning that the SNOM technique is not only applicable for optical fields, but also for microwaves and millimeter waves. This interesting method was reviewed in [59]. Many configurations are the long-wave analogues of optical nanosized apertures or apertureless structures.

REFERENCES

[1] Stowe, T.D., et al., "Attonewton force detection using ultrathin silicon cantilevers," *Appl. Phys. Lett.*, Vol. 71, No. 2, 1997, pp. 288-290.

[2] Harris, J.G.E., et al., "Fabrication and characterization of 100 nm-thick GaAs cantilevers," *Rev. Sci. Instrum.*, Vol. 67, No. 10, 1996, pp. 3591-3593.

[3] Chand, A., et al., "Microfabricated small metal cantilever with silicon tip for atomic force microscopy," *J. Microelectromechanical Systems*, Vol. 9, No. 1, 2000, pp. 112-116.

[4] Petersen, K.E., "Dynamic micromechanics on silicon; techniques and devices," *IEEE Trans. Electron Devices*, Vol. 25, No. 10, 1978, pp. 1241-1249.

[5] Elwenspoek, M. and R. Wiegerink, *Mechanical Microsensors*, Berlin: Springer, 2001, pp. 72-77.

[6] Dequesnes, M., S.V. Rotkin, and N.R. Aluru, "Calculation of pull-in voltages for carbon-nanotube-based nanoelectromechanical switches," *Nanotechnology*, Vol. 13, No. 1, 2002, pp. 120-131.

[7] Sapmaz, S., et al., "Carbon nanotubes as nanoelectromechanical systems," *Phys. Rev. B*, Vol. 67, No. 2, 2003, pp. 235414/1-7.

[8] de Los Santos, H.J., "Nanoelectromechanical quantum circuits and systems," *Proc. IEEE*, Vol. 91, No. 11, 2003, pp. 1907-1921.

[9] Chan, H.B., et al., "Nonlinear micromechanical Casimir oscillator," *Phys. Rev. Lett.*, Vol. 87, No. 21, 2001, 211801/1-4.

[10] Buks, E. and M.L. Roukes, "Metastability and the Casimir effect in micromechanical systems," *Europhys. Lett.*, Vol. 54, No. 2, 2001, pp. 220-226.

[11] Serry, F.M., D. Walliser, and G.J. Maclay, "The anharmonic Casimir oscillator (ACO)–The Casimir effect in a model microelectromechanical system," *J. Microelectromechanical Systems*, Vol. 4, No. 4, 1995, pp. 193-205.

[12] Zalalutdinov, M., et al., "Frequency-tunable micromechanical oscillator," *Appl. Phys. Lett.*, Vol. 77, No. 20, 2000, pp. 3287-3289.

[13] Weigert, S., M. Dreier, and M. Hegner, "Frequency shifts of cantilevers vibrating in various media," *Appl. Phys. Lett.*, Vol. 69, No. 19, 1996, pp. 2834-2836.

[14] Zhu, R., "Theoretical and experimental studies of vibrations of optical fiber cantilevers for atomic force microscopy," *Rev. Sci. Instrum.*, Vol. 69, No. 4, 1998, pp. 1753-1756.

[15] Poncharal, P., et al., "Electrostatic deflection and electromechanical resonances of carbon nanotubes," *Science*, Vol. 283, No. 5407, 1999, pp. 1513-1516.

[16] Roukes, M.L., "Nanoelectromechanical systems," *Technical Digest of the 2000 Solid-State Sensor and Actuator Workshop*, Cleveland, OH, Hilton Head Island, 2000, pp. 1-10.

[17] Yasumura, K.Y., et al., "Quality factors in micron- and submicron-thick cantilevers," *J. Microelectromechanical Systems*, Vol. 9, No. 1, 2000, pp. 117-125.

[18] Hoummady, M. and H. Fujita, "Micromachines for nanoscale science and technology," *Nanotechnology*, Vol. 10, No. 1, 1999, pp. 29-33.

[19] Rast, S., et al., "The noise of cantilevers," *Nanotechnology*, Vol. 11, No. 2, 2000, pp. 169-172.

[20] Mehta, A., et al., "Manipulation and controlled amplification of Brownian motion of microcantilever sensors," *Appl. Phys. Lett.*, Vol. 78, No. 11, 2001, pp. 1637-1639.

[21] Blencowe, M.P. and M.N. Wybourne, "Sensitivity of a micromechanical displacement detector based on the radio-frequency single-electron transistor," *Appl. Phys. Lett.*, Vol. 77, No. 23, 2000, pp. 3845-3847.

[22] Yang, J., T. Ono, and M. Esashi, "Energy dissipation in submicrometer thick single-crystal silicon cantilevers," *J. Microelectromechnical Systems*, Vol. 11, No. 6, 2002, pp. 775-783.

[23] Srikar, V.T. and S.M. Spearing, "Materials selection in micromechanical design: an application of the Ashby approach," *J. Microelectromechnical Systems*, Vol. 12, No. 1, 2003, pp. 3-10.

[24] Cho, H.J. and C.H. Ahn, "A bidirectional magnetic microactuator using electroplated permanent magnet arrays," *J. Microelectromechnical Systems*, Vol. 11, No. 1, 2002, pp. 78-84.

[25] Dragoman, D. and M. Dragoman, "Optical actuation of micromechanical tunneling structures with applications in spectrum analyzing and optical computing," *Appl. Opt.*, Vol. 38, No. 32, 1999, pp. 6773-6778.

[26] Vogel, M., C. Mooser, K. Karrai, and R.J. Warburton, "Optically tunable mechanics of microlevers," *Appl. Phys. Lett.*, Vol. 83, No. 7, 2003, pp. 1337-1339.

[27] Dragoman, D. and M. Dragoman, "Micro/Nano-Optoelectromechanical systems," *Prog. Quantum Electron.*, Vol. 25, No. 6, 2001, pp. 229-290.

[28] Heinzelman, R., et al., "Force microscopy," in *Scanning Tunneling Microscopy and Related Methods*, R.J. Behm et al. (eds.), Dordrecht: Kluwer Academic, 1990, pp. 443-467.

[29] Bhushan, B. (ed.), *Springer Handbook of Nanotechnology*, Berlin: Springer, 2004, pp. 330-331.

[30] Sarid, D., *Scanning force microscopy with applications to electronic, magnetic and atomic forces*, New York: Oxford University Press, 1994.

[31] Manalis, S.R., et al., "Interdigital cantilevers for atomic force microscopy," *Appl. Phys. Lett.*, Vol. 69, No. 25, 1996, pp. 3944-3946.

[32] Dai, H.J., et al., "Nanotubes as nanoprobes in scanning probe microscopy," *Nature*, Vol. 384, No. 6605, 1996, pp. 147-150.

[33] Beck, R.D., et al., "GaAs/AlGaAs self-sensing cantilevers for low temperature scanning probe microscopy," *Appl. Phys. Lett.*, Vol. 73, No. 8, 1998, pp. 1149-1151.

[34] Kawakatsu, H., et al., "Millions of cantilevers for atomic force microscopy," *Rev. Sci. Instrum.*, Vol. 73, No. 3, 2002, pp. 1188-1192.

[35] Kawakatsu, H., et al., "Towards atomic force microscopy up to 100 MHz," *Rev. Sci. Instrum.*, Vol. 73, No. 6, 2002, pp. 2317-2320.

[36] Dragoman, D. and M. Dragoman, "Time-frequency modeling of atomic force microscopy," *Optics Communications*, Vol. 140, No. 2, 1997, pp. 220-225.

[37] Girard, P., "Electrostatic force microscopy: principles and some applications to semiconductors," *Nanotechnology*, Vol. 12, No. 4, 2001, pp. 485-490.

[38] Nechay, B.A., et al., "Applications of an atomic force microscope voltage probe with ultrafast time resolution," *J. Vac. Sci. Technol.*, Vol. B 13, No. 3, 1995, pp. 1369-1374.

[39] Hou, A.S., et al., "Scanning probe microscopy for testing ultrafast electronic devices," *Opt. Quantum Electron.*, Vol. 28, No. 12, 1996, pp. 819-841.

[40] Leyk, A., et al., "104 GHz signal measurement by high frequency scanning force microscope test system," *Electronics Letters*, Vol. 31, No. 13, 1995, pp. 1046-1047.

[41] Grütter, P., H.J. Mamin, and D. Rugar, "Magnetic Force Microscopy," in *Scanning tunneling microscopy II*, R. Wiesendanger and H.-J. Güntherodt (eds.), Berlin: Springer, 1995, pp. 151-207.

[42] Saito, H., et al., "High resolution MFM: simulation of the tip sharpening," *IEEE Trans. Magnetics*, Vol. 39, No. 5, 2003, pp. 3447-3449.

[43] Garcia, J.M., et al., "Characterization of cobalt nanowires by means of force microscopy," *IEEE Trans. Magnetics*, Vol. 36, No. 5, 2000, pp. 2981-2983.

[44] Zhu, X. and P. Grutter, "Magnetic force microscopy studies of patterned magnetic structures," *IEEE Trans. Magnetics*, Vol. 39, No. 5, 2003, pp. 3420-3425.

[45] Hansma, P.K. and J. Tersoff, "Scanning tunneling microscope," *J. Appl. Phys.*, Vol. 61, No. 2, 1987, pp. R1-R23.

[46] Binnig, G. and H. Rohrer, "Scanning tunneling microscopy," *IBM J. Res. Develop.*, Vol. 44, No. 1/2, 2000, pp. 279-292.

[47] Pope, K.J., J.L.P. Smith, and J.G. Shapter, "Imaging molecular adsorbates using scanning tunneling microscopy and image processing," *Smart Mater. Struct.*, Vol. 11, No. 2, 2002, pp. 679-685.

[48] Sagisaka, K., et al., "Scanning tunneling microscopy in extreme fields: very low temperature, high magnetic fields, and extreme high vacuum," *Nanotechnology*, Vol. 15, No. 1, 2004, pp. S371-S375.

[49] Takeuchi, K., A. Mizuhara, and Y. Kasahara, "Application of scanning tunneling microscope to high-speed optical sampling measurement," *IEEE Trans. on Instrumentation and Measurement*, Vol. 44, No. 3, 1995, pp. 815-818.

[50] Nejo, H. (ed.), *Nanoelectrodynamics. Electrons and Electromagnetic Fields in Nanometer-Scale Structures*, Berlin: Springer, 2003.

[51] Banno, I., "Classical theory on electromagnetic near field," in *Progress in Nano-Electro-Optics II: Novel Devices and Atom Manipulation*, M. Ohtsu (ed.), Berlin: Springer, 2004, pp. 1-57.

[52] Ohtsu, M. and K. Sawada, "High-resolution and high-throughput probes," in *Nano-Optics*, S. Kawata, M. Ohtsu, and M. Irie (eds.), Berlin: Springer, 2002, pp. 61-74.

[53] Ono, T., et al., "Integrated and functional probes," in *Nano-Optics*, S. Kawata, M. Ohtsu, and M. Irie (eds.), Berlin: Springer, 2002, pp. 111-135.

[54] Heisig, S., O. Rudow and E. Oesterschulze, "Scanning near-field optical microscopy in the near-infrared region using light emitting cantilever probes," *Appl. Phys. Lett.*, Vol. 77, No. 8, 2000, pp. 1071-1073.

[55] Fujihira, M., et al., "Near-field imaging of molecules and thin films," in *Nano-Optics*, S. Kawata, M. Ohtsu, and M. Irie (eds.), Berlin: Springer, 2002, pp. 151-190.

[56] Yanagida, T., et al., "Near-field microscopy for biomolecular systems," in *Nano-Optics*, S. Kawata, M. Ohtsu, and M. Irie (eds.), Berlin: Springer, 2002, pp. 191-236.

[57] Gonokami, M., H. Akiyama, and M. Fukui, "Near-field imaging of quantum devices and photonic structures," in *Nano-Optics*, S. Kawata, M. Ohtsu, and M. Irie (eds.), Berlin: Springer, 2002, pp. 237-286.

[58] Flück, E., et al., "Amplitude and phase evolution of optical fields inside periodic photonic structures," *IEEE J. Lightwave Technology*, Vol. 21, No. 5, 2003, pp. 1384-1393.

[59] Rosner, B.T. and D.W. van der Weide, "High-frequency near-field microscopy," *Rev. Sci. Instrum.*, Vol. 73, No. 7, 2002, pp. 2505-2525.

Chapter 3

Carbon Nanotube Devices

Carbon nanotubes (CNTs) of different forms and shapes are the most studied nanomaterials because it is hoped that either the electrical or the optoelectronic devices based on them will exceed the performances of similar devices based on Si and GaAs semiconductors. The CNT-based electrical devices are studied in this chapter, while the corresponding optoelectronic devices will be presented in Chapter 6. Before detailing various CNT-based electrical devices we analyze their physical properties, which reveal astonishing behaviors never encountered before in any known material.

3.1 PHYSICAL PROPERTIES

3.1.1 Band Structure and Band Modulation

The CNTs, as well as the electronic, optical, and NEMS (nano-electro-mechanical systems) devices based on them, represent one of the most studied topics in modern nanoelectronics. Theoretical, experimental and advanced technological processes are involved in the study of CNT properties and applications. CNTs have a series of amazing electrical, thermal, optical, or mechanical characteristics, which are either not encountered in other materials or prevail over similar characteristics of any existing material with few orders of magnitude. These properties justify the huge interest in CNT devices.

CNTs are empty cylinders, which can be viewed as rolled up sheets of a single or several concentric layers of graphene. As displayed in Figure 3.1, the graphene is a 2D honeycomb structure of sp^2 carbon atoms. The CNT with a single rolled up sheet of graphene is called single-walled CNT (SWCNT), while the CNT having several concentric graphene layers is termed multiwalled CNT (MWCNT). Very often the physical properties of SWCNTs differ significantly from those of MWCNTs, and thus care must be taken in choosing the type of CNT involved in a certain application.

Depending on how the graphene layer (or layers) is (are) rolled we can get CNTs with metallic or semiconducting behavior. This remarkable possibility, of tailoring the fundamental material properties by folding graphene sheets around different directions, is unique among any known material. The way in which a graphene sheet is folded is described by two parameters: the chirality or chiral vector $\boldsymbol{C}$ and the chiral angle θ (see Figure 3.1). The chiral vector of a CNT, which links two equivalent crystallographic sites, is given by

$$\boldsymbol{C} = n\boldsymbol{a}_1 + m\boldsymbol{a}_2 , \tag{3.1}$$

where $\boldsymbol{a}_1$ and $\boldsymbol{a}_2$ are unit vectors of the graphene lattice, and the numbers n and m are integers.

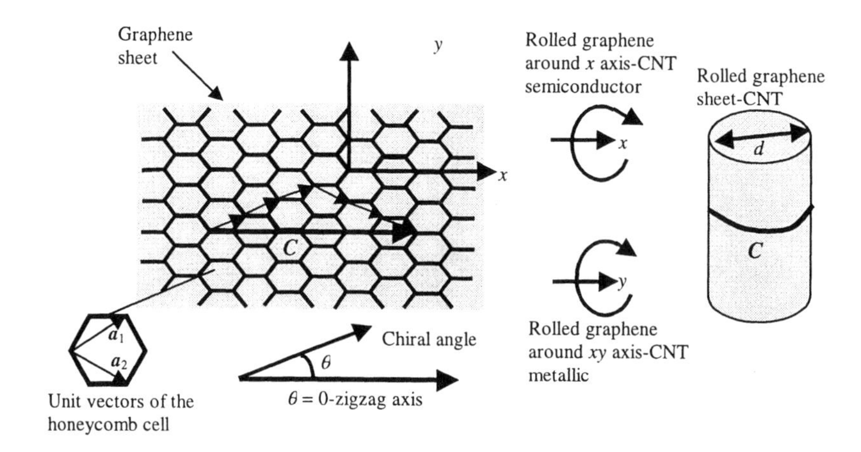

Figure 3.1 Schematic description of the CNT structure.

The set of integer numbers (n, m) describes entirely the metallic or the semiconducting character of any CNT. In general, the CNT is metallic if $n = m$, it shows a small bandgap (has allegedly a semimetallic character) if $n - m = 3i$, with $i = 1, 2, 3, \ldots$, and is semiconducting when $n - m \neq 3i$ [1]. The most encountered situations are the (n, n) metallic CNTs, also called armchair CNTs, and the CNTs characterized by $(n, 0)$, which are semiconducting and are termed zigzag CNTs [2].

There is a direct link between the set (n, m) and the geometrical characterristics of the CNT. In particular, the CNT diameter is given by

$$d = a_{C-C}[3(m^2 + mn + n^2)]^{1/2} / \pi = |C| / \pi, \qquad (3.2)$$

where $a_{C-C} = 1.42$ Å is the carbon bond length, and $|C|$ is the length of the chiral vector. The above formula illustrates the significance of the chiral vector: its modulus is equal to the circumference of the CNT.

The chiral angle, defined by

$$\theta = \tan^{-1}[\sqrt{3}\, n / (2m + n)] \qquad (3.3)$$

takes the value $\theta = 30°$ for (n,n) armchair CNTs and is equal to $\theta = 60°$ for $(n,0)$ zigzag CNTs. It is common, however, to limit the domain of θ to the range (0, 30°); then, as shown in Figure 3.1, due to symmetry, we assign $\theta = 0°$ for zigzag CNTs and consider $\theta = 0°$ as the reference axis or the zigzag axis. Instead of the chiral vector and the chiral angle, the set of integers (n,m), for example (10,10), (9,0), or (4,2), can be alternatively used to specify a CNT, the diameter and chiral angle of which can then be calculated using (3.2) and (3.3), respectively.

The dispersion relation of graphene can be obtained by considering that the complex energy structure of sp^2 bonded carbon atoms can be reduced to only two bands: the conduction band π^* and the valence band π. Then, a simple tight-binding model [3] based on the nearest-neighbor Hamiltonian

$$H = \begin{pmatrix} E_{2p} & -\gamma_0 g(\boldsymbol{k}) \\ -\gamma_0 g^*(\boldsymbol{k}) & E_{2p} \end{pmatrix} \qquad (3.4)$$

associated to carrier doping between the two bands can be used to calculate the dispersion relation, provided that the overlap of carbon atoms in the graphene structure is also taken into account. In the above Hamiltonian $\gamma_0 > 0$ is the carbon-carbon energy, E_{2p} is the site energy of the $2p$ atomic orbital, and the function $g(\boldsymbol{k})$, with $\boldsymbol{k} = (k_x, k_y)$ a vector in the reciprocal space of the graphene lattice, is given by

$$g(\boldsymbol{k}) = \exp[i(k_x a)/3^{1/2}] + 2\exp[-i(k_x a/2)/3^{1/2}]\cos[(k_y a)/2], \qquad (3.5)$$

where $a = |a_1| = |a_2| = 3a_{C-C}$. With an overlap matrix associated with two carbon atoms contained in any graphene honeycomb structure of the form

$$S_o = \begin{pmatrix} 1 & sg(\boldsymbol{k}) \\ sg^*(\boldsymbol{k}) & 1 \end{pmatrix}, \qquad (3.6)$$

where s is the overlap of the electronic wavefunction over adjacent sites, the band structure $E_{2D}(\boldsymbol{k})$ of the graphene is determined from the equation

$$\det(H - E_{2D}S_o) = 0, \tag{3.7}$$

which has the solution

$$E_{2D}^{\pm}(\boldsymbol{k}) = \frac{E_{2p} \pm \gamma_0\omega(\boldsymbol{k})}{1 \mp s\omega(\boldsymbol{k})}, \tag{3.8}$$

with

$$\omega(\boldsymbol{k}) = |g(\boldsymbol{k})|^{1/2} = [1 + 4\cos(3^{1/2}k_x a/2)\cos(k_y a/2) + 4\cos^2(k_y a/2)]^{1/2}. \tag{3.9}$$

The dispersion relation of the two branches of the energy band structure of (3.8) is represented in Figure 3.2 for $E_{2p} = 0$, $\gamma_0 = 3\,\text{eV}$, and s = 0.13, while Figure 3.3 shows the band structure of the graphene for the same parameters when the two branches E_{2D}^{+} and E_{2D}^{-} are superimposed.

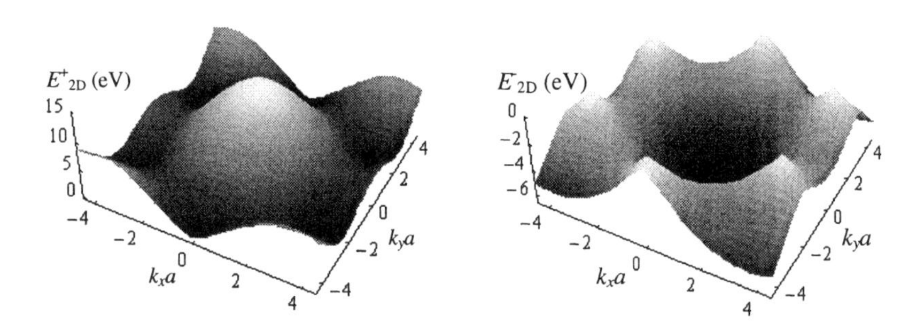

Figure 3.2 E_{2D}^{+} and E_{2D}^{-} dependence on the wave vector.

From Figure 3.3 it follows that when E_{2D}^{+} and E_{2D}^{-} are superimposed six common points named Fermi points appear, which are denoted by K and K' in Figure 3.4 and are located at the Fermi level E_F and form a hexagon that defines the limit of the graphene Brillouin zone (see Figure 3.4). The Fermi energy is assigned for $E = 0$ at K points but, since the corresponding density of states (DOS) of allowed states at the Fermi level is zero, the graphene is a zero bandgap semiconductor. Inside the hexagon the iso-energy lines are circles around $K(0, 4\pi/3a)$ and $\Gamma(0,0)$ and are straight lines that connect the $M(2\pi/3^{1/2}a, 0)$ points.

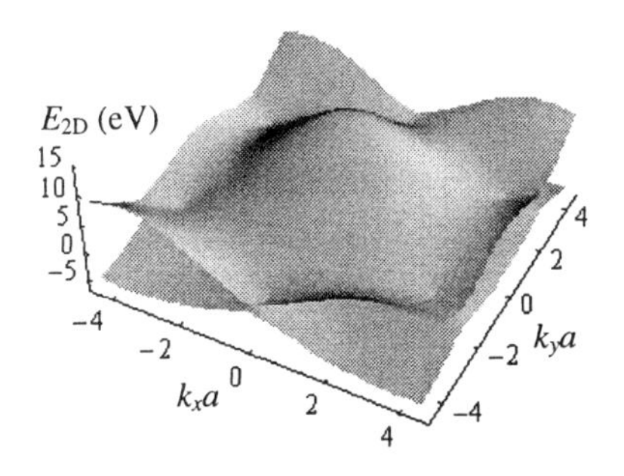

Figure 3.3 The band structure of graphene.

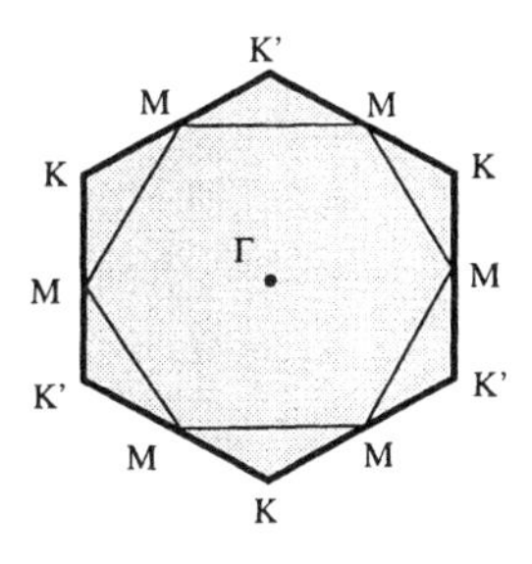

Figure 3.4 The Brillouin zone of graphene.

The CNT band structure is obtained by considering that the folding of the 2D graphene sheet into a nanotube produces confinement of carriers in the circumferential direction of the tube, which can be expressed by the quantization rule

$$\boldsymbol{k}_c \boldsymbol{C} = 2\pi j, \tag{3.10}$$

where $\boldsymbol{k}_c$ is the circumferential component of the wave vector $\boldsymbol{k}$ and $j = 1, 2, 3, \ldots, N$ is an integer with N the number of hexagons in a nanotube unit cell. The requirement (3.10) implies that the allowed energy band structure of the CNT is obtained by cutting the 2D band structure of the graphene into 1D slices. In the vicinity of a K point, for $ka \ll 1$ with $k = |\boldsymbol{k}|$ measured from K, the function $\omega(k)$ given by (3.9) is linearly dependent on the wave vector k:

$$\omega(k) \cong 3^{1/2} ka/2 + \ldots. \tag{3.11}$$

Further, for $E_{2p} = 0$ and $s = 0$ the graphene energy (3.8) becomes

$$E_{2D} \cong 3^{1/2} \gamma_0 k a_{C-C}, \tag{3.12}$$

from which it follows that the CNT energy dependence on the wave vector can be expressed as [4].

$$E_{\mathrm{CNT},j} = E_{\mathrm{2D}}(k\boldsymbol{K}_{\mathrm{a}} / |\boldsymbol{K}_{\mathrm{a}}| + j\boldsymbol{K}_{\mathrm{c}})\,. \tag{3.13}$$

Here $\boldsymbol{K}_{\mathrm{c}}$ is a unit vector along the circumferential direction and $\boldsymbol{K}_{\mathrm{a}}$ is the reciprocal lattice vector along the axis of the tube. The wave number k takes values in the interval $(-\pi/T, \pi/T)$, where T is the length of the translational vector $\boldsymbol{T}$. Relation (3.13) implies the existence of N discrete values of the CNT wave vector in the $\boldsymbol{K}_{\mathrm{c}}$ direction while in the $\boldsymbol{K}_{\mathrm{a}}$ direction the wave vector is continuous. This means that N lines of the form $k\boldsymbol{K}_{\mathrm{a}} / |\boldsymbol{K}_{\mathrm{a}}| + j\boldsymbol{K}_{\mathrm{c}}$ cut the hexagonal Brillouin zone of graphene. As illustrated in Figure 3.5, the CNT is metallic if the lines pass through the K point, and semiconductor otherwise. In other words, the CNT is metallic if $n - m = 3i$ and semiconductor when $n - m \neq 3i$, with i an integer. At low energies the band structure of a metallic CNT consists of two bands with linear dispersion that intersect at the K point, while for semiconductor CNTs an energy gap E_{g} opens between the two bands.

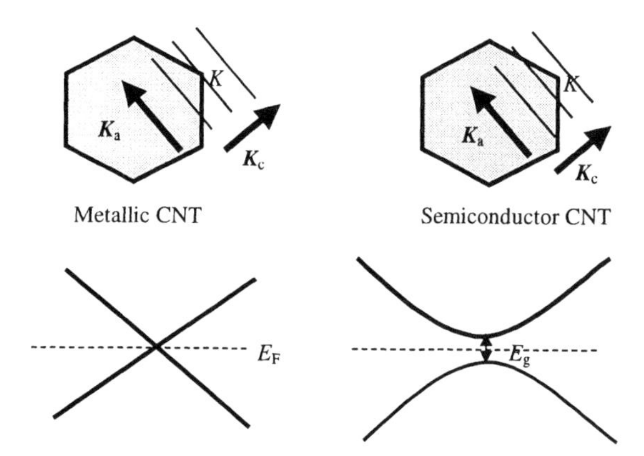

Figure 3.5 CNT band structure (*After*: [4]).

The bandgap of the semiconducting CNT is given by [1]

$$E_{\mathrm{g}} = 4\hbar v_{\mathrm{F}} / 3d\,, \tag{3.14}$$

and takes the value

$$E_{\mathrm{g}}(\mathrm{eV}) \cong 0.9 / d(\mathrm{nm}) \tag{3.15}$$

for the Fermi velocity $v_F = 8\times10^7$ m/s. Because SWCNTs have diameters ranging from a fraction of a nanometer to several nanometers, the semiconducting CNTs have a bandgap in the range 20 meV–2 eV. Bandgap engineering is accomplished in the case of CNT by simply changing the nanotube diameter. Although bandgap engineering can also be achieved in AIII-BV semiconductor heterostructures by changing the chemical composition of the heterostructure, the variation of the energy gap cannot attain such a large value as for CNTs.

The quite simple model presented above, based on the tight-binding Hamiltonian applied for the two-band structure of graphene, is applicable for the large majority of CNTs, even for nanotubes with quite large diameters, but fails for CNTs with a very small diameter of 2–6 Å. For this case various methods for energy band structure calculation (such as first-principles, tight-binding, LDA) provide contradictory results [5]. For example, the zigzag SWCNTs with $n < 7$, which are semiconductors according to the theory above, should be metallic if first-principle calculations are performed. On the contrary, the tight-binding model predicts bandgaps for small-diameter semiconducting tubes, which are, however, slightly larger than in reality. In this case, the situation can be alleviated by considering the involvement of chirality θ in the calculation of E_g. In the tight-binding model E_g is independent of θ and given by $E_g = 2\gamma_0 a_{C-C} / d$, with γ_0 the doping matrix element $V_{p\pi\pi}$, but a more appropriate formula for semiconducting CNTs with small diameters $(4 < n < 9)$ is [5]

$$E_g = 2\gamma_0 a_{C-C}[1 + (-1)^p 2\beta a_{C-C} \cos(3\theta) / d] / d , \tag{3.16}$$

with β a constant and p an integer that satisfies $n - 2m = 3i + p$. By choosing $\gamma_0 = 2.53\,\text{eV}$, $\beta = 0.43$, $\theta = 0$, we get

$$E_g = 2\gamma_0 a_{C-C} / d \pm 4\gamma_0 \beta a_{C-C}^2 / d^2 . \tag{3.17}$$

Although the above results have been obtained for defect-free graphene and thus for a resulting crystalline CNT, amazingly, the $1/d$ dependence of the bandgap is preserved even when the graphene is disordered [6]. The amorphous semiconducting CNT has a bandgap given by

$$E_g \cong 3\times(2)^{-1/2} m_C \overline{\omega^2} a_{C-C} / d , \tag{3.18}$$

where m_C is the mass of carbon (2×10^{-23} g) and $\overline{\omega}$ is the average phonon energy (1600 cm^{-1}). The disorder, produced by defects in graphene, yields a faster variation of the bandgap with the CNT diameter compared to the crystalline case, which can be useful in certain applications. This is an example of CNT functionalization, that is, of a drastic and controllable change of the physical properties of CNTs under the action of an external excitation. The variation of the bandgap for crystalline and amorphous CNTs is represented schematically in Figure 3.6.

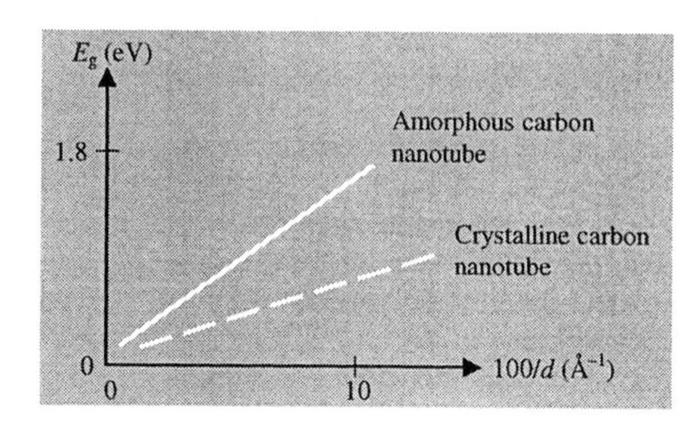

Figure 3.6 Diameter dependence of the bandgap of amorphous and crystalline semiconducting CNTs (*After*: [6]).

By changing the physical properties of CNTs one can induce new properties of the CNT devices. In the above situation, we functionalized the crystalline CNT by introducing defects in it, the consequence being a significant change of the bandgap, but CNTs can be functionalized in many ways including by doping, by adsorption of individual atoms or molecules (hydrogenation, oxygenation), by radial mechanical deformations, and by application of external or magnetic fields.

Irrespective of the functionalization method the energy band structure of the CNT is profoundly modified. In particular, a reversible semiconductor-isolator transition occurs in some cases, which changes entirely the material properties of a single CNT or of an array of CNTs, with important consequences in CNT devices. In what follows, we analyze qualitatively the changes in the energy band produced by some of the functionalization techniques mentioned above. The reader is advised to peruse reference [5] for an updated and in-depth review of this subject.

In the case of hydrogenation, the geometry, band structure, and binding energies of CNTs strongly depend on the pattern of hydrogenation, called decoration. For example, when zigzag CNTs are uniformly exohydrogenated at half-coverage ($\Theta = 0.5$, where Θ is a measure of different isomers at different hydrogen coverage), the (7,0) CNT changes its shape from circular to rectangular, while the (n,0) CNTs, with n = 8, 9, 10, become squared. Moreover, at half-exohydrogenation (Θ = 0.5) the semiconducting tubes show two double-degenerated dispersionless conduction and valence bands separated by a bandgap that decreases as d increases and that becomes zero for large n. The effect of full exohydrogenation (Θ = 1) on CNTs is also very pronounced. The bandgap of semiconducting CNTs is increased by 2 eV, attaining a maximum value of about 5 eV, while metallic armchair CNTs are transformed into semiconducting CNTs with a slightly greater bandgap than semiconducting nanotubes with a similar diameter.

Adsorbing oxygen molecules again induce significant transformations of the band energy structure. For example, low-diameter semiconducting CNTs, such as (8,0), are transformed into metallic CNTs when oxygen molecules are physisorbed.

The mechanical pressure exerted on metallic CNTs squashes them [7], while a hydrostatic pressure [8] induces radial deformations on CNTs, as indicated in Figure 3.7. The shape deformation from circle to ellipse is dominated by the competition between compression, which reduces the perimeter, and bending, which increases the curvature of the CNT. Beyond a certain pressure P_1 the bending process is dominant, this threshold pressure between the circular and the ellipse shape being given by

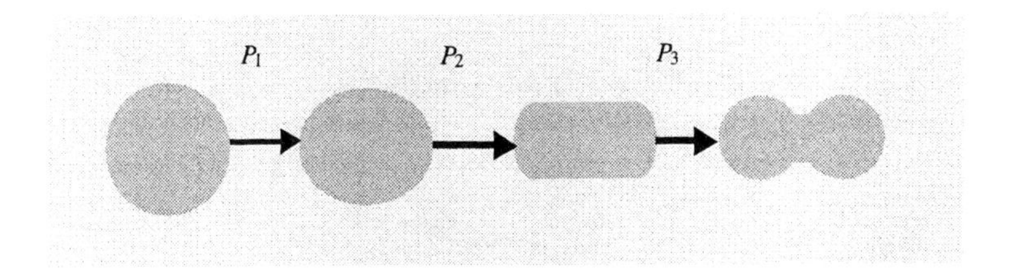

Figure 3.7 Deformation of the cross-section of a metallic CNT due to an increasing external pressure.

$$P_1 \cong 24R_F / d^3 , \tag{3.19}$$

where R_F is the flexural rigidity. The other threshold pressures for shape deformations are

$$P_2 = P_1 - B\ln(A_2 / A_1) , \tag{3.20}$$

$$P_3 = P_1 - B\ln(A_3 / A_1) , \tag{3.21}$$

where A_1, A_2, and A_3 are the cross-section areas of the CNT at pressures P_1, P_2, and P_3, respectively, and B is the radial modulus of the tube after the first shape transition. Because the ratio $A_2 / A_1 \cong 0.81$ is independent of the tube radius, we obtain $P_2 \cong 1.2P_1$. At the pressure P_3 the metallic CNT experiences a transition towards a semiconducting nanotube, the metal-semiconductor transition being identified by measuring the conductance dependence on the dc voltage applied to the CNT. At the metal-insulator transition a gap opens, manifested by the decrease of the CNT conductance with two orders of magnitude. This gap is about 0.2 eV for a (10,10) metallic CNT. The pressure P_3 decreases if the diameter of the CNT increases. For example, P_3 is 8.5 GPa for a (6,6) metallic nanotube, it becomes 2 GPa for a (10,10) metallic nanotube, while for a (100,100) metallic nanotube P_3 is only 3 kPa. The pressures P_1 and P_2 show the same behavior.

The axial strain applied on a semiconducting CNT increases or decreases the bandgap, depending on the type of the strain (see Figure 3.8). The gap variation as a function of the strain σ is given in this case by [9]

$$dE_g / d\sigma = \mathrm{sgn}(2p+1)3\gamma_0(1+\nu)\cos 3\theta \,, \tag{3.22}$$

where ν is the Poisson ratio, n and m satisfy $n - m = 3i + p$, with $p = -1$, 0, or 1 and i an integer. The semiconductor CNTs with $p = 1$ have $dE_g / d\sigma > 0$, while those with $p = -1$ have $dE_g / d\sigma < 0$. The controllable increase or decrease of the bandgap of semiconductor CNTs under strain can be used to bandgap-engineer heterostructures, as in the case of AIII-BV semiconductors.

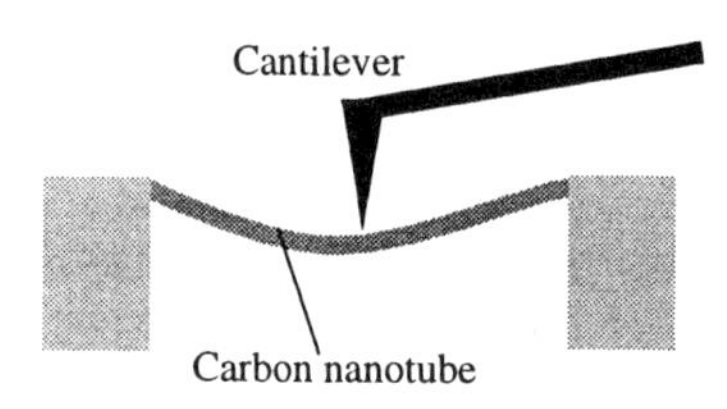

Figure 3.8 Tuning of the CNT bandgap via an applied strain.

The axial strain applied on a metallic CNT via an AFM cantilever generates a metallic-semiconductor transition, during which the conductance decreases with orders of magnitude, as in the case of CNT squashing. The conductance depends on strain as

$$G \cong \gamma_0(8e^2 / h)\{1+\exp[E_g(\sigma)/ k_B T)]\}^{-1}. \tag{3.23}$$

The electronic properties of individual semiconducting CNTs can be controlled by chemical doping, which tunes the Fermi energy level value in either the conduction band (for an *n*-type CNT) or in the valence band (for a *p*-type CNT). At room temperature and in a normal atmospheric pressure the semiconductor CNT is not intrinsic ($n = p$), but is of *p*-type, so that the electronic properties are dictated by holes, whose concentration is dominant. The natural *p*-type doping of semiconductor CNTs was first explained as a chemical contamination and was attributed either to the charge transfer from metal contacts, which are patterned over the CNT, or to adsorbed molecules, such as oxygen or chemical groups, with which the CNT is in contact when patterned on various types of substrates. Nevertheless, it was demonstrated that neither adsorbed molecules nor chemical groups induces the *p*-type behavior. Rather, the origin of the *p*-type behavior is the self-doping phenomenon [10], which is specific to nanoscale materials and is caused by the curvature of the CNT. More precisely, the intrinsic dominant hole concentration is due to a curvature-induced charge redistribution among the bonding orbitals, this rehybridization of atom orbitals depending strongly on the tube diameter. The rapid decrease of the number of holes per C atom when the CNT diameter increases is supported by various experiments. The *p*-type intrinsic concentration, which resides in the geometry of the CNT, increases the strength of π bond and induces a $\pi - \sigma$ charge transfer, manifested by electron depletion from the π valence band.

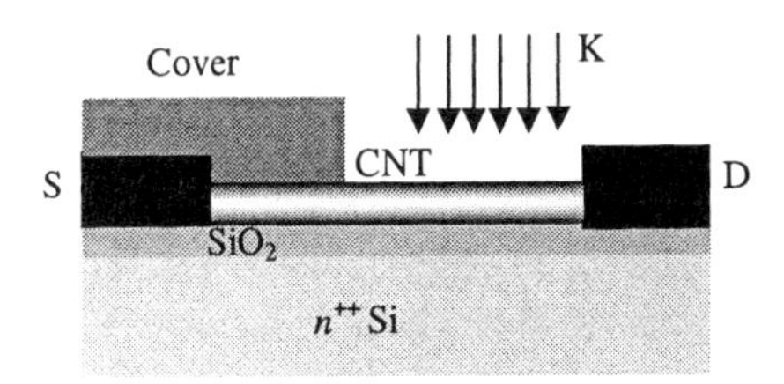

Figure 3.9 CNT device configuration used for doping (*After*: [12]).

The *n*-doping of semiconducting CNTs is crucial for the realization of electronic devices such as diodes or transistors. It was established that potassium (K) acts as an electron donor for CNTs, increasing significantly their conductance. The *n*-doping, which changes the type of majority carriers in semiconducting CNTs from holes to electrons [11] is realized in a device geometry shown in Figure 3.9, in which the CNT is patterned with two metallic electrodes (drain D and source S) and is positioned over a Si/SiO_2 substrate that plays the role of gate; the configuration in Figure 3.9 is known as CNT MOSFET (metal-oxide-semiconductor field-effect transistor). The CNT is n-doped when placed in a vacuum vessel that contains potassium vapors obtained by electrical heating of a potassium source; only a part of the CNT can be exposed to K doping if the CNT is covered with an organic material such as PMMA [12]. The doping levels are in the range $n_K = 100 - 700\ \mu m^{-1}$, which corresponds to a few K atoms/1000 C atoms.

As illustrated in Figure 3.10, the type of doping can be easily identified by measuring the conductance of the device. For p-doped CNT the conductance decreases when the gate voltage increases, indicating that holes are the predominant carriers and that the Fermi level E_F is located in the valence band VB, as shown in Figure 3.11. On the contrary, if the conductance increases with the gate voltage, the electrons dominate the conduction mechanism and E_F is located in the conduction band CB.

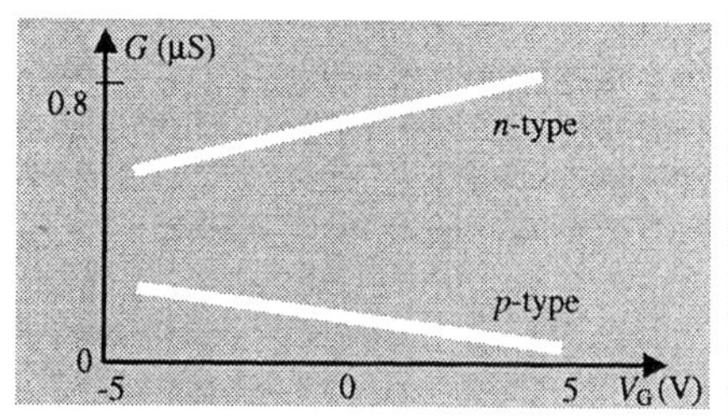

Figure 3.10 Identification of CNT doping via conductance measurement in the case of CNT ropes (*After*: [11]).

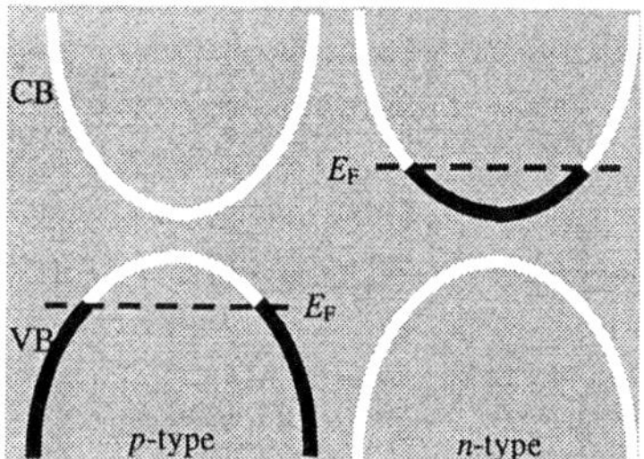

Figure 3.11 The location of the Fermi level in doped semiconductor CNTs (*After*: [11]).

Depending on the gate voltage applied on the n-doped CNT, all types of p-n doping can be achieved: p^+n, pn^+, and p^+n^+ [12]. Even the degenerate doped p^+n^+ region, which shows a negative resistance, was observed in Esaki tunneling diodes, when electrons tunnel between the conduction band and the valence band through a thin depleted layer. However, the ends of n-doped CNTs are still of p-type due to doping from the metal contacts. In this way, a small p-n junction is formed near the contact, which allows the creation of a small p-type quantum dot between the p-n junction and the contact. The zero-dimensional (0D) quantum dot formed at the end of an n-type 1D semiconducting CNT is the lower dimensional analog of the 2D inversion layer formed at the boundary of a gated 3D semiconductor [13].

Intrinsic CNT semiconductors are obtained when the CNT is suspended in air and is not in direct contact with the gate oxide. This can be done by micromachining a trench in the central portion of the silicon dioxide layer in Figure 3.9 [14].

The band structure modulation of a CNT can also be realized by applying dc electric and magnetic fields. An electric field transverse to the tube axis (a transverse electric field) narrows significantly the bandgap of a semiconducting CNT, whereas a metallic CNT is not very sensitive to the applied field. In contrast to the CNT MOSFET configuration described in Figure 3.9, when beyond a certain threshold gate voltage the Fermi level is shifted in the conduction or valence band, in a split-gate approach, in which the semiconducting CNT is placed in the gap between two metallic gates, the Fermi level is kept all the time at zero; the split-gate configuration is represented in Figure 3.12. In the split-gate configuration, the required SiO_2 thickness of the gate, which warrants a uniform field distribution around the CNT circumference, is comparable with the tube diameter [15], the corresponding dielectric thickness being quite large compared to the tube diameter in the CNT MOSFET configuration. In the split-gate configuration the bandgap varies from 1.1 eV (for a gate voltage $V_G = 0$ V) to 0.4 eV (for $V_G = 9$ V) for a (10,0) nanotube.

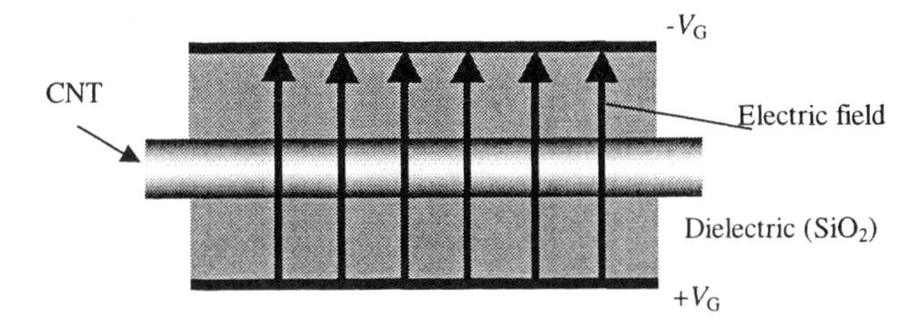

Figure 3.12 The semiconducting CNT in a split-gate configuration.

In contrast to the results reported in [15], recent studies demonstrate that (n,n) metallic CNTs undergo a reversible metallic-semiconductor transition beyond a certain field threshold [16]. The transverse electric field opens a gap in the metallic CNT and transforms it into a semiconductor, the gap of the (n,n) CNT increasing with increasing fields until it attains its maximum value

$$E_g(\text{eV}) = 6.89 / n \tag{3.24}$$

at the gate voltage

$$V_{G,max}(V) = 12.09 / n . \tag{3.25}$$

A further increase of V_G produces a decrease of the gap. For weak transverse fields, there is a universal relation between the increasing gap and the gate voltage given by

$$nE_g = \alpha(neV_G)^2 \tag{3.26}$$

where α is a constant equal to 0.07 $(eV)^{-1}$.

However, similar tight-binding methods showed that armchair tubes remain always metallic while semiconducting CNTs, in particular zigzag nanotubes, open and close their bandgap as a function of the applied gate voltage (or transverse dc electric field) [17]. The controversy about the behavior of CNTs in transverse dc electric fields cannot be solved until experimental results validate one prediction or the other.

When the CNT is placed in a magnetic field $\boldsymbol{B}$ directed along its axis, the wave vector in the circumferential direction acquires an additional phase due to the Aharonov-Bohm effect, and increases from its zero-field value $|\boldsymbol{K}_c| = 2(i - p/3)/d$, with i an integer and $p = 0$ or ± 1, to

$$|\boldsymbol{K}_c| = 2(i - p/3 + \phi/\phi_0)/d . \tag{3.27}$$

Here ϕ is the magnetic flux through the tube and $\phi_0 = h/e$ is the magnetic flux quantum. The change in the wave vector that corresponds to a change in the magnetic flux ϕ via a variation of $\boldsymbol{B}$ can produce a modification of the bandgap of semiconducting CNTs or even a reversible metal-semiconductor transition. For example, if the CNT is metallic at $\phi = 0$, it becomes semiconducting at $\phi = 0.5\phi_0$, then again metallic at $\phi = \phi_0$, and so on. The metal-semiconductor transitions in CNTs via an induced Aharonov-Bohm phase shift have been demonstrated in recent experiments reviewed in [18]. However, a sizeable change of the band gap even at very low temperatures requires quite high magnetic fields of 10–50 T [18].

Another effect associated with magnetic fields $\boldsymbol{B}$ parallel to the nanotube axis is a shift of the energy states $\Delta E = -\boldsymbol{\mu}_{orb}\boldsymbol{B} = \pm dev_F B/4$, where $\mu_{orb} = dev_F/4$ is the orbital magnetic moment induced by the movement of electrons with Fermi velocity v_F around the tube diameter. The value of the orbital magnetic moment estimated from measurements of these shifts is 0.7 meVT^{-1} for nanotubes with diameters of 2.6 nm, and 1.5 meVT^{-1} for nanotubes with 5-nm diameter [14]. The conductance of the nanotubes changes dramatically when fields of about 10 T are applied.

3.1.2 Electrical Properties of CNTs

For a given p value the allowed circumferential wave vectors are spaced by $2/d$ (see (3.27) when $\phi = 0$). Because each $|\boldsymbol{K}_c|$ inside the Brillouin zone produces a 1D subband, a set of subbands is generated by varying i. A large-diameter CNT will thus have many subbands, while a small-diameter CNT will have only a few subbands. In the vicinity of the Fermi energy, the subbands are given by

$$E_i(k) = \pm(2\hbar v_F / d)[(i - p/3)^2 + (kd/\pi)^2]^{1/2} = \pm E_0[(i - p/3)^2 + (kd/2)^2]^{1/2}. \tag{3.28}$$

The dispersion relation (3.28), which describes metallic CNTs when $p = 0$ and semiconducting CNTs when $p = \pm 1$, is represented in Figure 3.13. Near the Fermi level the energy band minimum for $p = 0$ becomes

$$E_i = \pm 2\hbar v_F i / d \tag{3.29}$$

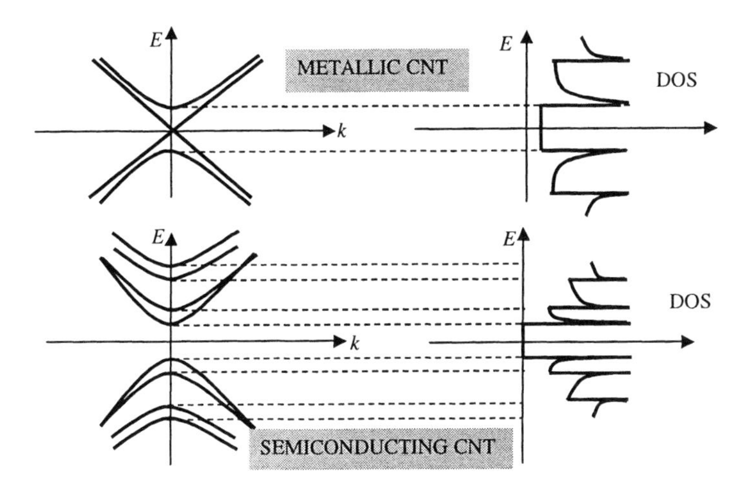

Figure 3.13 The dispersion relation of CNTs and the DOS.

and, as shown in Figure 3.13, there are two crossing levels for $i = 0$ and non-crossing levels for $i \neq 0$. According to (3.28) no crossing levels exists for semiconducting CNTs. The DOS is typical for 1D quantum systems (see, for example, the density of states of quantum wires in Chapter 1) and has a series of sharp peaks (van Hove singularities) with a mean level spacing of $E_0/2$, where $E_0 = 2\hbar v_F / d$.

The conductance of a CNT is expressed by the Landauer formula $G = (2e^2/h)T$, where T is the transmission of the nanotube. In metallic nanotubes two subbands are occupied and cross at E_F, so that in the ballistic regime, for which $T = 1$, the conductance of a metallic CNT at the Fermi level is

$$G = 2 \times G_0 = 2 \times (2e^2/h) = 2 \times (12.9\,\text{k}\Omega)^{-1} = 155\,\mu\text{S}. \tag{3.30}$$

The corresponding resistance is $R = 1/G = 6.5\,\text{k}\Omega$.

The ballistic regime is present in a CNT device when

$$L_{\text{fp}} >> d \quad \text{and} \quad L_{\text{fp}} >> L \tag{3.31}$$

where L is the tube length and L_{fp} is the mean free path of the carriers. In metallic CNTs the mean free path is a few micrometers and can attain even 10 μm in theory, while in semiconducting CNTs the mean free path is of the order of 300–400 μm [19].

If a metallic CNT is placed in a MOSFET configuration, such as that presented in Figure 3.9, its conductance is independent of the variation of the gate voltage, whereas for a semiconducting (p-type) CNT, the conductance drops to zero when the gate becomes positive.

The mobility is related to the mean free path via the relation

$$\mu = (2G_0)L_{\text{fp}} / Ne\,, \tag{3.32}$$

where N is the carrier density. Because the conductance of highly doped CNTs behaves like that of metallic CNTs and because beyond a certain amount of doping $E_F = E_g$, so that $k_{\text{F,met}} = k_{\text{F,semic}}$, we get $N/2 = 2k_F/\pi$ from which it follows that the carrier density is

$$N = 4 \times 3^{1/2} E_g / h v_F\,. \tag{3.33}$$

The mobility is in the range $10^4 - 5\times10^4\ \text{cm}^2\text{V}^{-1}\text{s}^{-1}$ for nanotubes with diameters varying from 1 nm to 10 nm. According to (3.32) and (3.33) the mobility increases with the diameter. On the other hand, the mobility of a CNT placed in a dc electrical field E applied along the nanotube axis is $\mu = v_d / E$. The data from [20] indicate that for a (59,0) tube of larger diameter the mobility can attain 120,000 $\text{cm}^2\text{V}^{-1}\text{s}^{-1}$ at an electric field of 10 kV/cm. At low values of the applied dc electric fields the mobility is given by $e\tau / m$, with τ the scattering time, and is found to be of the form

$$\mu = (n\gamma_0^{3/4} / 4d)^2[1 + (1 - \gcd(n+1,3) / n^{2/3})] \times 10^4 \text{cm}^2/\text{Vs} . \tag{3.34}$$

Here $\gcd(n+1,3)$ means the greatest common divisor between $n+1$ and 3. The dependence of mobility on the applied field can be expressed as [21]:

$$\mu_E = \mu / [1 + \lambda(E)E / v_{\max}], \tag{3.35}$$

where $v_{\max}$ is the peak of the drift velocity, given by

$$v_{\max} = (3n\gamma_0^2 / 8)^{1/3}[1 + (1 - \gcd(n+1,3) / 2n)] \times 10^7 \text{ cm/s} \tag{3.36}$$

and

$$\lambda(E) = \text{H}(E - E_0)\{\mu \exp[\log_{10}^2(E / E_{\text{cr}}) / S] - v_{\max} / E_{\text{cr}}\}, \tag{3.37}$$

with

$$E_{\text{cr}} = \gamma d^2\{1 + (8 / n)^2[\gcd(n+1,3) - 1]\} / 27n^{3/2} \times 10^6 \text{ V/cm} \tag{3.38}$$

the critical dc electric field where the drift velocity attains the highest value, $E_0 = 350 \times [1 + 3\gcd(n+1,3) / 7] / n^2$ and $S = 1.3 + \text{H}(E - E_{\text{cr}})n^{1/2} / 2$ with H the Heaviside function. Formulas (3.32)–(3.28) give a complete analytical description of the mobility of the semiconducting CNT. It is worth noting that when the applied dc electric field is lower than E_{cr}, the mobility is constant and almost independent of the field, while beyond E_{cr}, the mobility decreases very fast leading to a negative differential mobility. This is of paramount importance in high-frequency devices such as Gunn diodes or HEMTs (high electron mobility transistors).

According to (3.32) the mobility is also strongly dependent on the conductance of the nanotube, which has in principle only two conducting channels (see (3.30)). However, any change in the chemical potential $\Delta\mu_{\text{chem}}$ due to doping or oxygenation contributes to the opening of new conduction channels [22]. The dispersion relation of CNTs shows six Fermi points that link the conduction and valence bands but, if there is a small shift in the chemical potential $\Delta\mu_{\text{chem}}$, the Fermi points transform into circles with radii of $2|\Delta\mu_{\text{chem}}| / (6\gamma_0 a_{C-C})$. For metallic CNTs two new channels open at integer multiples of the lateral quantization energy $E_q = 6\gamma_0 a_{C-C} / d$ for each Fermi point that transforms in the center of a circle, whereas for semiconducting CNTs a new channel is opened at $\Delta\mu_{\text{chem}} = (i + p/3)E_{\text{q}}$. Hence, the maximum number of channels opened for metallic and semiconducting CNTs is, respectively,

$$n_{\text{met}} = 2[1 + 2\text{Int}(\Delta\mu_{\text{chem}} / E_{\text{q}})], \tag{3.39a}$$

$$n_{\text{semic}} = 2\{\text{Int}[2/3 + |\Delta\mu_{\text{chem}}| / E_{\text{g}}] + \text{Int}[1/3 + |\Delta\mu_{\text{chem}}| / E_{\text{g}}]\}. \tag{3.39b}$$

The behavior of the conductance of doped CNTs is presented in Figure 3.14, where G/G_0 represents the number of channels calculated above.

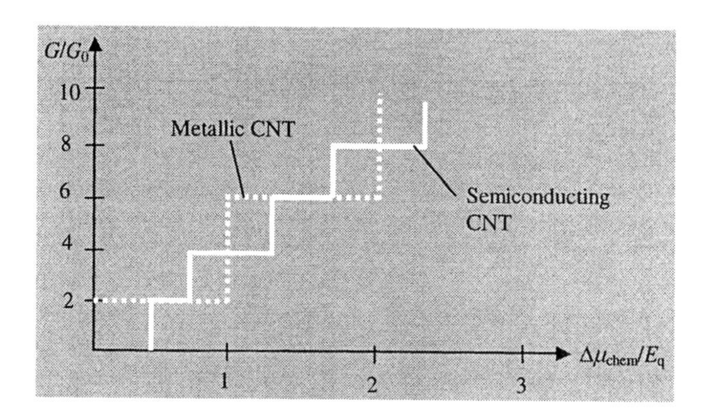

Figure 3.14 Opening of multiple conduction channels of CNTs (*After*: [22]).

As explained in Section 3.1.1, the number of channels can also be changed by applying a radial pressure on a metallic CNT. The transitions at each of the pressures P_1, P_2, and P_3 (see Figure 3.7) induce a staircase-like decrease of the conductance in steps of $2G_0$. The conductance vanishes around the Fermi energy, considered to be the zero energy level. This is the indication that the metal-semiconductor transition has occurred and a gap has opened. The energy interval over which the conductance is zero (or very small) is a measure of the bandgap of the semiconducting CNT, which is obtained by applying a pressure on a metallic CNT. So, the conductance of a functionalized CNT shows a staircase-like behavior, as in a 2D quantum confined structure. The number of conduction channels is universal in the sense that at a certain chemical potential shift the total number of conducting channels depends only on the tube radius and is independent of its chirality.

For measurements of transport characteristics both metallic and semiconducting CNTs are placed in a MOSFET configuration, similar to that in Figure

3.9. In this configuration, represented in Figure 3.15, the conductance of metallic CNTs is independent of the gate voltage, whereas a strong dependence on V_G is observed in semiconducting CNTs.

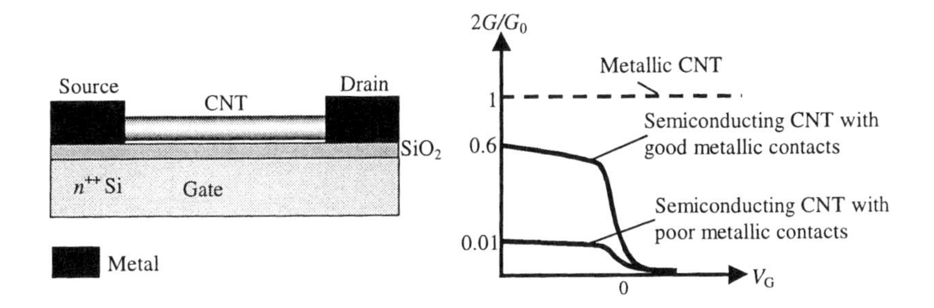

Figure 3.15 CNT MOSFET device geometry for transport measurement.

One of the thorniest problems with the CNT MOSFET geometry is the contact resistance between the CNT and the metal electrodes called source and drain. The total resistance of the CNT,

$$R_{tot} = h/4e^2 + (h/4e^2)(L/L_{fp}) + R_c \ , \tag{3.40}$$

is composed of three components: a ballistic component, the Drude resistance due to scattering, and the contact resistance R_c [23]. In the ideal situation, the intrinsic resistance of the metallic nanotube is $h/4e^2 = 6.5\,\mathrm{k\Omega}$. However, due to the parasitic tunnel barriers between the CNT and the metallic contacts resistances over 1 MΩ can be encountered. This huge resistance can be lowered up to 8–10 kΩ by evaporating gold on the top of CNT via a CVD technique or by using planarized and annealed gold electrodes, so that intrinsic transport phenomena in metallic CNT can be eventually observed. In semiconducting CNTs the problem is more complex since the contacts play the role of Schottky barriers or are purely resistive (ohmic), depending on the work function difference between the metal and the CNT and on the carrier type (n or p). Since the CNT work function is about 4.5 eV, we get ohmic contacts if

$$W_{met}(\mathrm{eV}) > 4.5 + E_g/2 \ , \tag{3.41}$$

where W_{met} is the metal work function. Palladium (Pd) has a large work function, which satisfies (3.41), and therefore is a very good contact electrode for semiconducting CNTs. The conductance of Pd-contacted semiconducting CNTs attains more than $0.5\times(2G_0)$, indicating that no parasitic barrier is formed [24].

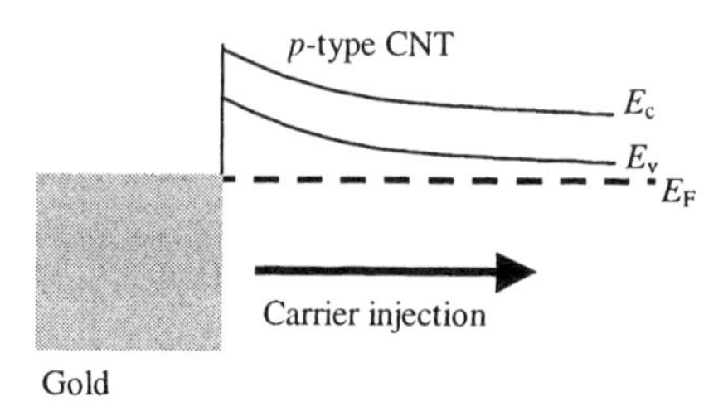

Figure 3.16 Ohmic contact between the p-type CNT and gold (*After*: [25]).

The AFM and STM, described in detail in Chapter 2, are extensively used to experimentally investigate the physical properties of CNTs, and, in particular, to measure the conductance of CNTs at different positions along the tube. Many important physical properties of CNTs can be extracted from these conductance measurements at the nanoscale. For example, the AFM was used as a movable electrical nanoprobe to locally sense the electrical properties of semiconducting CNTs, especially in order to reveal the physical nature of the metallic contacts to semiconducting CNTs [25]. Using a metallized tip as a voltage nanoprobe it was demonstrated that there is ohmic contact between a *p*-type CNT and Au, the conductance, after a short time annealing at 600°C in an argon environment, almost attaining the theoretical value in the ON state, which is $1.5G_0$. The energy band diagram of the *p*-type CNT with Au contacts is represented in Figure 3.16, in which E_c and E_v denote the edges of the conduction and valence band energies, respectively. This figure shows that the carriers are directly injected in the valence band of the CNT, the high quality contact between gold and the *p*-type CNT being characterized by typical resistances of 10–50 kΩ, independent of the gate voltage, and approaching the theoretical limit of $h/8e^2 = 3.2$ kΩ. The mean free path of a *p*-type CNT, determined from the relation $R_{CNT}/L = (h/4e^2)/L_{fp}$ was found to be 300 nm at room temperature for $R_{CNT}/L = 20$ kΩ/μm.

On the contrary, large resistances are encountered in *n*-type CNTs with gold contacts, either due to the parasitic *p-n* junction, or the Schottky contact that forms near the contacts. In a parasitic *p-n* junction, as that in Figure 3.17(a), the gold makes an ohmic contact with the *p*-type region of the *n*-type tube, which forms near the contact when the Fermi level is located in the valence band; the *p-n* junction is entirely situated in the CNT. The contact resistance due to the parasitic *p-n* junction or the Schottky contact is of the order of MΩ. To reduce the barrier effects and thus to lower the conductance from MΩ to a value close to $2G_0$, Pd contacts are used to destroy the parasitic Schottky barrier. Thus, room temperature conductance that attains the limit of the ballistic conductance $2G_0$ can be obtained in metallic CNT, as well as in semiconducting CNT.

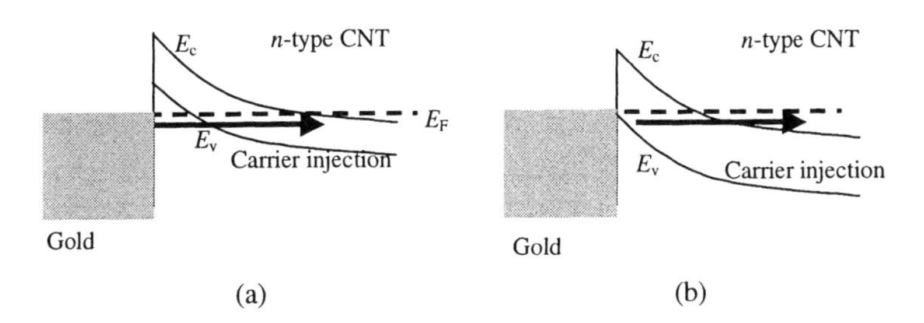

Figure 3.17 Contact between the *n*-type CNT and gold: (a) parasitic *p-n* junction, and (b) Schottky contact (*After*: [25]).

The STM is also an important tool to analyze CNTs. Electronic wavefunctions associated with the quantized energy levels in short metallic CNTs (with lengths of a few tens of nm) were reconstructed from the low-temperature (4–5 K) periodic oscillations in the differential conductance at different positions along the nanotube [26,27]. In Chapter 2, we have shown that the conductance dI/dV is proportional to the local density of states (LDOS) of the device under test. The electron in a short-length CNT behaves like a "particle in a box," with quantized energy levels given by $\Delta E = hv_F/4L \approx 1\,\text{meV}/L(\text{nm})$. In this case, discrete electronic wavefunctions $\psi_i(\boldsymbol{r})$ with corresponding discrete energy levels E_i, with i = 1, 2, 3,... are characteristic for the short nanotube and the measured STM conductance at various bias voltages V and positions $\boldsymbol{r}$ is given by

$$dI/dV \propto \sum_{|eV-E_i|<\delta E} |\psi_i(\boldsymbol{r})|^2 , \tag{3.42}$$

where δE is the energy resolution. If ΔE is less than the difference between two consecutive energy levels $|E_{i+1} - E_i|$, the sum in (3.42) reduces to a single term and the mapping of $|\psi_i(\boldsymbol{r})|^2$ becomes possible through the measurement of the conductance at energy E_i. The observed steps in current in the measured $I-V$ characteristic correspond to different discrete energy states. Near the Fermi level the wavefunction consists of a sum of damped exponentials, which matches the experimental results.

The same STM method can be used to determine experimentally the form of the dispersion relation, which is linear for metallic CNTs. The sharp peaks in the dI/dV curve as a function of the bias voltage correspond to the van Hove singularities in the DOS, which diverge as $E^{-1/2}$. The number of these singularities indicates the number of occupied subbands, while the width of the bandgap is determined from the distance between two consecutive peaks. Finally, the set of integers (n,m) can be determined also from the van Hove singularities [26].

The conductance curve $G(V_G)$ of metallic CNTs with a large contact resistance at very low temperature consists of a periodic series of spikes, which originate in the Coulomb blockade effect. The CNT plays the role of a metallic island weakly coupled to the source and drain contacts via tunnel barriers. The electrostatic capacitance $C = 2\pi\varepsilon_0\varepsilon_r / \ln(2z/d)$ of the CNT, where z is the gate height, has very small values, of about 50 aF/μm or less, so that adding a single electron requires a charging electrostatic energy of $E_{ch} \approx 5\text{meV}/L(\mu\text{m})$. The poor contacts between the CNT and the electrodes allow us not only to observe the Coulomb blockade effect, but also to study the Luttinger-liquid behavior due to the electron-electron interaction in the metallic CNT. The Luttinger-liquid is a ground state with the following properties: 1) low-energy charge, 2) spin excitation that propagates with different velocities, and 3) a suppression of the DOS according to the law $\rho(E) \propto |E - E_F|^\alpha$ [28]. The Luttinger liquid is characterized by the parameter

$$g = (1 + 4E_{ch}/\Delta E)^{-1/2}, \tag{3.43}$$

which indicates the level of electron correlation, with ΔE the energy level spacing. The value $g = 0.22$ indicates a strong electron correlation in CNTs. In the linear response regime $(eV << k_B T)$ the conductance of the CNT has a power-law variation with the temperature T of the form

$$G(T) \propto T^\alpha, \tag{3.44}$$

where the exponent α is related to g; the above dependence is opposite to that found in normal metals. Additionally, at large biases $(eV >> k_B T)$ we have $dI/dV \propto V^\alpha$. It is interesting to note that the Luttinger-liquid behavior in CNTs persists even at room temperatures.

Other physical properties of CNTs, which are not analyzed in detail here, are summarized in Table 3.1. They refer mainly to SWCNTs, which have benefited from many experiments that tried to elucidate their physical properties; the MWCNTs have a more complicated structure. The most developed review on the physical properties of MWCNTs is [29]. We point out here only that the MWCNT, having a Russian-doll structure with each doll a rolled SWNCT, has a contact resistance many times lower than the SWCNT and has a better mechanical stability. The electrical behavior of the MWCNT is similar to that of the SWCNT since the current is confined in the outer SWCNT. Fundamental phenomena, such as quantum interference and the Aharonov-Bohm effect, can be studied more easily in MWCNTs than in SWCNTs. For example, the applied magnetic field intensity for which the Aharonov-Bohm effect is observable is five times lower in MWCNTs than in SWCNTs.

Table 3.1

Physical Properties of Carbon Nanotubes

Parameter	*Value and units*	*Observations*
Length of the unit vector	$a = \sqrt{3}a_{C-C} = 2.49$ Å	$a_{C-C} = 1.44$ Å is the carbon bond length
Current density	$>10^9$ A/cm^2	-1000 times larger than the current density in copper - Measured in MWCNTs
Thermal conductivity	6600 W/mK	More thermally conductive than most crystals
Young modulus	1 Tpa	Many orders of magnitude stronger than the steel
Mobility	10,000-50,000 cm^{2}V^{-1}s^{-1}	Simulations indicate motilities beyond 100,000 cm^{2}V^{-1}s^{-1}
Mean free path (ballistic transport)	300-700 nm semiconducting CNT 1000-3000 nm metallic CNT	- Measured at room temperature - At least three time larger than the best semiconducting heterostructures
Conductance in ballistic transport	$G = 4e^2 / h = 155\mu S$; $1/G = 6.5 k\Omega$	
Luttinger parameter g	0.22	The electrons are strongly correlated in CNTs
Orbital magnetic moment	0.7 meVT^{-1} (d = 2.6 nm) 1.5 meVT^{-1} (d = 5 nm)	The orbital magnetic moment depends on the tube diameter

3.2 CNT-BASED ELECTRONIC DEVICES

In what follows, we analyze a series of devices based on CNTs, which work at different frequencies and have different purposes. We start with the most studied device: the CNT transistor.

3.2.1 The CNT Transistor

Almost all CNT transistors are of the FET (field-effect transistors) type, with different configurations. The design and the realization of CNT transistors (CNTFETs) is an emerging area of research, a lot of effort being invested by many companies for the implementation of reliable CNTFETs and of integrated circuits based on them. The reason is that recent CNTFET configurations, like MOSFET CNTFETs, have a figure of merit at a room temperature 20 times larger than that of the best complementary metal oxide (CMOS) transistor, which is the key component in modern computers, communication systems, or devices. Thus, justified by the much better performances of CNTFET transistors, it is hoped that carbon technology will, in the future, replace the existing worldwide Si CMOS technology. Although the design and the technological implementation of CNTFETs is at the very beginning, the progress is extremely rapid.

The first CNTFET, represented in Figure 3.18, was similar to that in Figure 3.15; it had a back gate of doped Si followed by a thin SiO_2 layer, the semiconducting CNT with a diameter of a few nm (with a bandgap of 0.6–0.8 eV) being terminated by two metallic (gold) electrodes with a thickness of 100–300 nm. These electrodes, defined by electron beam lithography, play the role of source and drain, respectively. If the gate voltage V_G is zero, the source-drain current I_{DS} has a linear dependence on the source-drain voltage V_{DS}, the linear $I_{DS} - V_{DS}$ characteristic being preserved for negative gate voltages. However, when $V_G >> 0$, the $I_{DS} - V_{DS}$ curve becomes nonlinear and decays to negligible values, the CNT suffering a transformation from a quasi-metallic state to an insulator state. The behavior of this CNTFET is analogous to that of a *p*-MOSFET transistor, which demonstrates that the device in Figure 3.18 is indeed a FET-like transistor and that the transport is dominated by holes [30].

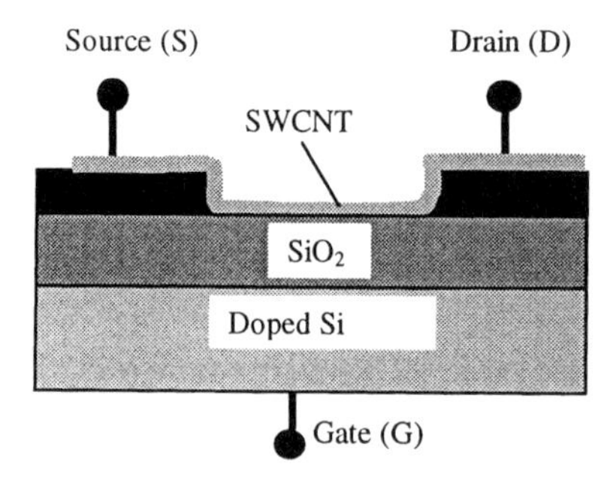

Figure 3.18 CNTFET device (*After*: [30]).

This first rudimentary CNT-based FET transistor consists simply of a semiconducting SWCNT linking two metallic electrodes deposited on a thin silicon dioxide layer, that, in turn, is deposited on a doped silicon that acts as a gate. Nonetheless, the conductance dI_{DS}/dV_G is modulated by more than 5 orders of magnitude when the gate voltage ranges between 0 V and 5 V. When the gate voltage is negative, the source-drain current I_{DS} is almost constant, its saturation indicating that the contact resistance of the two electrodes prevails over the CNT resistance, which depends on the gate voltage. Practically, for $V_G = 0$ the CNTFET is in the ON state and the Fermi energy is located near the valence band. If the band bending length is comparable to the CNT length L, and if the gate-CNT distance is shorter than the distance between the two electrodes, a barrier is raised in the middle of the CNT for positive gate voltages. Beyond a certain threshold $V_{G,th}$ the holes are depleted and suppressed in the center of the tube, this state corresponding to the OFF state of the device. The hole density can be determined from $I_{DS} - V_G$ measurements as [30]:

$$p = Q / eL = CV_{G,th} / eL, \tag{3.45}$$

where

$$C / L = 2\pi\varepsilon_r\varepsilon_0 / \ln(4h / d) \tag{3.46}$$

is the capacitance/unit length of the CNT, with h the thickness of the silicon dioxide substrate and $\varepsilon_r \approx 2.5$ the relative electrical permittivity of the CNT.

For example, at a gate threshold of 6 V, the hole density reaches $9\times10^6\,\text{cm}^{-1}$, which means that there is one hole for each 250 atoms of carbon. This is a very high hole density compared to that in graphite, where there is a hole at each 10000 atoms. For a transconductance of $dI_{DS}/dV_G = \mu_h (C/L)^2 V_{SD} \approx 2\times10^{-9}\,\text{A/V}$ the hole mobility μ_h is 20 cm^2/Vs, which is a very low value compared to the 10^4 cm^2/Vs value in grapite. Such a low mobility indicates a diffusive transport at room temperature in this first CNTFET. However, a couple of years later ballistic transport at room temperature was evidenced in CNTFET transistors with improved performances, based on better quality nanotubes with low contact resistance. If the SWCNT in Figure 3.18 is replaced by an MWCNT no transistor behavior is detected until structural deformations are applied on the MWCNT.

A device similar to that presented in Figure 3.18 is the TUBFET [31]. The TUBFET has Pt electrodes with a work function of 5.7 eV, which is larger than the work function of the CNT, so that the carriers are injected in the CNT via tunneling. A polarization layer forms at the electrode-CNT interface until the valence band is aligned to the Fermi level of the metallic electrode, producing shallow barriers for holes even when no gate voltage is applied. The height of these barriers, which are caused by the difference in the work function between the CNT and the electrodes, is modulated by the applied gate voltage as follows: for $V_G < 0$, the valence band bends upwards and flattens until it gives rise to a metal-like conductance (i.e., to a constant value of conductance), and for $V_G > 0$, the valence band bends downwards and the barrier height for holes increases, suppressing the hole transport between the two electrodes. This particular band structure is similar to that of a BARITT diode, which consists of two Schottky diodes connected back-to-back.

It is interesting to note that the TUBFET, which is still a rudimentary FET transistor, has a traversal time of only 0.1 ps, which corresponds to 10 THz. For a CNT with a capacitance of about 1 aF, the resulting RC time is 100 GHz when R is of the order of 1–2 MΩ. However, the resistance R is of about 10 kΩ for CNTFETs with Pd contacts that show ballistic transport at room temperature, the cutoff frequency being about 10 THz. The gain of the TUBFET is low, of about 0.35, but can be increased beyond 1 by thinning the silicon dioxide layer.

Unlike the above transistors, which have a diffusive transport, the CNTFET transistor with palladium contacts shows ballistic transport at room temperature [24]. The conductance in the ON state attains almost the ballistic limit $4e^2/h$ at ambient temperature, similar to the ohmic metallic nanotubes. The explanation resides in the suppression of the Schottky barrier at the metal-CNT interface because palladium has a high work function and a good wetting interaction with the CNT. The freely injected carriers in the valence band of the semiconducting CNT are characterized by a conductance G, which attains in the ON state the value

$$G_{\text{ON,Pd}} = 0.4 - 0.7 \times (2G_0) . \tag{3.47}$$

The ballistic transport in CNTFETs with Pd electrodes allows a huge ratio between the conductances in the ON and OFF states, of six orders of magnitude, and high current carrying capabilities (of 25 μA/tube).

Another type of CNT transistor, displayed in Figure 3.19, is the Schottky-barrier transistor (SB-CNTFET). It consists of a nanotube embedded in a dielectric layer, which is positioned between a top gate and a ground plane and is terminated with two metal electrodes that act as source and drain. Unlike in the above configurations, where the transistor action (switching) is produced by varying the channel conductance, in the SB-CNTFET, this action is caused by variations in the contact resistance [32]. The switching is realized by modulating the tunneling current via a change in the top gate voltage.

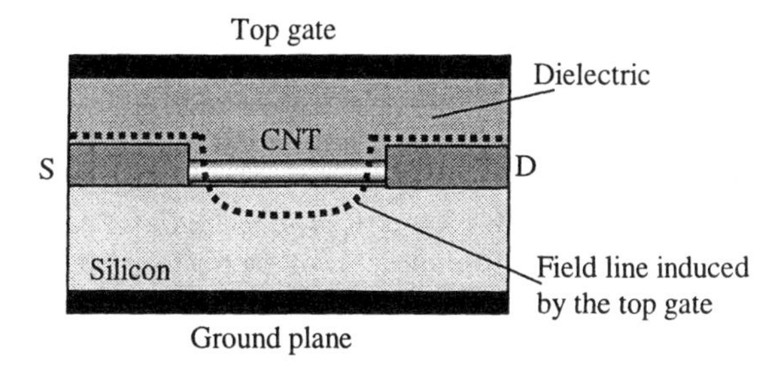

Figure 3.19 Schematic representation of a SB-CNTFET (*After*: [32]).

The conductance behavior of the SB-CNTFET with thin gate oxides suggests an ambipolar conduction (see Figure 3.20 and Figure 3.21) in contrast to all the CNT transistors studied up to now, where the conduction is unipolar, of *p*-type. The huge difference between the two types of CNT transistors studied up to now can be seen from Figure 3.20. The conductance of the SB-CNTFET is discontinuous (see the case when the barrier height ϕ_b is 1/6 of the CNT bandgap). By pinning the Fermi level of the metal contacts at $E_g/2$ this transistor shows electron conduction at high voltages and hole conduction at low voltages, the result being an ambipolar conduction. This is a major drawback, since the hole current is a leakage current with a sizable value in the OFF state. In particular, due to the ambipolar conduction, negative gate-to-source voltages and thus high leakage currents are produced when the SB-CNTFETs are stacked [33].

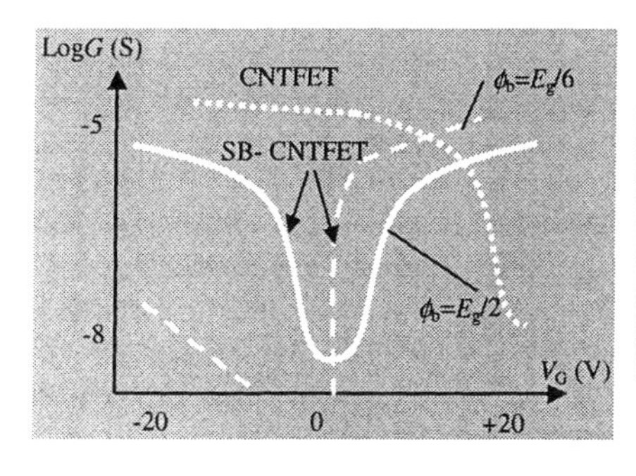

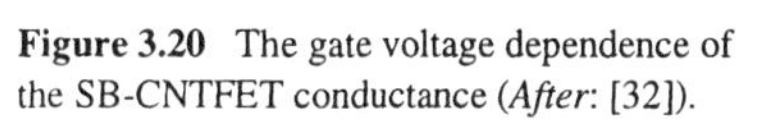

Figure 3.20 The gate voltage dependence of the SB-CNTFET conductance (*After*: [32]).

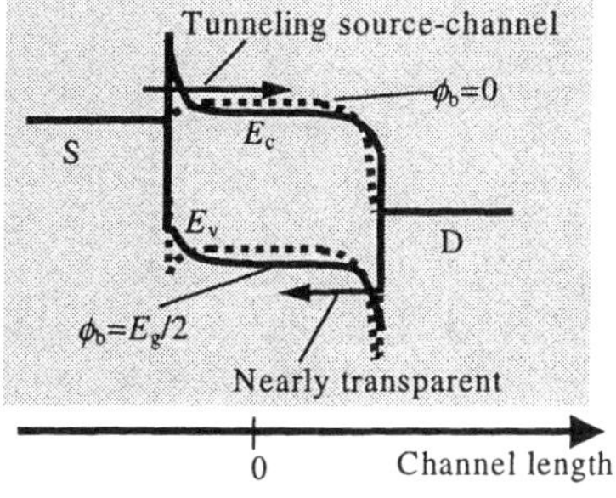

Figure 3.21 SB-CNTFET band diagram (*After*: [33]).

While in the CNTFET the conductance is changed by varying the electrostatic potential in the channel, in the SB-CNTFET this is achieved by controlling the electric field variation at the contact, as shown in the band diagram in Figure 3.21. The variation of the geometry of contacts is due to the different heights of the Schottky barrier ϕ_b beneath the gate. A sharper contact induces field focusing, which means a larger field at the contact. A reduction of the Schottky barrier height by a 10 times decrease of the thickness of the top metal gate electrode from its initial value of 50 nm strongly improves the ratio between the ON and OFF conductance values.

As evident from Figure 3.21, the SB-CNTFET is a tunneling transistor when $\phi_b = E_g / 2$, the carriers tunneling into the source-channel junction as the height of the barrier is modulated by the gate voltage. Unfortunately, even when the barrier height is zero, the conduction is still strongly ambipolar. The resulting leakage current increases exponentially with the drain voltage and, hence, limits the applications of this device. Moreover, this transistor offers no scaling advantage over existing silicon MOSFETs. Its scaling limit is 5–10 nm due to source-drain tunneling and is determined by the CNT bandgap. Larger bandgaps (larger tube diameters) reduce the height of the barrier, which provides higher current in the ON state, but in the same time generates larger leakage currents [33].

Dielectrics with high electrical permittivity, called high-κ dielectrics, can improve the current in the ON state and allow the realization of very performing CNTFETs. SiO_2, which is the traditional dielectric in MOSFETs, has a low electrical permittivity, of 3.2, that limits its use in transistors with gate lengths smaller than tens of nanometers. Instead, using high dielectric films such as ZrO_2 with an electrical permittivity in the range 20–30 [34], large capacitances are obtained without the need of very thin thicknesses. In this manner the electrical charges are efficiently injected in transistor channels and the direct-tunneling leakage current is reduced. The transconductance of the *p*-CNTFET transistor, given by

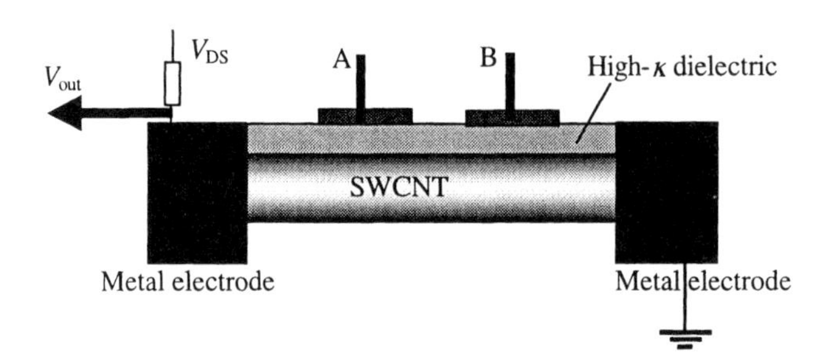

Figure 3.22 OR logic gate using a CNTFET with high-κ dielectric substrate.

$$g_{DS} = I_{DS} / V_{DS} = 2(V_{GS} - V_{G,th})\mu_h C_{GS} / 2L^2 \tag{3.48}$$

where $V_{G,th}$ is the threshold gate voltage, is 3000 Sm^{-1} in CNTFETs with ZrO_2 dielectric gates, a value that exceeds about 4 times that of the best Si MOSFETs. Logic gates with high performances can be implemented by applying multiple gates on a CNT terminated with two metal electrodes, as shown in Figure 3.22.

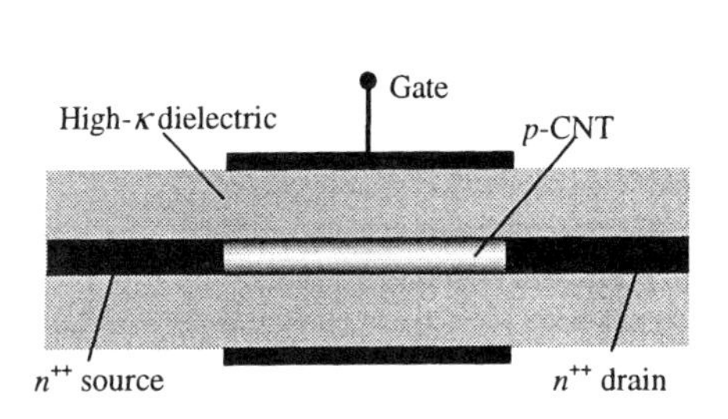

Figure 3.23 Schematic representation of the MOSFET-CNT.

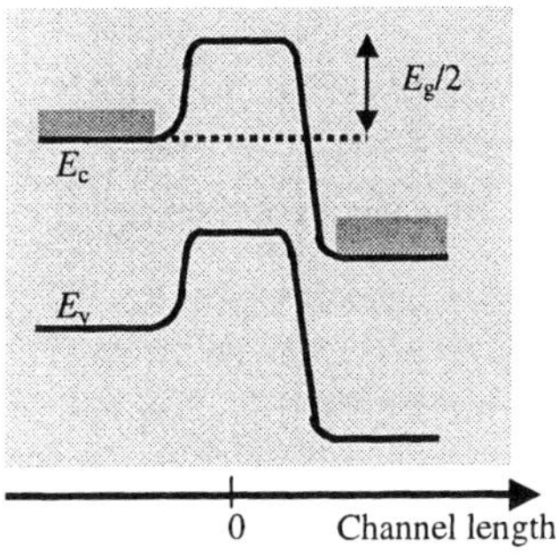

Figure 3.24 The band diagram of the MOSFET-CNT.

A very promising transistor, which mimics a normal MOSFET, has highly doped source and the drain ungated regions. This MOSFET-CNT, represented in Figure 3.23, works on the same principle as the SB-CNTFET, namely barrier-height modulation via the gate voltage. However, the ambipolar character of the conduction, specific to the SB-CNTFET, is suppressed in the MOSFET-CNT due to the high doping concentrations of the source and the drain, and the Schottky barrier between the source and the channel no longer exists (see the band diagram in Figure 3.24). Thus, because in the ON state the MOSFET-CNT works like a SB-CNTFET but with a zero or even negative barrier height, the current in the ON state increases and the leakage current decreases. In the OFF state the MOSFET-CNT still has a leakage current, but it is controllable by engineering the bandgap of the CNT and the band-to-band tunneling.

Simple models can be developed for the MOSFET-like CNT transistor by considering that the gate-source voltage V_{GS} produces charges in the channel and also modulates the top of the source-drain energy band [35]. When the source-drain barrier decreases, a ballistic current flows between the source and the drain. The voltage gate V_G lowers the channel potential with a certain amount ψ_S and thus induces an accumulation of charges in the channel, this extra charge producing as a result a voltage drop $(V_G - \psi_S)$ across the insulator, which lowers the energy band by ψ_S. Therefore, by modifying V_G the band profile in Figure 3.24 can be moved up or down. The total charge n in the CNT channel is the sum of the charges in the source and the drain, each of them being given by

$$n_i = (1/2)\int_0^\infty D(E - E_c) f(E - E_c - \mu_i)\, dE\,, \tag{3.49}$$

where $i = \text{S}, \text{D}$ denotes drain or source, respectively, $D(E)$ is the DOS of the CNT, $f(E)$ is the Fermi-Dirac distribution function, with E_c the bottom of the conduction band and μ_i the energy levels of source or drain.

The calculation of the DOS of CNTs, represented in Figure 3.13, is not an easy task, but we can use a simplified model, not taking into account the mixing of states. (This model is not applicable for MOSFET-CNTs with a bias exceeding some hundreds of mV.) In this case the DOS near the Fermi level is [36]

$$D_l(E) = D_0 E (E^2 - \Delta E_l^2)^{-1/2}\,, \tag{3.50}$$

where $D_0 = 8/3\pi\gamma_0 a_{\text{C-C}}$ and ΔE_l is the minimum of the l-th conduction subband (l = 1, 2, 3…), which can be written as

$$\Delta E_l \cong (E_g/2)[6l - 3 - (-1)^l]/4\,. \tag{3.51}$$

For $\mu_S = 0$ and $\mu_D = -eV_{DS}$, denoting

$$\alpha_{i,l} = (\psi_S - \Delta_l - \mu_i)/k_B T\,, \tag{3.52}$$

we can determine the charge in the l-th subband as

$$n_{i,l} = \int_0^\infty (4k_B T/3\pi\gamma_0 a_{\text{C-C}})\{1 + \exp[((z^2 + \Delta_l^2)^{1/2} - \Delta_l)/k_B T - \alpha_i]\}^{-1/2} dz\,, \tag{3.53}$$

the total charge on all subbands being

$$n_{\text{CNT}} = \sum_l (n_{S,l} + n_{D,l}) \tag{3.54}$$

with $n_{i,l}$ given by (3.53). Using a mean-value of the conduction band minimum for the first subband and dropping the contribution of other subbands, the above sum is finally reduced to two terms corresponding to the source and drain, respectively. The drain current I_D is given by

$$I_D = (4ek_BT/h)\sum_l \ln\{[1+\exp(-\alpha_{S,l})]+\ln[1+(-\alpha_{D,l})]\}\,. \tag{3.55}$$

The basic characteristic of the transistor, $I_D(V_{GS}, V_{DS})$, is determined by repeating the above procedures for various (ψ_S, V_{DS}) points. The equivalent circuit of the MOSFET-CNT, depicted in Figure 3.25, can be found by following the steps described in [35], which determine in an iterative way the quantum capacitances C_{Gi} and the surface potential ψ_S related to the drain current.

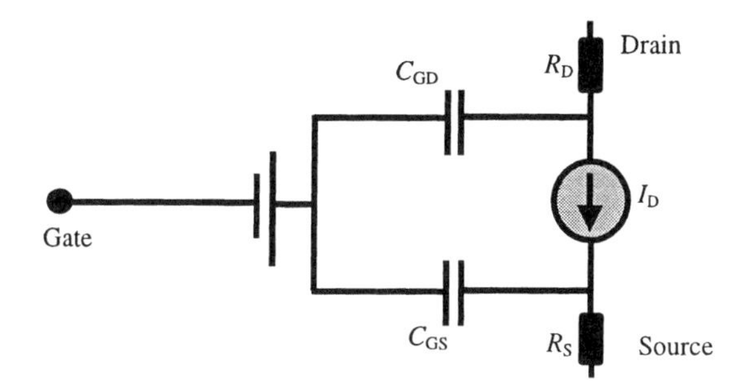

Figure 3.25 The equivalent circuit of MOSFET-CNT.

The latest tendency is to realize CNTFETs in a vertical position, not in a horizontal position shared by all the transistors previously described. The vertical orientation of the CNTFET is preferred because it can significantly increase the density of an array of transistors. For example, in [37] an ultrahigh-density vertical transistor array was realized starting from a CNT array. The vertical transistor VERT-CNTFET depicted in Figure 3.26 was fabricated performing the following technological steps: 1) nanopore formation in high-purity aluminum by anodization using a sulfuric acid solution; a template is created in this way; 2) electron-beam patterning for selective growth; the SiO_2 layer is deposited first, then the electron-beam lithography is employed, and subsequently the structure is reactive ion etched; 3) the CNT is synthesized in the nano-pores using thermal CVD; 4) the surface of the template is etched, SiO_2 is deposited followed by electron-beam lithography, and the Au/Ti metal electrodes are deposited; and 5) the gate oxide (SiO_2) and the gate-electrode (Au) are deposited and patterned by electron-beam lithography. In the final product, the cell size occupied by the CNT was $20\times20\,\text{nm}^2$, with a pitch of 40 nm, the density of VERT-CNTFET being

$2\times10^{11}/\text{cm}^2$. This teralevel density of nanotransistors is one of the most encouraging achievements of nanoelectronics. Moreover, despite the huge density of nanotransistors that act as switches, each transistor in the array can be individually contacted since it is placed at the crossing point of a top electrode (drain) and the gate electrode. The first experiments with such transistor arrays were made at 5 K, but the real struggle is to demonstrate VERT-CNTFET arrays at room temperature with currents in the ON state higher than 10 nA for a bias of a few volts and at a gate voltage not exceeding 5 V.

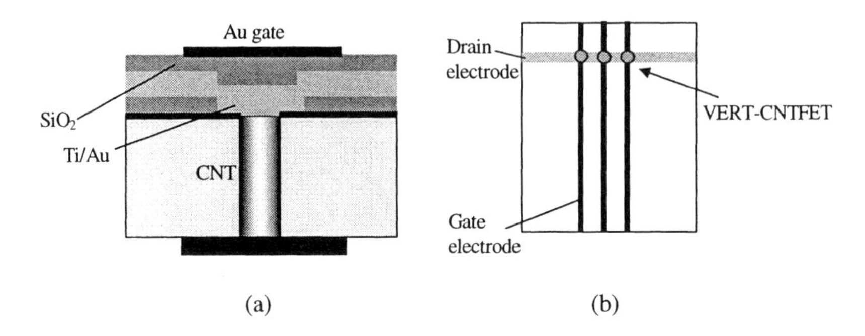

Figure 3.26 (a) VERT-CNTFET, and (b) an array of VERT-CNTFET (*After*: [37]).

Besides CNT-based FET transistors, room-temperature single-electron transistors based on metallic CNT were recently reported [38]. When the tip of an AFM in tapping mode is pressed down onto a portion of the CNT, it creates two buckles, which constitute two junctions realized by forming two tunneling barriers. The resulting structure, consisting of a conducting island (the CNT) connected by tunnel barriers to metal electrodes, is a single-electron transistor. Conductance oscillations typical for the Coulomb-blockade effect were observed in such structures by measuring $G(V)$.

All transistor configurations described above, and the NEMS nanoswitches described in Chapter 2, Figure 2.6, are promoted as new building blocks for ultra-high-density devices such as memories or processors. Terascale integration implies ultrahigh density of switches, of 10^{11}–10^{12} switches per square centimeter, low power consumption, and high speed. These requirements cannot be satisfied by non-CNT MOSFET transistors, which encounter some problems in ultrahigh-density applications regarding 1) thermal dissipation, 2) power consumption, 3) the fluctuation of the electrical parameters, and 4) leakage.

Although the CNTFETs are in their infancy, it is hoped that they will replace the existing MOSFETs in terascale integration. This hope is based on the fact that the ratio between the ON and OFF states attains 10^5–10^6 in even rudimentary CNTFETs, as well as on the high thermal conductivity and the impressive density currents carried by CNTs. In particular, the search of CNT-based logical circuits and memories is directly linked to the development of CNTFETs. The first logic circuits based on CNTFETs have used a semiconducting CNT with a bandgap of 0.7 eV, which was contacted by two gold electrodes that acted as source and drain. An Al wire beneath the semiconducting CNT, which was covered with few nanometers of Al_2O_3, ensured a good capacitive coupling between the gate and the CNT [39]. This transistor, which has a transconductance of 0.3 μS, an ON/OFF ratio higher than 10^5 at room temperature, a gain greater than 10, and a maximum operating current of 0.1 μA, was used to demonstrate basic binary logic circuits such as inverters (which convert a logical 1 into 0 and vice versa), NOR, or flip-flops (see Figure 3.27).

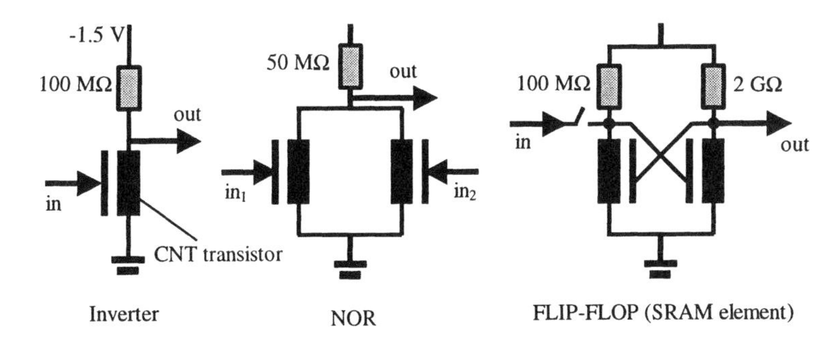

Figure 3.27 Basic binary logic gates based on CNT transistors.

The CNTFETs could also be key elements of the multiple-valued logic (MVL), which involves more than the two logical states of binary logic. Ternary (3 logical values) and quaternary (4 logical values) are especially envisaged. The MVL is a possible solution for high-density integration since the circuits based on MVL are able to reduce significantly the number of elementary operations necessary to implement certain mathematical functions. In particular, the chip area and the power dissipation can be reduced with up to 50% [40]. For example, the simplest MVL operator, called complement and defined as $\bar{n} = (m-1) - n$, where m is the base number assigned for the MVL takes the following values in ternary logic ($m = 3$): $\bar{n} = 2$ when $n = 0$, $\bar{n} = 1$ for $n = 1$, and $\bar{n} = 0$ when $n = 2$ (in binary logic $m = 2$, and the complement operator is transformed into an inverter operator). The complement operator in ternary logic can be implement by two CNTFETs with slightly different diameters and thus slightly different threshold voltages, which switch the transistors from the OFF state to the ON state, as shown in Figure 3.28. This threshold voltage is given by:

$$V_{\text{th}}(\text{eV}) = 0.5 / d(\text{nm}) = E_{\text{g}} / 2 \tag{3.56}$$

The above relation shows, once again, how powerful for various applications is the engineering of the nanotube bandgap via the geometrical dimensions of the CNT.

The two CNT transistors with different nanotube diameters are both in the OFF state at low input voltages so that the output is at voltage V_{DD}. When the input voltage increases and reaches the threshold voltage of the first transistor T1 this transistor is switched in the ON state, and the output voltage decreases to $V_{\text{DD}}/2$. A further increase of the input voltage causes the second transistor to switch in the ON state, the output voltage becoming zero. The output voltage is thus given by

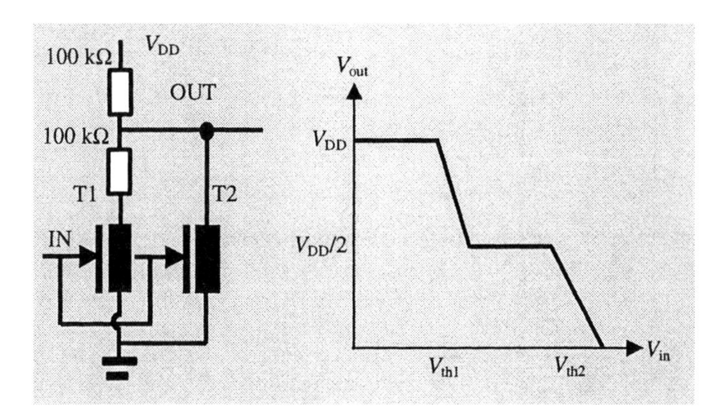

Figure 3.28 The implementation of the MVL complement operator using CNTFETs.

$$V_{\text{out}} = \begin{cases} V_{\text{DD}}, & V_{\text{in}} < V_{\text{th1}} \\ V_{\text{DD}}/2, & V_{\text{th1}} < V_{\text{in}} < V_{\text{th2}} \\ 0, & V_{\text{in}} > V_{\text{th2}} \end{cases}, \tag{3.57}$$

so that three stable states with three logical values are implemented. Other MVL operators such as the truncated sum or the cycle of an MVL variable, which is an

MVL register, can be quite easily implemented with a couple of CNTFETs. These simple logic gates are key elements for future processors.

Memories based on CNTs are also a hot topic nowadays, because the first attempts have already demonstrated a teralevel density. The first proposal of a CNT-based nonvolatile random access memory involved a crossbar CNT array, each bistable element of the memory being represented by suspended and crossed CNTs, which are electrically switched in the ON or OFF state; such a configuration is shown in Figure 3.29. The crossbar CNT memory consists of a set of parallel metallic CNTs on a substrate and another set of perpendicular metallic CNTs suspended on a periodic array of posts [41]. This memory is able to attain the teralevel of integration (i.e., 10^{12} elements could be integrated on a square centimeter) using 5-nm-long device elements made of metallic CNTs and 5-nm-long supports, the addressability being achieved with 10-μm-long SWCNTs. For a 5-nm long element the switching time of this elementary CNT memory is about 5 ps, corresponding to an operation frequency of 200 GHz, more than 60 times higher than the operation frequency of present-day computers, which slightly exceeds 3 GHz.

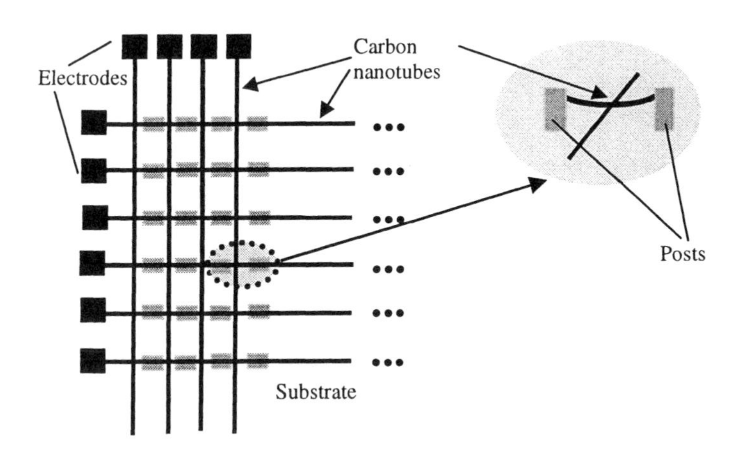

Figure 3.29 CNT-based nonvolatile RAM memory.

CNTFET memories were also developed. For example, the CNTFET consisting of quite long SWCNT bundles, with a diameter of 2–4 nm, deposited on highly Sb-doped silicon wafer on which a 100-nm-thick SiO_2 layer is thermally grown, shows memory effect (i.e., a reproducible hysteresis as that in Figure 3.30) in the drain current when the gate voltage is sweeping the range 0–3 V. The 15 nm

AuPd source and drain contacts were separated in this case at a distance of 150 nm, and the silicon substrate was used as a back gate. The significant memory effect, manifested by the fact that the conductance states at zero gate voltage differ by two orders of magnitude [42], originates from the preparation method of the CNT bundles: thermal annealing in air for several hours or controlled oxygen plasma exposure. The CNT bundles are a mixture of semiconducting and metallic tubes, only the metallic nanotubes being oxidized when heated. Therefore, the gate voltage dependence of the drain current is more pronounced in this case than for unheated bundles, due to the remaining intact semiconducting nanotubes. Moreover, oxidation-related defects, which play the role of charge storage traps, are created at the surface of amorphous carbon.

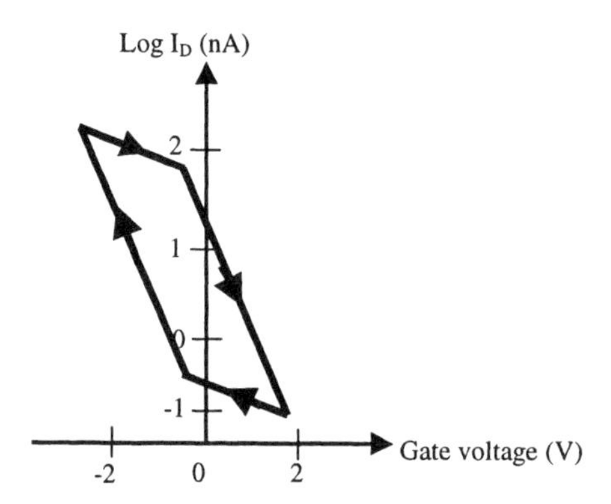

Figure 3.30 Hysteresis effect in SWCNT bundles transistors (*After*: [42]).

More advanced nonvolatile memory devices were fabricated using SiO_2-Si_3N_4-SiO_2 (ONO) layers as a storage node in the CNTFET configuration with a top gate. These CNT flash memory devices consist of a semiconducting CNT with two metallic source and drain electrodes patterned at its ends. The CNT is placed over a silicon substrate, the ONO layer and the top metallic gate [43] being deposited over the CNT. The ONO layer, which has a high breakdown voltage and a high charge retention capability and is almost defect-free, can be viewed together with the CNT as a charge nanochannel, forming the basic element of the flash memory. The $I_{DS} - V_G$ dependence of this device is typical for a *p*-type CNTFET. However, when the gate voltage is swept to a positive threshold value, the charges tunnel from the surface of the CNT and are captured by the free traps of ONO. The stored charges increase the threshold voltage with a fixed amount because ONO has traps with quantized energy states. (The threshold voltage is closely linked to the localized electric field in the CNT channel.) Sweeping up and down the gate voltage reveals a hysteresis behavior similar to that in Figure 3.30.

3.2.2 CNT-Based Field Emission Devices

The field emission phenomenon consists of the emission of electrons into vacuum due to a very high electric field. When an intense electric field is applied to a solid, the surface potential discontinuity, which confines the electrons in the solid, is transformed into a potential barrier. If the thickness of this barrier is only a few nm, so that it has the same order of magnitude as the tunneling distance of low energy electrons (of about 2–3 nm), the electrons tunnel from the highest occupied states in the solid into the vacuum and then travel ballistically in vacuum without any scattering. The Fowler-Nordheim model (see Chapter 1, Section 1.1.6) describes the dependence of the emitted current on the applied electric field assuming a triangular-shaped barrier. Because the emitted current depends exponentially on the applied electric field, a small variation of the slope of the $\ln I - V$ curve, of the surrounding of the emitter, or of the chemical state of the surface can change significantly the emitter current. Among the vacuum devices that employ field electron emitters, we mention field emission displays, CRT displays, switches, klystrons, or traveling wave amplifiers. Any vacuum device consists of a cathode, which emits electrons and an anode, which collects them, both embedded in a high vacuum enclosure. For an in-depth review of field emission and vacuum microelectronic by devices we recommend the book of Zhu [44].

The electric field, which generates a measurable emission current in metals with a work function of 5 eV, is about 2500–3000 V/μm. Such high fields can only be generated using sharp emitters, because the electric fields tend to concentrate around sharp objects. The field concentration of a tip-like structure is expressed by the enhancement factor $\beta = h / r$, where h is the height of the tip and r the radius curvature at the apex. CNT thin films show current densities of up to 1 mA/cm^2 for applied fields of less than 5 V/μm; these fields are 500 times lower than in metals [45] since high field concentrations develop at the sharp nanotube tips. Therefore, CNTs are among the best electron field emitters available today, their enhancement factor of β = 10^2–10^3 exceeding by far the performances of field emission arrays consisting of sharp metal tips obtained via micromachining.

The emission properties of CNTs depend on the type of CNT, its size, method of preparation, and the filling fraction of nanotubes in the case of thin films. Irrespective of these parameters CNTs have excellent field emission properties despite their quite high work function of about 5 eV. Individual CNTs are able to emit 1 μA, while the current density for thin films containing randomly oriented CNT arrays can easily reach 1–3 A/cm^2, much greater current densities being emitted by ordered arrays of CNTs. In all cases no structural degradation is apparent after many hours of continous emission. The threshold electric field for electron emission varies from 2 V/μm in the case of SWCNTs up to 4.5 V/μm for MWCNTs grown via CVD [44].

The $I - V$ characteristic of CNT-based field emitters shows unique features; a typical characteristic for a single SWCNT is schematically represented in Figure 3.31. In region 1 the electron emission follows the Fowler-Nordheim dependence, as in the majority of known field emitters. However, at a higher voltage, $V_1 \cong 700$ V, the current shows a region of saturation (region 2), which is very useful for device applications because in this region the field emitter regime is stable and less perturbed by other factors. Several mechanisms have been suggested to explain the occurrence of the saturation region, the most plausible being that the saturation is due to an adsorbate effect at the surface of CNTs. This explanation is supported by experimental facts, which show that clean CNTs do not show saturation effects, and assumes that the CNT emits electrons through the water-related surface adsorbate states, enhancing the emission with 3–4 orders of magnitude. When the voltage increases the adsorbate states are strongly distorted, and produce the saturation current. Region 3, which occurs at very high voltages, for V_2 over 1200 V, is typical for a clean-SWCNT field-emission behavior.

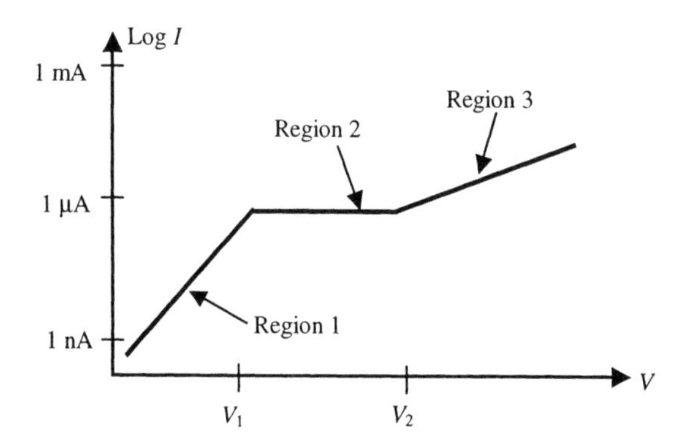

Figure 3.31 The field emission I-V characteristic for a single SWCNT (*After*: [44]).

The applications of CNT-based field emission devices cover various areas of applied sciences. In particular, display applications that employ CNT cold cathodes in CRT lighting elements with a diameter of 2 cm are very appealing. These devices have demonstrated twofold brightness over conventional CRTs and have a lifetime greater than 10^4 hours [46]. A vacuum-sealed 5-inch color field emitter display was fabricated by Samsung using aligned arrays of CNT emitters [47].

An array of CNT-based field emitters can also generate a very large current density, of 100 A/cm^2, in a nanoklystron device designed to produce 3 mW at 1.2 THz. The reflex nanoklystron is made using Si micromachining techniques and is composed from two parts, which are sealed in vacuum. The entire vacuum microtube is very small; its dimensions do not exceed 100 μm. Figure 3.32 presents the THz klystron device, which uses a CNT ordered array as a cold cathode [48].

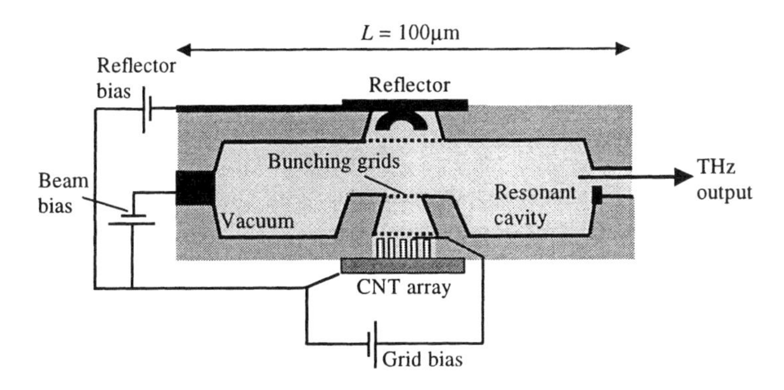

Figure 3.32 THz klystron with an array of ordered CNTs as a cold cathode (*After*: [48]).

The device in Figure 3.32 is only a member of a larger class of devices that mimic vacuum tubes and are based on a combination of micromachining techniques and field emitters based on CNT arrays. The resulting vacuum tube devices are micron-scale miniaturized versions of vacuum-tube triodes, klystrons, or traveling wave amplifiers. For example, a fully integrated on-chip microtriode was manufactured via silicon micromachining [49], the triode being fabricated laterally using the MEMS (microelectromechanical systems) principles explained in Chapter 2. More precisely, the polysilicon electrode panels of the triode are etched and released on a silicon-nitride-coated Si substrate. Then, the CNT array is selectively grown on the cathode and finally the electrode panels are rotated and locked into their vertical position. The triode, represented in Figure 3.33, has a conductance of 2.7 μS and a cutoff frequency of about 200 MHz, which can be increased with orders of magnitude by reducing the distance between the anode and the cathode. Such a device could be an interesting solution for an amplifier in the frequency range beyond 150 GHz, where semiconductor amplifiers cannot work.

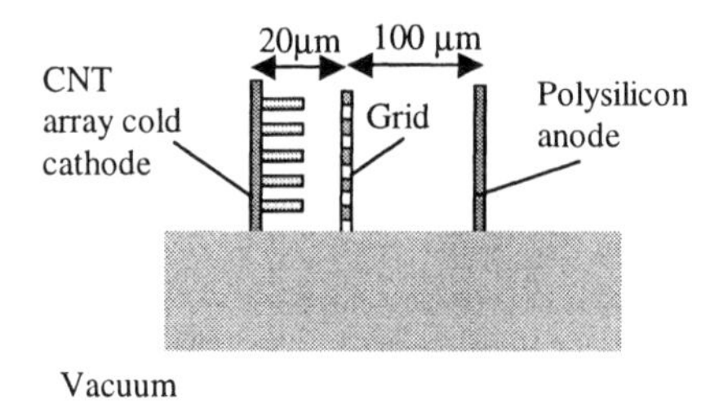

Figure 3.33 Vacuum triode with a CNT array as a cold cathode (*After*: [49]).

The aligned CNT field emitter arrays can also be grown on flexible plastic substrates [50]. Free-standing nanotubes with a 20–30 nm diameter grown onto a chromium-covered polymide foil show a 3.2 V/μm turn-on field and a 4.5 V/μm threshold field (where the current density is 1 μA/cm^2). Flexible field emitters for display and plastic electronics applications can be fabricated using this technique.

3.2.3 Junctions, Heterojunctions, and Quantum Confined Structures Based on Carbon Nanotubes

CNT junctions consist of two or more CNTs connected in different manners. These junctions can be made from the same type of CNT, or from different (semiconducting or metallic) types. Figure 3.34 presents the most encountered CNT junctions.

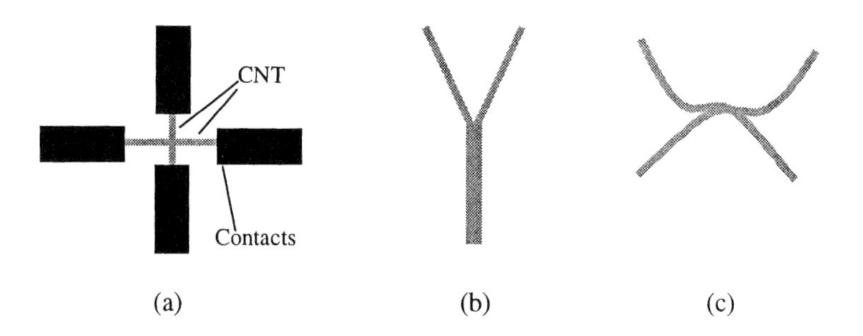

Figure 3.34 Different CNT junctions: (a) crossed, (b) Y-shaped, and (c) formed from buckled CNTs.

The crossed nanotube junction, represented in Figure 3.34(a), is made from two SWCNTs, each of them ended with metallic contacts. Metallic and/or semiconducting SWCNTs were studied in the crossed configuration [51], in which the crossed junction and the contacts are deposited on a doped silicon substrate, which plays the role of a gate that is able to change the charge densities in both nanotubes. The results showed that metallic-metallic and semiconducting-semiconducting junctions have a quite high conductance, of $G = 0.1e^2 / h$, and that the conductance at different temperatures has a transmission probability $T = G/(4e^2 / h)$, which ranges from 0.01 to 0.06, indicating that some percents of electrons that arrive at the junction tunnel from a nanotube to the other. Because the nanotubes are deformed at the junction by van der Waals forces (see Chapter 2) the tunneling probability is 4% for junctions between metallic nanotubes, a similar behavior being observed in semiconducting-semiconducting SWCNT junctions.

The metallic-semiconducting junction behaves in a different manner from the junctions made from the same type of CNTs. The metallic-semiconducting junction is a rectifying diode because the semiconducting SWCNT is depleted at the junction with the metallic SWCNT. A Schottky barrier appears, since at the junction the Fermi level of the metallic SWCNT is aligned with that of the semiconducting SWCNT, depleting it, while far from the junction, the Fermi level is located in the valence band of the semiconducting SWCNT, assumed of *p*-type. Therefore, there are two tunneling barriers at the junction: the tunneling barrier at the junction of the two nanotubes, similar to that in metallic-metallic or semiconducting-semiconducting CNT junctions, and the Schottky tunneling barrier with a height equal to $E_g / 2$. In consequence, the conductance is with some orders of magnitude smaller than that obtained in the case of crossed CNTs of the same kind and the transmission probability is about 2×10^{-4}.

The Y-shaped CNT junction in Figure 3.34(b) consists from a large-diameter CNT (the stem), which branch into two smaller-diameter CNTs [52]. The $I - V$ characteristic of an individual Y junction shows a rectifying behavior, the current increasing rapidly at negative bias and becoming zero at positive voltages. The rectification action of the Y junction takes place at room temperature. An array of parallel Y junctions, with all junctions contacted by a bottom and a top electrode, shows the same nonlinear behavior of the $I - V$ characteristic as a single Y junction, described by the equation

$$I \propto \exp(eV / \gamma T) . \qquad (3.58)$$

Here $\gamma = 1 + \varepsilon_2 N_2 / \varepsilon_1 N_1$, where N_1, N_2, and ε_1, ε_2 are the carrier concentrations and the dielectric permittivity in the stem and branch side, respectively. The ratio

$N_2 / N_1 \propto (d_1 / d_2)^4$ indicates the direct relation between the current and the CNT diameters at the Y junction.

A crossed nanotube junction, as that indicated in Figure 3.34(c), is formed by manipulating two buckled SWCNTs using an AFM [53]. If the buckle angle is greater than 60° the buckles modify significantly the transport properties by acting as tunnel barriers. Thus, the conductance is much smaller than the $4e^2 / h$ ballistic conductance, this junction being the simplest way to implement an SET transistor.

There are not only horizontally oriented CNT junctions, supported (in principle) by a substrate, but also vertically aligned CNT junctions, such as the multiple MWCNT/MWCNT junctions reported in [54], which have a density of 10^2–10^3 junctions/array and are suitable for multilevel interconnections.

The simplest heterojunction between two CNTs is formed by connecting in series two CNTs, one metallic and the other semiconducting (see Figure 3.35). In this way, a straight Schottky barrier is formed [55] enclosed by a metallic cylinder gate of radius R. This structure shows rectifying properties and a characteristic step-like increase of the current because the long-range Coulomb interactions play an important role in band shaping and reshaping during charge transfer.

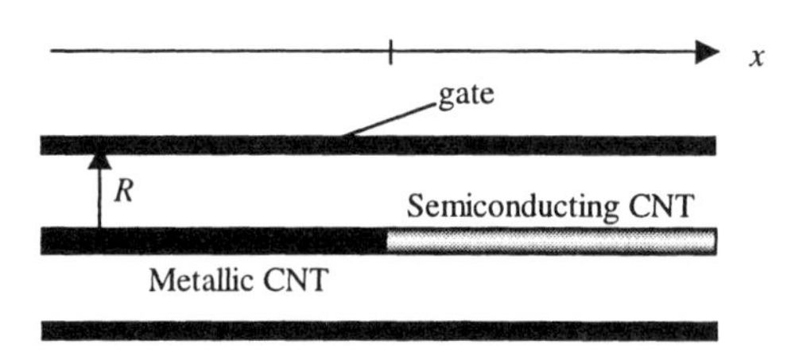

Figure 3.35 The CNT Schottky barrier formed by a metallic and semiconducting CNT (*After*: [55]).

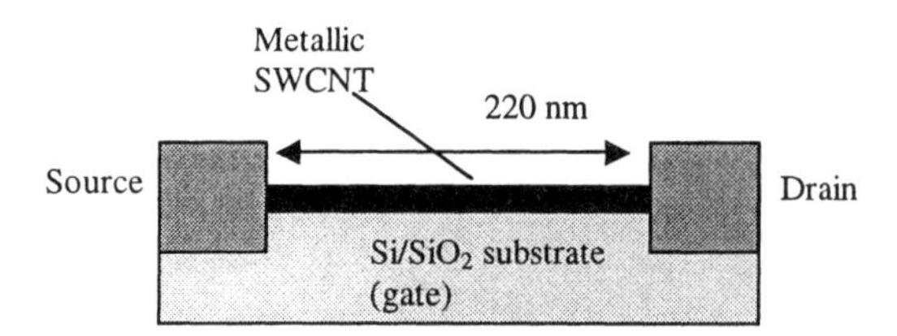

Figure 3.36 The SWCNT-based electron Fabry-Perot cavity.

A step further in the development of CNT-based devices is the Fabry-Perot cavity for electrons, which is a double heterojunction between a metallic SWCNT and the two metallic contact electrodes at its end, which play the role of partially reflective barriers. In this way, the electrons are longitudinally confined between the two electrodes, the metallic SWCNT acting as an electron waveguide [56]. The realization of this device, represented schematically in Figure 3.36, is possible only using metallic CNTs with a measured resistance of 7 kΩ (a value very close to the theoretical ballistic limit of 6.5 kΩ) terminated by nearly ohmic contacts to two source and drain Au/Cr electrodes. The entire nanostructure is supported by a Si/SiO_2 substrate, which acts as a gate. At low temperature the conductance (i.e., the $\partial I / \partial V$ curve) shows strong oscillations when the gate voltage is varied, the almost sinusoidal variation of the current between the drain and source contacts with the gate voltage indicating a similar behavior to electromagnetic waves in a Fabry-Perot cavity. However, in the above device the multiple reflections of electrons at the contacts are different from the reflections of electromagnetic waves because, due to its dispersion relation depicted in Figure 3.13, electrons can propagate in a metallic SWCNT in two modes. The electrons acquire different phase shifts in the two propagating modes during the traversal of the nanotube, the periodic pattern observed in experiments originating in the change of the phase as a function of the electron energy.

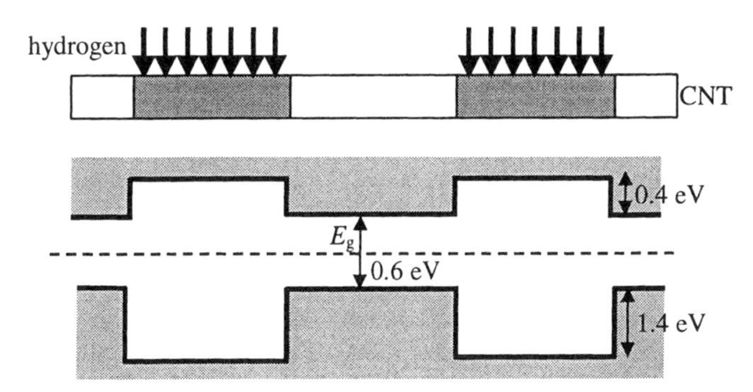

Figure 3.37 A quantum resonant tunneling structure realized by selective hydrogen absorption in a CNT (*After*: [57]).

More complex heterostructures, such as quantum structures similar to those obtained via bandgap engineering of III–V semiconductors, can be realized with CNTs modulated by hydrogen absorption [57]. By selectively exposing a CNT to hydrogen, the bandgap of the exposed zone widens with respect to the non-exposed zones, due to hydrogen absorption, leading to conduction and valence band offsets. In this way, the bandgap of the CNT can be modulated along its axis, and a series of quantum wells or dots, in particular a resonant tunneling diode such as that displayed in Figure 3.37, can be implemented. The regions where the hydrogen is absorbed are barriers, the regions between them being quantum wells.

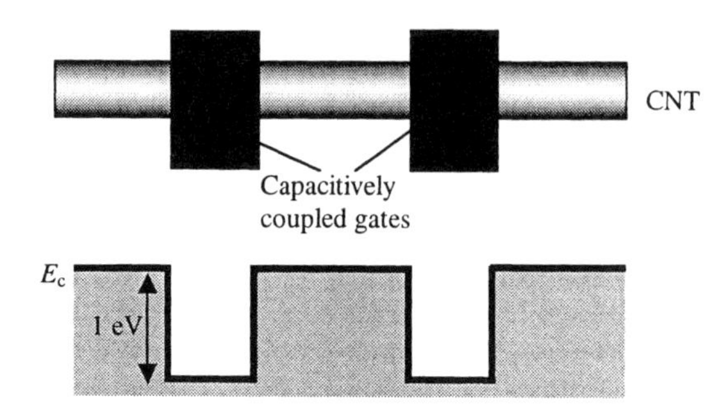

Figure 3.38 Parametric electron pump created via capacitively coupled gates patterned onto a CNT.

A quantum structure similar to that in Figure 3.37 can be created by patterning periodical metal gates on a CNT. The formation of barriers beneath metallic gates is a known effect in 2D and 1D electron gases, various devices, in particular split-gate quantum wire electron waveguides, being based on this effect. The possibility of well creation under the gates deposited above a CNT was used to implement a parametric electronic pump, represented schematically in Figure 3.38 [58], and a resonant tunneling diode for high-frequency applications [59], which will be explained in the next section.

The parametric electron pump consists of a metallic CNT capacitively coupled to two electrode gates, on which sinusoidal voltages with a certain phase shift between them are applied. When the Fermi level is varied, the pumped current along the CNT oscillates due to the resonant competition in the two wells. The resonant pumping is also the cause of the nonsinusoidal, but periodic dependence of the pumped current on the phase difference between the two pumps. The generation of a current that depends periodically on the phase and that can be controlled by the phase variation is of paramount importance for many NEMS applications.

Even in simple CNTFET configurations quantum dot behavior, as, for example, single electron charging and 0D confinement of electrons, manifests itself at low temperatures, not exceeding a few K. The simplicity of the quantum CNT-based structures is in deep contrast with the elaborated technological procedures used to create quantum dots in III–V semiconductor heterostructures.

As we pointed out earlier, a *p*-type quantum dot forms at the end of an *n*-type CNT [13], so that a *p-n* junction is created near the contact, producing a small quantum dot between the junction and the contacts. The quantum dot characteristics involved in single electron charging phenomena are the charging energy

$$E_{\text{ch}} = e^2 / C , \tag{3.59}$$

where C is the quantum dot capacitance, and the separation between quantized energy levels in the dot

$$\Delta E = h v_{\text{F}} / 2L . \tag{3.60}$$

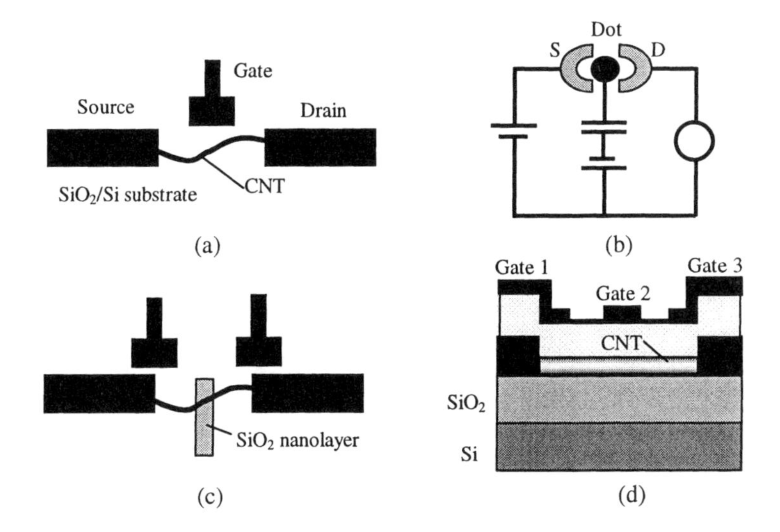

Figure 3.39 CNT-based quantum dot structures: (a) a single quantum dot, (b) its equivalent circuit, (c) double quantum dot, and (d) double quantum dots with local gates (*After*: [62]).

In general, a semiconducting CNT terminated with two metallic contacts and placed in close proximity with a gate electrode that modulates its potential, behaves at low temperatures as a quantum dot coupled via tunneling barriers to two electrodes; Figures 3.39(a) and 3.39(b) illustrate this behavior [60,61]. This CNT quantum dot has $C = 9.4$ aF, $E_{\text{ch}} = 8.6\ \text{meV}$ and $\Delta E = 2.3\ \text{meV}$. Coupled quantum dots can be implemented by adding more contacts to the CNT; for example, a CNT with six metallic contacts and two gate electrodes is equivalent to two capacitively coupled quantum dots. The same configuration can be realized introducing a SiO_2 nanolayer between the source and the drain contacts and using two gates, as shown in Figure 3.39(c).

A very interesting configuration of two coupled quantum dots, especially designed for logic operations at the nanoscale and quantum computation, is presented in Figure 3.39(d) [62]. This configuration consists of three gates, the middle one controlling locally the field in the CNT. In this way, the regime between the two dots is electrically tuned from weak to strong coupling.

In conclusion, it is much easier to produce quantum dots using CNTs than employing the AIII–BV technology, although CNT devices still work at very low temperatures. The applications of CNT-based quantum dots in logical devices, quantum computing, high-frequency devices, and optoelectronics are not matured enough, their development being the major task of future CNT nanoelectronics.

3.2.4 Microwave Devices Based on Carbon Nanotubes

A metallic or semiconducting CNT suspended over a ground plane, as shown in Figure 3.40(a), can be modeled as a transmission line, so that an equivalent circuit as that displayed in Figure 3.40(b) can be easily developed [63]. However, the circuit elements of the equivalent circuit of a CNT transmission line differ significantly from those of any microscopic microwave transmission line.

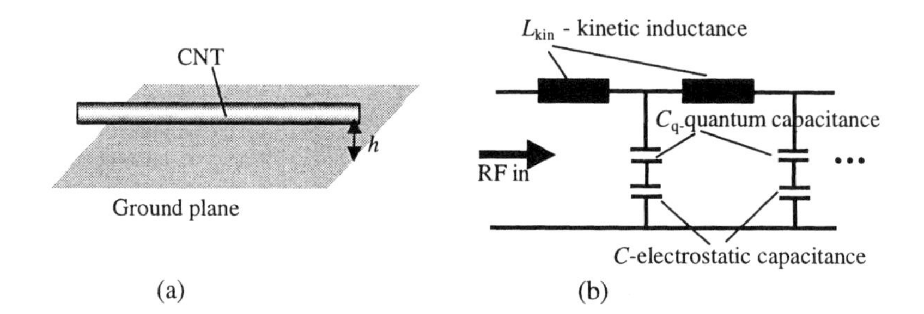

Figure 3.40 (a) The CNT transmission line, and (b) its equivalent circuit (*After*: [63]).

The CNT transmission line model is used when the CNT connects devices or circuits working at high frequencies, or when a long CNT is excited with high-frequency radiation. The CNT length can nowadays reach 1 cm [64], which corresponds to the free-space wavelength of a 30 GHz radiation, so that CNT applications in the millimeter range of frequencies and beyond could be very promising. The inductance in the equivalent circuit of the CNT transmission line is not a magnetic inductance, as in usual microwave transmission lines characterized by 3D electron transport, but is a kinetic inductance that originates in the inertia of electrons. In a CNT transmission line, where we have 1D electron transport, the electrons do not respond instantaneously to the applied field, but there is a delay between the current and its phase, expressed by the kinetic inductance

$$L_{kin} = \hbar\pi / e^2 v_F . \qquad (3.61)$$

For $v_F = 8\times10^5$ m/s, L_{kin} is about 16 nH/μm, predominating over the magnetic inductance. The kinetic inductance in (3.61) is proportional to the Planck constant, which underlines the quantum nature of L_{kin}, the corresponding reactance being proportional to the electron energy.

In a usual microwave transmission line it is possible to add electrons with arbitrary energies, whereas in a CNT transmission line adding of electrons with energy lower than the Fermi energy E_F is not possible. The electrons are added in available quantum states above E_F, which have energies separated by $\Delta E = 2\pi\hbar v_F / L_{kin}$. By re-writing the energy separation as $\Delta E = e^2 / C_q$, it follows that the quantum capacitance is given by

$$C_q = 2e^2 / hv_F , \qquad (3.62)$$

its value being of about 100 aF/μm.

The electrostatic capacitance in Figure 3.40(b) is the well-known capacitance $C = 2\pi\varepsilon / \ln(h/d)$ between a wire and a ground plane and has a typical value of about 50 aF/μm. The electrostatic capacitance can be neglected if we ignore the screened Coulomb interaction, a case in which the electron wave velocity is $v_{CNT} = (L_{kin}C_q)^{-1/2} = v_F << c$. When the electron-electron interaction is accounted for, the wave velocity becomes $v_{CNT} = (L_{kin}C_{tot})^{-1/2} > v_F$, where $C_{tot}^{-1} = C_q^{-1} + C^{-1}$, but is still smaller than the speed of light. Thus the CNT transmission line is a slow wave structure. In the ideal case, when electron-electron interactions are neglected, the characteristic impedance of the CNT line is $Z = (L_{kin}C_q)^{1/2} = h/2e^2 = 12.5\,\text{k}\Omega$, while when the electron-electron interactions are considered the characteristic impedance becomes gZ, where g is the Luttinger parameter. The damping in the transmission line is introduced via a resistance distributed along the CNT, with values in the range 1–10 kΩ/μm; the transmission line is quite lossy. The ways in which the very high characteristic impedance of the CNT line can be matched is an open question.

The cutoff frequency of CNT transistors can be estimated from the behavior of the CNT transmission line. Such a transmission line can be a matching circuit for a CNTFET transistor, the equivalent circuit of which is displayed in Figure 3.41 [65]. The RC cutoff frequency for a CNTFET with a length of 100 nm is $f_{RC} = (2\pi RC)^{-1}$ = 6.5 THz for R = 6.25 kΩ and C = 4 aF. The gate-source capacitance $C_{GS}^{-1} = C^{-1} + (4C_q)^{-1}$ can be estimated at about 50 aF/μm. Another cutoff frequency, $f_{g_m} = (g_m / 2\pi C_{GS}) \approx 0.5\,\text{THz}$, is determined by the gate-source capacitance above and the transconductance g_m, which has a typical value of 10 μS extracted from experiments. However, since transconductances of 60 μS or higher are possible, f_{g_m} is beyond 1 THz, which means that a CNTFET has a cutoff frequency in the terahertz range. The cutoff frequency of a CNTFET can be increased by reducing the parasitic capacitances C_G and C_{GD}, some solutions being indicated in [64] for electrodes with special geometries.

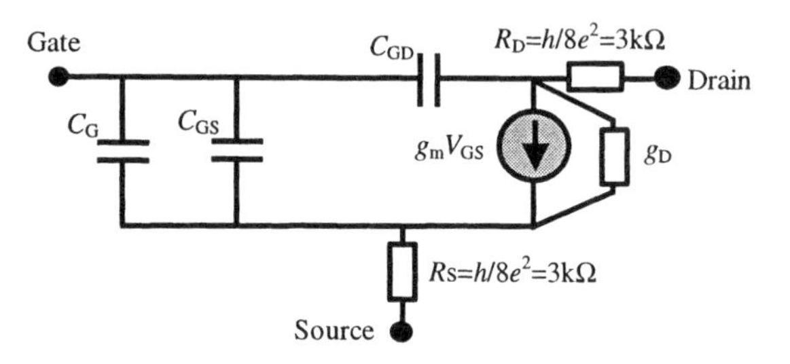

Figure 3.41 The simplified equivalent circuit of a CNTFET (*After*: [65]).

If the parasitic capacitances are negligible, the cutoff frequency of a CNTFET can be calculated via Burke's formula [65]

$$f_{\text{cut}} = g_m / 2\pi C_{GS} = 80 / L_G\,(\text{GHz}), \tag{3.63}$$

where L_G is the gate length in nm, which is identical to, or smaller than, the CNT length. Thus, the ideal cutoff frequency of the CNT, which is in the THz range, can decrease dramatically up to a few GHz if the parasitic capacitances are important, as is the case in many CNTFETs. Therefore, the first operation in microwaves of a CNT transistor was observed at 2.6 GHz [66]. This CNTFET was similar to the back-gate transistors described above, but had an added matching circuit to transform the high-value impedance of the transistor at 50 Ω. The transistor was placed on a microstrip test fixture and was contacted via a gold wire, which played the role of inductors and which, together with the capacitance to ground from the electrical contacted pads, formed a rudimentary matching circuit. At room temperature the losses in the silicon substrate prevailed over the microwave behavior, and so measurements were made at only at few K. The observation of a resonance of about –12 dB at 5 GHz in the S_{11} modulus indicated that the matching condition was accomplished and offered the first hint that CNTFETs will work at microwaves at least as resonators. The source-drain voltage at 2.6 GHz and at dc is different, indicating a possible transistor action at GHz. However, the CNTFETs are too rudimentary at this stage to achieve THz cutoff frequencies; major improvements are expected in the next years.

A SWCNT-based resonant tunneling diode (RTD) at THz frequencies, which is displayed in Figure 3.42, was proposed in [59]. As indicated in the previous section, lower-potential regions can be implemented by applying bias voltages on a series of gate electrodes. The conduction energy band structure of the semiconducting SWCNT in Figure 3.42 changes underneath the gates on which the voltages V_{G1}, V_{G2}, V_{G3} are applied such that the potential energies in these regions are lowered with V_1, V_2, V_3, respectively, in comparison with the non-gated regions. Because the gate electrodes are separated from the semiconducting SWCNT through a thin dielectric layer, the potentials V_i, $i = 1, 2, 3$ are different from the gate voltages V_{Gi} but are determined by the latter. For an appropriately chosen, sufficiently thick dielectric layer no tunneling between the gates and nanotube occurs and no significant parasitic capacitance besides that of the CNT itself appears. If a dc voltage V_{dc} is applied along the axis of the SWCNT, the region on which V_{G1} is applied acts as emitter for electrons, that region on which the gate voltage V_{G3} is applied acts as a collector, and the region on which V_{G2} is applied is a quantum well. The width of the well region L is given by the length of the inner gate and the barrier widths d_1 and d_2 are determined by the separation between the inner and outer gate electrodes. The planar gates create rectangular-shaped potential wells and barriers if the applied voltage is less than $E_g(d_G / d)$, where d_G is the distance between the gates and the SWCNT with diameter d. The quantum well and the surrounding barriers form the RTD structure.

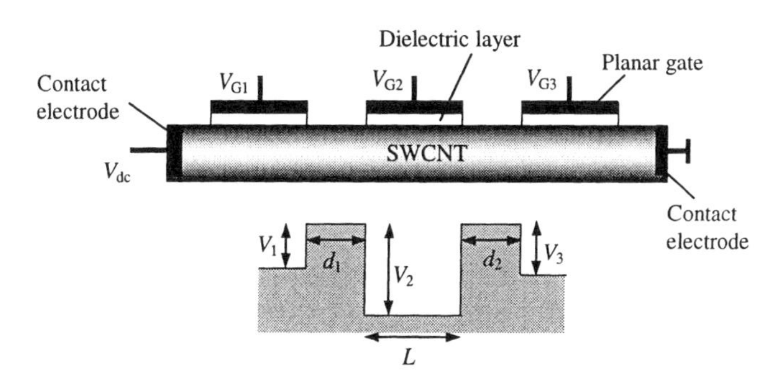

Figure 3.42 The THz CNT RTD (top) and its conduction energy band diagram (bottom) (*After*: [59]).

The simulations of the RTD in Figure 3.42 performed in [59] for a semiconducting (17,0) SWCNT with an effective mass of $0.216\,m_0$, $d_1 = d_2 = 10$ Å, L = 20 Å, and $V_1 = V_2 = V_3 = 0.8$ eV, demonstrated that the transmission coefficient and the transit time in the RTD depend on the electron energy E as shown in Figure 3.43(a). The reference for the electron energy was taken at the bottom of the well. Figure 3.43(a) shows that the transmission coefficient is significant only around two resonant energies of the quantum well: $E_1 = 0.193$ eV and $E_2 = 0.72$ eV, the corresponding transit times being $\tau_1 = 1.5 \times 10^{-13}$ s and $\tau_2 = 1.35 \times 10^{-14}$ s, respectively. Because of the high value of τ_1, the high-frequency capabilities of the device can be enhanced by suppressing the contribution of the electrons that tunnel through E_1. This can be achieved by choosing V_1 so that the bottom of the conduction band in the emitter region is raised above E_1, a case in which the electron states around E_1 are not available for the electrons in the emitter. For example, if the bottom of the conduction band in the emitter region is raised with 0.35 eV above the bottom of the quantum well, the transmission and transit time dependences on E modify as displayed in Figure 3.43(b). The solid lines in this figure indicate the case when the bias along the SWCNT axis is absent, while the dashed lines are drawn for V_{dc} = 0.5 V. From Figure 3.43(b) it follows that the position of the highest resonant state in the well, as well as the transit time at the corresponding energy, shift towards lower values when the SWCNT is biased. The transit time in the biased device attains 10 fs, which is consistent with a cutoff oscillation frequency of about 16 THz.

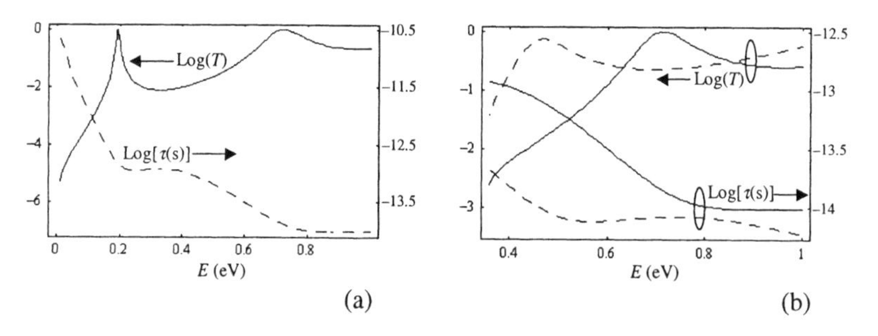

Figure 3.43 Transmission coefficient and transit time energy dependence of electron energy in the CNT-based RTD when (a) $V_1 = V_2 = V_3 = 0.8$ eV, (b) $V_1 = 0.45$eV, $V_2 = V_3 = 0.8$eV, no bias (solid line), and $V_{dc} = 0.5$eV (dashed line).

The $I - V_{dc}$ dependence of the device at room temperature, displayed in Figure 3.44 for three values of the Fermi level calculated from the bottom of the conduction band in the well: 0.6 eV (dashed line), 0.65 eV (solid line), and 0.7 eV (dotted line) was computed using the Landauer formula $I(V) = (2e/h) \times\int T(E,V)[f_L(E) - f_R(E)]dE$ (see Section 1.1.4). A strong region of negative differential resistance is observed, which indicates that oscillations with THz frequencies are produced along the SWCNT axis. When the Fermi level is at 0.65 eV, the ratio between the peak and valley values of the currents, indicated by the p and v subscripts, respectively, is $I_p / I_v = 2.23$. Negative differential conductivity in CNTs was evidenced in [67] and [68]. The oscillation condition of the SWCNT-based RTD is $R_N + R_c + R_A < 0$, with R_N the negative RTD resistance, R_c the contact resistance, and R_A the radiation resistance of an antenna that is coupled to the RTD to quasi-optically deliver the THz radiation. This condition is fulfilled for $R_c = 12.9\,\mathrm{k\Omega}$, $R_A = 80\ \Omega$, and $R_N = -99.2\,\mathrm{k\Omega}$ for the Fermi level of 0.65 eV, the cutoff frequency of oscillations being $f_{RC} = (2\pi C_{RTD})^{-1} \times[-G_N /(R_c + R_A) - G_N^2]^{1/2}$ = 10.3 THz, where C_{RTD}= 0.4 aF is the capacitance of the SWCNT-based RTD. This value of f_{RC} is very close to the oscillation frequency estimated from the transit time computation. The output power $\Delta I \Delta V = (I_p - I_v)(V_p - V_v)$ of the RTD oscillator matched to the antenna is 2.5 μW for the data in Figure 3.44. The p and v subscripts of V indicate the peak and valley voltages, respectively. The quite high impedance of the SWCNT-based RTD oscillator in the THz domain, of about 20 kΩ, can be fully matched to a bowtie antenna if the antenna is supported on a 1-μm-thick GaAs membrane (see [59] for details and further references on matching).

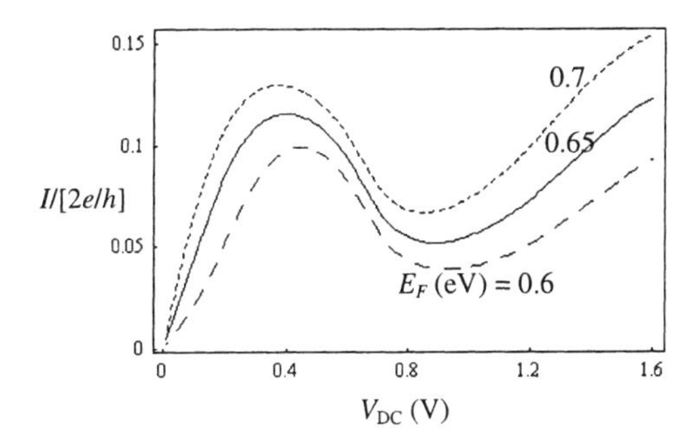

Figure 3.44 Room temperature $I - V_{DC}$ characteristics with various Fermi energy levels measured from the bottom of the quantum well: 0.6 eV (dashed line), 0.65 eV (solid line), and 0.7 eV (dotted line).

CNT-based terahertz signal amplification was recently proposed in [69] based on the Monte-Carlo simulations in [20], which demonstrate that undoped semiconductor CNTs display a region of negative mobility for a large interval of n values, from 10 to 59. The negative mobility for a (34,0) CNT is schematically displayed in Figure 3.45.

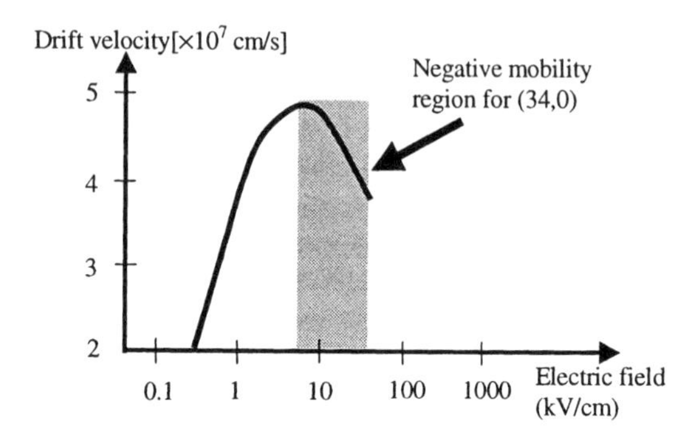

Figure 3.45 Negative mobility region for the (34,0) semiconducting CNT (*After*: [20]).

The physical origin of negative mobility in undoped 1D semiconductor CNTs is similar to that encountered in connection to the Gunn effect in doped bulk semiconductors. When a dc electric field is applied on a semiconducting CNT the electrons jump from the first subband, where they have a low effective mass and a high mobility, to the second subband, characterized by a higher effective mass and a lower mobility. However, unlike in bulk semiconductors, where electrons are transferred between unequivalent valleys in the electronic bandstructure, in semiconducting CNTs the transfer occurs within the same or within different but equivalent bandstructure valleys and originates in electron-phonon scattering. Therefore, unlike the above RTD or CNTFETs, the devices based on the Gunn effect are not ballistic, but require that the scattering time is much smaller than the electron transit time. For scattering times of about 10 fs in semiconductor CNTs the Gunn effect can occur at THz frequencies, whereas the corresponding energy relaxation time in bulk GaAs and InP semiconductors is in the 1–2 ps range. However, it is not possible to build a THz Gunn oscillator with an undoped semiconductor CNT since Gunn domains and domain oscillations occur only in heavily doped materials. But an undoped semiconductor with a negative carrier mobility that originates in the Gunn effect can amplify at frequencies equal to the inverse of the transit time and its integer multipliers [70]. If a bias voltage V is applied between two metallic contacts patterned at a distance L on a semiconductor CNT, the transit time between them is $t_t = L / v_d$, with v_d the drift velocity of electrons. The transit time is 1 ps (much longer than the electron intersubband transfer rate) for a (34,0) semiconductor CNT with $L = 400$ nm and $v_d = 4\times10^5$ m/s, which corresponds to the frequency $f_t = 1$ THz. So, according to Kroemer's theory, amplification occurs at the frequencies $f = (p + 0.25) / f_t$, where p = 1, 2, 3,.... The input impedance of this amplifier is $Z(f) = (L^2 / \varepsilon_0 \varepsilon_r A v_d)\{[\exp(-\gamma) + \gamma - 1] / \gamma^2\}$, where γ is related to the frequency f via $\gamma = (L / v_d)[(n_0 e\mu / \varepsilon_0 \varepsilon_r) - i2\pi f]$, with $\varepsilon_r = 2.5 + 1.19i$ the relative permittivity of the CNT at 1 THz, A the CNT cross-sectional area, and n_0 the carrier concentration. The considered (34,0) semiconducting CNT, with a diameter of 2.64 nm, must be biased at V = 0.8 V [20] to obtain an applied electric field $F = 2 \times 10^6$ V/m, situated in the middle of the negative mobility region, where $\mu = -440$ cm^2/(Vs) and $v_d = 4\times10^5$ m/s. The voltage swing associated to V = 0.8 V is about 0.4 V, the resulting output power being 1.6 mW for an impedance of 100 Ω. The current density $j = V / AR$ of the CNT in this example is 0.36×10^6 A/cm^2 for an intrinsic CNT resistance of 40 MΩ, the corresponding carrier concentration being $n_0 = 0.6 \times 10^{17}$ cm^{-3}.

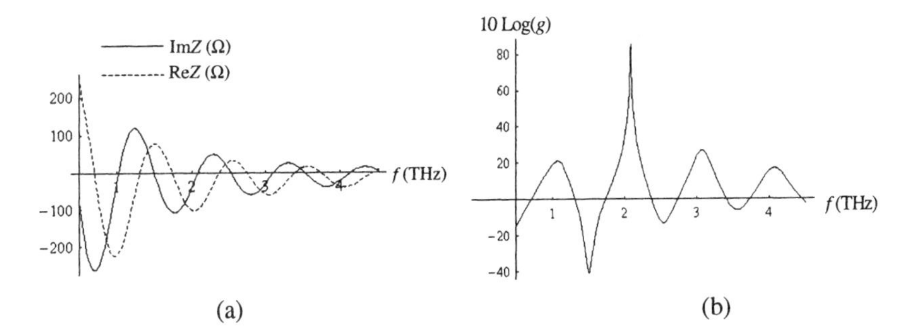

Figure 3.46 (a) The real and imaginary parts of the input impedance of the THz Gunn CNT, and (b) the gain in dB within the THz range.

Because it is very difficult to propagate THz radiation through CNTs with a tiny cross-sectional area, an array of parallel well-aligned CNTs should be used instead. The impedance of an array of CNTs connected in parallel equals the value for a single nanotube divided by the number of CNTs in the array. The input impedance of an array of 2×10^6 CNTs, which can be produced by self-assembly and which can be contacted by two metallic electrodes since the array acts like a single crystal, is represented in Figure 3.46(a). The real part of the impedance in Figure 3.46(a) is negative, as, for example, in the intervals 0.7–1.32 THz and 1.73–2.37 THz, where the CNT array amplifies the THz signal, attenuating the incoming signals in the frequency ranges for which Re(Z) is positive. $g = [(Z_0 + R)^2 + X^2]/[(Z_0 - R)^2 + X^2]$ is the gain of the array with impedance $Z = -R + iX$, Z_0 being the characteristic impedance of the feeding line that excites the CNT array. The frequency dependence of $10 \times \text{Log}(g)$ is represented in Figure 3.46(b). The maximum gains in the 0.7–1.32 THz and the 1.73–2.37 THz ranges, of about 20 dB and 80 dB, respectively, for $Z_0 = 100\ \Omega$, are unexpectedly high for the THz range. These calculated gain values are lower in real CNT arrays due to a filling factor smaller than 1, but they are still impressive because the filling factors can be as high as 80%. The importance of the theoretical demonstration of an amplifier at THz frequencies based on CNT arrays is enhanced by the fact that no electronic room-temperature THz amplifier has been yet reported.

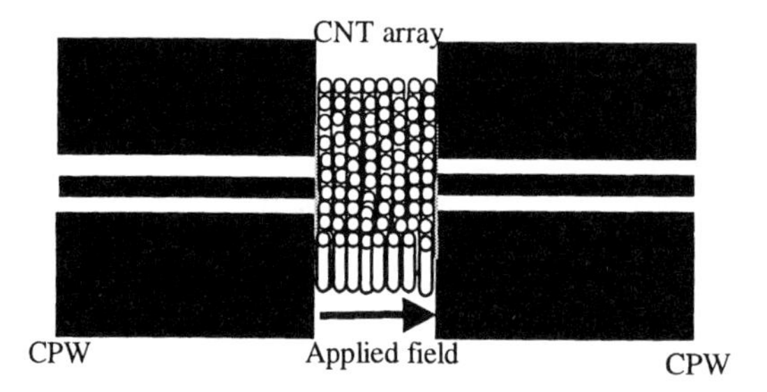

Figure 3.47 RF filter based on CNT array.

An interesting approach for a microwave resonator and filter working in the 1–3 GHz range is to combine CNT mechanical resonators, which have quality factors of 10^3 in this frequency range, with coplanar waveguide (CPW) lines [71]. The CPW line couples the microwave signal to the array of metallic CNTs (see Figure 3.47), which in turn reflects any microwave signal, except that with a frequency equal to the mechanical frequency of the CNT array, f_0. Coupling at f_0 takes place due to the Coulomb forces between the microwave signal and the electric charges that exist on the metallic CNT array. These charges can be produced via a dc field applied on the array or with an electrode located beneath the array, which is in contact with the grounded electrodes of the CPW. The resonance frequency f_0 of the resonator or filter can be tuned by changing the bias of the field applied on the CNT array. The vibration of the excited tubes located near the input electrode propagates along the entire array like an acoustic excitation.

CNT films were coated over Cu and used as microwave resonators at 4 GHz. Because the CNTs change their electrical permittivity when adsorbing various molecules (and thus a frequency shift of the microwave resonator indicates the adsorbance of a certain gas molecule or biomolecule) a simple resonator configuration containing the CNT resonator and a microstrip feedline can be used as a gas detector [72]. For example, the ammonia gas produces a frequency shift of 10 MHz of the 4-GHz resonator at room temperature, with a recovery time of 10 minutes, while isopropanol, which is a biomaterial, produces a shift of 2 MHz of the same resonator [73]. A more complex gas sensor was fabricated by depositing an MWCNT-SiO_2 composite layer on a planar inductor-capacitor resonant circuit that works at 20 MHz. The sensing device was able to sense CO_2, oxygen, and ammonia, the response to these gases being linear and reversible [74].

3.2.5 CNT-Based NEMS

The general properties of NEMS were described in Chapter 2. Many NEMS devices are related to RF devices in the sense that in RF devices the electromagnetic frequencies are in the MHz and GHz range, while in the CNT-based NEMS (CNT-NEMS) devices the mechanical frequencies are in the MHz and GHz range. In consequence, the combination of CNT-NEMS mechanical devices with RF devices can produce a wealth of new physical phenomena and devices. The static and dynamic mechanical deflections are produced electrostatically in a CNT cantilever (a biased CNT clamped at one end and free at the other), following the same rules and equations as described in Chapter 2. The experimental analysis performed in [75] has demonstrated that the electrical force exerted on a CNT, and hence the charge, is located entirely on the tip. The CNT-NEMS cantilever deflection is proportional to V_s^2, where V_s is the static (dc) voltage applied on the cantilever, βV_s being the corresponding static electric field, with β a constant. Hence, the static force applied on the CNT cantilever tip is $F_s = \alpha\beta V_s^2$, with α a constant that depends on the nanotube geometry. Extreme bending, with a radius of curvature of 80 nm was demonstrated for a CNT cantilever with a diameter of 20 nm, the cantilever returning afterwards to the original straight configuration.

The application of a dynamical electrical signal, which varies periodically in time according to $V(t) = V_0 \cos(\omega t)$, produces a force on the CNT cantilever of the form

$$F(t) = \alpha\beta[\Delta V + V_s + V_0 \cos \omega t]^2 = \alpha\beta[(\Delta V + V_s)^2 + 2(\Delta V + V_s)V_0 \cos(\omega t) + (V_0^2/2)\cos 2\omega + (1/2)V_0^2] \tag{3.64}$$

where ΔV is the part of the bias voltage, which is necessary to neutralize the static charge accumulated on the cantilever and the bottom electrode even when they are in mutual electric contact. By tuning the electrical frequency ω we can excite the cantilever at any of its mechanical resonances, which in the case of a multiwalled CNT are given by

$$\nu_j = \beta_j[(E/\rho)(d_{\text{out}}^2 + d_{\text{in}}^2)]^{1/2} / 8\pi L^2 , \tag{3.65}$$

where d_{out} and d_{in} are the outer diameter and inner diameters of the CNT, respectively, $\beta_1 = 1.875$ and $\beta_2 = 4.694$. The amplitude of the CNT cantilever has a maximum value at the resonant frequency, which could range from a few MHz to a few GHz depending on the nanotube diameter and length. Experiments made on various CNT geometries using the harmonic excitation of their fundamental mode have demonstrated that the Young modulus varies as a function of the nanotube diameter, ranging from 1 TPa for CNTs with diameters less than 10 nm to 0.1 TPa for CNTs with large diameters, of 25–35 nm [75]. By modeling the frequency dependence of the amplitude with a Lorentzian shape around the mechanical resonances, it was found that for nanotubes with high Young modulus the quality factor Q was very high, greater than 500. From (3.64) it can be seen that the mechanical vibration amplitudes at resonance show two regimes: the linear regime, in which the mechanical amplitude is a linear function of V_s, and the quadratic mode, in which the amplitude is proportional to V_0^2 but insensitive to the static voltage. By monitoring the shift of the CNT resonant frequency in the linear regime, it is possible to measure the mass of a particle adsorbed on its tip and kept there due to van der Waals forces (see Figure 3.48). This nanobalance is able to detect particle masses in the picogram-femtogram range.

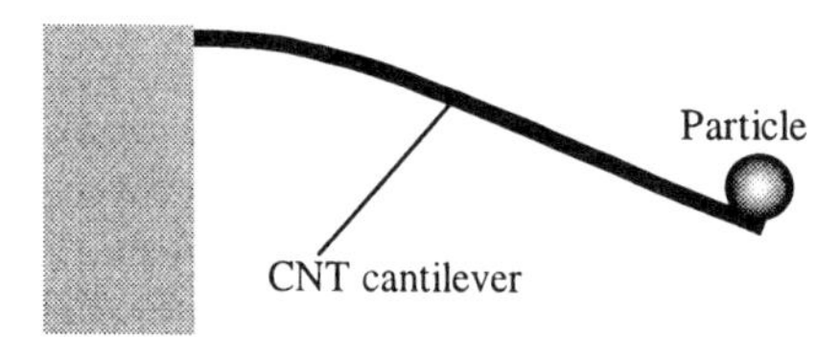

Figure 3.48 CNT nanobalance.

According to the simulations presented in [76], the sensitivity of the nanobalance in Figure 3.48 can attain 10^{-21} grams, but this value is overestimated since the influence of temperature, the environment (air or vacuum), and other factors was not included in [76]. The same precautions must be taken into consideration when assessing the claims in [77] regarding the performance of

CNT-based strain and pressure sensors. For example, the CNT in [77] has a fundamental mechanical frequency of 300–400 GHz, but the frequency shift due to axial strain does not exceed 5 kHz, while the maximum frequency shift due to an applied pressure of 1 MPa is about 120 kHz. Such small frequency shifts are almost impossible to detect with present millimeter-wave technologies. NEMS performances, which are indeed impressive, must be always correlated with the present capabilities of microwave and millimeter-wave instrumentation.

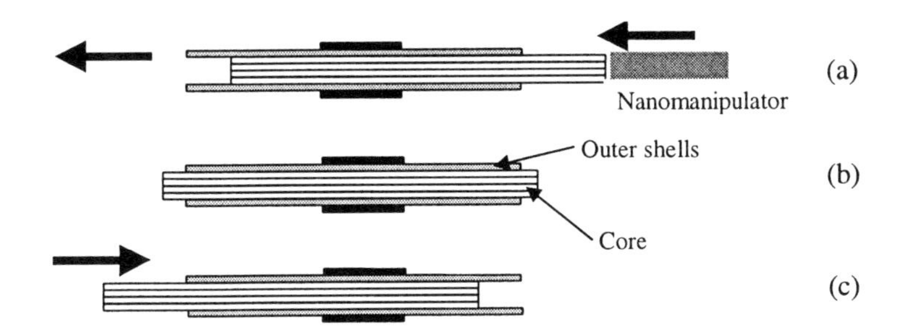

Figure 3.49 MWCNT-NEMS mechanical oscillator in the GHz range: (a) releasing of the core, (b) retraction of the core due to the van der Waals force, (c) reversed motion of the core (*After*: [80]).

Proposed CNT-NEMS switches and oscillators can oscillate mechanically in the GHz range using MWCNT with extruded cores. These proposals are based on the experimental observation that after peeling open one end of the nanotube, the inner tubes slide in and out with very low friction, behaving like a switch with a switching time in the range of ns [77]. If both ends of MWCNTs are peeled away the inner tubes oscillate back and forth, as indicated in Figure 3.49, with a mechanical frequency in the range 1–30 GHz depending on the nanotube diameter [78]. In both cases, the van der Waals force is the restoring force responsible for pulling back the inner tubes inside the outer tube.

In the shown MWCNT-NEMS switches and oscillators, the internal forces responsible for the spectacular telescoping of the inner shells are, mainly, the static and dynamic frictional forces and the van der Waals forces (F_{vdW}). The retraction force F_{vdW} is given by

$$F_{vdW} = -dU(x)/dx = C_1 , \quad (3.66)$$

where $U(x)$ is the excess of the van der Waals potential energy, x is the length of the overlap between the inner and outer walls of the MWCNT, and the constant

C_1 is 0.16C [78] with $C \cong 60$ nm the circumference of the nanotube bearing cylinders. If the initial extrusion is 330 nm, the complete retraction of inner walls is made in 4 ns, for $F_{\rm vdW}$ of 9 nN, which implies a very low friction force, of 2×10^{-14} N/atom or 4×10^{-14} N/nm. An alternative expression for C_1 is [79]

$$C_1 = -2.4 d_{\rm in} \varsigma \Delta E / a_{\rm C-C}^2 , \qquad (3.67)$$

where ς is a numerical factor that depends on the number of inner cores and outer shells of the MWCNT, and $-\Delta E$ is the energy difference between a carbon atom on the inner core shell and the atoms of the outer shells. In the mechanical oscillator presented in Figure 3.49 the core shells have kinetic energy in position (b), where the van der Waals energy is minimized, so the core passes through the point of minimal energy, reverses its motion direction, and is telescoped again. The frequency of this back and forth mechanical motion of the core is

$$f = (\beta / 4)(\varsigma \Delta E / 2 m_{\rm C} \eta L_{\rm c} \Delta L)^{1/2} , \qquad (3.68)$$

where $L_{\rm c}$ is the core length, ΔL is the initial extrusion, $m_{\rm C}$ is the mass of the carbon atom, and $\beta = (1-\delta)^{1/2}(1-\delta/2)$, with $\delta = | L_{\rm o} - L_{\rm c} | / 2\Delta L$ and $L_{\rm o}$ the length of the outer shells. The mechanical frequency of oscillation for an initial extrusion of 100 nm, with $\delta \le 0.48$ and $\Delta E \cong 10^{-20}$ J, is 1.2 GHz, a value that can be increased up to 120 GHz depending on the geometry of the MWCNT [80].

Rotational actuators based on CNTs were implemented by attaching a metal plate to an MWCNT anchored to two electrodes. By placing other two electrodes around the plate to induce rotation, the entire ensemble comprising the plate and the MWCNT can be turned around with an arbitrary angle in the range 0–360°, depending on the applied voltages [81].

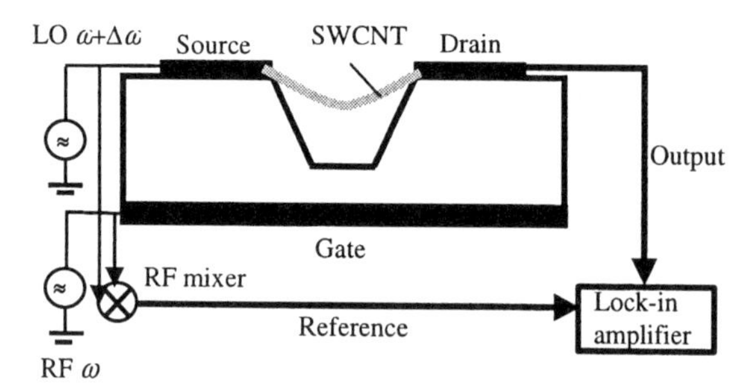

Figure 3.50 Schematic representation of the tunable SWCNT mechanical resonator (*After*: [82]).

A double clamped SWCNT patterned over a trenched gate geometry [82] was used as a mechanical resonator in an RF mixer configuration shown in Figure 3.50. The mechanical frequency of the resonator depends on the gate voltage according to $f_0 \propto V_G^{2/3}$ and thus can be tuned in a wide range by changing V_G. This resonator senses extraordinary low forces, of 1 $\text{fF} \times \text{Hz}^{-1}$, at room temperature.

This chapter has demonstrated a wealth of new concepts and devices based on CNTs. This trend will continue in Chapter 4, where the spintronic applications of the CNTs will be presented, and in Chapter 6, which will describe optoelectronic devices based on CNTs.

REFERENCES

[1] Avouris, Ph., et al., "Carbon nanotube electronics," *Proc. IEEE*, Vol. 91, No. 11, 2003, pp. 1772-1784.

[2] Dresselhaus, M.S. and Ph. Avouris, "Introduction to carbon material research," in *Carbon Nanotubes*, M.S. Dresselhaus, G. Dresselhaus, and Ph. Avouris (eds.), Berlin: Springer, Topics in Applied Physics 80, 2001, pp. 1-9.

[3] Hamada, N., S.-I. Sawada, and A. Oshiyama, "New one-dimensional conductors: graphitic microtubules," *Phys. Rev. Lett.*, Vol. 68, No. 10, 1992, pp. 1579-1581.

[4] Saito, R. and H. Kataura, "Optical properties and Raman spectroscopy of carbon nanotubes," in *Carbon Nanotubes*, M.S. Dresselhaus, G. Dresselhaus, and Ph. Avouris (eds.), Berlin: Springer, Topics in Applied Physics 80, 2001, pp. 213-247.

[5] Ciraci, S., et al., "Functionalized carbon nanotubes and device applications," *J. Phys.: Condens. Matter*, Vol. 16, No. 7, 2004, pp. R901-R960.

[6] Rakitin, A., C. Papadopoulos, and J.M. Xu, "Electronic properties of amorphous carbon nanotubes," *Phys. Rev. B*, Vol. 61, No. 8, 2000, pp. 5793-5796.

[7] Lammert, P.E., P. Zhang, and V.H. Crespi, "Gapping by squashing: metal-insulator and insulator-metal transitions in collapsed carbon nanotubes," *Phys. Rev. Lett.*, Vol. 84, No. 11, 2000, pp. 2453-2456.

[8] Wu, J., et al., "Computation design of carbon nanotube electromechanical pressure sensors," *Phys. Rev. B*, Vol. 69, No.15, 2004, pp. 153406/1-4.

[9] Minot, E.D., et al., "Tuning carbon nanotube band gaps with strain," *Phys. Rev. Lett.*, Vol. 90, No. 15, 2004, pp. 156401/1-4.

[10] Rakitin, A., C. Papadopoulos, and J.M. Xu, "Carbon nanotube self-doping: calculation of the hole carrier concentration," *Phys. Rev. B*, Vol. 67, No. 3, 2003, pp. 033411/1-4.

[11] Bockrath, M., et al., "Chemical doping of individual semiconducting carbon-nanotube ropes," *Phys. Rev. B*, Vol. 61, No. 16, 2000, pp. R10606-R10608.

[12] Zhou, C., et al., "Modulated chemical doping of individual carbon nanotubes," *Science*, Vol. 290, No. 5496, 2000, pp.1552-1555.

[13] Park, J. and P.L. McEuen, "Formation of a p-type quantum dot at the end of an n-type carbon nanotube," *Appl. Phys. Lett.*, Vol. 79, No. 11, 2001, pp. 1363-1365.

[14] Minot, E.D., et al., "Determination of electron orbital magnetic moments in carbon nanotubes," *Nature*, Vol. 428, No. 6892, 2004, pp. 536-529.

[15] O'Keeffe, J., C. Wei, and K. Cho, "Bandstructure modulation for carbon nanotubes in a uniform electric field," *Appl. Phys. Lett.*, Vol. 80, No. 4, 2002, pp. 676-678.

[16] Zhou, X., H. Chen, and O.-Y. Zhong-can, "Can electric field induce energy gaps in metallic carbon nanotubes?," *J. Phys.: Condens. Matter*, Vol. 13, No. 1, 2001, pp. L635-L640.

[17] Li, Y., S.V. Rotkin, and U. Ravaioli, "Electronic response and bandstructure modulation of carbon nanotubes in a transverse electric field," *Nano Lett.*, Vol. 3, No. 2, 2003, pp. 183-187.

[18] Kong, J., L. Kouwenhoven, and C. Dekker, "Quantum change for nanotubes," *Physics World*, No. 7, 2004, pp. 17-18.

[19] Dürkop, T., B.M. Kim, and M.S. Fuhrer, "Properties and applications of high-mobility semiconducting nanotubes," *J. Phys.: Condens. Matter*, Vol. 16, No. 18, 2004, pp. R553-R580.

[20] Pennington, G. and N. Goldsman, "Semiclassical transport and photon scattering of electrons in semiconducting carbon nanotubes," *Phys. Rev. B*, Vol. 68, No. 4, 2003, pp. 045426/1-11.

[21] Pennington, G. and N. Goldsman, "Electrical mobility of semiconducting carbon nanotubes," Int. Semiconductor Device Research Symposium, 2003, pp. 412-413.

[22] Kleiner, A., "Universal number of conducting channels in doped carbon nanotubes," *Phys. Rev. B*, Vol. 69, No. 8, 2004, pp. 081405(R)/1-2.

[23] McEuen, P.L., M.S. Fuhrer, and H. Park, "Single-walled carbon nanotube electronics," *IEEE Trans. Nanotechnology*, Vol. 1, No. 1, 2002, pp. 78-85.

[24] Javey, A., et al., "Ballistic carbon nanotube field-effect transistors," *Nature*, Vol. 424, No. 6949, 2003, pp. 654-657.

[25] Yaish, Y., et al., "Electrical nanoprobing of semiconducting carbon nanotubes using an atomic force microscope," *Phys. Rev. Lett.*, Vol. 92, No. 4, 2004, pp. 046401/1-4.

[26] Venema, L.C., et al., "Imaging electron wave functions of quantized energy levels in carbon nanotubes," *Science*, Vol. 283, No. 5398, 1999, pp. 52-55.

[27] Lemay, S.G., et al., "Two-dimensional imaging of electronic wavefunctions in carbon nanotubes," *Nature*, Vol. 412, No. 6847, 2001, pp. 617-620.

[28] Yao, Z., C. Dekker, and Ph. Avouris, "Electrical transport thorough single-wall carbon nanotubes," in *Carbon Nanotubes*, M.S. Dresselhaus, G. Dresselhaus, and Ph. Avouris (eds.), Berlin: Springer, Topics in Applied Physics 80, 2001, pp. 147-171.

[29] Forró, L. and C. Schönenberger, "Physical properties of multi-walled carbon nanotubes," in *Carbon Nanotubes*, pp. 329-391, M.S. Dresselhaus (eds.), Berlin: Springer, Topics in Applied Physics 80, 2001.

[30] Martel, R., et al., "Single- and multi-wall carbon nanotube field-effect transistors," *Appl. Phys. Lett.*, Vol. 73, No. 17, 1998, pp. 2447-2449.

[31] Tans, S.J., A.R.M. Verschueren, and C. Dekker, "Room-temperature transistor based on a single carbon nanotube," *Nature*, Vol. 393, No. 6680, 1998, pp. 49-51.

[32] Heinze, S., et al., "Carbon nanotubes as Schottky barrier transistors," *Phys. Rev. Lett.*, Vol. 89, No. 10, 2002, pp. 106801/1-4.

[33] Guo, J., S. Datta, and M. Lundstrom, "A numerical study of scaling issues for Schottky-barrier carbon nanotube transistors," *IEEE Trans. Electron Devices*, Vol. 51, No. 2, 2004, pp. 172-177.

[34] Javey, A., et al., "High-κ dielectrics for advanced carbon-nanotube transistors and logic gates," *Nature Mater.*, Vol. 1, No. 4, 2002, pp. 241-246.

[35] Raychowdhury, A., S. Mukhopadhyay, and K. Roy, "A circuit-compatible model of ballistic carbon nanotube field-effect transistors," *IEEE Trans. on Computer-Aided Design of Integrated Circuit and Systems*, Vol. 23, No. 10, 2004, pp. 1411-1420.

[36] Mintmire, J.W. and C.T. White, "Universal density of states for carbon nanotubes," *Phys. Rev. Lett.*, Vol. 81, No. 12, 1998, pp. 2506-2509.

[37] Choi, W.B., et al., "Ultrahigh-density nanotransistors by using selectively grown vertical carbon nanotubes," *Appl. Phys. Lett.*, Vol. 79, No. 22, 2001, pp. 3696-3698.

[38] Postma, H.W.Ch., et al., "Carbon nanotube single-transistors at room temperature," *Science*, Vol. 293, No. 5527, 2001, pp. 76-79.

[39] Bachtold, A., et al., "Logic circuits with carbon nanotube transistors," *Science*, Vol. 294, No. 5545, 2001, pp. 1317-1320.

[40] Raychowdhury, A. and K. Roy, "A novel multiple-valued logic design using ballistic carbon nanotube FETs," Proc. 34th International Symposium on Multiple-Valued Logic (ISMVL'04), 2004, pp. 1-6.

[41] Rueckes, T., et al., "Carbon nanotube-based nonvolatile random access memory for molecular computing," *Science*, Vol. 289, No. 5476, 2000, pp. 94-97.

[42] Cui, J.B., et al., "Carbon nanotube memory devices of high charge stability," *Appl. Phys. Lett.*, Vol. 81, No. 17, 2002, pp. 3260-3262.

[43] Choi, W.B., et al., "Carbon-nanotube-based nonvolatile memory with oxide-nitride-oxide film and nanoscale channel," *Appl. Phys. Lett.*, Vol. 82, No. 2, 2003, pp. 275-277.

[44] Zhu, W. (ed.), *Vacuum Microelectronics*, New York: John Wiley & Sons, 2001.

[45] Gröning, O., et al., "Field emission properties of carbon nanotubes," *J. Vac. Sci. Tehnol.*, Vol. B 18, No. 2, 2000, pp. 665-678.

[46] Saito, Y., et al., "Cathode ray tube lighting elements with carbon nanotube field emitters," *Jpn. J. Appl. Phys.*, Vol. 37, Part 2, No. 3B, 1998, pp. L346-L348.

[47] Choi, W.B., et al., "Fully-sealed, high-brightness carbon-nanotube field display," *Appl. Phys. Lett.*, Vol. 75, No. 20, 1999, pp. 3129-3131.

[48] Manohara, H.M., et al., "Fabrication and emitter measurement for a nanoklystron: a novel THz micro-tube source," Far IR, Submillimeter and Millimeter Detector Technology Workshop, Monterey, 2002.

[49] Bower, C., et al., "On-chip vacuum microtriode using carbon nanotube field emitters," *Appl. Phys. Lett.*, Vol. 80, No. 20, 2002, pp. 3820-3822.

[50] Hofmann, S., et al., "Direct growth of aligned carbon nanotube field emitter arrays onto plastic substrates," *Appl. Phys. Lett.*, Vol. 83, No. 22, 2003, pp. 4661-4663.

[51] Fuhrer, M.S., et al., "Crossed nanotube junctions," *Science*, Vol. 288, No. 5465, 2000, pp. 494-497.

[52] Papadopoulos, C., et al., "Electronic transport in Y-junction carbon nanotubes," *Phys. Rev. Lett.*, Vol. 85, No. 16, 2000, pp. 3476-3479.

[53] Postma, H.W.Ch., et al., "Electrical transport through carbon nanotube junctions created by mechanical manipulation," *Phys. Rev. B*, Vol. 62, No. 16, 2000, pp. R10653-R10656.

[54] Cassel, A.M., et al., "Vertically aligned carbon nanotube heterojunctions," *Appl. Phys. Lett.*, Vol. 85, No. 12, 2004, pp. 2364-2366.

[55] Odintsov, A.A., "Schottky barriers in carbon nanotube heterojunctions," *Phys. Rev. Lett.*, Vol. 85, No. 1, 2000, pp. 150-153.

[56] Liang, W., et al., "Fabry-Perot interference in a nanotube electron waveguide," *Nature*, Vol. 411, No. 6889, 2002, pp. 665-669.

[57] Gülseren, O., T. Yildirim and S. Ciraci, "Formation of quantum structures on a single nanotubes by modulating hydrogen adsorption," *Phys. Rev. B*, Vol. 68, No. 11, 2003, pp. 115419/1-6.

[58] Wei, Y., et al., "Carbon nanotube parametric electron pump: a molecular device," *Phys. Rev. B*, Vol. 64, No. 11, 2001, pp. 115321/1-4.

[59] Dragoman, D. and M. Dragoman, "Terahertz oscillations in semiconducting carbon nanotube resonant-tunneling diodes," *Physica E*, Vol. 24, No. 1, 2004, pp. 282-289.

[60] Ishibashi, K., et al., "Quantum dots in carbon nanotubes," *Jpn. J. Appl. Phys.*, Vol. 39, No. 12B, 2000, pp. 7053-7057.

[61] Suzuki, M., et al., "Carbon nanotubes as a building block of quantum dot devices," *Physica E*, Vol. 24, No. 1, 2004, pp. 10-13.

[62] Mason, N., M.J. Biercuk, and C.M. Marcus, "Local gate control of a carbon nanotube double quantum dot," *Science*, Vol. 303, No. 5638, 2004, pp. 655-658.

[63] Burke, P.J., "A RF circuit model for carbon nanotubes," *IEEE Trans. Nanotechnology*, Vol. 2, No. 1, 2003, pp. 55-58.

[64] Huang, S.M., et al., "Ultralong, well-aligned single-walled carbon nanotubes architectures on surfaces," *Adv. Mater.*, Vol. 15, No. 19, 2003, pp. 1651-1655.

[65] Burke, P.J., "AC performance of nanoelectronics: towards a ballistic THz nanotube transistor," *Solid-State Electronics*, Vol. 48, No. 5, 2004, pp. 1981-1986.

[66] Li, S., et al., "Carbon nanotube transistor operation at 2.6 GHz," *Nano Lett.*, Vol. 4, No. 4, 2004, pp. 753-756.

[67] Maksimenko, A.S. and G.Ya. Slepyan, "Negative differential conductivity in carbon nanotubes," *Phys. Rev. Lett.*, Vol. 84, No. 2, 2000, pp. 362-365.

[68] Léonard, F. and J. Tersoff, "Negative differential resistance in nanotube devices," *Phys. Rev. Lett.*, Vol. 85, No. 22, 2000, pp. 4767-4770.

[69] Dragoman, D. and M. Dragoman, "Terahertz continuous wave amplification in semiconductor carbon nanotubes," *Physica E*, Vol. 25, No. 4, 2005, pp. 492-496.

[70] Kroemer, H., "Detailed theory of the negative conductance of the bulk negative mobility amplifiers, in the limit of zero ion density," *IEEE Trans. Electron Devices*, Vol. 14, No. 9, 1967, pp. 476-491.

[71] Davies, J.F., et al., "High-Q mechanical resonator arrays based on carbon nanotubes," 3rd IEEE Conference on Nanotechnology, IEEE-NANO 2003, Vol. 2, 2003, pp. 635-638.

[72] Chopra, S., et al., "Development of RF carbon nanotube resonant circuit sensors for gas remote sensing applications," IEEE Microwave Theory and Techniques-Symposium Digest, 2002, pp. 639-642.

[73] Aihara, K., et al., "GHz carbon nanotube resonator bio-sensors," 3rd IEEE Conference of IEEE on Nanotechnology, Vol. 2, 2003, pp. 612-614.

[74] Ong, K.G., K. Zeng, and C.A. Grimes, "A wireless, passive carbon nanotube-based gas sensor," *IEEE Sensors Journal*, Vol. 2, No. 2, 2002, pp. 82-88.

[75] Poncharal, P., et al., "Electrostatic deflections and electromechanical resonances of carbon nanotubes," *Science*, Vol. 283, No. 5407, 1999, pp. 1513-1516.

[76] Li, C. and T.-W. Chou, "Mass detection using carbon nanotube-based nanomechanical resonators," *Appl. Phys. Lett.*, Vol. 84, No. 25, 2004, pp. 5246-5248.

[77] Li, C.-Y. and T.-W. Chou, "Strain and pressure sensing using single-walled carbon nanotubes," *Nanotechnology*, Vol. 15, No. 11, 2004, pp. 1493-1496.

[78] Cumings, J. and A. Zettl, "Low-friction nanoscale linear bearing realized from multiwall carbon nanotubes," *Science*, Vol. 289, No. 5479, 2000, pp. 602-604.

[79] Guo, W., et al., "Energy dissipation in gigahertz oscillators from multiwalled carbon nanotubes," *Phys. Rev. Lett.*, Vol. 91, No. 12, 2003, pp. 125501/1-4.

[80] Zheng, Q. and Q. Jiang, "Multiwalled carbon nanotube as gigahertz oscillators," *Phys. Rev. Lett.*, Vol. 88, No. 4, 2002, pp. 045503/1-4.

[81] Fennimore, A.M., et al., "Rotational actuators based on carbon nanotubes," *Nature*, Vol. 424, No. 6947, 2003, pp. 408-410.

[82] Sazonova, V., et al., "A tunable carbon nanotube electromechanical oscillator," *Nature*, Vol. 431, No. 7006, 2004, pp. 284-287.

Chapter 4

Spintronics

Spintronics is a generic name for spin-based electronic devices, which use the spin degree of freedom instead of the charge for information processing. The interest in spintronic devices is motivated by the anticipated lower power consumption and higher degree of functionality than in electronic devices. In addition, spintronic devices that rely on semiconducting materials can incorporate novel functionalities and are more adaptable, since the spin-dependent properties are, in many cases, controlled through electric fields. However, viable spintronic devices require efficient injection of spin-polarized electrons into nonmagnetic materials, minimization of spin dephasing, ability of spin control during transport, and, finally, efficient spin detection. Although some of these requirements have been successfully demonstrated, spintronic devices still await an experimental confirmation of the predicted theoretical expectations.

Because spin polarization arises naturally in ferromagnetic (FM) materials, the first spin-based devices were all-metallic structures, which consisted of layers of magnetic and nonmagnetic materials that form magnetic tunnel junctions and display hysteresis and giant magnetoresistance (GMR). In the GMR effect, the resistance changes from a small to a large value when the magnetizations of the FM layers change from parallel to antiparallel. These devices function as GMR sensors, circuit isolators, read heads, or magnetoresistive memory cells used as nonvolatile magnetic random access memories, which store trillions of bits defined as magnetization directions established by an external magnetic field (see overviews in [1–5]). The devices based on magnetoresistance effects and on the manipulation of ensemble magnetizations imposed the field of magnetoelectronics. Modern spintronic devices, on the other hand, are based on semiconductor materials, in which spin-dependent transport properties can be altered also by electric fields and in which observable effects can be generated by the manipulation of only a small number of spins or even of one spin in quantum computing devices. (FM ordering can also be switched on and off by electric fields via magnetoelectric interactions in some nonsemiconducting materials, but this property is uncommon, and the materials in which it appears, such as hexagonal $HoMnO_3$ [6] or $Pr_{1-x}Ca_xMnO_3$ manganites [7], have not yet shown relevant applications in magnetoelectronics.) Moreover, hybrid metal/semiconductor structures can display unique properties, such as the extremely high room temperature magnetoresistance change of 320, 000% at a magnetic field of $H = 2$ kOe displayed by granular films containing nanoscale FM metal clusters of MnSb grown on a GaAs substrate by MBE, which is caused by the magnetic-field dependence of avalanche breakdown at the MnSb/GaAs interface [8].

Because of recent interest, we focus in this chapter on semiconductor-based spintronic devices that, unfortunately, in most cases, are still at the proposal stage. In addition to these, several spintronic device proposals based on organic semiconductors, organic ferromagnets, or carbon nanotubes (CNT) wait for a successful implementation. The technology and materials aspects of spintronic devices based on semiconductors are outlined in [9,10], while the physical principles of spin generation, manipulation, dynamics, and detection are reviewed in [1]. A recent book that provides an overview of semiconductor spintronics, including the quantum computation applications, is [11]. Among the reviews on optical, electrical and magnetic manipulation of spins we mention [12] and [13].

4.1 PHYSICAL PRINCIPLES OF SPINTRONIC DEVICES

The applications of modern spintronic devices depend on the ability to generate, control, and detect the spin polarization of charge carriers. The coupling between the electron spin degree of freedom s, with magnetic moment $-g\mu_B s$ (g is the Landé factor and μ_B the Bohr magneton) and the charge transport appears at both the microscopic level, through a spin-orbit term in the Hamiltonian that expresses the dependence of electron dynamics on spin orientation, and at the level of statistical ensembles of electrons. In the latter case, the Pauli principle limits the occupation of states to single spins and thus imposes spin-dependent transport properties of electrons due to the different density of states, Fermi velocities, scattering times, and so on, for electrons with different spin orientations.

Spin injection in a nonmagnetic material is, in most cases, achieved by the creation of a nonequilibrium spin population (called spin accumulation) at the interface with a magnetic electrode. In this case, the rate of spin injection depends on the spin relaxation and dephasing mechanisms in the nonmagnetic material, which tend to restore the equilibrium in the accumulated spin population; the relatively long lifetimes of nonequilibrium electronic spins in semiconductors and metals, of about 1 ns, are essential for spintronic devices [1]. The spin lifetimes can however increase to hundreds of ns in confined semiconductor heterostructures [14], which imply transport of coherent spin packets over hundreds of μm.

The spin relaxation time T_1 (also called longitudinal or spin-lattice time) and the spin dephasing time T_2 (called transverse or decoherence time) are defined and calculated in the theory of magnetization dynamics. In an ensemble of mobile electrons the spin population and the lattice reach a thermal equilibrium in T_1, while in T_2 an ensemble of transverse electron spins, with precessing phases about the longitudinal field, lose their phase due to temporal and spatial fluctuations of the precessing frequencies. In isotropic and cubic solids $T_1 = T_2$ for magnetic fields up to several tesla, this common value being denoted τ_s (τ_s is often used for either spin relaxation or dephasing as a relevant characteristic time in spintronic applications), while in anisotropic systems, in general, $T_2 \leq 2T_1$ [1]. τ_s is determined, for example, from the transverse magnetic field dependence of the circularly polarized luminescence, which results from the recombination of spin-polarized electrons and holes.

Spin polarization of electrons in several semiconductors as a function of time is recorded by time-resolved Faraday rotation, a pump-probe spectroscopy technique, in which the optical selection rules impose that a circularly polarized pump pulse excites preferentially electrons with spins polarized along the beam direction. The longitudinal spin lifetime T_1 is obtained by fitting with $\theta \propto \exp(-\Delta t / T_1)$ the time dependence at zero magnetic field of the polarization rotation θ of a transmitted linearly polarized probe beam delayed with Δt. The transverse spin lifetime T_2 is determined from measurements of time-dependent polarization rotation of the probe beam $\theta \propto \exp(-\Delta t / T_2)\cos(\omega_L \Delta t / 2\pi)$ in the presence of a magnetic field B; $\omega_L = g\mu_B B / \hbar$ is the Larmor frequency of electron spin precession in the plane perpendicular to B [13]. When measurements are performed in reflection instead of transmission geometry the technique is known as time-resolved Kerr rotation [13]. Unlike spin polarization of holes, which relaxes very quickly (in less than 1 ps) due to the strong mixing of the valence bands, the nonequilibrium spin polarization of the electrons generated by excitation with circularly polarized light has a much slower decay time and lasts even after the excited carriers relax, the recombination of spin-up and spin-down electrons being equally probable.

Electron spin coherence times in lightly *n*-doped semiconductors are typically longer by several orders of magnitude than carrier lifetimes and can exceed 100 ns in optimally doped GaAs (with electron density $n = 10^{15}$ cm^{-1}); values up to 300 ns at liquid-helium temperature were measured from photoluminescence data of GaAs-based heterostructures optically excited with circularly polarized light [14]. Spin coherence depends on crystalline orientation; for example, the lack of D'yakonov-Perel' relaxation mechanism (see Section 4.1.1) in GaAs/AlGaAs (110) quantum wells enhances the spin coherent time with an order of magnitude compared to (100) quantum wells [15].

Note that besides spin polarization due to optical or electrical excitation, spontaneous spin polarization in less than 50 ps occurs in semiconductors in close proximity with an FM layer, electron spins being polarized along the direction of magnetization. This FM proximity polarization effect is observed when the incident light is linearly polarized and its magnitude is about 1/3 of the polariza-tion induced by circular polarized light.

In spin-based quantum computing devices, the relevant time constant is the single-spin decoherence time (or single-spin correlation time) τ_{sc}; this time should be sufficiently long in fault tolerant algorithms to perform 10^4–10^6 gate operations. In quantum computing it is important to maintain, as long as possible, the coherent superposition of spin states, allowing them at the same time to interact. This requirement is satisfied for spin systems well isolated from the environment. An equally important condition in quantum computing is to move spin packets with minimal decoherence. Electric field transport of coherently precessing electrons is possible over hundreds of μm and across interfaces of dissimilar materials, such as GaAs/ZnSe (see references in [13]). Coherence is not that important in spintronic devices, where the transport of polarization (coherent or not) with minimal loss is crucial. The relation between τ_s and τ_{sc} is complex and depends on many factors.

4.1.1 Spin Relaxation Mechanisms

There are several relevant mechanisms for electron spin relaxation and dephasing in metals and semiconductors, among which the most important are the Bir-Aronov-Pikus, the hyperfine interaction, the Elliott-Yafet, and the D'yakonov-Perel' mechanisms [1]. All of these can also be employed for controlling the spin; D'yakonov-Perel' is the most widely used mechanism in this sense since it requires electrical fields to tune the strength of the spin-orbit interaction.

The Bir-Aronov-Pikus relaxation mechanisms are significant in *p*-doped semiconductors, where fluctuating local magnetic fields that flip electron spins are caused by electron-hole exchange interaction governed by the Hamiltonian

$$H_{\mathrm{BAP}} = A\boldsymbol{S} \cdot \boldsymbol{J}\delta(\boldsymbol{r})\,. \tag{4.1}$$

Here A is proportional to the exchange integral between conduction and valence states, $\boldsymbol{S}$ is the electron spin operator, $\boldsymbol{J}$ the angular momentum operator for holes, and $\boldsymbol{r}$ the relative position of electrons and holes.

The hyperfine interaction mechanism is important in quantum heterostructures (quantum wells and dots) based on semiconductors with a nuclear magnetic moment. The associated Hamiltonian term is

$$H_{\mathrm{hyp}} = (2/3)\mu_0 g_0 \mu_{\mathrm{B}} \Sigma_i \hbar\gamma_i \boldsymbol{S} \cdot \boldsymbol{I}_i \delta(\boldsymbol{r} - \boldsymbol{R}_i)\,, \tag{4.2}$$

where μ_0 is the vacuum permeability, $g_0 \cong 2$ is the free-electron g factor, γ_i is the nuclear gyromagnetic ratio of the nuclei at positions $\boldsymbol{R}_i$, and $\boldsymbol{S}$ and $\boldsymbol{I}_i$ are the electron and nuclear spin operators, respectively, expressed in units of $\hbar$. The hyperfine interaction induces spin dephasing of localized electrons in the presence of nuclear dipole–dipole interactions when there are strong orbital or spin correlations between neighboring electron states, or when the orbit and spin correlation between separated electron states and nuclear spin states is small. The hyperfine interaction is not very effective for spin relaxation of free electrons in metals or bulk semiconductors, but becomes essential in spin-based computing devices as a means of coupling nuclear and electron spins in a controllable way.

In the Elliott-Yafet mechanism spin relaxation via momentum scattering (by phonons at high temperatures or impurities at low temperatures) is caused by the admixture of opposite-spin states induced by spin-orbit coupling. The Hamiltonian of this mechanism, which couple, electron states in different subbands with the same wave vector $\boldsymbol{k}$ but opposite spins, is

$$H_{\mathrm{EY}} = (\hbar / 4m^2c^2)\nabla V \times \boldsymbol{p} \cdot \boldsymbol{\sigma}\,, \tag{4.3}$$

where V is the spin-independent scalar periodic lattice potential, $\boldsymbol{p} = -i\hbar\nabla$ is the momentum operator and $\boldsymbol{\sigma}$ are the Pauli matrices. The Elliott-Yafet mechanism is significant in small-gap semiconductors with large spin-orbit splitting, like InSb. The spin-orbit coupling H_{EY} leads to coupling of spin-up and spin-down states (and hence to spin relaxation) only in combination with momentum scattering.

In the D'yakonov-Perel' mechanism, spin dephasing occurs due to an effective magnetic field that appears in solids with no (intrinsic or induced by asymmetric confining potentials, for example) center of symmetry in the presence of spin-orbit interaction. This $\boldsymbol{k}$-dependent effective magnetic field $\boldsymbol{B}_{\mathrm{eff}}(\boldsymbol{k})$ is randomized when electrons scatter to different momentum states. Electron spin precession with Larmor frequency $\boldsymbol{\Omega}(\boldsymbol{k}) = (e/m)\boldsymbol{B}_{\mathrm{eff}}(\boldsymbol{k})$ around $\boldsymbol{B}_{\mathrm{eff}}(\boldsymbol{k})$ is described by the Hamiltonian

$$H_{\mathrm{DP}}(\boldsymbol{k}) = (1/2)\hbar\boldsymbol{\sigma} \cdot \boldsymbol{\Omega}(\boldsymbol{k})\,, \tag{4.4}$$

where $\boldsymbol{\sigma}$ are the Pauli matrices. H_{DP} produces spin-splitting of the Fermi surface for spins parallel or antiparallel to $\boldsymbol{\Omega}(\boldsymbol{k})$. The spin-orbit interaction described by (4.4) leads to spin dephasing only in combination with momentum scattering; otherwise it can be used to control the orientation of electron spins. While in the Elliott-Yafet mechanisms the precession frequency of electrons is preserved between collisions and the loss of phase occurs during collision (such that the associated τ_{s} is proportional to the momentum relaxation time τ_{p}), in the D'yakonov-Perel' dephasing mechanism the phases of spins are randomized between collisions since electrons with different momenta precess with different frequencies (τ_{s} is inversely proportional to τ_{p}). In addition, the spin diffusion length $L_s = (D\tau_s)^{1/2}$ depends on momentum scattering as $L_s \propto \tau_p$ in the Elliott-Yafet mechanism, since $D \propto \tau_p$, while in the D'yakonov-Perel' mechanism it does not depend on τ_p and should be independent of temperature in degenerate electron systems. In a 2DEG, for instance, the spin diffusion lengths for these cases are $L_{s,EY} = L_e / (2^{1/2}\hbar\alpha_{EY}k_F^2)$ and $L_{s,DP} = \hbar^2\pi / (2m\alpha_{DP})$, respectively (see [16] and the references therein), where L_e is the elastic mean free path, k_F is the Fermi momentum, and α_{EY}, α_{DP} are the respective spin-orbit coupling coefficients. $L_{s,EY}$ depends on L_e (on impurities), while $L_{s,DP}$ is controlled only by the structural asymmetry. The importance of the D'yakonov-Perel' mechanism increases as the bandgap and temperature increase. For example, the Bir-Aronov-Pikus mechanism dominates in heavily *p*-doped samples at low temperatures, whereas the D'yakonov-Perel' mechanism can become more important at high temperatures even for large acceptor densities.

In zinc-blende semiconductors with an intrinsic lack of inversion symmetry the D'yakonov-Perel' mechanism is also known as Dresselhaus, or bulk inversion asymmetry, and manifests in a k^3 spin splitting of the conduction band for small electron wave numbers *k*, but generates a splitting linear in *k* in 2D systems. In structural asymmetric systems the D'yakonov-Perel' mechanism is known as Rashba spin splitting, for which the in-plane precession vector is parallel to $\boldsymbol{\Omega}(\boldsymbol{k}) = 2\alpha_R(\boldsymbol{k} \times \boldsymbol{n})$ with $\boldsymbol{n}$ a unit vector along the confinement direction and α_R the Rashba coefficient. Large relaxation times for spins oriented along $\boldsymbol{\Omega}(\boldsymbol{k})$ are expected in quasi-1D channels. In fact, an applied electric field in (110)-oriented GaAs/AlGaAs quantum wells can increase tenfold the spin relaxation rate; room temperature spin-relaxation times up to 10 ns, along with high electron mobility, could be achieved through the manipulation of conduction band spin splitting via the Rashba effect [17].

The Rashba effect [18], which introduces an energy splitting $\Delta E = 2\alpha_R |k|$ between spin-up and spin-down electrons, even in the absence of a magnetic field, originates in the perpendicular electric field that appears at heterojunction interfaces, determined by ∇V with *V* the potential energy. The net electric field in a layer is a sum of the fields at the two interfaces, which are generally not equal. The Rashba splitting vanishes at the zone center, for zero in-plane wave vector components $\boldsymbol{k}_{\parallel}$. States with $\boldsymbol{k}_{\parallel}$ and $-\boldsymbol{k}_{\parallel}$ in a spin-split subband have opposite spins, the ensemble of transmitted electrons in local thermal equilibrium having no net spin polarization unless a lateral electric field (perpendicular to the growth direction) shifts the incident electron distribution towards one direction in the $\boldsymbol{k}_{\parallel}$ space. The degree of spin polarization of the current increases with the spin splitting between the two subbands. The Rashba effect is particularly strong in narrow-gap semiconductor heterostructures, for example, in InGaAs-based heterostructures, but also in wurtzite GaN/AlGaN heterostructures [19]. Because transition-metal-doped GaN shows, in addition, room temperature ferromagnetism, wide bandgap GaN-based semiconductors are becoming one of the most interesting materials for spintronic applications [20].

The Dresselhaus and Rashba effects can be of comparable magnitude. For example, in zinc-blende 2DEG [001]-grown semiconductor heterostructures lying in the (x,y) plane the total Hamiltonian that accounts for both bulk inversion asymmetry and structural inversion asymmetry is given by [21]

$$H = H_0 + \alpha_R(\sigma_x k_y - \sigma_y k_x) + \alpha_D(\sigma_x k_x - \sigma_y k_y)\,. \tag{4.5}$$

Here H_0 is the spin degenerate Hamiltonian, σ_i are the Pauli matrices, and k_i are the in-plane components of the electron wave vector $\boldsymbol{k}$. The strength of the Dresselhaus spin-orbit interaction α_D is determined by the parameters of the constituent materials and the thickness of the 2DEG, whereas α_R can be tuned by applying an electric field normal to the 2DEG plane.

4.1.2 Spin Injection

Spin injectors provide a spin-polarized electron population, which is generally a nonequilibrium spin population. Initially spin polarization was achieved by illumination with circularly polarized photons, which transfer angular momenta to electrons (see [1] and the references therein) and thus excite preferentially electrons (and holes) with particular spin orientations due to optical selection rules. However, electrical spin injection from a magnetic semiconductor or metal electrode with a spin-split density of states is more desirable. Several spin injector devices based on different physical principles and involving different heterojunctions (for example, FM metal/superconductor metal, which rely on the Andreev reflection, FM/normal metals, which exhibit GMR, FM semiconductor/semiconductor) have been proposed [1]. The first successful spin injection experiment was performed in an all-metal device consisting of permalloy spin injector and detector pads deposited at different distances on a "bulk wire" of aluminum [22]. The spin diffusion length in aluminum can be measured by varying the position of the FM detector pad (see Section 4.1.3). A review of the theoretical models and experimental achievements of spin injection into semiconductors can be found in [23].

Spin injection from an FM contact into a 2DEG was predicted to generate an unexpectedly low spin-polarization efficiency, of less than 0.1 [24]. If spin scattering occurs much more slowly than electron scattering this counterintuitive situation can be modeled by defining two (generally different) electrochemical potentials $\mu_\uparrow$ and $\mu_\downarrow$ for the opposite spin directions at any point of a device, which consists of a 2DEG of length L patterned with two identical FM contacts. In the diffusion theory of spin injection, these potentials are connected to the 1D current along x via the spin-dependent conductivities $\sigma_{\uparrow,\downarrow}$, the weighted average of the diffusion constants for both spin directions D, and the spin-flip time constant τ_{sf} through

$$\frac{\partial \mu_{\uparrow,\downarrow}}{\partial x} = -\frac{ej_{\uparrow,\downarrow}}{\sigma_{\uparrow,\downarrow}}, \qquad \frac{\mu_{\uparrow} - \mu_{\downarrow}}{\tau_{sf}} = D\frac{\partial^2(\mu_{\uparrow} - \mu_{\downarrow})}{\partial x^2}. \tag{4.6}$$

The current densities $j_{\uparrow,\downarrow}$ and electrochemical potentials $\mu_{\uparrow,\downarrow}$ are continuous for perfect ohmic interfaces (without spin scattering or interface resistance), $(\mu_{\uparrow} - \mu_{\downarrow})$ decaying exponentially from the interface over a spin-flip length $(D\tau_{sf})^{1/2}$. In semiconductors the spin-flip length L_{sc} is several orders of magnitude larger than the value L_{fm} in FM materials and hence it is taken as $L_{sc} \to \infty$. The spatial dependence of the electrochemical potentials is obtained from (4.6) using symmetry considerations and the assumption that there is no spin-flip in the 2DEG, which imply $\mu_{\uparrow}(0) - \mu_{\downarrow}(0) = \pm[\mu_{\uparrow}(L) - \mu_{\downarrow}(L)]$, the + sign referring to parallel magnetizations of the FM contacts, while the – sign refers to antiparallel magnetizations; this spatial dependence is depicted in Figure 4.1.

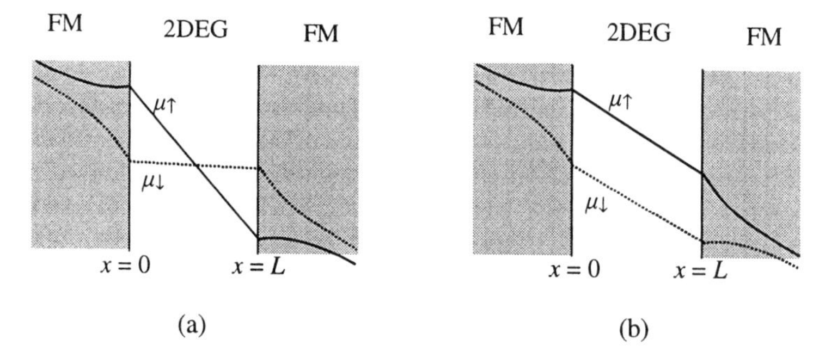

Figure 4.1 Spatial dependence of the electrochemical potentials for the opposite directions when the magnetizations of the FM contacts are (a) parallel and (b) antiparallel (*After*: [24]).

As seen from Figure 4.1(a), when the identical FM contacts have parallel magnetizations, the electrochemical potentials cross in the middle of the 2DEG and the different voltage drop for the two spin directions over the 2DEG leads to a spin-polarized current. The spin-polarization of the injected current, defined at an x = const. plane as $\eta(x) = [j_{\uparrow}(x) - j_{\downarrow}(x)]/[j_{\uparrow}(x) + j_{\downarrow}(x)]$, is

$$\eta(0) = \beta \frac{L_{\text{fm}}}{\sigma_{\text{fm}}} \frac{\sigma_{\text{sc}}}{L} \frac{2}{(1 + 2L_{\text{fm}}\sigma_{\text{sc}} / L\sigma_{\text{fm}}) - \beta^2}, \quad (4.7)$$

where β is the magnetization of the contacts. On the contrary, for antiparallel magnetization of the FM contacts the current is unpolarized; this case is illustrated in Figure 4.1(b). The relative difference in resistance between antiparallel and parallel configurations $(R_{\uparrow\downarrow} - R_{\uparrow\uparrow}) / R_{\uparrow\uparrow}$ takes the maximum value $\beta^2 /(\beta^2 - 1)$.

The spin injection efficiency η in (4.7) cannot exceed β and is dominated by the mismatch at the ohmic contact between the relevant part of the FM resistance $R_{\text{fm}} = L_{\text{fm}} / \sigma_{\text{fm}}$ and the resistance of the semiconductor $R_{\text{sc}} = L / \sigma_{\text{sc}}$, characterized by the ratio $(L_{\text{fm}} / \sigma_{\text{fm}}) /(L / \sigma_{\text{sc}})$. Typically, for diffusive spin injection from an FM metal into a semiconductor (in general, a normal conductor) $\eta \approx R_{\text{fm}} / R_{\text{sc}} << 1$, due especially to the low value of L_{fm}; characteristic values are $\eta < 0.1\%$. (Spin injection from an FM into a paramagnetic metal is achieved with spin injection efficiency $\eta \approx R_{\text{fm}} / R_{\text{para}} \cong 1$.)

Large values of η can be obtained only for $\beta \cong 1$, magnetization value encountered in semimagnetic semiconductors (or DMS, for diluted magnetic semiconductors) with large Zeeman splitting. DMS, in which magnetic elements substitute a small fraction of host elements in a semiconductor, can turn into ferromagnets when doped [25,26]. The enhancement of spin injection efficiency from a magnetic into a nonmagnetic semiconductor is also justified by the comparable resistivity values in the two materials. Efficient spin injector materials are Cr- and Eu-based chalcogenide FM semiconductors, or more recently (II,Mn)VI paramagnetic semiconductors [27] such as CdMnTe, BeMnZnSe,and ZnMnSe, which at low Mn concentrations have a giant Zeeman splitting $\Delta E = g^* \mu_B H$ caused by the *sp-d* exchange between the spins of conduction electrons and the spins of localized Mn^{2+} ions. Here $g^* = g + \alpha M /(g_{\text{Mn}} \mu_B^2 H)$, with g the effective electron Landé factor at $H = 0$, M the magnetization and α the exchange integral. In addition, (II,Mn)VI compounds can be integrated with optically active AIII-BV nonmagnetic semiconductors that generate circularly polarized luminescence under electric excitation, which can be used as spin detection method. Recent experiments show that FM metals such as MnAs (MnAs is ferromagnetic at room temperature) can be also epitaxially grown by the MBE method on AIII-BV semiconductor substrates such as GaAs(001), GaAs(111)B, and even on Si(001), forming integrated FM/semiconductor hybrid structures [28].

Experimentally, 50% spin injection efficiency from a non-lattice-matched II-VI $Zn_{1-x}Mn_xSe$ semimagnetic semiconductor into a III-V GaAs-based quantum well light-emitting diode (LED) was reported at 4.2 K [29]; this value was extracted from measurements of the degree of polarization P of the circularly polarized light emitted by the LED when excited by spin-polarized electrons. Selection rules imply that P is related to the spin injection efficiency as $P = -\eta / 2$

for bulk GaAs and as $P = -\eta$ for GaAs quantum wells in which the degeneracy between light and heavy holes in the valence band is removed. Low-temperature 90% spin-injection efficiency was demonstrated in a GaAs/AlGaAs LED from a lattice matched *n*-type $Be_xMn_yZn_{1-x-y}Se$ spin aligner, the giant *g*-factor of which produces large Zeeman splitting in an applied magnetic field that aligns the injected carriers to the energetically favorable lower Zeeman level [30]. (The *g* factor is up to 100 for low Mn concentrations, but decreases significantly in this material with increasing temperature.) The polarization degree of the injected electrons depends on the thickness of the spin aligner layer, which must be sufficiently large to completely relax all spins into the lower Zeeman level; the 90% value was obtained for a thickness of 300 nm. 70% spin injection efficiency at 2 K from a semimagnetic BeMnZnSe spin aligner into a single self-assembled CdSe/ZnSe quantum dot was reported in [31]; the DMS spin aligner polarizes electron-hole pairs that are created optically by illumination with an Ar^+ laser, the spin injection of the carriers transferred into the ground state of the quantum dot being confirmed by the polarization degree of the luminescence of the quantum dot exciton.

Another class of efficient spin injector materials are the (III,Mn)V FM semiconductors, such as (In,Mn)As, with a transition temperature of 30 K, or (Ga,Mn)As. In these materials, however, the magnetic ion often acts as an acceptor impurity, providing simultaneously a hole and a localized magnetic moment. Thus, the magnetic interaction is mediated by holes and the main carriers are generally spin-polarized holes, which lose their polarization very fast after injection into nonmagnetic semiconductors due to spin-orbit coupling. A non-negligible technological advantage of (III,Mn)V FM semiconductors is that they can be epitaxially grown within nonmagnetic semiconductor heterostructures. 80% electron spin injection efficiency in GaAs from a (Ga,Mn)As spin injector was obtained in a Zener diode structure, detected through an oblique Hanle effect (see reference in [1]).

Spin injection efficiency from FM contacts can be substantially improved if nonohmic contacts are used. In particular, in [32] it was demonstrated that spin-selective tunnel contacts, with different conductivities $\sigma_\uparrow$ and $\sigma_\downarrow$ for spin-up and spin-down electrons, placed between FM metals and normal conductors significantly enhance spin injection and solve the problem of conductivity mismatch if the tunnel resistance $R_c \geq L_{fm} / \sigma_{fm}, \min(L_n, W) / \sigma_n$. Here the effective contact resistance is $R_c = (\sigma_\uparrow + \sigma_\downarrow) / 4\sigma_\uparrow\sigma_\downarrow$, L_n and σ_n are the spin diffusion length and conductivity of the normal metal, respectively, and W is the width of the normal conductor; the spin injection coefficient in an FM/tunnel/normal metal junction is controlled by the element with the largest effective resistance. Examples of tunnel contacts are STM tips in vacuum or air, resonant tunneling structures and Schottky barriers. Moreover, spin injection into a ballistic semicon-ductor through a diffusive interface [33] is characterized by the efficiency $\eta \approx R_{fm} / R_{sc}^{*}$ rather than the expression $\eta \approx R_{fm} / R_{sc}$ obtained in the diffusion theory, with $R_{sc}^{*} = (h/e^2)(\pi / k_F^2)$ the Sharvin resistance of the semiconductor, which is determined by the electron concentration (πk_F^2 is the Fermi-surface cross-section) and by the resistance quantum h/e^2. In this case, again, spin-selective contacts, such as tunnel or Schottky barriers, with large resistance $R_c \geq R_{fm}, R_{sc}^{*}$ increase the efficiency of spin injection.

In one of the earliest spin injection experiments into semiconductors a magnetized FM metal STM Ni tip was used to inject electrons into a GaAs(110) surface across vacuum [34]. The negative spin polarization associated with the magnetic state of the tip $\eta = (\eta^{+} - \eta^{-})/2$, which indicates that minority spin electrons dominate the tunneling current, was extracted from the degrees of circular polarization P^{+} and P^{-} of the recombination luminescence in GaAs for both polarities of the magnetizing electromagnet as $\eta = (P^{+} - P^{-})/(2\rho_r \cos\theta)$. Here $\eta^{\pm} = (n_{\uparrow}^{\pm} - n_{\downarrow}^{\pm})/(n_{\uparrow}^{\pm} + n_{\downarrow}^{\pm})$ with $n_{\uparrow,\downarrow}^{\pm}$ the number of majority and minority electrons in the ferromagnet for both polarities of the electromagnet, θ is the angle between the spin polarization vector of electrons and the direction of light propagation towards the detector, and $\rho_r = 0.5/(1 + \tau / \tau_s)$ is the spin-polarization detection sensitivity, with τ the electron lifetime at the bottom of the conduction band. The efficiency η is about -30%, and is maximum at very low values of electron injection energies. If a STM Ni tip injects spin polarized electrons into an $Al_{0.06}Ga_{0.94}As$(110) surface the spin polarization across the vacuum potential barrier between the tip and the semiconductor surface increases from 20% to about 50% as the tunneling current through the barrier varies between 20 pA and 500 pA [35]. This increase of spin polarization is explained by the decrease of the barrier thickness with 0.18 nm and a corresponding increase of the contribution to the tunneling current of the localized and highly polarized 3*d*-like states in the tip with respect to the delocalized and low-polarized 4*sp*-like states.

Room-temperature spin injection from Fe (also a FM metal) into the GaAs layer of a GaAs/$In_xGa_{1-x}As$ quantum well LED, detected through the degree of circular polarization of electroluminescence from the LED, was observed in [36] with an efficiency of 2%. This efficiency value, which is higher than the 0.1% limit imposed in [24], is due to the Schottky-type contacts that form at the Fe/GaAs interface, the electron tunneling process through the Schottky barrier being independent of temperature and unaffected by resistance mismatch. The room temperature spin injection efficiency can be increased up to 30% if efficient tunnel barriers are created at the interface between epitaxial Fe films and the AlGaAs layer of a AlGaAs/GaAs quantum well LED [37]. A review on spin injection into GaAs from FM metals Fe and MnAs (with a Curie temperature of 40°C) can be found in [38].

A detailed theoretical analysis of electron spin injection through tunneling at a Schottky contact between a spin-polarized electrode and a nonmagnetic semiconductor [39] shows that the electron density in the nonmagnetic region is essential for efficient spin injection: the electron spin population in the nonmagnetic semiconductor changes only slightly even for highly spin-polarized injected current if the electron density in the nonmagnetic semiconductor is high or the injection current is low. Moreover, spin injection is impeded by a significant depletion region at the Schottky contact, but an appropriate doping profile in the semiconductor (with a heavily doped region near the interface and a small potential drop in the depletion region) maximizes spin injection. For a review on spin injection from magnetic semiconductors and FM metals into nonmagnetic semiconductors through a tailored Schottky barrier, see [40].

Spin injection across an FM/2DEG interface can also be observed by measuring the change in interface resistance either at the reversal of the magnetization orientation of the ferromagnet for a given polarization of 2DEG carriers or at the reversal of the polarization of 2DEG carriers for a fixed magnetization [41]. In this case, the spin-polarized current in the ferromagnet is first projected onto the spin-split density of states of the 2DEG, the gradient of the confining potential playing the role of an effective magnetic field that induces a Zeeman splitting of the conduction band into spin subbands through spin-orbit coupling. The relative difference of the interface resistances is given by $\beta\eta/2$, where $\eta = 2(G_{s\uparrow} - G_{s\downarrow})/(G_{s\uparrow} + G_{s\downarrow})$ is the net spin polarization in the 2DEG, in which the conductance of each spin subband $G_{s\uparrow,\downarrow}$ is proportional to the carrier density, and $\beta = 2(G_{F\uparrow} - G_{F\downarrow})/(G_{F\uparrow} + G_{F\downarrow})$ is the net spin polarization in the FM contact, proportional to the respective density of states near the Fermi level. In Permalloy/InAs systems $\eta = 10\%$, $\beta = 40\%$, the relative 0.91% change in interface resistance being independent of temperature but depending on the relative orientation between the magnetization of the FM contacts and the direction of the equivalent magnetic field in the 2DEG.

The ac theory of diffusive spin injection from an FM contact into a normal conductor through a tunnel or Schottky contact is developed in [42] based on frequency-dependent complex impedances. These express the diffusion and relaxation of nonequilibrium spins and are controlled by the spin relaxation rates and the resistances. A drift-diffusion equation for spin polarization shows that electric field effects that dominate carrier motion in nondegenerate semiconductors can enhance spin injection from a ferromagnet into a nonmagnetic semiconductor in the high-field regime [43], this spin injection being further increased by the presence of interfacial barriers. The efficiency η is enhanced since, unlike in the low-field injection regime, high electric fields create a significant density difference between spin-up and spin-down electrons, which becomes comparable to the spin polarization of the current. The spin injection efficiency from DMS into a nonmagnetic semiconductor decreases strongly, however, in the nonlinear regime, for voltage drops across the interface (ohmic contact) larger than a few mV. In this case band bending at interface causes repopulation of the minority spin level in the magnetic semiconductor and hence a decrease in the spin polarization β close to the interface [44].

A model for spin injection across a heterojunction between two electrodes in thermodynamic equilibrium within the ballistic regime, which includes exchange interaction or large Zeeman splitting in one electrode, and Rashba spin-orbit interaction in the other, as well as interface scattering effects, is developed in [45]. This model is suitable for FM metal/normal metal and FM metal/semiconductor heterojunctions when the appropriate band-structure parameters are used, and explains qualitatively the behavior of DMS/semiconductor junctions. The model predicts that the spin-injection rate across the heterojunction at $x = 0$, $\eta = (G_\uparrow - G_\downarrow)/(G_\uparrow + G_\downarrow)$, where the conductance at temperatures $k_B T << E_F$ is given by the Landauer formula

$$G_\sigma = (e^2 / h) \sum_{n=0}^{N} \sum_{\sigma'=\uparrow,\downarrow} t_{\sigma,\sigma'}^* t_{\sigma,\sigma'} \tag{4.8}$$

with $\sigma = \uparrow, \downarrow$ the spin index is determined in the linear response regime by only five dimensionless parameters. These are: the ratio of effective masses in the right (R) and left (L) electrodes, $m = m_R / m_L$, the ratio of the corresponding Fermi energies $\mu = E_{FR} / E_{FL}$, $\eta_0 = h_0 / E_{FL}$ the spin polarization in the left electrode, with h_0 the exchange interaction or the Zeeman shift in the left electrode, $r_0 = E_{so} / E_{FR}$ the spin polarization in the right electrode, with E_{so} the Rashba spin-orbit energy, and $Z_0 = (U / \hbar)(2m_L / E_{FL})^{1/2}$, with $U\delta(x)$ the δ-function potential that determines the elastic scattering at the interface. In particular, in a FM/normal metal or DMS/semi-conductor junction with matched electrodes (m = 1, μ = 1) and $r_0 = 0$, η increases with increasing η_0, while the interface scattering reduces the total conductance but increases η. In junctions with mismatched electrodes η varies only slowly with increasing η_0 except for $\eta_0 \cong 1$. The scattering strength also reduces the total conductance for mismatched electrodes, but it can either increase or decrease $|\eta|$ depending on its strength; the Rashba effect does not influence η significantly.

Spin-polarized transport in a ballistic 2DEG channel contacted with interdigital FM source and drain contacts with different coercive fields was demonstrated in [46]; these contacts suppress effects related to the formation of random domains. The estimated 4.5% spin injection across the FM/2DEG interface produces a strongly temperature-dependent relative increase in device resistance when the source/drain magnetization is switched from parallel to anti-parallel, which decreases continuously from 0.2% at low temperatures and vanishes for T > 10 K. By increasing the channel length the influence of spin precession, dephasing and momentum scattering can be estimated.

Recently, it was revealed that spin-polarized charge transfer at the surface between an FM metal substrate and a multiwalled CNT induces contact magnetization of the CNT corresponding to a spin transfer of $0.1\mu_B$ per contact carbon atom [47]. The prospect of CNT spintronics is encouraged by a large spin diffusion length, which can exceed 100 nm in graphitic structures.

4.1.3 Spin Detection

As mentioned in the previous section, spin polarization can be determined from the polarization degree of the photoluminescence, which is generated by electron-hole recombination. This method is especially suited when spin dynamics is studied in electron-hole systems. In particular, the time-resolved right-circularly polarized photoluminescence component of a GaAs quantum well excited with left-circularly polarized laser pulses at different magnetic fields envisions the electron spin precession perpendicular to the magnetic field and offers information about spin dynamics and the spin relaxation time [13]. The exact relation between spin polarization and the polarization degree of photoluminescence depends on the band structure of the emitting material.

Optical methods for detection of spin polarization in semiconductors, which results from injection from FM materials, include also the polarization of emitted electroluminescence. This shows a hysteresis behavior as a function of an applied magnetic field, with the two parts of the curve corresponding to spin-up and spin-down polarization for temperatures lower than the Curie temperature of the ferromagnet [13] (see Figure 4.2). For higher temperatures the injector has no FM behavior, the emitted electroluminescence has no polarization and the hysteresis behavior with H disappears.

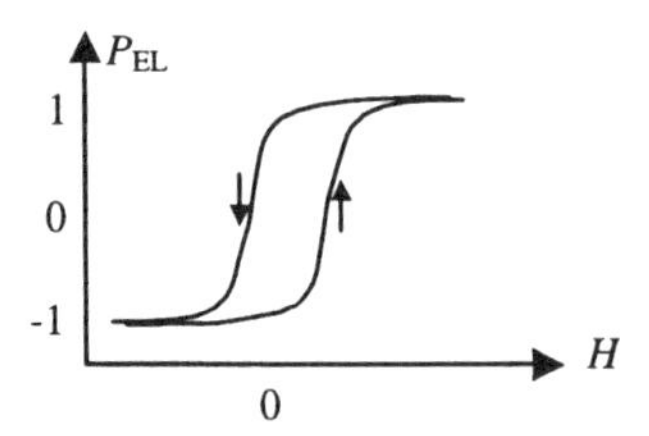

Figure 4.2 Typical dependence of the polarization of electroluminescence on the applied magnetic field.

Spin injection from an FM electrode into a normal conductor (including a semiconductor) or a paramagnetic metal can be also detected by electrical means, in particular by measuring the open circuit voltage (floating potential) on the FM electrode. The spin accumulation in the non-FM region induces a spin-electromotive force signal, similar to the photo-electromotive force signal. When a sample is placed between two FM electrodes, FM1 and FM2, these act, respectively, as injector and detector of electron spin. In FM1/N/FM2 structures the nonequilibrium density polarization in the normal region N is the source for the spin electromotive force and generates a measurable spin-coupled voltage across the N/FM2 detector electrode. This detection technique is called the polarizer/analyzer method, the reversal of magnetization in one ferromagnet from parallel to antiparallel inducing a change in voltage (in open circuits with large impedance) or resistance (in circuits with small impedance).

In the polarizer/analyzer configuration in Figure 4.3(a) [22, 48] the spin polarization $\eta = (n_\uparrow - n_\downarrow)/(n_\uparrow + n_\downarrow)$ of the current I injected from the FM1 electrode, with $n_{\uparrow,\downarrow}$ the spin-dependent density of states at the Fermi level of the electrons in the electrode, generates an unbalanced density of states for the spin-up and spin-down electrons in the N strip, which is transported by diffusion to the FM2 electrode, the detected output voltage V being proportional to η^2. More precisely,

$$V/I = \pm(1/2)\eta^2 \lambda_{sf} \exp(-L/\lambda_{sf})/(\sigma_n A)\,, \tag{4.9}$$

where $\lambda_{sf} = (D\tau_{sf})^{1/2}$ is the spin-flip length in the N strip, σ_n is the conductivity of the N strip and A its cross-sectional area, the positive and negative signs corresponding to parallel and antiparallel magnetizations of FM1 and FM2, respectively. By fitting the experimental data for different inter-electrode distances L to this formula it is possible to find both η and λ_{sf}.

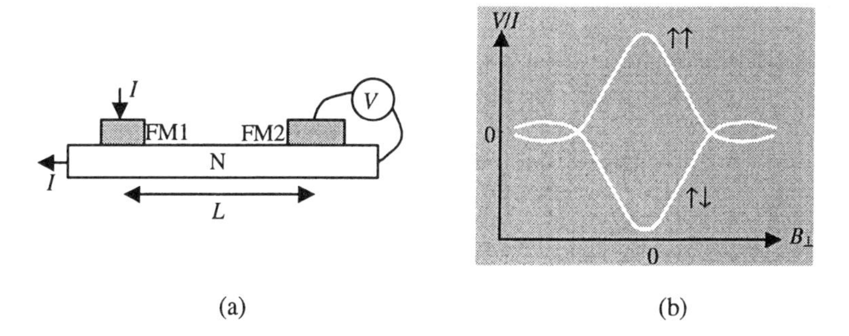

Figure 4.3 Electrical detection methods of spin polarization: (a) polarizer/analyzer method, and (b) Hanle method (*After*: [48]).

Spin precession in the N strip between FM1 and FM2 is analyzed by applying a $B_\perp$ field perpendicular to the N strip plane. Then, electron spins in N precess around $B_\perp$ with the Larmor frequency $\omega_L = g\mu_B B_\perp / \hbar$, their spin direction being altered by an angle $\varphi = \omega_L t$ with t the diffusion time between FM1 and FM2. Because FM2 only detects the spin projection onto its magnetization direction, the measured signal is

$$V(B_\perp) = \pm I \frac{\eta^2}{e^2 \rho_n A} \int_0^\infty \Theta(t) \cos(\omega_L t) \exp(-t / \tau_{sf}) dt \tag{4.10}$$

where $\Theta(t) = (4\pi Dt)^{-1/2} \exp(-L^2 / 4Dt)$ is the distribution of electron diffusion times and ρ_n is the density of states in N at the Fermi energy. The signal decreases with $B_\perp$ and changes sign when electrode magnetizations change from parallel to antiparallel; a typical signal of this so-called Hanle effect is represented in Figure 4.3(b). The parameters η, λ_{sf}, and D are obtained by fitting (4.10) with experimental data. If the magnetization direction of FM electrodes is tilted out of the substrate plane with an angle ϑ the signal is $V(B_\perp, \vartheta) = V(B_\perp)\cos^2\vartheta + |V(B_\perp = 0)| \sin^2\vartheta$ [48].

Single spins can be directly detected by magnetic resonance force microscopy, which measures the magnetic force between an FM tip and the spin [49]. Although this force is a million times smaller than typical forces in an atomic force microscope, spins as deep as 100 nm below the surface of the sample can be detected with a spatial resolution that attains 25 nm for an unpaired spin in SiO_2. Spin detection of low-density unpaired electron spins is performed with a 150-nm-wide Si cantilever with SmCo magnetic tip, positioned at about 125 nm above the sample. The experiments must be carried out at 1.6 K and in a vacuum chamber in order to minimize the force noise and reduce the spin relaxation rate. The microwave magnetic field and the inhomogeneous field $\boldsymbol{B}_{tip}$ from the magnetic tip (see Figure 4.4) establish a resonant slice within the sample, defined as the points where $|\boldsymbol{B}_{tip} + \boldsymbol{B}_0| = \omega_{rf} / \gamma$, with $\boldsymbol{B}_0$ a static external field perpendicular to the sample surface, ω_{rf} the frequency of the microwave field, and γ the gyromagnetic ratio. The cantilever vibrations induce a swing of the resonant slice through the sample, which generates a cyclic adiabatic inversion of the spin observed through a frequency shift of the cantilever. This frequency shift is caused by the alternating magnetic force exerted by the unpaired spin on the tip, which mimics a variation of the cantilever stiffness. Substantial response is obtained when the spin is located slightly in front of or behind the cantilever, along the line of cantilever vibration. In the experiment, for ω_{rf} = 2.96 GHz, $\boldsymbol{B}_0 \cong 30$ mT, ω_{rf} / γ = 106 mT, and for the amplitude of the microwave magnetic field of 0.3 mT, the fundamental frequency of the cantilever, 5.5 kHz, was shifted with 4.2

mHz. This shift is detected with an analogue frequency discriminator followed by a digital lock-in amplifier.

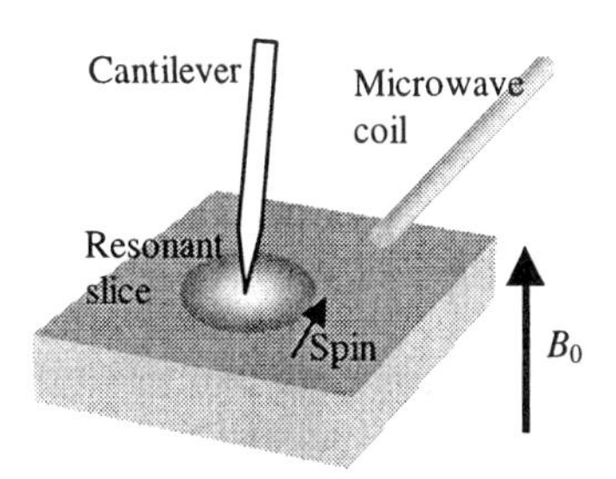

Figure 4.4 Schematic representation of the magnetic resonant force microscopy method for detecting single spins (*After*: [49]).

A 2D spin density map with nanoscale resolution can be recorded if the magnetic particle tip in the magnetic resonance force microscopy is shuttled in synchrony with an rf pulse sequence, which causes precession of spins in a constant-field slice near the tip with a rate proportional to the spin coordinates. The spin density map is obtained by a linear Fourier transform of a set of integrated force signals, the pixel size depending on the nanometer scale amplitude of orthogonal shuttle motions and the thickness of the region of detected spins. The method can also be used to phase encode spin positions [50].

Indirect detection of individual spins, which is crucial in quantum computation read-out, is usually done with single-electron-transistors (SETs). These devices, which will be presented in detail in Chapter 5, detect single electron charges, which encode the spin state of electrons after a spin-to-charge conversion. Spin-to-charge conversion of a single electron confined in a dot, and a subsequent single-shot detection of the single electron charge in a quantum point contact, which renders spin measurement visibility of 65%, was experimentally demonstrated in [51]. The twin-SET architecture described in [52] was also designed for single-spin detection purposes and has the advantage of a considerable reduction of random charge noise through the correlation of two detector outputs. The SET in [53] is especially interesting for quantum computation applications since its sensitivity is so high that it is in principle able to detect with optimized circuitry 1% of an electron in less than 1 μs. Qubit states can thus be read during computation, without destroying their state.

Electrical detection of the magnetic resonance spin flip of a single electron paramagnetic spin center situated in the vicinity of the channel of a Si field-effect transistor was demonstrated in [54]. The detection method is based on conversion of the spin orientation into an electric charge, which changes the value of the source-drain channel current. This is achieved by aligning the Fermi level in the channel E_F at the middle between the Zeeman split energy levels of the paramagnetic trap. In this case, the trap occupied with one electron on the lower Zeeman level cannot accept an additional electron from the Fermi sea in the channel (this additional electron cannot occupy the lower Zeeman level since the Pauli principle forbids it, and cannot occupy the upper Zeeman level because it doesn't have sufficient energy), whereas when the upper Zeeman level in the trap is occupied an additional electron can be transferred from the channel to the lower Zeeman state. Therefore, when a paramagnetic spin flips under the action of a resonant microwave field and thus jumps from the lower to the upper Zeeman level, the lower Zeeman level becomes empty and can trap an additional electron from the channel producing a decrease of the source-drain channel current. This minute variation of the source-drain channel current can be experimentally observed at temperatures below 1 K.

Many proposals of solid-state computers rely also on nuclear spins of impurities that can be manipulated by electron spins through the hyperfine interaction. Coherent manipulation and detection of about 10^7 nuclear spins, which have lifetimes of up to hours and thus are suitable for long-term storage of quantum information, can be achieved by exciting locally and periodically precessing electron spins with an electromagnetic radiation. All-optical nuclear magnetic resonance is observed when the frequency of excitation is in resonance with the gyromagnetic ratio (see the references in [13]). Nuclear spins are polarized when the electron spins generated by optical pumping relax through the hyperfine interaction; Faraday rotation at a fixed delay in a sweeping magnetic field detects nuclear magnetic resonance.

4.2 SPINTRONIC DEVICES

Many unipolar or bipolar spintronic devices have been proposed in recent years. Their modeling is not always easy but advantage can be taken from the fact that unipolar diodes and transistors made from FM semiconductors with different magnetization directions, such as the majority spin carriers on one side are the minority carriers on the other side, behave similarly to nonmagnetic bipolar *p-n* diodes or transistors [55]. The major difference between these two classes of devices is that the interface between layers in a *p-n* diode with two types of oppositely charged carriers is a charge depletion layer, whereas in a unipolar spin diode it is a spin depletion layer. Another difference is that, whether the barriers for both electrons and holes moving across the junction are reduced under forward bias and increased for reverse bias, leading to rectification of charge current, in spin diodes under both forward and reverse biases the barrier for one spin polarization is increased and for the other is decreased.

4.2.1 Spin Filters

Spin filters are devices that filter spins with a definite orientation from an otherwise unpolarized spin distribution, and can therefore act as spin-polarized sources for other devices. The vast majority of spin filter configurations are based on electron tunneling through resonant structures, although it was theoretically demonstrated that atomically ordered and suitably oriented interfaces between some FM metals and some semiconductors can act as ideal spin filters. They can transmit electrons only from the majority spin bands or only from the minority spin bands of the ferromagnet to the semiconductor at Fermi energy and vice versa, even when both majority and minority bands of the ferromagnet are at the Fermi level [56]. Disorder-induced intermixing between the ferromagnet and semiconductor significantly reduces the spin polarization property of the interface, reverting it into a spin antifilter, characterized by a much weaker spin polarization of transmitted electrons in semiconductor than that of the ferromagnet at the Fermi energy level.

Resonant tunneling structures generate spin-polarized electronic currents when spin splitting of electronic bands occurs in asymmetric heterostructures, with different left and right barriers (with build-in electric field) or subject to an external electric field. Spin splitting is caused in this case by spin-orbit coupling and can be controlled by an applied electric field; spin splitting is determined by band-edge discontinuity at interfaces, which imply spin-dependent boundary conditions, and by the additional electrostatic potential, which introduces a spin-dependent term in the effective mass Hamiltonian. In general, transport through a resonant tunneling structure with spin-split energy levels in the well material leads not only to a spin-dependent transmission coefficient, but also to spin-dependent tunneling times [57].

The polarization of the transmitted electron current is $\eta = (I_{\uparrow} - I_{\downarrow})/(I_{\uparrow} + I_{\downarrow})$, where the tunnel current for electrons with polarization $\sigma = \uparrow, \downarrow$ and wave vector $\boldsymbol{k}$ is

$$I_{\sigma} = (e/8\pi^3)\int T_{\sigma}(E,\boldsymbol{k})[f_{\mathrm{E}}(\boldsymbol{k}) - f_{\mathrm{C}}(\boldsymbol{k})]\boldsymbol{v}d\boldsymbol{k} \,, \tag{4.11}$$

with T_{σ} the spin-dependent transmission coefficient calculated using the transfer matrix approach (see, for example, [58] for AIII-BV asymmetric nanostructures in the Kane model), f_{E} and f_{C} the electron distribution functions in the emitter and collector and $\boldsymbol{v}$ the electron velocity component along the tunneling direction in the emitter. η is different from zero only in the presence of an in-plane (perpendicular to the tunneling direction) asymmetry in the electron distribution.

In symmetric resonant tunneling structures this asymmetry can be generated by an in-plane electric field F, case in which the dependence of η on F shows two extrema (one with $\eta < 0$ and the other with $\eta > 0$) that appear when the Fermi level crosses the lowest spin-split resonant level in the well and, respectively, when the highest spin-split resonant level (with opposite spin polarization) crosses the bottom of the conduction band in the emitter. For an InAs/GaAs resonant tunneling structure the spin polarization obtained in this way is lower than 0.25.

Higher spin polarization values (up to 40%) are achieved in asymmetric InAs/GaAs/InAs/AlAs/InAs resonant tunneling structures, with a built-in spin splitting that amplifies the spin splitting of the external electric field [59]. Larger filtering efficiencies, of more than 99.9%, have been observed in a nonmagnetic InAlAs/InGaAs triple barrier resonant tunneling diode (RTD), which enhances the Rashba spin-splitting resonant tunneling (for which $\eta < 90\%$) by combining it with the spin blockade effect between the outer regions separated by the middle barrier [60]. In this device the potential profile peaks in the middle barrier since the outer barriers are impurified with n-type impurities while p-type impurities are introduced in the middle barrier. Therefore the built-in electric field felt by electrons as well as the corresponding Rashba constants in the two wells are opposite. Because of the mirror symmetry of the device and the opposite sign of the Rashba coefficients α_R in the two wells the emitter-collector voltage that matches the spin-split resonance levels in the wells is different for the two spin orientations, so that two peaks are observed in the $I-V$ curve. The separation of the peaks decreases when $|\alpha_R|$ decreases.

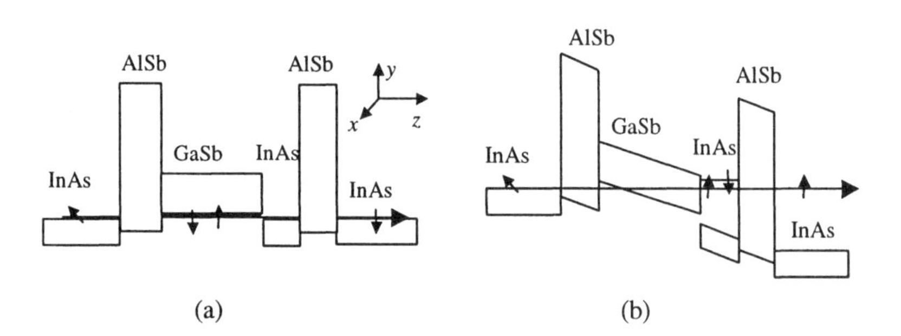

Figure 4.5 Energy band diagram of the resonant (a) interband and (b) intraband tunneling spin filters (rectangles represent the band gap) (*After*: [61]).

The fact that in Rashba resonant tunneling spin filters the spin of the transmitted electron aligns with that of the resonant state in the well can be exploited to design a spin filter that can operate either in the resonant interband tunneling regime for low biases or in the resonant intraband tunneling regime for moderate biases; the corresponding energy band diagrams are represented in Figures 4.5(a) and 4.5(b), respectively [61]. In the first case, the electron passes through heavy-hole valence band states in the GaSb layer, whereas in the second case it traverses conduction band states in the InAs well. The interband resonant tunneling spin filter has an enhanced spin-orbit interaction, the fast spin relaxation in the valence band being avoided by collecting the electrons in the InAs conduction band.

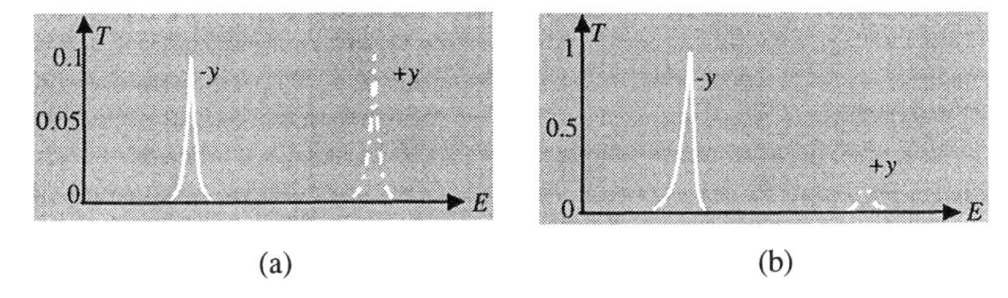

Figure 4.6 Energy dependence of the spin-polarized transmission coefficient for in-plane electrons in (a) an intraband and (b) an interband InAs/AlSb/GaSb/InAs/AlSb/InAs RTD (*After*: [61]).

In the InAs/AlSb/GaSb/InAs/AlSb/InAs RTD, the two peaks of the transmission coefficient T as a function of the electron incident energy E for in-plane spin-up and spin-down polarized electrons (spins along $+y$ and $-y$ and $\boldsymbol{k}_{\parallel} \neq 0$) that tunnel through the InAs quantum well in the intraband tunneling regime have comparable magnitudes, as shown in Figure 4.6(a), but the calculated T is quite small, of about 0.1, since the tunneling structure is asymmetric (the barriers are AlSb and GaSb on one side and AlSb on the other). In the interband tunneling regime through the structure that consists from the AlSb barrier, the GaSb well, and the InAs and AlSb layers as the other barrier the peaks in the transmission coefficient differ strongly in magnitude, as illustrated in Figure 4.6(b). The explanation is that the spin-orbit interaction tends to make the structure more symmetric for spin-down electrons (∇V, which is proportional to the spin-orbit interaction, has opposite signs at the two interfaces of the quantum well) and thus enhances the transmission coefficient as the in-plane component of the electron wave vector increases. On the contrary, for spin-up electrons the structure becomes more asymmetric as $\boldsymbol{k}_{\parallel}$ increases and the corresponding transmission coefficient decreases; the difference in magnitude between the two peaks increases also with the strength of the spin-orbit interaction. The current polarization is negative, with $|\eta| > 50\%$ for an applied lateral electric field of 70 V/cm, as opposed to intraband operation where η reaches a much smaller value at the onset of resonant tunneling, of 10%, for a lateral field higher than 100 V/cm due to the comparable magnitudes of the transmission coefficient peaks. The lateral field generates opposite net spin currents for each of the heavy-hole spin-split subbands in GaSb and the imbalance of the magnitudes of the transmission coefficient for the two polarizations produces a spin-dependent tunneling. The Rashba effect in interband resonant tunneling spin filters has also been analyzed in [62]. Experimentally, control gates along the sides of an asymmetric resonant interband tunneling structure must be patterned to provide a controllable electric field perpendicular to the resonant tunneling current [63]. The two separate side gates must be biased independently for the device to work as a Rashba spin filter; modulation of the tunneling current with the lateral electric field was observed in [63].

Simulations of the symmetric InAs/AlSb/GaMnSb/AlSb/InAs resonant interband heterostructure, in which conduction electrons from the nonmagnetic n-InAs propagate through the valence band of the FM p-GaMnSb layer, which is magnetized to saturation by a magnetic field parallel to the growth axis, have also demonstrated that unpolarized emitted electrons reach the collector with spin polarization greater than 75% [64]. By modulating the emitter-collector bias with 0.1 V the spin polarization of the current changes sign since transmission through different resonant spin split heavy- and light-hole states in the well can be selected by the bias.

Spin filtering can be also achieved by placing a 2DEG in an inhomogeneous magnetic field created by FM stripes or by sets of conductive bars traversed by currents. In this case the electron transmission coefficient depends on both the wave vector component parallel to the barrier, q, and on the spin direction $s = \pm 1$. The 2DEG electrons are subject to the potential

$$V(x,q,s) = U(x,q) + s(m_0 / m) g\mu_B B \,, \tag{4.12}$$

with $U(x,q) = [eA(x) - \hbar q]^2 / 2m$, $A(x)$ the component of the vector potential along y and $B(x)$ the magnetic field component along z [65]. It must be distinguished between odd vector potentials (o-structures), for which $A(x) = -A(-x)$, $B(x) = B(-x)$, and $U(x,q) \neq U(-x,q)$ for $q \neq 0$, and even vector potentials (e-structures), for which $A(x) = A(-x)$, $B(x) = -B(-x)$, $U(x,q) = U(-x,q)$, and $V(x,q,s) = V(-x,q,-s)$. These structures are illustrated in Figures 4.7(a) and 4.7(b), respectively. The equality of tunneling probabilities for electrons moving in opposite directions imposes that e-structures posses no spin filtering properties (i.e., the tunneling probability is independent of the spin direction), whereas in a double-barrier o-structure the resonant energy levels

$E_n(q,s)$ for which the transmission probability is 1 depend on the spin direction. In particular, if the electron energy is lower than $E_{\max}$ and $E_0(q,1) < E_{\max} < E_0(q,-1)$, only electrons with $s = 1$ tunnel through (are filtered out by) the resonant structure.

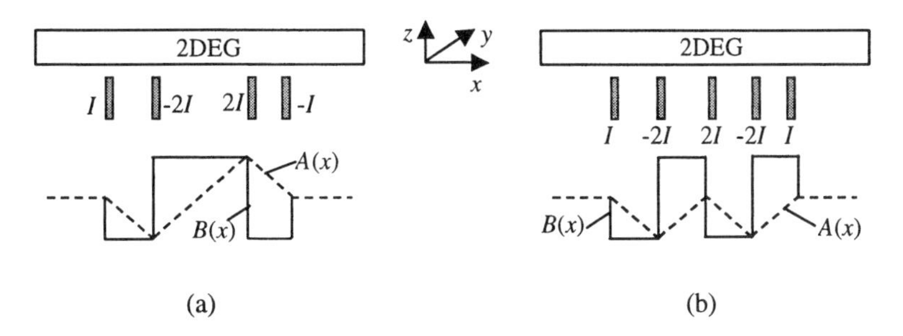

Figure 4.7 Vector potentials and magnetic fields of the (a) o-structure and (b) e-structure created by a set of conductive bars traversed by currents and placed in close proximity with a 2DEG (*After*: [65]).

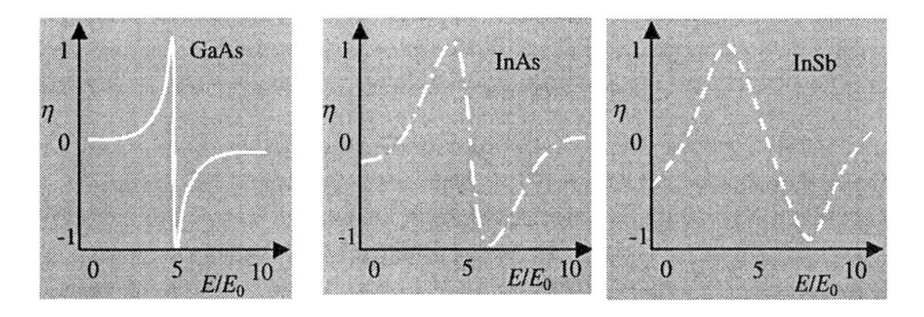

Figure 4.8 Energy dependence of the spin polarization of transmitted electrons in GaAs, InAs, and InSb 2DEGs (*After*: [66]).

In [66] it was theoretically demonstrated that 100% spin polarization can be achieved in an o-structure in which one or two FM stripes are patterned on top of a 2DEG. The spin-filtering property is due to electron tunneling through the spin-dependent magnetic barrier of the stripe with magnetization perpendicular to the 2DEG, the degree of spin polarization varying with the electron energy and with the 2DEG semiconductor material, as shown in Figure 4.8 in dimensionless units $E_0 = \hbar eB / m$. From Figure 4.8 it follows that the transmitted electron beam is fully spin polarized (η = 1 or $\eta = -1$) at two resonant energies separated by a distance that increases with the gap $(m/m_0)gB$ of spin splitting levels and thus with the effective g factor of the 2DEG (the g factors in GaAs, InAs, InSb are –0.44, –15, and –51, respectively). The position of the resonant energies can be further tuned by applying an electric field on the stripes, which shifts the polarization extremes towards lower energies for an electric well or towards higher energies for an electric barrier. The spin filter is thus controlled by the voltage applied on the stripe.

When a second FM metallic stripe is patterned on the 2DEG with perpendicular magnetization, electrons tunnel through two barriers, which can form a RTD if the spacing between them is sufficiently small. In this case the resonance peaks in η split into triplets and almost 100% spin polarization can be obtained in an energy plateau with a position and width that depend on the distance between the stripes and the voltage applied on them. The operating temperature of the device must satisfy $T << (m / m_0) gB / k_B$, which means that it should be about 100 mK in GaAs 2DEG and 1 K in InSb systems. The operating temperature can be increased if materials with larger g factors (for example, EuTe antiferromagnetic epilayers with g = 1140 [67] or metamagnetic EuSe epilayers with g factors up to 18000 [68]) are used.

Another efficient spin filtering device is RTD based on (Zn,Mn,Be)Se, in which the quantum well is a dilute magnetic material. In this case, the resonant energy level in the quantum well splits into spin-up and spin-down states in the presence of an external magnetic field, as shown in Figure 4.9(a). The separation between these energy levels is 15 meV at 1.3 K for fields of 1–2 T, and grows with the applied magnetic field as $\Delta E = N_0 \alpha x s_0 \mathrm{Br}[sg\mu_B B / k_B (T + T_{\mathrm{eff}})]$, where $N_0 \alpha$ is the s-d exchange integral, x and s are the Mn concentration and spin, g the Mn Landé factor, Br[] is the Brillouin function, and s_0 and T_{eff} denote the Mn effective spin and temperature, respectively. (Unlike in the diffusive transport regime, where spin-polarized currents in semimagnetic semiconductors originate in spin alignment caused by spin-flip processes, in the ballistic or quantum-coherent limit spin filtering is caused by the selective spin-polarized transmission of electrons.) The transmission coefficient of the structure in the presence of an applied bias is resonant whenever the spin-up or spin-down states are selectively brought into resonance with the Fermi energy level into the emitter; the spin orientation can be controlled via the bias V across the resonant tunneling structure. Therefore, the transmission coefficient as a function of V has two distinct peaks, which correspond to the two spin-polarized resonance levels, and the $I - V$ characteristic of the structure looks like that in Figure 4.9(b) [69]. The peaks move closer and eventually merge if the magnetic field decreases or the temperature increases since the giant Zeeman splitting in semimagnetic $Zn_{1-x}Mn_xSe$ depends on the temperature. (Zeeman-based spin splitting is not always convenient since it often needs large magnetic fields and requires on-chip placement of micro-magnets.)

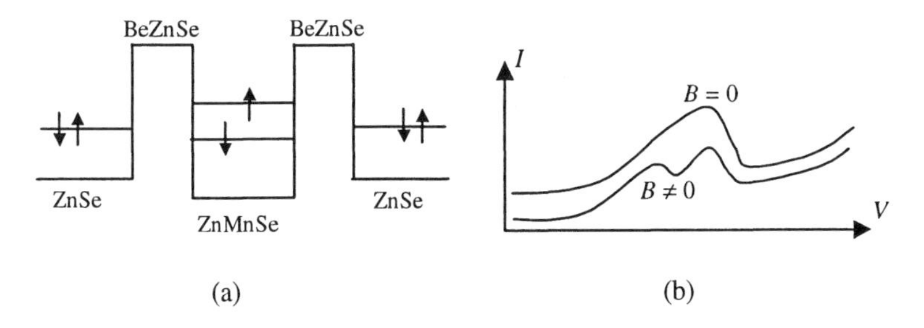

Figure 4.9 (a) Schematic diagram of the RTD and its spin-polarized levels, and (b) typical current-voltage characteristics, which are offset for clarity (right) (*After*: [69]).

In a similar BeTe/$Zn_{1-x}Mn_xSe$/BeTe resonant tunneling structure grown above an AlGaAs/GaAs LED, the spin polarization of the tunneling current was deduced from the degree of circular polarization of the electroluminescence emitted by the LED [70]. It was found that at 1.6 K the degree of circular polarization increases with the magnetic field and saturates at 80% for magnetic fields above 2–3 T, while an increase of the bias from 1.8 to 2.3 V at a constant current of 50 μA generates a reduction of the polarization degree over the first resonance down to 38%, explained by the increased contribution at transport from the second spin-split subband.

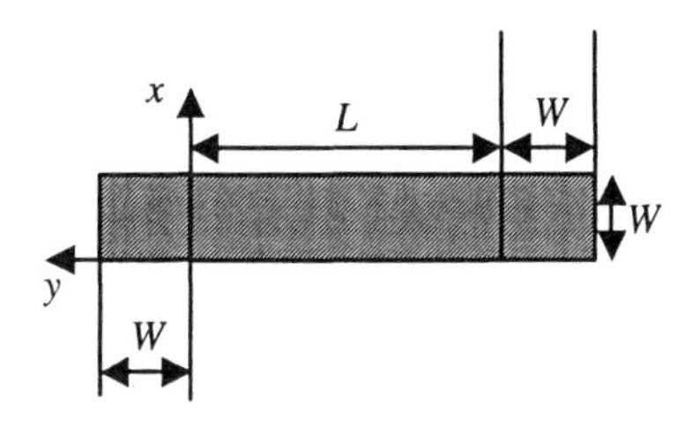

Figure 4.10 Geometry of a double-bend structure (*After*: [71]).

An alternative for spin filtering to resonant tunneling through electric or magnetic heterostructures is the double-bend structure represented in Figure 4.10 [71]. The double-bend discontinuity, as treated by the mode-matching theory, causes strong resonances in the energy dependence of the transmission coefficient, the width and spacing of which depend on the cavity length L. If, in addition, a uniform weak magnetic field is applied over the striped area in Figure 4.10 by using magnetic semiconductors or by sticking a magnetic strip on top of the sample, the spin-dependent potential $V_s(x, y) = sV_0(x, y)$ in the double-bend region with $s = \pm 1$ for spin-up and -down electrons causes spin-polarized currents because electrons with different spins experience different potentials. In particular, tight-binding calculations show that if $E_F >> V_0$ extremely large spin-polarized currents (of the order of μA) can be obtained, which increase with increasing V_0.

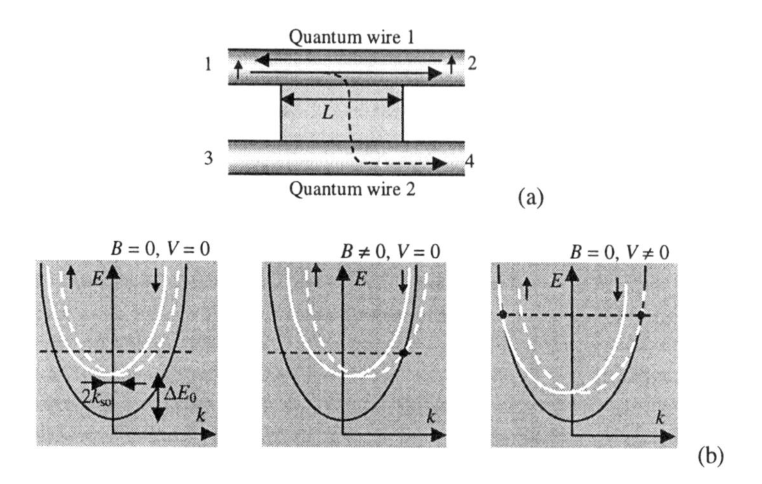

Figure 4.11 (a) Schematic diagram of a spin-filter device based on tunneling between parallel quantum wires and (b) the energy band diagram of the upper wire (spin-split) and lower wire (degenerate) in the absence and presence of electric and magnetic fields (*After*: [72]).

A three-terminal spin polarizer or spin filter configuration, which relies on the interplay between the Rashba effect and the selectivity of wave number at tunneling between parallel electron waveguides (quantum wires) was proposed in [72]. The quantum wires, each connected to electron reservoirs with equal chemical potential and biased with a gate voltage that controls the strength of the Rashba spin-orbit interaction, are coupled via tunneling through a nonmagnetic barrier with uniform height and length L (see Figure 4.11(a)). An electron can tunnel from one wire to the other if its momentum is conserved, the spin state remaining unchanged during the process; spin filtering is achieved by a coupling of spin to momentum, such that the electron canonical momentum $\hbar \boldsymbol{k}$ becomes spin dependent. We assume that the Rashba spin-orbit coupling vanishes in the lower wire so that it has a spin-degenerate dispersion relation as in Figure 4.11 and that the energy dispersion in the upper wire is given by $E_u = E_{0u}$

$+\hbar^2(k - sk_{so})^2/2m - \Delta_{so}$, where $s = \pm 1$ for spin-up and spin-down electrons, and $\Delta_{so} = \hbar^2 k_{so}^2/2m$ with k_{so} a wave number that characterizes the spin-orbit coupling; k_{so} can be tuned by the bias voltage applied on the gate. Electrons from one wire can tunnel to the other only if the dispersion curves of the upper and lower wires cross at the electron wave number. As seen from Figure 4.11(b), tunneling of electrons with spin s can occur in the presence of a magnetic field perpendicular to the plane of the two wires whenever its strength B satisfies the condition

$$-eBd/\hbar = \pi(\gamma_u n_u - \gamma_l n_l)/2 - sk_{so} \quad (4.13)$$

with d the wire separation, $\gamma_{u,l} = \pm 1$ and $n_{u,l}$ the electron densities in the upper and lower wires. If $\pi L << k_{so}$ simultaneous tunneling of electrons with opposite spins is forbidden and the device acts as spin-polarizer. In Figure 4.11, spin-polarized electrons from reservoir 1 enter reservoir 4 for $\gamma_u = \gamma_l = 1$, while electrons from reservoir 2 with the same spin but opposite momentum can only reach reservoir 1. The magnetic field required to tune the wave number selectivity is generally smaller than that needed to filter spins via Zeeman splitting, but the device works for temperatures lower than about 10 K. Alternatively, a finite bias V between the two wires can tune the wave number, case in which electrons with spin s and wave number k tunnel if

$$eV = \Delta E_0 - \hbar^2 k_{so} sk/m. \quad (4.14)$$

Unlike in the previous case electrons with opposite spin and wave number tunnel simultaneously, ending in opposite leads of the wire in which they tunnel, such that the currents in the leads are fully spin polarized. For example, the currents in leads 1 and 3 have the same polarization, opposite to that in 2 and 4, the selection of spin-up or spin-down electrons being achieved by adjusting V. Spin filtering is achieved by a redistribution of spin-polarized electrons between the leads.

Rashba spin-orbit interaction can lead to spin filtering in a transverse 2DEG focusing configuration displayed in Figure 4.12 [73]. Transverse focusing of 2DEG electrons emitted from an electron source (quantum wire 1 in Figure 4.12) into an electron collector (quantum wire 2) in the presence of a magnetic field normal to the 2DEG plane takes place since electron trajectories become circular due to the Lorentz force. Magnetic electron focusing is a typical example of electron optics experiments, which can only be performed in ballistic 2DEG [74]. Electrons are collected in wire 2 (the current in wire 2 peaks) if the distance R is an integer multiple of the cyclotron diameter $2\hbar k/eB$. In the presence of spin-orbit coupling, however, electrons with the same energy follow one of two semiclassical orbits according to their spin, since their energy can be approximated with $E_j^{\pm} \cong \hbar\omega_c j \pm \alpha_R (2j)^{1/2} / l_c$ for large Landau index j, where $\omega_c = eB/m$ is the cyclotron frequency and $l_c = (\hbar / m\omega_c)^{1/2}$ [73]. The difference between the radii of the two orbits is $\Delta R \cong 2\alpha_R / \hbar\omega_c$, and spin filtering can occur if only electrons with one spin are collected by wire 2. Significant spin splitting of electron beam occurs only at the first focusing peak, splitting at odd focusing points becoming observable for large Rashba coupling; at even focusing peaks no spin splitting occurs. Splitting of the first focusing peak has been experimentally observed in a 2D hole gas (see the reference in [73]). A review of the Rashba effect in electron optics experiments can be found in [75].

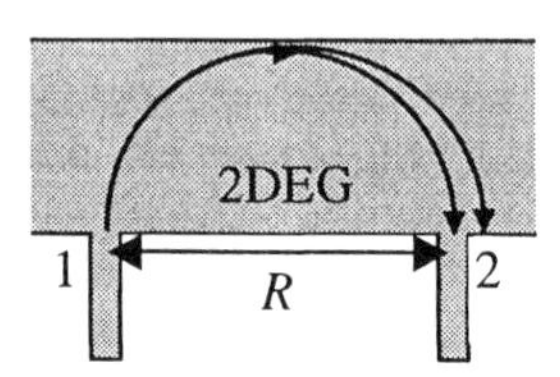

Figure 4.12 Electron trajectories in the transverse focusing experiment (*After*: [73]).

Theoretically, it was demonstrated that a nanotube double junction consisting of two metallic CNTs joined by a semiconducting CNT through transition regions made up of sets of five and seven member carbon rings can also act as a spin filter. The central semiconducting CNT region is a quantum well, which is surrounded by Schottky barriers, spin currents appearing when Coulomb interaction lifts the spin degeneracy of resonant levels in the well; the spin direction can be controlled by a gate voltage as long as the length of the semiconducting CNT region is appropriately chosen (it must be larger than a threshold value and lower than an upper limit) and the Fermi level is near the band gap edges at the operating gate voltage [76]. CNT junctions can be constructed by allowing the open ends of $(3m,0)$ zigzag CNTs and $(3m \pm 1,0)$ CNTs to react.

Other proposed configurations that could act as spin filters include a system of two capacitively coupled quantum dots, each of them connected to a common source (a normal metallic lead) and to a distinct outer current lead. This spin splitter device, which works in a local magnetic field in the Kondo regime and requires only nonmagnetic semicondutor materials, filters and spatially separates oppositely spin polarized currents in the two outer leads [77]. Spin filtering is obtained by energy filtering. Another proposal is a 1D electron wire with a strong Rashba spin-orbit coupling and Zeeman splitting, which is covered on one side with a FM gate biased with a voltage that pins the Fermi energy in the middle of the local gap induced by the local magnetic field, the ungated part lacking Zeeman splitting [78]. This device is an anti-symmetric spin filter in the sense that electrons with opposite spins are filtered out for opposite current directions. Also, a T-shaped quasi-1D channel with a control electrode placed over/under it can split an incoming unpolarized electron beam into highly polarized output beams, electrons with opposite spins being redirected to the two opposite output leads of the structure. The spin splitting efficiency is significantly enhanced if the difference between the energy of the unpolarized electron beam and the quasilocalized states in the intersection region becomes comparable to the spin-orbit interaction induced by the asymmetry of the confining potential [79].

4.2.2 Spin Valves

Spin valves are devices based on the GMR effect, namely on the significant change in resistance for different magnetization orientations of two FM layers, which can be changed by an applied magnetic field. Traditionally GMR is found in layered structures consisting of a nonmagnetic material sandwiched between FM layers, in which the resistance or the conductance of the device depends on the relative orientation of the magnetizations of the two FM layers because the density of states for the two spin components at Fermi level depend on the relative orientation of magnetizations.

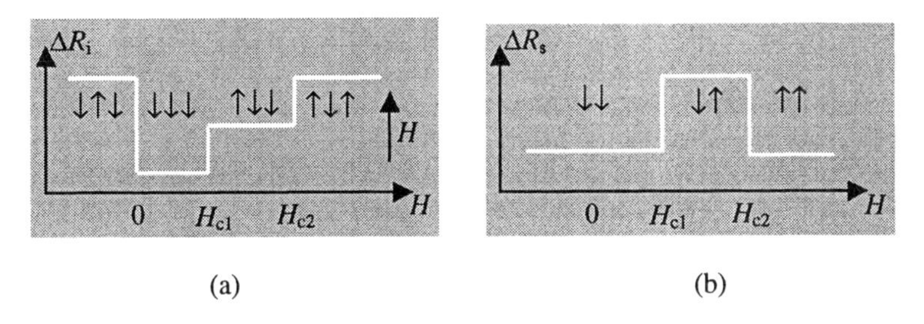

Figure 4.13 (a) Magnetic-field dependence of the interface spin-valve effect and (b) direct spin-valve effect (*After*: [80]).

Spin-valve effects have also been observed in field-effect transistor configurations in which the AlSb/InAs 2DEG channel is contacted by magnetic permalloy thin films with different coercivities H_{c1} and H_{c2} and with magnetization directions perpendicular to the current flow direction; the contacts can be magnetized parallel or antiparallel by a sweeping magnetic field [80]. It was found that two different spin-valve effects are involved: one at the FM/2DEG interface and the other at spin propagation in the 2DEG. The behavior of the components of the magnetoresistance $\Delta R = R(H) - R(0)$ corresponding to the interface, ΔR_i, and the direct spin-valve effects, ΔR_s, are represented in Figures 4.13(a) and 4.13(b), respectively. The total magnetoresistance varies with only 0.2% at low temperatures, the spin-valve effect disappearing at 10 K due to thermal activation that smears the zero-field spin splitting. The interface spin-valve effect depends on the relative orientations of the magnetization direction in contact 1, the spin orientation in the 2DEG, and the magnetization direction in contact 2, represented in this order by the arrows, as well as on the orientation of the magnetic field H, while the direct spin-valve effect depends only on the magnetization directions of the contacts, represented by arrows.

For a multiwalled CNT placed between FM Co contacts or between NiFe and Co contacts a spin valve effect has been observed in [81]. More precisely, when the magnetization of the FM contacts was switched from parallel to antiparallel by a magnetic field, which is swept back and forth from a positive to a negative value, the CNT resistance switched from a low-resistance to a high-resistance state, as shown in Figure 4.14. The hysteretic behavior of the magnetoresistance of the small-diameter CNT is characteristic for FM contacts with different coercivities, and is due to the misalignment of the magnetic moments of the small number of magnetic domains in the two electrodes that are in contact with the CNT. The width of the resistance peak is commensurate with the coercive field strength of the FM contact, and the relative difference between the tunnel resistance in the parallel and antiparallel states is

$$(R_{\uparrow\downarrow} - R_{\uparrow\uparrow}) / R_{\uparrow\downarrow} = 2\beta_1\beta_2 / (1 + \beta_1\beta_2)\,, \tag{4.15}$$

with $\beta_{1,2}$ the spin polarization in the majority spin band in the FM contacts. At 4.2 K the maximum relative resistance change was 9% for Co contacts.

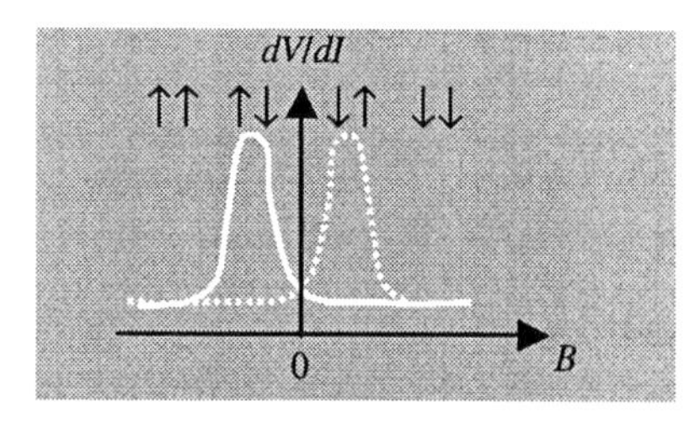

Figure 4.14 Typical dependence of the differential resistance of a multiwalled CNT with FM contacts on the magnetic field. The arrows indicate the magnetization directions of FM contacts (*After*: [81]).

An enhanced spin-valve effect, characterized by a magnetoconductance ratio $(G_{\uparrow\uparrow} - G_{\uparrow\downarrow}) / G_{\uparrow\uparrow}$ of up to 20%, was demonstrated for metallic armchair single-walled CNTs (SWCNT) [82]. The spin valve is a magnetic tunnel junction FM/SWCNT/FM, in which the transport is dominated by resonant transmission and in which the magnetizations of the FM contacts point in different direction. The tunnel junction has a minimum resistance when the contact magnetizations are parallel (and perpendicular to the SWCNT axis) and a maximum resistance when they are antiparallel. Different transport behaviors are observed for tubes with lengths commensurate with $3N+1$-unit cells, which have resistances with one order of magnitude smaller at the Fermi level than other tubes.

A spin valve configuration consisting of a 2DEG on which two FM metal gates FM1 and FM2 are patterned, placed between source S and drain D nonmagnetic contacts has been proposed in [83]; the device is represented in Figure 4.15 The magnetizations of the FM contacts are perpendicular to the direction of current flow, a configuration typical for spin valve devices, and the magnetization direction of FM2 can be changed. If the interface barrier between the 2DEG and the gates are sufficiently thin the FM proximity effect yields spin-dependent electron transport in the 2DEG. The tunneling coupling of the 2DEG electron wavefunction to the metal exchange-split band states generates spin-dependent broadening and shift of the energy levels in the 2DEG and a corresponding spin-dependent conductance

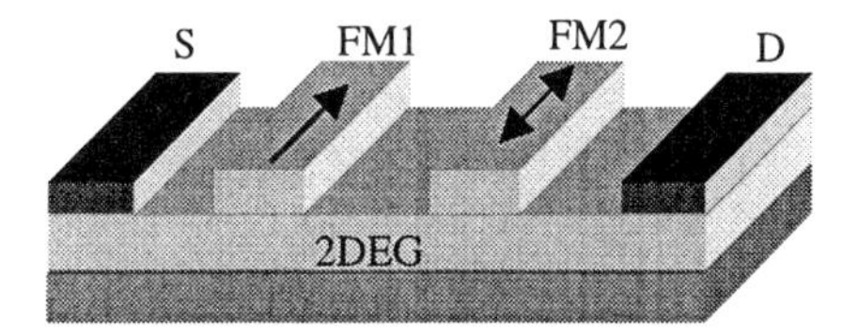

Figure 4.15 Schematic diagram of a spin-valve configuration with two FM metal gates (*After*: [83]).

$$\frac{1}{\sigma_{\uparrow,\downarrow}} = \frac{m}{n_{\uparrow,\downarrow} e^2} \left(\frac{1}{\tau_0} + \frac{1}{\tau_{\uparrow,\downarrow}} \right). \tag{4.16}$$

Here the subscripts $\uparrow$ and $\downarrow$ refer to the majority and minority spin channels, respectively, $n_{\uparrow,\downarrow}$ is the spin-dependent density in 2DEG due to spin-splitting, τ_0 is the spin-independent Drude scattering time, and $\tau_{\uparrow,\downarrow}$ the escape time of quasibound 2DEG electrons into the FM metal band. Additionally, a spin-dependent leakage probability $\alpha_{\uparrow,\downarrow}$ for nonequilibrium channel electrons into the gate appears, the magnetoresistance ratio $MR = (I_{\uparrow\uparrow} - I_{\uparrow\downarrow}) / I_{\uparrow\uparrow}$, with $I_{\uparrow\uparrow}$ and $I_{\uparrow\downarrow}$ the drain currents for parallel and antiparallel magnetizations of the gates, being approximately given by

$$MR \cong \left[\frac{\sigma_{\uparrow}(1-\alpha_{\uparrow}) - \sigma_{\downarrow}(1-\alpha_{\downarrow})}{\sigma_{\uparrow}(1-\alpha_{\uparrow}) + \sigma_{\downarrow}(1-\alpha_{\downarrow})} \right]^2 . \tag{4.17}$$

For a $Si/SiO_2/Fe$ heterostructure with gate length 0.1 μm, in-plane field 0.2 V/μm, $\tau_0 = 1$ ps, $\tau_{\uparrow} = 1$ ps, and $\tau_{\downarrow} = 2$ ps, it was found that $\alpha_{\uparrow} = 40\%$, $\alpha_{\downarrow} = 20\%$ and *MR* = 10%. A decrease of the drift velocity in 2DEG via a decrease of the source-drain bias increases the leakage probability and the *MR*, but decreases the drain current.

An absolute spin valve effect, which involves devices with only one magnetic contact, based on the spin-orbit interaction, was shown to exist in a hybrid system consisting of a 2DEG placed between a FM lead and a normal metallic lead NM [16]. In this case the conductance of the system depicted in Figure 4.16 changes with the angle θ between the magnetization vector μ and the z axis, but the current is always perpendicular to μ. In this absolute spin valve the conductance depends on the absolute direction of magnetization because of the spin-dependent scattering of electrons due to the spin-orbit interaction. Calculations for both the Elliot-Yafet and the Rashba spin-orbit interactions show that the total conductance G depends symmetrically on θ, having maxima for $\theta = \pi/2$ and $\theta = 3\pi/2$ and minima for $\theta = 0$, $\theta = \pi$, and $\theta = 2\pi$, but the magnitudes of the effects are different (the Rashba effect is more efficient) since the spin diffusion lengths are different (the spin diffusion length for Rashba is ten times smaller in the example considered than for Elliot-Yafet). The absolute spin valve coefficient, defined as $\gamma = 2[G(\mu \parallel z) - G(\mu \parallel y)]/[G(\mu \parallel z) + G(\mu \parallel y)]$, where $G(\mu \parallel z)$ and $G(\mu \parallel y)$ are the conductances for $\mu \parallel z$ and $\mu \parallel y$, respectively, was found to be negative and quite small (0.1%) for weak spin-orbit interaction since in the example the system size was small with respect to the diffusion length, but increased to about 1% for large spin-orbit coupling.

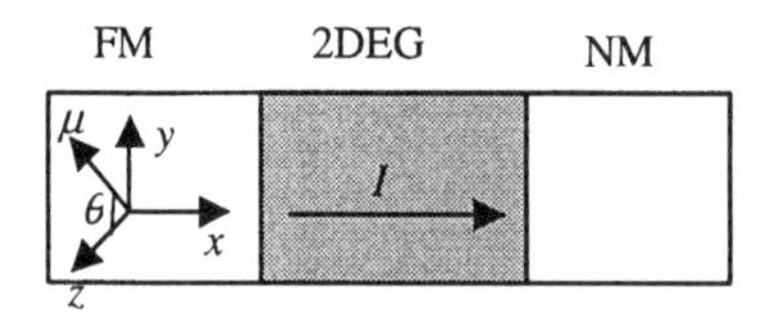

Figure 4.16 Schematic representation of the absolute valve, with the magnetization direction μ of the FM lead in the yz plane and the current flowing along x (*After*: [16]).

Calculations based on the tight-binding model predict that spin valve effects can occur in short DNA molecules sandwiched between FM contacts. The corresponding magnetoresistance values $(R_{\uparrow\downarrow} - R_{\uparrow\uparrow}) / R_{\uparrow\downarrow}$, where $R_{\uparrow\downarrow}$ and $R_{\uparrow\uparrow}$ are the resistances for the antiparallel and the parallel spin configuration of the contacts, respectively, equal 26% for Ni contacts and 16% for Fe contacts [84]. Spin dependent transport can thus be observed in molecular structures.

4.2.3 Spin Pumps

A spin pump generates net spin currents with negligible charge currents by driving spin-up and spin-down electrons in opposite directions. The device in Figure 4.5 can act also as a spin pump at zero magnetic field when a lateral electric field is applied in the emitter region but no bias is applied between the emitter and collector of the resonant interband tunneling structure [85]. For electrons with energies close to the spin-split subband in the InAs-GaSb well with clockwise pinwheel spin states in the $\boldsymbol{k}_{\parallel}$ plane (which is designed such that the resonant tunneling through it dominates over the resonant tunneling through the other spin-split subband with counterclockwise states) the in-plane field F creates an excess of carriers on the $+k_x$ side, which tunnel into the collector with spins polarized along $-y$, and a deficit of carriers is generated on the $-k_x$ side, which receive electrons that tunnel from the collector with spins along $+y$. As a consequence the device is a bidirectional spin pump, which generates significant oppositely spin-polarized currents in opposite directions with very little net electrical current through the tunneling structure, from emitter to collector. There is however a net electrical current along x induced by the in-plane E_x electrical field. The current spin polarization $\eta = [I(+y) - I(-y)]/[I(+y) + I(-y)]$ can take, in this case, values greater than 1 since $I(+y)$ and $I(-y)$ have opposite signs. The spin pump efficiency decreases with temperature mainly due to the decrease of the momentum relaxation time τ, which causes a decrease of the Fermi sphere shifts $\Delta k_x = -eE_x\tau / \hbar$ in the k space.

Another proposed spin pump device, which works on the principles of quantum adiabatic transport based on dynamical destruction of time reversal invariance, consists of FM stripes patterned on a 2DEG [86]. This device acts as a spin pump when the magnetic barriers are adiabatically modulated by an external magnetic field that modulates the magnetization strength of the stripes and by modulating the distance between the stripes and the 2DEG through gate voltages.

4.2.4 Spin Diodes

In a usual *p-n* junction consisting of nonmagnetic materials spin-polarized currents can be generated by external spin excitation [87]. In particular, if the *p* region, for example, is illuminated by circularly polarized light, the minority electrons become spin polarized (hole spin polarization is neglected) and diffuse toward the depletion layer where they are swept to the *n* side by the built-in field. The electrons now become majority carriers and can diffuse away from the depletion layer or can relax on a time scale T_1. Calculations show that the nonequilibrium spin population $s = n_\uparrow - n_\downarrow$ in the majority region is given by $s = \eta R_{gen}(T_1\tau_s)^{1/2}$, where R_{gen} is the generation rate of electrons with spin polarization $\eta = (n_\uparrow - n_\downarrow)/(n_\uparrow + n_\downarrow)$ by illumination in the *p* regions and τ_s is the effective spin relaxation time in the *p* region. Because *s* is typically larger than the spin in the minority region, a spin amplification process occurs, in which the spin of majority carriers is pumped by the minority channel, similarly to spin pumping of majority carriers in semiconductors by circularly polarized light. The spin-polarized *p-n* junction acts thus as a spin-polarized solar cell, in which the amount of injected nonequilibrium spin is controlled by a reverse bias applied to the junction. This bias controls the width of the depletion layer and hence the spin that arrives at the *p* side of the depletion layer through the so-called spin capacitance effect. Spin injection by majority carriers, upon illuminating the *n* side with circularly polarized light is also very effective.

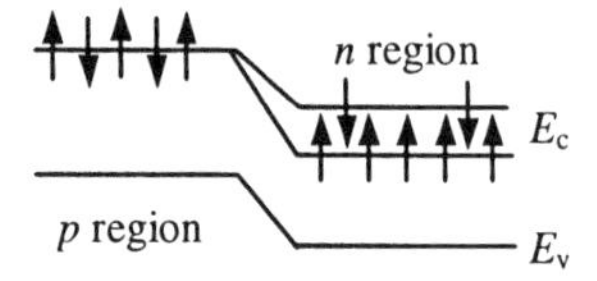

Figure 4.17 Energy band diagram in a magnetic *p-n* junction with a magnetic *n* region (*After*: [87]).

Spin-dependent transport properties are expected to be significantly enhanced in magnetic *p-n* junctions, which contain magnetic impurities that lead to spin splitting of carrier energy bands in the presence of an applied magnetic field [87]. However, this is not usually the case, unless nonlinear spin-charge coupling is involved. In a magnetic *p-n* diode where only the *n* region is magnetic (see Figure 4.17), spin injection, manifested by the increase of spin polarization in the *p* region, is not possible at small biases *V*. For small *V* the depletion layer is in (quasi)equilibrium, and the exponential rectification $I - V$ characteristics is evident because the unbalanced density of spin-up and spin-down electrons is compensated by the unequal barrier heights for the two spin orientations. Spin injection is observable only at higher biases at which nonequilibrium spins accumulate in the magnetic region and are then injected to the minority nonmagnetic side; on the contrary, spin injection from the minority magnetic side appears as spin extraction.

Significant spin injection in magnetic *p-n* junctions at low biases occurs only in the presence of nonlinear spin-charge coupling, when a nonequilibrium source spin is introduced into the junction. In this case spin polarization is efficiently injected through the depletion layer, but the charge transport across the junction is affected. For example, the $I-V$ characteristic of a magnetic *p-n* junction with equilibrium spin polarization η_0 on the *p* side of the depletion layer and nonequilibrium spin polarization $\Delta\eta$ on the *n* side is

$$I = I_0[(1+\eta_0\Delta\eta)\exp(eV/k_BT)-1], \tag{4.18}$$

with I_0 the material-dependent generation current of the diode.

A spin diode based on two semiconducting quantum wires coupled by a tunnel barrier, which can rectify the ballistic spin current when a bias and a magnetic field gradient are applied between the quantum wires, was studied in [88]. The structure consists of two leads with a periodic potential, in which a gap of width Δ is opened at the Fermi energy, the chemical potentials in the left and right leads being given by $\mu_L^\sigma = -(eV+\sigma g\mu_B H)/2$ and $\mu_R^\sigma = (eV+\sigma g\mu_B H)/2$, respectively, where $\sigma = +1/2$ (or $-1/2$) corresponds to spin-up (or spin-down) electrons, V is the applied voltage, and it was assumed that the magnetic field in the right (left) lead is $H/2$ ($-H/2$); a magnetic field gradient at nanometer scale can be applied if the semiconductor quantum wires have different gyromagnetic ratios. In this case the spin-polarized current obtained from a scattering state formalism is

$$I_\sigma \propto \Gamma\sinh[(eV+\sigma g\mu_B H)/(2k_BT)], \tag{4.19}$$

where Γ is determined by the tunneling matrix element and by the Boltzmann factor $\exp(-\Delta/k_BT)$. The factor $\sinh[(eV+\sigma g\mu_B H)/(2k_BT)]$ expresses the asymmetry of the current and the rectification effect. In particular, the current is completely polarized if $eV = g\mu_B H$. For strong tunneling the charge and spin conductances are $G_e \cong (e^2/h)\exp(-\Delta/k_BT)/[1+\exp(-\Delta/k_BT)]$ and $G_s = 2(g\mu_B/e)^2 G_e$, respectively. The Fermi-Dirac distribution function is replaced in the calculation of conductances by the Boltzmann function, if $|\Delta-\mu_{L,R}^\sigma| > k_BT$.

4.2.5 Spin Transistors

In spin transistors, the spin-dependent transport properties are controlled through a signal applied on a gate contact in unipolar devices or a base contact in bipolar devices. The first proposal for a field-effect spin transistor, known as the Datta-Das transistor, was based on the analogy with an electro-optic light modulator [89]. A schematic representation of its structure is represented in Figure 4.18. The polarizer and analyzer of an electro-optic modulator are replaced in this case by FM source and drain iron contacts, the direction of polarization selectivity being now the spin direction in the contacts.

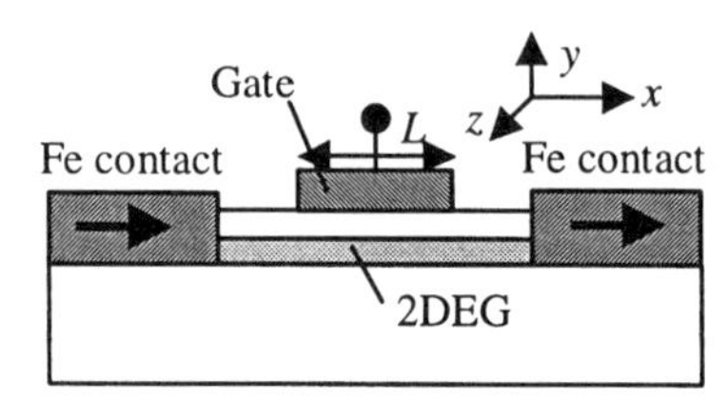

Figure 4.18 Schematic representation of the Datta-Das transistor (*After*: [89]).

In the Datta-Das transistor the spins in the FM contacts are parallel aligned along x and 100% spin injection/detection efficiency in 2DEG is assumed. The FM source contact launches a linear combination of $+z$ -and$-z$ -spin-polarized electrons along the current direction x. The gate voltage introduces a phase shift between these oppositely polarized electron spins in the 2DEG through the Rashba term in the effective mass Hamiltonian, which takes in this case the form $H_R = \alpha_R(\sigma_z k_x - \sigma_x k_z)$.

For electrons launched along x with $k_z = 0$ this term raises the energy of $+z$ -polarized electrons with $\alpha_R k_x$ and lowers with the same amount the energy of $-z$ -polarized electrons. Therefore, oppositely spin-polarized electrons with the same energy E have different wave vector components along x, k_{x+}, and k_{x-}, respectively, given by $E = \hbar^2 k_{x+}^2 / 2m - \alpha_R k_{x+}$ and $E = \hbar^2 k_{x-}^2 / 2m + \alpha_R k_{x-}$. Spin-precession due to spin-orbit coupling is expected, the differential phase shift (spin precession angle)

$$\theta_R = (k_{x+} - k_{x-})L = 2m\alpha_R L / \hbar^2 \tag{4.20}$$

between the $+z$ - and $-z$ -polarized electrons after propagation along a distance L in the 2DEG being proportional to α_R. At the drain contact the transmission probability of electrons depends on the angle between the incoming spins and the fixed magnetization direction of the contact, the spin-resolved conductance

$$G_{\pm} = e^2(1 \pm \cos\theta_{\mathrm{R}})/h \tag{4.21}$$

for electrons spin-polarized along x oscillating as a function of the spin-orbit coupling strength α_{R}. The state of the transistor is ON when current flows through it or OFF when the gate prohibits current flowing.

The spin-orbit coefficient, and hence the current modulation, can be controlled via the gate voltage, which changes the macroscopic electric field in the 2DEG. Since in an InGaAs/InAlAs heterostructure $\alpha_{\mathrm{R}} = 3.9 \times 10^{-12}$ eV m, phase differences of π can be obtained for $L = 0.67$ μm; a value that can be made lower than the mean free path in high-mobility semiconductors at low temperatures. Gate voltage control of the spin-orbit coefficient α_{R} was demonstrated in a gated inverted $In_{0.53}Ga_{0.47}As/In_{0.52}Al_{0.48}As$ heterostructure [90]: the spin precession angle changed from π to 1.5π as the gate voltage changed from 1.5 V to –1 V for a gate length L = 0.4 μm. 100% modulation of α_{R} in a gated $In_xGa_{1-x}As/In_xAl_{1-x}As$ quantum well was demonstrated in [91]. Moreover, in InAs square asymmetric quantum wells with an applied positive back-gate voltage the Rashba parameter increased with the electron density and decreased (for a constant electron density) by a factor of 2 when an additional positive front gate voltage was applied, due to gate-control position of the electron wave function [92].

The differential phase shift θ_{R} in (4.20) is maximum for electrons with $k_z = 0$, whereas for $k_x = 0$ the effect is absent and no current modulation is expected. To restrict the angular spectrum of electrons, an electron waveguide along x instead of a 2DEG should be used, the mixing between different waveguide subbands being negligible for a sufficiently strong lateral confinement. In this case the differential phase shift is the same as in (4.20) for all subbands and for all energies, which eliminates quantum interference between different subbands and thus guarantees large current modulation even in multimode electron waveguides as long as electron transport between the source and drain Fe contacts is ballistic. The Rashba spin-orbit coupling parameter in quantum wires increases when the quantum confinement is stronger: it increases from 6.45×10^{-12} eV m to 9.94×10^{-12} eV m when the width of an InGaAs/InP quantum wire decreases from 1 μm to 600 nm [93].

In the presence of two weakly coupled Rashba bands in the electron waveguide, an additional spin rotation of injected electrons occurs at energies near the band crossing E_{cr}, as illustrated in Figure 4.19. In this case, the spin-resolved conductance in the two channels of the Datta-Das transistor with enhanced spin control becomes [94]

$$G_{\pm} = (e^2 / h)[1 \pm \cos(\theta_{cr} / 2) \cos \theta_R] , \tag{4.22}$$

where $\theta_{cr} = \theta_R d / k_{cr}$, with d the interband matrix element and k_{cr} the wave vector at band crossing. In the Datta-Das transistor with enhanced spin control the angle θ_{cr} can, in principle, be varied independently of θ_R through lateral gates that, however, alter the transverse confinement strength of the electron waveguide (or channel).

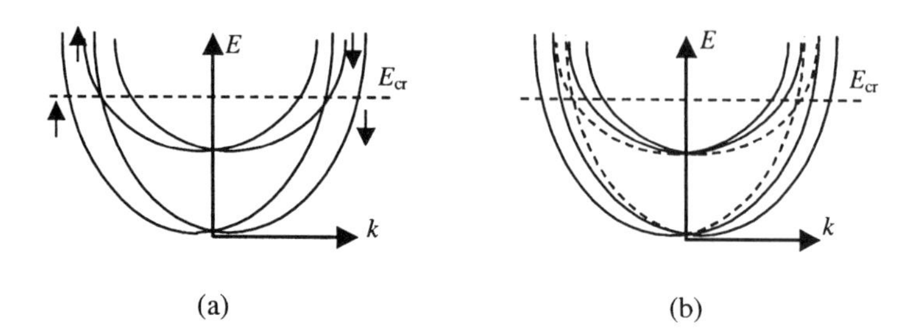

Figure 4.19 Band dispersion diagram (solid line) in the (a) absence and (b) presence of interband mixing in a Datta-Das transistor with enhanced spin control, which contains two bands split by spin-orbit (Rashba) coupling; the dashed lines in (b) are the energy bands in (a) (*After*: [94]).

When, instead of a static voltage, a time-dependent gate bias is applied on the Datta-Das transistor the time variation of the Rashba spin-orbit interaction generates a force $(m/2)[d\alpha_R(t)/dt]$ that acts in opposite directions on opposite electron spins. This force can generate a spin current but does not induce an electric current. On the other hand, the spin current in the electron gas provokes a variation of the gate voltage, a phenomenon that can be used to detect spin currents by measuring the voltage on a gate coupled to a nearby 2DEG, which carries a time-dependent spin current [95].

The theoretical modeling of the Datta-Das spin transistor configuration has received a great deal of attention. For example, when besides the Rashba interaction the influence of bulk inversion asymmetry in zinc-blende semiconductors is accounted for, the conductance is found to depend significantly on the crystallographic orientation of the channel [96]. In particular, numerical simulations of ballistic electron transport in semiconductor quantum wires show that optimal current modulation in the Datta-Das transistor is observed in channels oriented along $[\bar{1}20]$ for GaAs devices and along [100] for InAs devices. In [97] it is demonstrated that the variation of gate voltage moves the position of the Fermi level with respect to the conduction band edge, which causes conductance modulation when the Fermi level sweeps through the resonant energy levels above the barrier between the contacts. This so-called Ramsauer effect, which was not considered in the original proposal of the Datta-Das transistor, can overcome the conductance modulation due to the Rashba effect unless the contacts are properly engineered. Other simulations show that an almost square-wave modulation of spin transmission is achieved in a quantum wire with only one occupied subband and with periodically modulated Rashba interaction when spin-orbit subband mixing of the two lowest subbands is considered [98].

Efforts have also been made to implement the Datta-Das spin transistor, but, up to now, no experimental realization of this device has been reported. In this respect, single domain FM electrodes are not suitable because, although they show a defined magnetization direction for all magnetic fields, they also generate strong stray fields that cause a strong local Hall effect in the neighboring semiconductor, which overshadows small spin-polarized injection effects. However, appropriately tailored permalloy electrodes with a multidomain structure are suitable for spin injection in semiconductor/FM hybrid devices with a high degree of polarization and small stray fields [99].

In [100], a similar configuration to the Datta-Das transistor was proposed, in which the magnetizations of the source and drain contacts are transverse to the 1D channel instead of parallel and in which the current is modulated by the Dresselhaus instead of the Rashba spin-orbit interaction. In this configuration, the magnetic field in the channel caused by the FM source and drain electrodes in the original Datta-Das configuration is absent (except eventually the fringing field). The magnetic field in the channel is undesirable since it causes Zeeman spin splitting that affects the dispersion relation and hence generates spin mixing in each subband. This induces spin flipping (nonballistic transport) at nonmagnetic scattering events and/or energy dependence of the phase shift, which decreases the current modulation by ensemble averaging. In the device in [100] the Rashba interaction is negligible since the 2D [100] GaAs channel has no structural inversion asymmetry and the conditions for ballistic transport are relaxed since no Dyakonov-Perel' spin relaxation mechanism exists in the absence of a channel magnetic field. The eigenspinors in this case are $+x$ and $-x$ polarized states, with x the source-drain current direction, the corresponding eigenenergies being $E(+x) = \varepsilon + \hbar^2 k_{x+}^2 / 2m + \alpha_D k_{x+}$ and $E(-x) = \varepsilon + \hbar^2 k_{x-}^2 / 2m - \alpha_D k_{x-}$, with ε the energy of the lowest subband and $\alpha_D = 2\beta[m\omega/(2\hbar) - (\pi / W)^2]$ the strength of the Dresselhaus interaction expressed in terms of a material constant β, the curvature ω of the parabolic potential in the z direction defined by the split-gate Schottky contact that defines the 1D channel and W the channel width in the other transverse direction to the current, y. The parallel magnetizations of the source and drain contacts are along z. In these conditions the phase shift between orthogonal spin eigenstates is $\theta_D = 2m\alpha_D L / \hbar^2$ and the source-to-drain conductance is modulated by changing the split-gate voltage, which modulates the strength of the Dresselhaus interaction α_D through ω (in the Datta-Das transistor a top gate is used for modulating the Rashba interaction). The phase shift is independent of the electron energy or wave vector so that conductance modulation survives ensemble averaging over energies at high temperatures. In GaAs channels, with $\beta = 2.9 \times 10^{-29}$ eV m^3 and L = 10 μm a voltage swing of 70 mV can switch the transistor from the ON to the OFF state.

Variant configurations of the Datta-Das transistor have also appeared in the literature. For example, in [21] a Datta-Das-like spin transistor was proposed that works in the less technological demanding drift-diffusive regime, under the combined action of bulk inversion asymmetry and structural inversion asymmetry on a [001]-grown 2DEG. The gate bias controls, in this case, the lifetime of spins in the 2DEG, randomizing them or aligning them with the spins in the FM collector. From the corresponding Hamiltonian in equation (4.5) it follows that when $\alpha_{\mathrm{D}} = \alpha_{\mathrm{R}}$ the spin eigenstates point along $[1\bar{1}0]$ or $[\bar{1}10]$ depending on the angle between the electron wave vector and the [100] direction; also, for spins along $[\bar{1}10]$ the spin lifetime due to the D'yakonov-Perel' relaxation mechanism is proportional to $(\alpha_{\mathrm{D}} - \alpha_{\mathrm{R}})^{-2}$, having a resonant behavior for $\alpha_{\mathrm{D}} = \alpha_{\mathrm{R}}$. If the FM source and drain contacts have magnetizations along $[\bar{1}10]$, the gate bias can control the lifetime of spins injected in the 2DEG by tuning α_{R} into equality (or not) with α_{D}. In the first case (ON state) the spins arrive aligned with the magnetization of the FM collector and the resulting resistance is low, while in the second case (OFF state) the spins are randomized and the resistance has a high value. Because both turn-on and turn-off times are of the order of a few picoseconds, the operating frequency can reach a few hundred GHz. This spin transistor configuration can work as a flash memory if the gate is of the charged/uncharged floating type, as a different type of nonvolatile memory configuration if the gate bias is set to the resonance condition $\alpha_{\mathrm{D}} = \alpha_{\mathrm{R}}$ during read cycle only and is off resonance otherwise, or as a magnetic information readout head. The nonballistic Datta-Das-like spin transistor works even if the channel is a quantum wire instead of a 2DEG, current modulation in the presence of both Dresselhaus and Rashba effects occurring also for ballistic transport [101].

A nonmagnetic spin transistor can be made using a configuration similar to that in Figure 4.18, in which the FM source and drain contacts are replaced by (110) InAs/AlSb/GaSb heterostructures with a large bulk inversion asymmetry spin splitting. In this case, spin selection occurs through spin-dependent resonant interband tunneling in the presence of lateral electric fields that align the energy of incident electrons from the bulk InAs to the spin-split resonant states in the GaSb well [102]; almost 100% spin-polarized injection is expected. The transistor action is again accomplished via gate control through the Rashba effect of spin decay time associated with precessional relaxation in the symmetric InAs 2DEG. More precisely, the gate bias generates an in-plane magnetic field component that induces rapid precessional relaxation of the electron spins polarized along the InAs/AlSb/GaSb heterostructure growth direction and a dramatic decrease of the spin relaxation time. (This relaxation time is very long in the absence of the gate bias since both the injected electron spins and the bulk inversion asymmetry crystal magnetic field are directed along the growth direction.) Thus, at the collector resonant interband tunneling heterostructure the voltage has a finite value when no gate bias is applied (the spins do not relax) and no voltage is detected for small gate fields ($F < 5$ kV/cm).

All devices described so far are spin field-effect transistors, or SPINFETs. In SPINFET devices with 1D channel (such that the OFF conductance is zero) that rely on the gate-controlled Rashba interaction to modulate the source-to-drain current, the major spin relaxation mechanism in the channel (the D'yakonov-Perel' mechanism) can be completely eliminated; these are the only spin devices that could compete with present day Si CMOS devices, which monopolize the technology. However, detailed calculations show that InAs Datta-Das transistors with 1D channels have lower switching voltages than traditional FETs only if their channels are longer than 4.88 μm [103]; the low-power expectations of SPINFETs are not supported by calculations. Moreover, the transconductance per unit channel width, which determines the amplification and bandwidth of field-effect transistors, have lower values for the Datta-Das transistor than for high-electron-mobility GaAs transistors. The modified Datta-Das configurations in [94], [101], and [21], presented above, have even worse performances. Therefore, SPINFET devices can only become competitive with their electronic counterparts in memory devices due to their better noise margin (spins are less prone to couple with stray electric fields) or in nonconventional quantum computing applications.

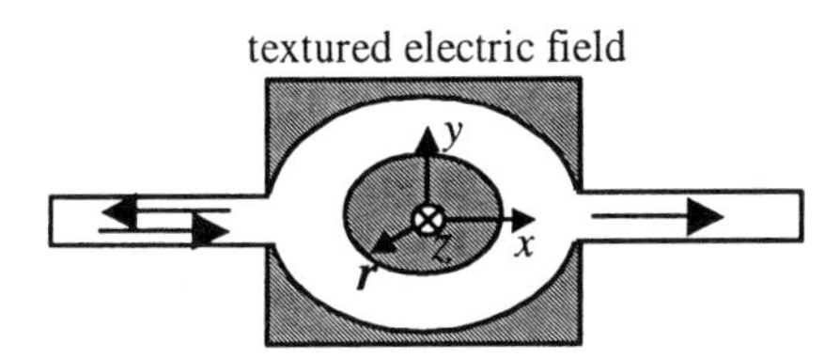

Figure 4.20 Schematic representation of a conducting ring in a transistor configuration (*After*: [104]).

Spin precession, which leads to current modulation via the Aharonov-Casher effect, can also be achieved in a conducting ring placed in a symmetrically textured electric field, as shown in Figure 4.20 [104]. The conducting ring is connected to the FM source and drain through two leads, the textured electric field being implemented by electrodes placed above and below the ring. This textured electric field influences the magnetic moment of charge carriers in the ring through the spin-orbit coupling, causing energy splitting between spin-up and spin-down electrons. The orientation of local spins in the ring is governed by the spin cyclic evolution along the ring, the quantum interference at the output lead giving rise to oscillations in conductance as a function of the electric field. Unlike in the Datta-Das transistor, where the phase difference between two spin states determines spin precession, in this device the electricfield-induced Aharonov-Casher phase difference of two spin eigenstates between the source and drain does not affect the spin polarizability but controls the transmission coefficients of the spins up- and down-polarized with respect to the z axis:

$$\begin{pmatrix} t_\uparrow \\ t_\downarrow \end{pmatrix} = \frac{i\sin(k\pi a)\sin(\phi/2)}{\sin^2(\phi/2)-[\cos(k\pi a)-(i/2)\sin(k\pi a)]^2}\begin{pmatrix} \cos(\alpha-\beta) \\ \sin(\alpha-\beta) \end{pmatrix}. \tag{4.23}$$

Here a is the radius of the ring, k the electron wave vector, ϕ the angle made by the radial component of the electric field $F = F(\hat{r}\cos\chi - \hat{z}\sin\chi)$ with the x axis, $\tan\beta = (\mu Fa/\hbar c)\sin\chi/[1+(\mu Fa/\hbar c)\cos\chi]$ with $\mu = g\mu_{\mathrm{B}}$ the magnetic moment of the charge carrier, and $(\cos\alpha, \sin\alpha)$ denotes the spin state of the injected electrons from the FM source. After going through the ring, this spin state evolves into a different one: $[\cos(\alpha-\beta), \sin(\alpha-\beta)]$, the source-drain current under an applied bias V is $I_{\mathrm{e}} = I_\uparrow + I_\downarrow = V(e^2/h)(|t_\uparrow|^2 + |t_\downarrow|^2)$ and the spin current along z is given by $I_{\mathrm{s}} = (\mu_{\mathrm{B}}/e)(I_\uparrow - I_\downarrow) = V(e\mu_{\mathrm{B}}/h)(|t_\uparrow|^2 + |t_\downarrow|^2)$ $\times\cos[2(\alpha-\beta)]$. The electron current is independent of the incident spin state α or of β (related to the Aharonov-Casher phase), while the transmission polarized spin current is controlled by both α and β. The spin polarization along z of the transmitted current is analogous to the spin injection in FM/semiconductor/FM heterostructures.

Ballistic spin transport through a finite chain of rings in the presence of Rashba spin-orbit coupling was studied in [105]. It was found that gaps in the conductance as a function of the spin-orbit strength α_{R} or of the electron wave vector k appear due to destructive interference of electrons propagating in opposite directions. These gaps widen and the conductance becomes almost square-wave-like if α_{R} or the ring radius are periodically modulated along the chain.

Another quantum interference device, in which the spin-dependent conductance can be modulated by the Rashba spin-orbit interaction manipulated by front and back gates is represented schematically in Figure 4.21 [106]. It is a Mach-Zender interferometer with quantum wires and perfectly reflecting mirrors M as interferometer arms and incident and output spin-polarized electron components $A = (A_\uparrow, A_\downarrow)$, $B = (B_\uparrow, B_\downarrow)$, and $C = (C_\uparrow, C_\downarrow)$, and $D = (D_\uparrow, D_\downarrow)$, respectively. If only the lowest subband in the quantum wires is occupied and the width W of the wire is much smaller than the spin-orbit-induced spin precession length L_{so} the spin eigenstates are perpendicular to the wire and in the plane of the device, the corresponding electron wave numbers for a given energy E being $k_s = k_E$

$-s\pi / L_{so}$ with $s = \pm 1$ and $k_E = [2mE/\hbar^2 + (\pi / W)^2]^{1/2}$. Electron transport from A to C and from B to D is characterized by the spin-independent conductance $G = (1/8)[1 - \cos(2\pi w / L_{so})][1 - \cos(2\pi h / L_{so})]$ (normalized to e^2 / h), while horizontal electron transport from A to D and vertical electron transport from B to C are characterized by spin-dependent normalized conductances

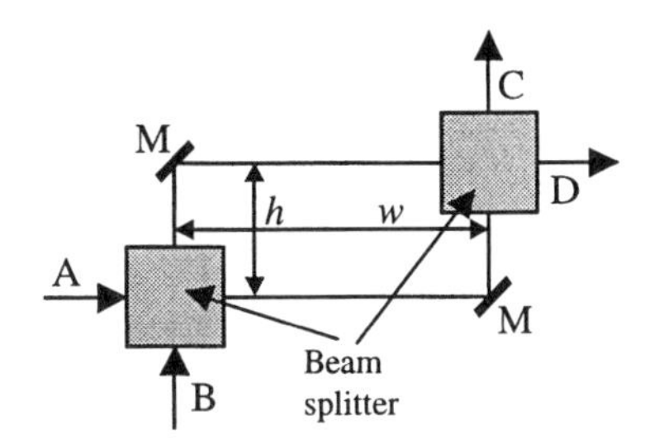

Figure 4.21 Schematic representation of the spin-dependent electronic Mach-Zender interferometer (*After*: [106]).

$$G_{H\uparrow\uparrow} = (1/2)[1 + \cos(2\pi h / L_{so})], \tag{4.24a}$$

$$G_{H\uparrow\downarrow} = (1/4)[1 + \cos(2\pi w / L_{so})][1 - \cos(2\pi h / L_{so})], \tag{4.24b}$$

$$G_{V\uparrow\uparrow} = (1/2)[1 + \cos(2\pi w / L_{so})], \tag{4.24c}$$

$$G_{V\uparrow\downarrow} = (1/4)[1 - \cos(2\pi w / L_{so})][1 + \cos(2\pi h / L_{so})], \tag{4.24d}$$

respectively, which have different values for conserved or flipped spins. As in the Datta-Das transistor, the conductances are independent of the electron energy. From formulas (4.24a)–(4.24d) it follows that the Mach-Zender interferometer has no effect if both the width w and the height h of the interferometer are integer multiples of L_{so}, the spin in the vertical channel is flipped for w a half-integer multiple of L_{so} and h an integer multiple, spin flip in the horizontal channel occurs for w an integer multiple of L_{so} and h a half-integer multiple, while pure reflection takes place when w and h are half-multiples of L_{so}. The device can act either as a voltage-controlled switch or as a spin-invertor, depending on the relation between w and h, without the need of magnetic contacts or applied magnetic fields.

In a metal-insulator-semiconductor field-effect transistor [107] as that shown in Figure 4.22(a) the FM properties of the (In,Mn)As semiconductor layer can be controlled by the voltage applied on the metallic gate. The gate voltage modifies the hole concentration in the semiconductor layer and thus the hole-mediated FM

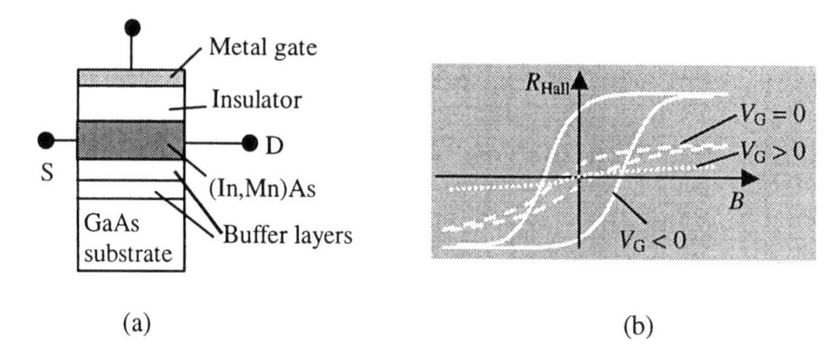

Figure 4.22 (a) Shematic representation of the metal-insulator-semiconductor field-effect transistor and (b) the dependence of the Hall resistance (magnetization) on the gate voltage (*After*: [107]).

exchange interaction between the localized Mn spins. This modification of the magnetic properties is evidenced by the variation of the sheet Hall resistivity

$$R_{\text{Hall}} = R_0 B / d + R_{\text{an}} M / d \tag{4.25}$$

with the gate voltage, shown in Figure 4.22(b). Here R_0 is the ordinary Hall coefficient, d is the thickness of the channel layer, R_{an} is the anomalous Hall coefficient and M is the magnetization perpendicular to the layer. At temperatures below the transition temperature the anomalous Hall effect dominates and a moderate hysteresis curve of the Hall resistivity of the (In,Mn)As FM layer with perpendicular easy axis can be observed at $V_G = 0$. A depletion of holes induced by a positive gate voltage determines a paramagnetic response with low susceptibility and no hysteresis, whereas a negative gate voltage produces a clear square hysteresis. Thus, the magnetic properties of the channel are modified reversibly and isothermally by a change in V_G. A gate bias swing of 125 V also changes the transition temperature $T_c \cong 25$ K with 4% (with 1 K), in this temperature range applied electric fields being able to turn on and off the ferromagnetism of the (In,Mn)As layer. Manipulation of spin states in semiconductors can be achieved in this way.

Among the various recent proposals for spin transistors, an interesting one is the magnetic bipolar transistor, schematically represented in Figure 4.23 [108]. It is a common *n-p-n* transistor configuration with a magnetic *p* region, in which the transport properties are determined by electrons, holes, and their spins. In particular, the current amplification when the emitter-base junction is forward polarized and the base-collector junction is reverse polarized can be controlled by spin. If, for simplicity, we consider that in the base region only the conduction band is split by 2Δ (either by the Zeeman effect in DMS with large g factors in a magnetic field or by exchange coupling in FM semiconductors), the equilibrium spin polarization in this region is $\eta_{0B} = \tanh(\Delta / k_B T)$. In addition to the equilibrium spins, a nonequilibrium spin population η_E can be induced in the emitter region by either electrical or optical external spin injection. Then, for emitter and collector widths within the spin diffusion length, in the narrow base limit and under the assumption of slow carrier recombination, as in Si-based transistors, the transistor gain is given by

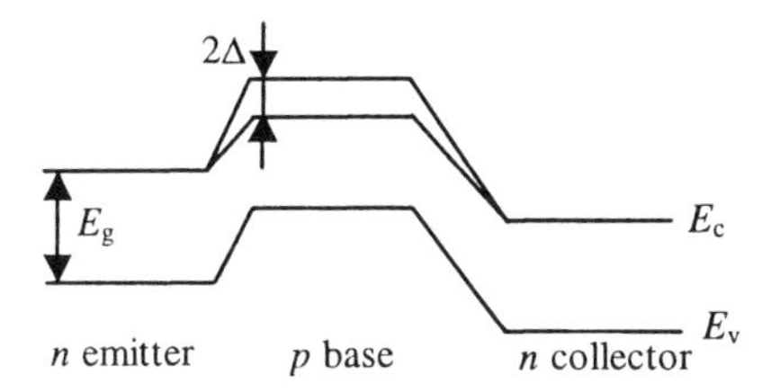

Figure 4.23 Schematic representation of a magnetic bipolar transistor with a forward polarized emitter-base junction and a reverse polarized base-collector junction (*After*: [108]).

$$\beta = \beta_0 (1 + \eta_E \eta_{0B}) / \sqrt{1 - \eta_{0B}^2} , \qquad (4.26)$$

where β_0 is the gain factor in a nonmagnetic transistor; the gain factor is controlled by both equilibrium spin population in the base region and by the nonequilibrium spin population in the emitter. Moreover, the flow of electrons in a magnetic bipolar transistor generates nonequilibrium spin accumulation in the collector region, indicating the ability of electrical spin injection. For typical material parameters of a Si-like transistor and for $\eta_E = 0.9$, β can vary with more than 50% for a variation of η_{0B} from 0 to 0.8, the amplification being larger for parallel orientation of the source and equilibrium spins. A detailed theoretical analysis of the magnetic bipolar transistor can be found in [109].

A transistor configuration very similar to that in Figure 4.23 can produce high current gains at room temperature [110]. It consists of a FM emitter that injects spin-polarized electrons into a (n- or p-type) Si base via a tunnel barrier, the spin-polarized minority carriers in the base traversing it diffusively and being collected by the half-metallic FM collector via another tunnel barrier. The back-biased collector assures spin selectivity through the density of final spin states, which differs from that of spin-polarized minority carriers in the base; the collector efficiency and the current gain (the differential ratio of the collector current to base current $\Delta I_C / \Delta I_B$) are controlled through switching the magnetic state of the collector. Experimental room temperature data indicate high current gain values, of 1.4 for an *n*-type Si base and 0.97 for a *p*-type Si base, while the magnetocurrents, defined as $MC = 100 \times (I_{C\uparrow\uparrow} - I_{C\uparrow\downarrow}) / I_{C\uparrow\downarrow}$ with $I_{C\uparrow\uparrow}$ and $I_{C\uparrow\downarrow}$ the collected currents in the parallel and antiparallel configurations, have corresponding values of 140% and 98%, respectively, at –110 Oe for a base current of 1 μA.

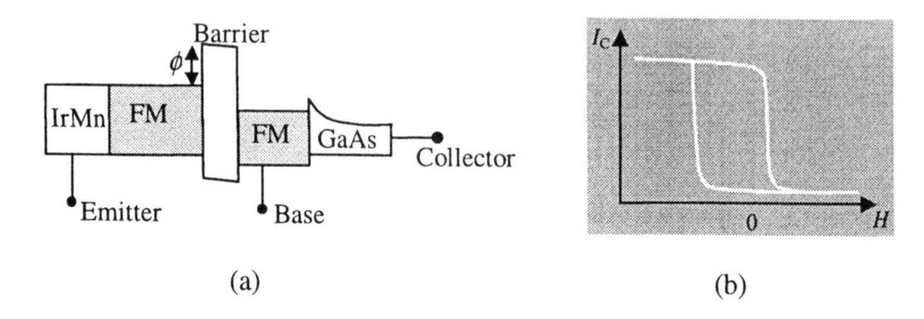

Figure 4.24 (a) Energy band diagram in a magnetic tunneling transistor and (b) a typical dependence of the collector current on the applied magnetic field (*After*: [3]).

A distinct type of spin-based transistor is the so-called magnetic tunneling transistor [1, 3], in which, for example, the spin polarized hot electrons emitted by the FM layer arrive at the FM base layer after tunneling across a barrier, as shown in Figure 4.24(a). The antiferromagnetic IrMn layer pins the magnetic moment in the emitter and thus allows independent switching of the magnetic moment of the base in an external magnetic field. The current in a semiconductor collector (GaAs for example) is determined by the electrons with a higher energy than the Schottky barrier height ϕ and depends on the magnetization directions of the emitter and base FM layers since spin-dependent scattering of hot electrons in the base preferentially scatters electrons with spin antiparallel to the majority spins in the base. A variation of the emitter-base voltage allows the exploration of hot electron transport over a large energy range of emitted electrons. This transistor is characterized by the magnetocurrent $MC = (I_{C\uparrow\uparrow} - I_{C\uparrow\downarrow}) / I_{C\uparrow\downarrow}$, where $I_{C\uparrow\uparrow}$, $I_{C\uparrow\downarrow}$ are the collector currents for parallel and antiparallel orientations of the emitter and base magnetizations; the dependence of the collector current on the applied magnetic field H is represented in Figure 4.24(b). The MC can reach 73% for an emitter-base bias of 1.8 V [3]; the MC has a nonmonotonic behavior with the emitter-base voltage, which depends on the conduction band structure of the collector material [1].

A hybrid Au/Co/Cu/NiFe/*n*-GaAs Schottky barrier structure similar to the one in Figure 4.24 but with no applied bias was used as a spin valve in [111]. Experiments show a 2400% increase in the helicity dependent photocurrent for photon energies above the Schottky barrier height accompanied by an increase in the spin polarization by a factor of more than 28 (from 0.2% to 5.9%) when the magnetization of the Co and NiFe FM layers are switched from parallel to antiparallel. Spin polarized ballistic electrons are thus also efficient filtered by this device. The measurements have been done for an in-plane magnetic field along the easy axis of magnetization under photoexcitation of the GaAs layer with laser light obliquely incident to the device, for which the photon helicity has an in-plane spin polarization component parallel to the magnetization direction of the ferromagnets.

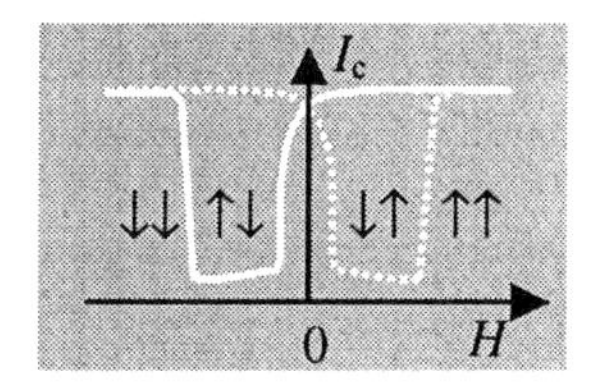

Figure 4.25 Typical dependence of the collector current on the magnetic field for a magnetic tunnel junction with a spin-valve base (*After*: [3]).

In a related but different magnetic tunneling transistor structure in which the base is a spin-valve consisting from a FM/normal metal/FM structure, the injector is a nonmagnetic Cu layer separated through a Schottky barrier from the base and the collector is GaAs, a variation of *MC* with more than 1200% of obtained by switching the magnetic moments of the two CoFe and NiFe FM layers of the base from parallel to antiparallel at an emitter-base voltage of 1.6 V [3]. The dependence of the collector current on the applied magnetic field is represented in Figure 4.25; arrows indicate the magnetization directions of the FM layers in the base.

A quite unusual configuration of a spin transistor is represented in Figure 4.26 and is known as the spin-torque transistor [112]. It consists of two spin-flip transistors (three-terminal devices in which the conducting channel of an antiparallel spin valve is in contact with a FM base) with a common base contact and source-drain contact magnetizations rotated with $\pi/2$. The FM base is in good electric contact with the normal metal nodes and is magnetically very soft. In this device, the source-drain current I_{DS} of the lower part of the device can be controlled by varying the base magnetization direction θ through a second, upper spin valve. For source and drain contacts made from high-coercivity metallic magnets biased by an electrochemical potential V_S, I_{DS} generates spin accumulation in the normal metal node N_1. If the magnetization angle of the FM base is not 0 or π the spin accumulation decreases due to the spin current that flows into the base and I_{DS} increases as θ increases up to $\pi/2$. The magnetization in the base is determined by the condition that the torque on the base exerted by the spin accumulation in N_1 is canceled by the torque exerted by the upper spin valve in N_2. If V is the applied potential on the upper spin valve the angle θ_0 for which the two torques cancel for upper and lower spin valves with the same parameters is determined from

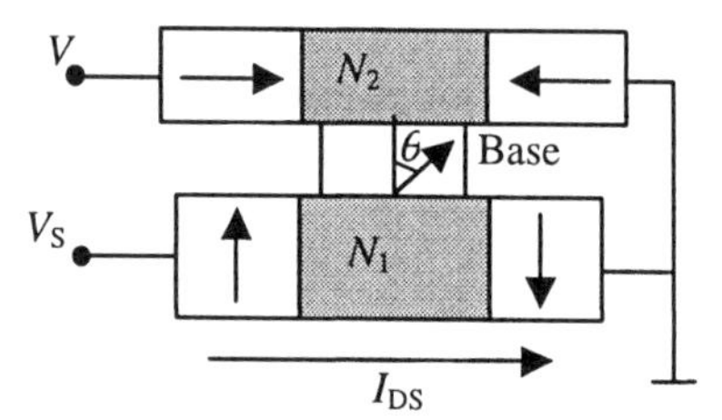

Figure 4.26 Schematic representation of the spin-torque transistor (*After*: [112]).

$$\frac{V}{V_S} = \frac{7+\cos(2\theta_0)}{7-\cos(2\theta_0)} \tan\theta_0 , \tag{4.27}$$

a variation in V modulating θ and, hence, I_{DS}. The source-drain differential conductance $G_{DS} = (\partial I_{DS}(\theta)/\partial V_s)_V$ defined for a constant V develops strong nonlinearities with increasing source polarization p_S, a negative differential resistance appearing at $V \cong V_S$ due to the competition between the increasing torque (which decreases the current) and the ohmic current (which increases with the bias). The differential current gain Γ, defined as the ratio between the differential transconductance $T = (\partial I_{DS}(\theta)/\partial V)_{V_S}$ calculated for a constant V_S and the channel conductance G_{DS}, is

$$\Gamma = \frac{\theta_0 / 2}{(1 - p_S^2)/(1 + p_S^2) - \theta_0^2 / 3} \tag{4.28}$$

for $V << V_S$. Γ is proportional to θ (to the control potential V) for small polarizations but becomes singular and changes sign for polarizations close to unity. This spin-torque transistor, based on the spin-transfer effect, can work for elevated temperatures with conventional FM contacts for source and drain and highly resistive base contacts fabricated from a magnetic insulator or from two magnetic films coupled through a thin insulating barrier, as long as the electron dwell time in the device is larger than the spin-flip relaxation time. The optimum situation is that of (nearly) half-metallic ferromagnets for source and drain contacts.

4.2.6 Spin-Based Optoelectronic Devices

As spin=polarized electrons can be created by left- and right-circular polarized light, σ^- and σ^+, the recombination of spin-polarized charges generates circularly polarized light. Besides these connections between spin polarized electrons and circularly polarized light, it was recently shown that the threshold current of the vertical-cavity surface-emitting laser (VCSEL) can be reduced if pumping is done with spin-polarized electrons, which can be generated by either optical or electrical injection [113]. Experimental data on a VCSEL with InGaAs quantum wells as active media and AlAs/GaAs Bragg reflectors as mirrors showed that the threshold current decreases from 0.65 A/cm^2 to 0.5 A/cm^2 when instead of producing randomly aligned electron spins (by optical excitation with linearly polarized light) spin-polarized electrons are generated by excitation with σ^+ polarized light, which generates electrons with 50% spin polarization. The reason is that spin-polarized electrons predominantly contribute to a circular polarized laser mode and drives it above threshold if the pump power is slightly below the laser threshold for randomly aligned electron spins. The influence of spin polarization on the laser threshold is exemplified by the fact that optical pumping with a polarized radiation that varies between σ^+, $\pi/4$ linear polarization, σ^-, $-\pi/4$ linear polarization, and so forth, produces an intensity modulation of 400% in the emitted laser radiation when the input pump intensity is between the threshold for +50% and –50% spin polarization (obtained by σ^+ and σ^- pumping). The maxima in emitted radiation correspond to σ^+ and σ^- pumping. The same behavior is expected for electrical pumping of spin-polarized electrons. The reduction of threshold currents indicates that spin lasers can work at higher temperatures than lasers with unpolarized electron spins.

In semiconductor GaAs/AlGaAs microcavities with embedded quantum wells working in the regime of strong light-exciton coupling, it was experimentally demonstrated that the emitted light becomes almost fully circularly polarized at the stimulation threshold of exciton-polaritons scattering under nonresonant optical pumping with either circular or linear polarized light [114]. Only the threshold power is influenced by the polarization of excitation (it is lower for linear pumping). The degree of circular polarization of the emitted remains almost 100% at higher pumping intensities with circular light but decreases for linear pumping. Thus, stimulated scattering of exciton-polaritons amplifies a selected polarization and inhibits all spin-relaxation processes, being useful in spin-dependent optoelectronic devices that manipulate light polarization on the mesoscopic scale. Under stimulated scattering the microcavity acts as a spin-polarization converter from linear to circular polarization.

Conversely, a solar cell illuminated with circularly polarized light produces charge and spin currents, the spin polarization of current greatly exceeding the spin polarization of carrier density for the majority carriers [115]. Generally, optical spin polarization of minority carriers is achieved by optical orientation while that of majority carriers is done by optical pumping. In a GaAs *p-n* junction with ideal ohmic contacts electron-hole pairs are created by circularly polarized light, the built-in field in the depletion region sweeping the spin-polarized electrons to the *n* region and the unpolarized holes, which lose their spin on the time scale of momentum relaxation, to the *p* region. The equation for charge currents in this spin-polarized solar battery is the same as in the unpolarized case since in nondegenerate semiconductors the diffusivities of spin-up and -down carriers are equal and hence spin polarization does not affect charge current. The equation for spin diffusion is

$$Dd^2s/dx^2 = (R_{\text{rec}}p + 1/T_1)s - \Delta R_{\text{gen}}, \tag{4.29}$$

where D is the electron diffusion length, $s = n_\uparrow - n_\downarrow$ with $n_\uparrow$, $n_\downarrow$ the spin-up and spin-down electron densities, R_{rec} is the electron-hole recombination rate, p is the hole density and $\Delta R_{\text{gen}} = R_{\text{gen}\uparrow} - R_{\text{gen}\downarrow}$ is the difference between the generation rates of spin-up and -down electrons. This equation must be solved together with the boundary conditions on the p side: $s = 0$ at the ohmic contact situated at $x = 0$, and at the edge of the depletion layer, where the generated carriers are immediately swept away by the built-in field. The solution for the spin density is

$$s(x) = s_p[1 - \cosh\xi + \sinh\xi(-1 + \cosh\xi_p)/\sinh\xi_p], \tag{4.30}$$

where $\xi = x/L_s^p$, $\xi_p = x_p/L_s^p$ with $L_s^p = (D\tau_s)^{1/2} \cong 0.8$ μm the spin decay length on the p side, $\tau_s \cong 0.067$ ns the electron spin decay time in the p region, and x_p the depletion layer boundary on the p side. The spin current density is $I_s = eDds/dx$, while the spin polarization is $\eta = s/n = (n_\uparrow - n_\downarrow)/(n_\uparrow + n_\downarrow)$. The 0.41 value of η near the ohmic contact is larger than the bulk value of 0.33. The current polarization $\eta' = (I_\uparrow - I_\downarrow)/(I_\uparrow + I_\downarrow)$ with $I_{\uparrow,\downarrow}$ the charge currents of spin-up and -down electrons is about –0.19 at the ohmic contact and takes the opposite values 0.19 at the edge of the depletion layer. Similarly, in the n region the boundary values of $s(L) = 0$ at the ohmic contact $x = L$ and $s(x_n) = s_0$ at the boundary x_n of the depletion layer give the solution

$$s(x) = s_n[1 - \cosh\varsigma + \sinh\eta(-1 + \cosh\varsigma_n + s_0 / s_n) / \sinh\varsigma_n] \tag{4.31}$$

with $\varsigma = (L - x) / L_s^n$, $\varsigma_n = (L - x_n) / L_s^n$, and $L_s^n = (DT_1)^{1/2} \cong 1.4$ μm the spin decay length on the n side ($T_1 = 0.2$ ns). The current spin polarization at the ohmic contact, $\eta'(L) \cong 0.33$ is smaller than the value 0.39 at x_n, but both are much larger than carrier polarization. At low temperatures and in long junctions illuminated only in the p region spin amplification can occur since $s_n / s_p = s_0 / s_p = (T_1 / \tau_s)^{1/2}$ and T_1 can exceed τ_s by orders of magnitude.

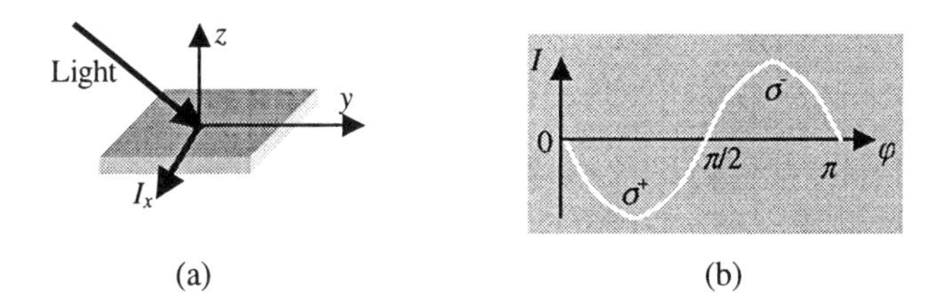

Figure 4.27 (a) Illumination with a circularly polarized light incident in the *yz* plane on a quantum well generates a perpendicular current, (b) which varies with the phase angle φ (*After*: [116]).

A quantum well, excited by circularly polarized radiation that generates nonequilibrium population of spin-up and spin-down states, acts as a spin battery that generates a spin-polarized current [116]. The direction of current is determined only by the predominant spin orientation. In particular, spin injection into quantum wells made of zinc-blende-type materials always generates an electric current in the quantum well plane due to the appearance of $\boldsymbol{k}$-linear terms in the reduced dimensionality Hamiltonian, which lift the spin degeneracy of energy bands inducing an unbalanced spin population and, hence, current flow. The spin photocurrent in the quantum well is perpendicular to the direction of illumination with circularly polarized light, as shown in Figure 4.27(a), and switches its direction when the helicity sign of the radiation and thus the spin orientation of free carriers is reversed. This phenomenon was observed in heterostructures with different symmetries, such as n-InAs/AlGaSb, n-GaAs/AlGaAs and p-GaAs/AlGaAs quantum wells with ohmic contacts illuminated with a high power infrared pulsed NH_3 laser. The radiation induces direct optical transitions between valence subbands in p-doped structures and indirect optical transitions in the lowest conduction subband in n-doped structures. Experimental data reveal that the current is normal to the direction of incident radiation and varies with the phase angle φ, which is connected to the helicity h of the incident light as $h = \sin 2\varphi$; left handed circular light σ^- corresponds to helicity –1 while right-handed light σ^+ corresponds to $h = +1$. A typical dependence of I on φ is represented in Figure 4.27(b). The angle φ can be varied by modifying the angle between the initial linear polarization plane of the laser beam and the optical axis of a subsequent $\lambda/4$ plate. The photocurrent is related to the helicity through $\boldsymbol{I} = \gamma \cdot \hat{\boldsymbol{e}} F^2 h$, where $\hat{\boldsymbol{e}}$ is the unit vector parallel to light propagation direction, F is the electric field of the radiation, and γ is a second-rank pseudotensor that depends (in both symmetry and sign) on the crystal symmetry and can have nonzero components in gyrotropic systems containing quantum wells from zinc-blende-type materials. The current generation upon illumination with circularly polarized light is a circular photogalvanic effect, which can be explained by transfer of the photon angular momentum into a directed motion of a free charge.

Unlike in the circular photogalvanic effect, where the spin polarized current is caused by the asymmetry of the momentum distribution of carriers excited in optical transitions sensitive to circular light polarization due to selection rules, unpolarized currents can be generated due to asymmetric $\boldsymbol{k}$-dependent spin-flip relaxation of spin-polarized electrons in systems with $\boldsymbol{k}$-linear terms in the electron Hamiltonian, which shift in $\boldsymbol{k}$-space the asymmetrically occupied spin-split energy bands; this is the spin galvanic effect [117]. The current is not spin-polarized if elastic scattering events occur but can become polarized for inelastic scattering. In the spin galvanic effect in a 2DEG situated in the (x,y) plane the current flows normal to the in-plane spin polarization of carriers, obtained by spin injection from FM contacts or by normal illumination of the 2DEG plane with circularly polarized light in the presence of an in-plane magnetic field. The incident light can cause intraband or interband optical transitions. For an illuminated 2DEG the spin-polarization η_{0z} induced by optical transitions is normal to the 2DEG plane (i.e., is along z) and is converted to an in-plane nonequilibrium spin polarization η_y by a magnetic field along x, B_x, which induces spin rotation with the Larmor frequency $\omega_L = g\mu_B B_x / \hbar$. Then,

$$\eta_y = -\omega_L \tau_{s,\perp} \eta_{0z} / [1 + (\omega_L \tau_s)^2] \tag{4.32}$$

with $\tau_s = (\tau_{s\parallel}\tau_{s\perp})^{1/2}$ and $\tau_{s\parallel}$, $\tau_{s\perp}$ the longitudinal and transverse electron-spin relaxation times, the polarity of the current, which flows along x, depending on both the direction of the excited spins (the left or right circular polarization of light) and on the direction of the applied magnetic field. A review of spin photocurrents generated by illuminating quantum wells with circular polarized radiation is reviewed in [118].

4.2.7 Spintronic Computation

Recent advances in spin-based computing systems have demonstrated amazing achievements. Magnetoresistive random access memories based on magnetoresistive tunneling through nanoscale barriers, which enhances the magnetoresistive response and allows the demonstration of a 256 kb memory chip [4], have been fabricated. In addition, ultrahigh-density ferroelectric data recording systems, which exceed 1 Tbit/inch^2 and which use scanning nonlinear dielectric microscopy technique with sub-nm resolution to read/write information in nanosized ferroelectric domains in a stoichiometric $LiTaO_3$ single-crystal film [119], have been demonstrated. In the latter case both read and write operations are performed by applying electric biases of 15 V between a conductive probe tip and an electrode placed behind the ferroelectric media for periods of tens or thousands of ns in order to produce inverted domain nano-dots that remain stable at least 24 h. The procedure is, however, time consuming. These technological performances enhance the applicability of magnetoelectronic devices in classical computation, based on Boolean logical operations that act upon registers of bits with only two logical values: 0 and 1. Boolean algorithms make free use of irreversible logical operations and the possibility of copying (multiplying) the logical value of the bit.

Impressive enhancements in computational speed have been predicted (at least in some cases) in quantum computing algorithms, in which the quantum bit (qubit) state ψ can be in any superposition

$$\psi = a\,|\,0\rangle + b\,|\,1\rangle \qquad (4.33)$$

of two eigenstates $|\,0\rangle$ and $|\,1\rangle$ of a quantum system that can be identified with the 0 and 1 logical values. This superposition principle, which is characteristic for quantum mechanical systems, allows parallel processing of the superposition states using algorithms based on reversible quantum operations, which are implemented by unitary Hamiltonian evolutions of the system. Unlike in irreversible logic, it is expected that, in principle, reversible logic operations can be performed with virtually no energy dissipation. Both irreversible logical operations and the possibility of copying (cloning) a quantum state are forbidden in quantum computa-tions. Moreover, the possibility of monitoring the evolution of a computation algorithm, which is a commonplace operation in classical computers, is forbidden in quantum computers because measurements of a quantum system alter its state; only the completion of the quantum algorithm can be eventually checked by measuring the state of a control qubit. More precisely, the superposition state in (4.33) cannot be recovered from measurements of the quantum state, which can only infer that the eigenstate $|0\rangle$ is occupied with probability $|a|^2$ and the eigenstate $|1\rangle$ with probability $|b|^2$; after the measurement the state of the quantum system is either $|0\rangle$ or $|1\rangle$. Thus, quantum and classical computing are based on entirely different physical principles; comprehensive reviews on this subject can be found in [74] and [120].

As in classical computing algorithms, quantum computation algorithms rely on universal operations or gates; any algorithm can be seen as a sequence of universal operations. The universal operations of quantum computing require the implementation of all one-qubit operations and of only a single two-qubit entangling gate, such as the $(\text{SWAP})^{1/2}$ or the CNOT gate. The potential increase in speed of quantum computers due to parallel processing, is hampered by decoherence. The preservation of quantum superposition (and hence parallel processing) implies lack of decoherence; so, the number of operations that can be performed by a quantum computer is limited by the decoherence time (often called coherence time since it represents the time in which coherence is preserved), which is unfortunately quite small in quantum systems that must interact with the environment in order to be able to implement quantum operations. The effects of decoherence can be (partially) corrected in fault tolerant algorithms if the computer can perform 10^4–10^6 gate operations.

As in any quantum computer, in solid-state computers it should be possible to identify a two-state quantum system with a qubit. It should also be feasible to initialize the state of the system, to drive this system from one state to another, to create superposition of qubit states, and finally, conditional logic operations, that is, the change of the state of one qubit only if another qubit is in a definite state. Isolation from environmental interactions is crucial for maintaining large dephasing times during which the quantum superposition states are not destroyed. The quest for solid-state computing systems originates from the difficulty of scaling up the number of qubits in the system without introducing decoherence. The most obvious two-state quantum system is a spin, which can have only two orientations in an applied magnetic field. Thus spin-based quantum computers have received considerable attention (see, for example, the review in [121]). Not all proposals of quantum computers based on magnetic materials encode the qubit onto the possible orientations of a single spin. For example, in [122] the ground and the first excited discrete spin states of nanometer-scale magnetic particles with large spin form the qubit states $|0\rangle$ and $|1\rangle$, while in high-anisotropy molecular clusters with significant tunneling through the anisotropy barrier the qubit states are the symmetric and antisymmetric combinations of the twofold degenerate ground state. We refer here only to spin-based quantum computers in which the spin orientations represent the qubit states. In this case, one-qubit operations can only be rotations of the spin direction; detection systems for reading the state of a single spin (eventually without destroying it) have been presented in Section 4.1.3.

Perhaps the most famous proposal of a spin-based quantum computer is the so-called Kane proposal [123], in which the qubits are the isolated (electron or nuclear) spins of donors in isotopically pure Si (with nuclear spin $I = 0$). The isolation ensures a long coherence time (up to hours) while individual control of spin dynamics is achieved via applied voltages on gates adjacent to the donors. The device must operate at low temperatures (up to 100 mK) and in strong magnetic fields $B \geq 2$ T to fully polarize the electron spins. To read out the state of single spin qubits, the spin quantum number is first converted to a charge polarization state and then the charge configuration is sensed with an exchange-coupled two-electron system. For example, P donors in Si have a nuclear spin $I = 1/2$ that is coupled with the spin $S = 1/2$ of the loosely bound electron through the hyperfine interaction energy expressed by the Hamiltonian term

$$H_{\text{hyp}} = A\boldsymbol{\sigma}_{\text{e}} \cdot \boldsymbol{\sigma}_{\text{n}} , \tag{4.34}$$

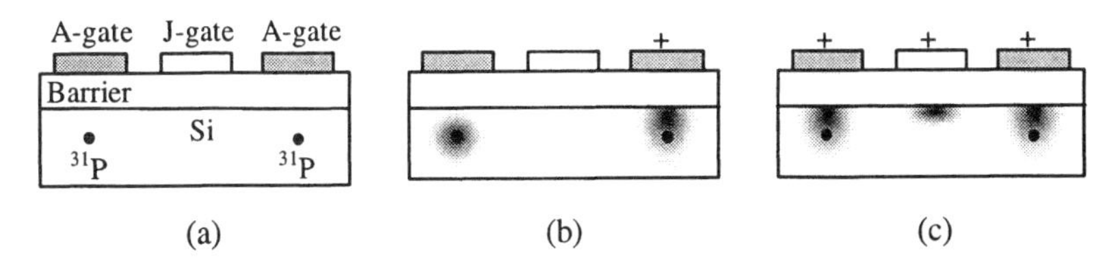

Figure 4.28 (a) The configuration of the Kane computer and the implementation (b) of one-qubit logical operations by biasing the A-gate, and (c) of two-qubit operations by biasing the A-gates and the J-gate (*After*: [124]).

where $\boldsymbol{\sigma}_{\text{e}}$ and $\boldsymbol{\sigma}_{\text{n}}$ are the Pauli spin matrices for the electron and nucleus with eigenvalues ± 1 and the contact hyperfine interaction energy A is proportional to the probability density of the electron wavefunction at the nucleus position. Because the nuclear spins in P are more weakly coupled to the environment than electron spins (these have coherence time of an hour instead of 1 ms for the electron spin coherence time) they act as qubits, while the delocalized and mobile electron spins in P couple to other electrons via the exchange interaction or to adjacent nuclei via hyperfine interaction. The electron spins are instrumental in implementing controlled quantum operations through biases applied on gates on the Si surface, which control the location and extent of the electron wavefunction. In a Kane computer, as that in Figure 4.28(a), one-qubit logical operations are performed by biasing the metal A-gate located above the donor side and thus by distorting the electron wavefunction around the donor, as shown in Figure 4.28(b). The change in the electron density at the nuclear site of the donor alters the hyperfine interaction energy A and also the A-dependent energy spacings of the spin levels. In this way, a single P donor can be selectively brought into or out of resonance with an applied rf external field B_{rf}. For $B_{\mathrm{rf}} = 10^{-3}$ T the single spin operation rate is 10–100 kHz in the presence of a magnetic field $B = 1$ T (in an applied magnetic field the electron and nuclear transitions become distinct but coupling between states with different nuclear spins decreases with increasing B, and so the computation speed decreases). Since the dephasing rate is about 0.1 s^{-1}, thousands of logical operations can be performed until the spins decohere [124].

Two-qubit logical operations involving adjacent spins, separated by a distance comparable to the Bohr radius, are performed by lowering the potential barrier between the donors via the positive bias on the J-gate and by a simultaneous turning on of the exchange coupling between the donors, as illustrated in Figure 4.28(c). On the contrary, a negative bias on the J-gate decouples the adjacent spins. The exchange coupling between electron spins S_1 and S_2 is described by the Hamiltonian

$$H_{\mathrm{exch}} = J(t)S_1 \cdot S_2, \tag{4.35}$$

with J the exchange energy. For a positively biased J-gate an exchange interaction between electron spins takes place and an indirect (electron mediated) nuclear spin exchange occurs between the donor nuclei if the electron spin exchange energy J is comparable to $\mu_{\mathrm{B}}B$ with B the applied magnetic field. This requirement imposes an upper limit of about 10–20 nm on the distance between the donors and the gates [124]. Larger separation between donor nuclei could be attainable if free electrons instead of electrons bound to the donor would transmit the quantum information. This is possible if the quantum information is first swapped between a nuclear spin and a bound electron at a donor site by a rf pulse, then the electron becomes ionized and is shuttled to a different donor site in a similar manner as in charge coupled devices. Finally, the electron is bound to a second donor and a second swap operation exchanges information between the electron and nuclear spin. Although technologically very demanding [125], it was shown that an atomically precise linear array of P on a Si surface can be fabricated by using a resist technology in which the resist is a layer of hydrogen atoms that terminate the Si surface. After a STM tip selectively desorbs individual hydrogen atoms, single phosphine (PH_3) molecules are adsorbed for the required placement of phosphorus atom in rows longer than 100 nm, with a pitch of 4 nm; however, the subsequent Si overgrowth that encapsulates the phosphorus has not been demonstrated. The successful fabrication of tiny metal gates suitable for the Kane computer architecture has been reported in [126].

A variant of the Kane computer, which is able to work at high temperatures and overcomes the tight technological requirements of the original proposal, is presented in [127]. The qubits are encoded in electron spins of deep donors, for example Bi in Si, which are not ionized at the working temperature and do not interact in the ground state, so that the spatial distance between them can be as large as 10 nm. Moreover, the gate electrodes located near the highly polarizable defects that are ionized except for very low temperatures are no longer required. A magnetic field is used to spin split the ground state of deep donors and to manipulate them. Initialization of qubit state can be made by spin injection or by polarization-selective optical pumping. Information transfer from a qubit A to a qubit B is controlled by optical excitation of a control impurity C (for example, Er), which excites the control electron from its ground state of energy E_{Cg} to an excited state E_{Ce} (see Figure 4.29) in which its wavefunction overlaps the qubit states of A and B, making exchange interaction between the qubit electrons via the control electron possible. Readout can also be performed via control electrons by tuning the laser frequency to the energy difference between the excited and ground states of the system formed from an extra qubit with a known state, a control atom and the qubit we wish to read, and by observing the appearance or not of scattered photons, situations that correspond to the two spin values of the unknown qubit state. Single qubits are manipulated by combining magnetic resonance and confocal optics. The qubit-qubit interaction is switched off by stimulated de-excitation of the control electron. To optically excite only one qubit pair with a radiation with a typical wavelength of 1 μm, high spatial resolution is, however, required.

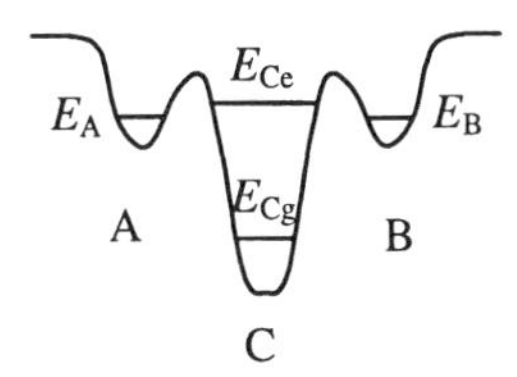

Figure 4.29 Energy diagram of the high-temperature quantum gate (*After*: [127]).

A quantum computer that does not require single-ion electrostatic gates, and in which the qubits are the nuclear-spins of $^{125}Te^{+}$ ions located on the surface of Si was proposed in [128]. In the presence of high magnetic fields $\boldsymbol{B}$ the localized electron spin of the ion precesses around $\boldsymbol{B}$ with the Larmor frequency, which depends on the nuclear spin state of the ion via the hyperfine interaction. A STM can then be used to detect the frequency of the Larmor precession since the interaction between the localized electron spin and the tunneling electrons cause a modulation of the tunneling current with the Larmor frequency. The nuclear spin of the ion is determined from the modulation frequency of the current, which is $f_{\uparrow} = (g\mu_B B + A/2)/2\pi\hbar$ if the nuclear magnetic moment is up and $f_{\downarrow} = (g\mu_B B - A/2)/2\pi\hbar$ if it points down; here g is the electron Landé factor and $A/2\pi\hbar \cong 3.5$ GHz is the constant of the hyperfine interaction. A selective excitation of a nuclear spin with Landé factor g_n is performed by flipping the nuclear spin through the application of a π-pulse (i.e., a pulsed rotating magnetic field B_{rot} of duration τ such that $\tau B_{rot} = \pi\hbar / g_n\mu_n$ with μ_n the nuclear magneton). Conditional logic operations can also be implemented by applying a series of π-pulses with definite frequency and making use of the dipole-dipole interaction between the electron and nuclear spins.

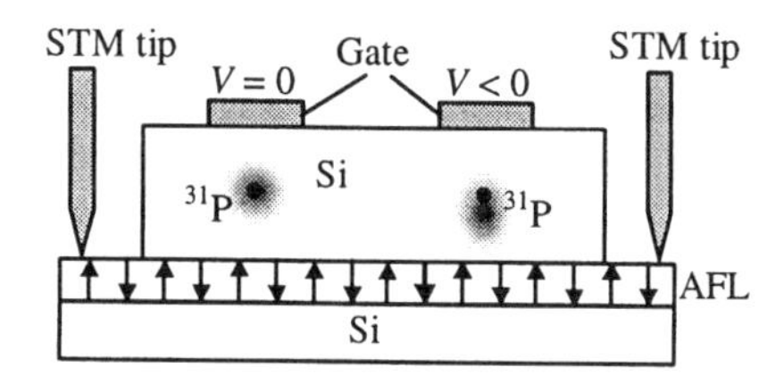

Figure 4.30 Schematic representation of a spin computer that utilizes spin waves for transmission of quantum information (*After*: [129]).

Yet another variant of the Kane proposal, which utilizes spin waves to transmit information between arbitrary nonadjacent qubits in the shortest time possible, has been described in [129] and is schematically represented in Figure 4.30. An antiferromagnetic layer (AFL) interacts selectively with the gated qubits on which the applied voltage is negative. A qubit with a spin state normal to the magnetization direction of the AFL can be rotated via exchange interaction between the electron wavefunctions of the donor and the AFL if a negative voltage is applied for a specific time duration. The originality of the proposal resides in the way in which two-qubit operations are performed: when the qubit electron couples to the AFL spin system, collective spin oscillations are excited, and propagate in the AFL as coherent cylindrical spin-waves, with group velocities of about 100 m/s. By negatively biasing two arbitrarily spaced qubits, they can interact and implement two-qubit operations via spin-wave exchange; spin waves act as a spin wave bus. It is thus possible to gain random access to any qubit and to entangle any two of them. Laterally applied STMs, which act as sources of local magnetic fields, and hence, of spin waves of large amplitude, can either initialize the qubits or read their logic values after the computational process is finished. To avoid thermal excitation of spin waves the operating temperature must be limited to 1 K. Because the decoherence time is about 10^{-6} s and a single computational step needs about 10^{-10} s, 10^4 operations can be performed with this computer.

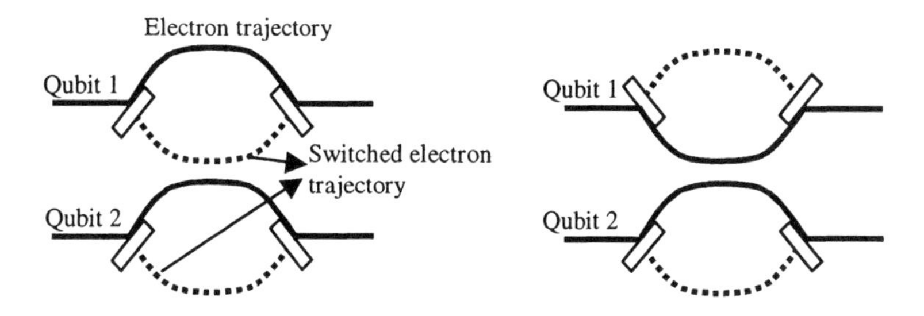

Figure 4.31 Switching of electron trajectory (solid line) by applying voltages on a pair of depleting gates (white rectangles) attached to each qubit. The qubits interact (right) or not (left) depending on the distance between the electrons in the two wires (*After*: [130]).

Another implementation of quantum logic circuits uses "flying" qubits, which are defined as the spin states of a single electron moving freely in a quantum wire. Flying qubits can be initialized via mesoscopic spin polarizing beamsplitters, efficient spin-coherent injectors, or via surface acoustic waves that trap electrons in the minima of a traveling potential, and can be detected by the same beamsplitter or by a spin filter accompanied by a SET [130]. One-quantum logic gates (controlled spin rotations along two independent axes) are implemented by static electric fields via spin-orbit interactions expressed by the Rashba Hamiltonian; this implementation takes into account the fact that all single-qubit transformations can be performed in maximum three steps via spin rotations around two axes. However, spin flip (rotation with π) requires lengths of the Rashba region (gate lengths) of 116 nm in InAs and 500 nm in InGaAs/InAlAs structures, which limits drastically the number of possible gates to 1500–2000 since in semiconductor substrates spin-coherent transport is possible over distances of at most a few hundred μm even at very low temperatures. Two-qubit gates are implemented via the isotropic exchange of two spins in adjacent quantum wires, described by (4.35), where the exchange coupling J can be switched on by reducing either the distance between the wires (via pairs of depleting gates that switch the electron trajectory, as depicted in Figure 4.31) or the potential (via electric fields) between the two electron sites. Up to an overall phase, the implementation of the universal $(\text{SWAP})^{1/2}$ gate requires that $\hbar^{-1}\int_{t_1}^{t_2} J(t)dt = \pi/2$, the universal CNOT or XOR gates being obtained from three one-qubit gates and two $(\text{SWAP})^{1/2}$ gates, while the quantum SWAP gate implies $\hbar^{-1}\int_{t_1}^{t_2} J(t)dt = \pi$, for which the spins of interacting electrons are swapped.

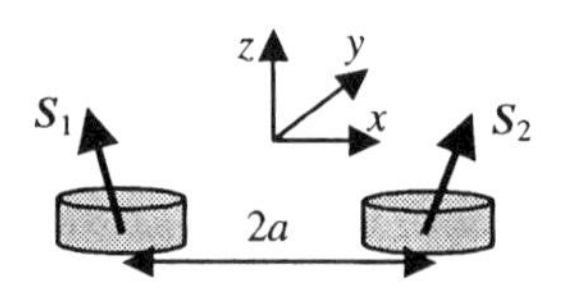

Figure 4.32 Electron spins in coupled semiconductor dots as qubit (*After*: [132]).

Electron spins in laterally coupled semiconductor quantum dots containing one electron each, as illustrated in Figure 4.32, can also act as qubits. The quantum dots can be electrostatically defined and the coupling between electron spins via the exchange interaction can be tuned by applying gate voltages that control the tunneling (the height of the tunneling barrier) between the dots [131]. A temporal coupling between the spins is necessary to implement two-qubit operations [132]. The ground state of two coupled electrons is a spin singlet (the next one being a spin triplet), the interaction between the two spins being described by the Heisenberg Hamiltonian (4.35). The variation of the exchange coupling J, which is equal in the low-temperature limit to the energy difference between the triplet and the single spin states, leads to the implementation of quantum operation as discussed in the previous paragraph. J can be controlled either by applying a magnetic field, which compresses the wavefunction, by applying an electric field, and hence, causing level detuning, by varying the interdot distance $2a$, or by varying the barrier height between the dots. For a magnetic field applied along z and a quartic confinement potential of the system of identical coupled dots in the (x, y) plane J is positive for small values of B, which indicates antiferromagnetic spin-spin coupling. With increasing magnetic field J changes sign for a B^* value and becomes negative, which suggests FM coupling, due to the long-range Coulomb interaction. The change of sign occurs for a wide range of values of a, the confining potential strength and the Coulomb strength interaction between electrons. Finally, for large B values J decays exponentially to zero due to the reduction in the electron wavefunction overlap with magnetic field compression of orbits. An additional electric field F applied along x also influences the exchange energy J, increasing it with a term proportional to F^2 as long as $J(B,F)-J(B,0) \leq J(B,0)$ and leading to suppression of J for larger F fields due to the reduced overlap of electron wavefunctions. A similar control of the J parameter can also be achieved in vertically coupled semiconductor quantum dots [133].

Single-qubit operations (i.e., spin rotations) are performed by exposing the selected spin to a time-varying Zeeman coupling $g\mu_B \boldsymbol{S} \cdot \boldsymbol{B}$ implemented by either a variable $\boldsymbol{B}(t)$ or $g(t)$, achieved by simply shifting the equilibrium position of an electron into a region with higher B or g through electrical gating. An electrical modulation at microwave frequencies (at several GHz) of the electronic g factor in undoped 100 nm GaAs/$Al_xGa_{1-x}As$ parabolic quantum wells grown by MBE, with a frequency modulation amplitude of up to 300 MHz, has been recently demonstrated in [134]; g control is achieved by displacing the electron wavefunction in regions with different material composition. Alternatively, a qubit can be flipped due to paramagnetic resonance, by applying an ac magnetic field with the Larmor frequency perpendicular to the spin eigenstate direction [133].

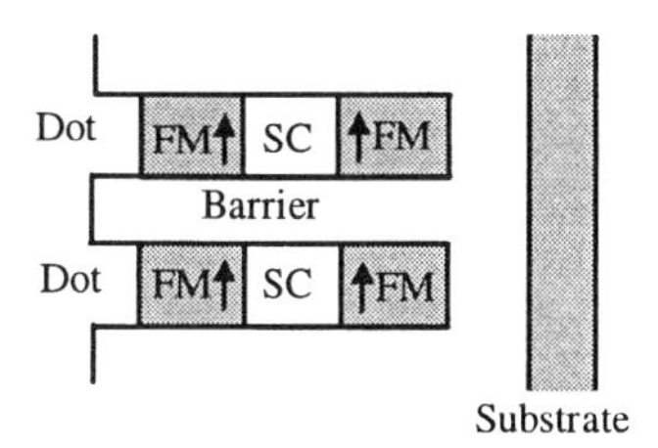

Figure 4.33 Schematic representation of two trilayered quantum dots separated by an insulating barrier and grown over a conducting substrate (*After*: [135]).

A self-assembled quantum computer fabricated by chemical synthesis, and based on exchange coupled multilayered quantum dots, was proposed in [135]. Each dot consists of FM/semiconductor/FM layers, Rashba interaction controlled by a potential applied between the outer layers causing spin splitting of the ground state of the dot. The qubit is encoded in the spin orientation of a spin-polarized electron injected from a FM layer and trapped by Coulomb blockade, as shown in Figure 4.33. One-qubit operations are implemented by bringing an ac magnetic field in resonance with the spin-splitting energy of a target quantum dot on which a potential pulse of appropriate duration and amplitude is applied between the FM layers. The pulse duration modulates the rotation angle and the pulse amplitude modulates the spin-splitting energy via the Rashba effect. Two-qubit operations are implemented via exchange coupling with the nearest-neighboring quantum dot since the spin-splitting energy in one dot depends, in this case, on the spin orientation in the other; the exchange coupling can be tuned by simultaneously applying potentials on both target and control qubits. The spin orientation is read through the current induced between the FM layers under an applied bias that overcomes the Coulomb blockade. The FM layers act as polarizer and analyzer of spin injection and detection; the current amplitude depends on the angle between the electron spin and the magnetization direction of the FM contacts.

A single-spin memory device for quantum computing operations with read-in or read-out capabilities can consist of a quantum dot in the Coulomb blockade regime connected to fully spin-polarized leads [136]. If the Zeeman splittings in the dot and the leads are different, Coulomb blockade peaks associated with a definite spin state in the dot appear, the spin-up or spin-down state of the electron in the dot being associated with a small cotunneling and a large sequential tunneling current, respectively. On the contrary, for unpolarized leads with negligible Zeeman splitting the quantum dot in the Coulomb blockade regime acts as an efficient spin filter at the single-spin level.

Other proposals for spin-based computing employ endohedral fullerenes, which are fullerenes that trap group-V paramagnetic atoms such as N or P, the qubit being encoded in the electron-nuclear spin system of the paramagnetic atom [137,138]. Coupling between qubits in a linear chain of endohedral fullerenes is accomplished by magnetic dipolar interactions between endohedral electron spins of adjacent qubits, and magnetic resonance pulses can implement universal quantum gates. An electron spin in a chain can be individually addressed if a magnetic gradient field is applied along the linear fullerene chain (for example, through a pair of micron-size wires), so that the resonance frequency of a qubit depends on the position in the chain and spin diffusion is suppressed. The spin-lattice relaxation time of electron spins in endohedral fullerenes can reach several seconds below 5 K, while the temperature-independent spin-spin relaxation time is about 20 μs. Electron spins have the advantage of much larger thermal spin polarization and higher sensitivity (due to a larger gyromagnetic ratio) with respect to nuclear spins [138]. Alternatively, electron-spin states of peapods inside single-walled CNTs, or of discrete units of the nanotubes, can act as molecular spin qubits in linear configurations [139].

A quantum Boltzmann machine neuron device, which uses stochastic architectures for processing, was proposed in [140]. It consists of a 2D array of quantum dots occupied by single electrons that are coupled by spin interaction. The spin-up or spin-down states of electrons are identified with the binary logic states 1 or 0, respectively. This quantum device mimics a neural network composed of identical interconnected neurons, which receive inputs from other neurons and imparts output states (0 or 1) to other neurons. The neuron produces an output state 1 with a probability $[1+\exp(-S/C)]^{-1}$, where $S=\sum_{i=1}^{n} x_i w_i + w_0$ with x_i the input signals, w_i the respective weight coefficients, w_0 the bias input, and C the control parameter. The neuron is replaced by an operating dot, which has two possible output states, 0 and 1, and the input signals are input dots related to the operating dots through input lines of coupled dots such that the states S_i of the input dots induce a local field $u = -4\sum_i^n J_i S_i \Delta S_0$ at the operating dot, where J_i is the interaction strength between the input dot and the operating dot for the ith input line and ΔS_0 is the change of spin polarization at the operating dot from up to down. The output state is determined by the spin polarization of the output dot, which is in turn connected to the operating dot through an output line. Because the spin polarization of the operating dots behave stochastically due to thermal agitation, the quantum Boltzmann machine generates an output of 1 with the same probability as a Boltzmann machine neuron if S is replaced by u and the control parameter is $k_B T$.

REFERENCES

[1] Žutić, I., J. Fabian, and S. Das Sarma, "Spintronics: fundamentals and applications," *Rev. Mod. Phys.*, Vol. 76, No. 2, 2004, pp. 323-410.

[2] Daughton, J., "Spin-dependent sensors," *Proc. IEEE*, Vol. 91, No. 5, 2003, pp. 681-686.

[3] Parkin, S., et al., "Magnetically engineered spintronic sensors and memory," *Proc. IEEE*, Vol. 91, No. 5, 2003, pp. 661-680.

[4] Engel, B.N., et al., "The science and technology of magnetoresistive tunneling memory," *IEEE Trans. Nanotechnology*, Vol. 1, No. 1, 2002, pp. 32-38.

[5] Johnson, M., "Overview of spin transport electronics in metals," *Proc. IEEE*, Vol. 91, No. 5, 2003, pp. 652-660.

[6] Lottermoser, T., et al., "Magnetic phase control by an electric field," *Nature*, Vol. 430, No. 6999, 2004, pp. 541-544.

[7] Asamitsu, A., "Current switching of resistive states in magnetoresistive manganites," *Nature*, Vol. 388, No. 6637, 1997, pp. 50-52.

[8] Akinaga, H., "Magnetoresistive switch effect in metal/semiconductor hybrid granular films: extremely huge magnetoresistance effect at room temperature," *Semicond. Sci. Technol.*, Vol. 17, No. 4, 2002, pp. 322-326.

[9] De Boeck, J., et al., "Technology and materials issues in semiconductor-based magnetoelectronics," *Semicond. Sci. Technol.*, Vol. 17, No. 4, 2002, pp. 342-354.

[10] Von Molnár, S. and D. Read, "New materials for semiconductor spin-electronics," *Proc. IEEE*, Vol. 91, No. 5, 2003, pp. 715-726.

[11] Awschalom, D.D., D. Loss, and N. Samarth (Eds.), *Semiconductor Spintronics and Quantum Computation*, Berlin: Springer, 2002.

[12] Young, D.K., et al., "Optical, electrical and magnetic manipulation of spins in semiconductors," *Semicond. Sci. Technol.*, Vol. 17, No. 4, 2002, pp. 275-284.

[13] Sih, V.A., E. Johnston-Halperin, and D.D. Awschalom, "Optical and electronic manipulation of spin coherence in semiconductors," *Proc. IEEE*, Vol. 91, No. 5, 2003, pp. 752-760.

[14] Dzhioev, R.I., et al., "Manipulation of the spin memory of electrons in n-GaAs," *Phys. Rev. Lett.*, Vol. 88, No. 25, 2002, pp. 256801/1-4.

[15] Ohno, Y., et al., "Spin relaxation in GaAs (110) quantum wells," *Phys. Rev. Lett.*, Vol. 83, No. 20, 1999, pp. 4196-4199.

[16] Pareek, T.P., "Absolute spin-valve effect: charge transport in two-dimensional hybrid systems," *Phys. Rev. B*, Vol. 70, No. 3, 2004, pp. 033310/1-4.

[17] Karimov, O.Z., et al., "High temperature gate control of quantum well spin memory," *Phys. Rev. Lett.*, Vol. 91, No. 24, 2003, pp. 246601/1-4.

[18] Bychkov, Yu.A. and E.I. Rashba, "Oscillatory effects and the magnetic susceptibility of carriers in inversion layers," *J. Phys. C: Solid State Phys.*, Vol. 17, No. 33, 1984, pp. 6039-6045.

[19] Litvinov, V.I., "Electron spin splitting in polarization-doped group-III nitrides," *Phys. Rev. B*, Vol. 68, No. 15, 2003, pp. 155314/1-6.

[20] Pearton, S.J., et al., "Wide bandgap GaN-based semiconductors for spintronics," *J. Phys.: Condens. Matter*, Vol. 16, No. 7, 2004, pp. R209-R245.

[21] Cartoixà, X., D.Z.-Y. Ting, and Y.-C. Chang, "A resonant spin lifetime transistor," *Appl. Phys. Lett.*, Vol. 83, No. 7, 2003, pp. 1462-1464.

[22] Johnson, M. and R.H. Silsbee, "Interfacial charge-spin coupling: injection and detection of spin magnetization in metals," *Phys. Rev. Lett.*, Vol. 55, No. 17, 1985, pp. 1790-1793.

[23] Schmidt, G. and L.W. Molenkamp, "Spin injection into semiconductors, physics and experiments," *Semicond. Sci. Technol.*, Vol. 17, No. 4, 2002, pp. 310-321.

[24] Schmidt, G., et al., "Fundamental obstacle for electrical spin injection from a ferromagnetic metal into a diffusive semiconductor," *Phys. Rev. B*, Vol. 62, No. 8, 2000, pp. R4790-R4793.

[25] Lee, B., T. Jungwirth, and A.H. MacDonald, "Ferromagnetism in diluted magnetic semiconductor heterojunction systems," *Semicond. Sci. Technol.*, Vol. 17, No. 4, 2002, pp. 393-403.

[26] Dietl, T., "Ferromagnetic semiconductors," *Semicond. Sci. Technol.*, Vol. 17, No. 4, 2002, pp. 377-392.

[27] Awschalom, D.D. and N. Samarth, "Spin dynamics and quantum transport in magnetic semiconductor quantum structures," *J. Magnetism and Magnetic Materials*, Vol. 200, No. 1-3, 1999, pp. 130-147.

[28] Tanaka, M., "Ferromagnet (MnAs)/III-V semiconductor hybrid structures," *Semicond. Sci. Technol.*, Vol. 17, No. 4, 2002, pp. 327-341.

[29] Jonker, B.T., et al., "Robust electrical spin injection into a semiconductor heterostructure," *Phys. Rev. B*, Vol. 62, No. 12, 2000, pp. 8180-8183.

[30] Fiederling, R., et al., "Injection and detection of a spin-polarized current in a light-emitting diode," *Nature*, Vol. 402, No. 6763, 1999, pp.787-790.

[31] Seufert, J., et al., "Spin injection into a single self-assembled quantum dot," *Phys. Rev. B*, Vol. 69, No. 3, 2004, pp. 035311/1-4.

[32] Rashba, E.I., "Theory of electrical spin injection: tunnel contacts as a solution of the conductivity mismatch problem," *Phys. Rev. B*, Vol. 62, No. 24, 2000, pp. R16267-R16270.

[33] Kravchenko, V.Ya. and E.I. Rashba, "Spin injection into a ballistic semiconductor microstructure," *Phys. Rev. B*, Vol. 67, No. 12, 2003, pp. 121310(R)/1-4.

[34] Alvarado, S.F and P. Renaud, "Observation of spin-polarized-electron tunneling from a ferromagnet into GaAs," *Phys. Rev. Lett.*, Vol. 68, No. 9, 1992, pp. 1387-1390.

[35] Alvarado, S.F., "Tunneling potential barrier dependence of electron spin polarization," *Phys. Rev. Lett.*, Vol. 75, No. 3, 1995, pp. 513-516.

[36] Zhu, H.J., et al., "Room-temperature spin injection from Fe into GaAs," *Phys. Rev. Lett.*, Vol. 87, No. 1, 2001, pp. 016601/1-4.

[37] Hanbicki, A.T., et al., "Efficient electrical spin injection from a magnetic metal/tunnel barrier contact into a semiconductor," *Appl. Phys. Lett.*, Vol. 80, No. 7, 2002, pp. 1240-1242.

[38] Ramsteiner, M., "Electrical spin injection from ferromagnetic metals into GaAs," *J. Superconductivity: Incorporating Novel Magnetism*, Vol. 16, No. 4, 2003, pp. 661-669.

[39] Albrecht, J.D. and D.L. Smith, "Electron spin injection at a Schottky contact," *Phys. Rev. B*, Vol. 66, No. 11, 2002, pp. 113303/1-4.

[40] Jonker, B.T., "Progress toward electrical injection of spin-polarized electrons into semiconductors," *Proc. IEEE*, Vol. 91, No. 5, 2003, pp. 727-740.

[41] Hammar, P.R. et al., Observation of spin injection at a ferromagnet-semiconductor interface," *Phys. Rev. Lett.*, Vol. 83, No. 1, 1999, pp. 203-206.

[42] Rashba, E.I., "Complex impedance of a spin injecting junction," *Appl. Phys. Lett.*, Vol. 80, No. 13, 2002, pp. 2329-2331.

[43] Yu, Z.G. and M.E. Flatté, "Spin diffusion and injection in semiconductor structures: electric field effects," *Phys. Rev. B*, Vol. 66, No. 23, 2002, pp. 235302/1-14.

[44] Schmidt, G., et al., "Spin injection in the nonlinear regime: band bending effects," *Phys. Rev. Lett.*, Vol. 92, No. 22, 2004, pp. 226602/1-4.

[45] Hu, C.-M. and T. Matsuyama, "Spin injection across a heterojunction: a ballistic picture," *Phys. Rev. Lett.*, Vol. 87, No. 6, 2001, pp. 066803/1-4.

[46] Hu, C.-M., et al., "Spin-polarized transport in a two-dimensional electron gas with interdigital-ferromagnetic contacts," *Phys. Rev. B*, Vol. 63, No. 12, 2001, pp. 125333/1-4.

[47] Céspedes, O., et al., "Contact induced magnetism in carbon nanotubes," *J. Phys.: Condens. Matter*, Vol. 16, No. 10, 2004, pp. L155-L161.

[48] Jedema, F.J., et al., "Electrical detection of spin precession in a metallic mesoscopic spin valve," *Nature*, Vol. 416, No. 6882, 2002, pp. 713-716.

[49] Rugar, D., et al., "Single spin detection by magnetic resonance force microscopy," *Nature*, Vol. 430, No. 6997, 2004, pp. 329-332.

[50] Kempf, J.G. and J.A. Marohn, "Nanoscale Fourier-transform imaging with magnetic resonance force microscopy," *Phys. Rev. Lett.*, Vol. 90, No. 8, 2003, pp. 087601/1-4.

[51] Elzerman, J.M., et al., "Single-shot read-out of an individual electron spin in a quantum dot," *Nature*, Vol. 430, No. 6998, 2004, pp. 431-435.

[52] Bühler, T.M., et al., "Single-electron transistor architectures for charge motion detection in solid-state quantum computer devices," *Smart Mater. Struct.*, Vol. 11, No. 5, 2002, pp. 749-755.

[53] van Ruitenbeek, J., "Noisy times ahead," *Nature*, Vol. 410, No. 6827, 2001, pp. 424-425.

[54] Xiao, M., et al., "Electrical detection of the spin resonance of a single electron in a silicon field-effect transistor," *Nature*, Vol. 430, No. 6998, 2004, pp. 435-439.

[55] Flatté, M.E. and G. Vignale, "Unipolar spin diodes and transistors," *Appl. Phys. Lett.*, Vol. 78, No. 9, 2001, pp. 1273-1275.

[56] Kirczenow, G., "Ideal spin filters: a theoretical study of electron transmission through ordered and disordered interfaces between ferromagnetic metals and semiconductors," *Phys. Rev. B*, Vol. 63, No. 5, 2001, pp. 054422/1-12.

[57] Wang, B., Y. Guo, B.-L. Gu, "Tunneling time of spin-polarized electrons in ferromagnet/insulator (semiconductor) double junctions under an applied electric field," *J. Appl. Phys.*, Vol. 91, No. 3, 2002, pp. 1318-1323.

[58] de Andrada e Silva, E.A. and G.C. La Rocca, "Electron-spin polarization by resonant tunneling," *Phys. Rev. B*, Vol. 59, No. 24, 1999, pp. R15583-R15585.

[59] Voskoboynikov, A., et al., "Spin-polarized electronic current in resonant tunneling heterostructures," *J. Appl. Phys.*, Vol. 87, No.1, 2000, pp. 387-391.

[60] Koga, T., et al., "Spin-filter device based on the Rashba effect using a nonmagnetic resonant tunneling diode," *Phys. Rev. Lett.*, Vol. 88, No. 12, 2002, pp. 126601/1-4.

[61] Ting, D.Z.-Y. and X. Cartoixà, "Resonant interband tunneling spin filter," *Appl. Phys. Lett.*, Vol. 81, No. 22, 2002, pp. 4198-4200.

[62] Ting, D.Z.-Y., et al., "Rashba effect resonant tunneling spin filters," *Proc. IEEE*, Vol. 91, No. 5, 2003, pp. 741-751.

[63] Moon, J.S., et al., "Experimental demonstration of split side-gated resonant interband tunneling devices," *Appl. Phys. Lett.*, Vol. 85, No. 4, 2004, pp. 678-680.

[64] Vurgaftman, I. And J.R. Meyer, "Ferromagnetic resonant interband tunneling diode," *Appl. Phys. Lett.*, Vol. 82, No. 14, 2003, pp. 2296-2298.

[65] Dobrovolsky, V.N., D.I. Sheka, and B.V. Chernyachuk, "Spin- and wave-vector dependent resonant tunneling through magnetic barriers," *Surface Science*, Vol. 397, No. 1-3, 1998, pp. 333-338.

[66] Xu, H.Z. and Y. Zhang, "Spin-filter devices based on resonant tunneling antisymmetrical magnetic/semiconductor hybrid structures," *Appl. Phys. Lett.*, Vol. 84, No. 11, 2004, pp. 1955-1957.

[67] Heiss, W., G. Prechtl, and G. Springholz, "Magnetic-field-tunable photoluminescence transitions in antiferromagnetic EuTe epilayers layers with an effective *g* factor of 1140," *Appl. Phys. Lett.*, Vol. 78, No. 22, 2001, pp. 3484-3486.

[68] Kirchschlager, R., et al., "Hysteresis loops of the energy band gap and effective *g* factor up to 18000 for metamagnetic EuSe epilayers," *Appl. Phys. Lett.*, Vol. 85, No. 1, 2004, pp. 67-69.

[69] Slobodskyy, A., et al., "Voltage-controlled spin selection in a magnetic resonant tunneling diode," *Phys. Rev. Lett.*, Vol. 90, No. 24, 2003, pp. 246601/1-4.

[70] Gruber, Th., et al., "Electron spin manipulation using semimagnetic resonant tunneling diodes," *Appl. Phys. Lett.*, Vol. 78, No. 8, 2001, pp. 1101-1103.

[71] Shi, Q.W., J. Zhou, and M.W. Wu, "Spin filtering through a double-bend structure," *Appl. Phys. Lett.*, Vol. 85, No. 13, 2004, pp. 2547-2549.

[72] Governale, M., "Filtering spin with tunnel-coupled electron wave guides," *Phys. Rev. B*, Vol. 65, No. 14, 2002, pp. 140403(R)/1-4.

[73] Usaj, G. and C.A. Balseiro, "Transverse electron focusing in systems with spin-orbit coupling," *Phys. Rev. B*, Vol. 70, No. 4, 2004, pp. 041301(R)/1-4.

[74] Dragoman, D. and M. Dragoman, *Quantum-Classical Analogies*, Berlin: Springer, 2004.

[75] Oliver, W.D., G. Feve, and Y. Yamamoto, "The Rashba effect within the coherent scattering formalism with applications to electron quantum optics," *J. Superconductivity: Incorporating Novel Magnetism*, Vol. 16, No. 4, 2003, pp. 719-733.

[76] Tamura, R., "Resonant spin current in nanotube double junctions," *Phys. Rev. B*, Vol. 67, No. 12, 2003, pp. 121408(R)/1-4.

[77] Feinberg, D. and P. Simon, "Splitting electronic spins with a Kondo double dot device," *Appl. Phys. Lett.*, Vol. 85, No. 10, 2004, pp. 1846-1848.

[78] Středa, P. and P. Šeba, "Rashba spin-orbit coupling and anti-symmetric spin filtering in one-dimensional electron systems," *Physica E*, Vol. 22, No. 1-3, 2004, pp. 460-463.

[79] Kiselev, A.A. and K.W. Kim, "T-shaped ballistic spin filter," *Appl. Phys. Lett.*, Vol. 78, No. 6, 2001, pp. 775-777.

[80] Gardelis, S., et al., "Spin-valve effects in a semiconductor field-effect transistor: a spintronic device," *Phys. Rev. B*, Vol. 60, No. 11, 1999, pp. 7764-7767.

[81] Alphenaar, B.W., K. Tsukagoshi, and H. Ago, "Spin electronics using carbon nanotubes," *Physica E*, Vol. 6, No. 1-4, 2000, pp. 848-851.

[82] Mehrez, H., et al., "Carbon nanotube based magnetic tunnel junctions," *Phys. Rev. Lett.*, Vol. 84, No.12, 2000, pp. 2682-2685.

[83] Ciuti, C., J.P. McGuire, and L.J. Sham, "Spin-dependent properties of a two-dimensional electron gas with ferromagnetic gates," *Appl. Phys. Lett.*, Vol. 81, No.2, 2002, pp. 4781-4783.

[84] Zwolak, M. and M. Di Ventra, "DNA spintronics," *Appl. Phys. Lett.*, Vol. 81, No. 5, 2002, pp. 925-927.

[85] Ting, D.Z.-Y. and X. Cartoixà, "Bidirectional resonant tunneling spin pump," *Appl. Phys. Lett.*, Vol. 83, No. 7, 2003, pp. 1391-1393.

[86] Benjamin, R. and C. Benjamin, "Quantum spin pumping with adiabatically modulated magnetic barriers," *Phys. Rev. B*, Vol. 69, No. 8, 2004, pp. 085318/1-10.

[87] Das Sarma, S., J. Fabian, and I. Žutić, "Spin-polarized bipolar transport and its applications," *J. Superconductivity: Incorporating Novel Magnetism*, Vol. 16, No. 4, 2003, pp. 697-705.

[88] Schmeltzer, D., et al., "Spin diode in the ballistic regime," *Phys. Rev. B*, Vol. 68, No. 19, 2003, pp. 195317/1-5.

[89] Datta, S., and D. Das, "Electronic analog of the electro-optic modulator," *Appl. Phys. Lett.*, Vol. 56, No. 7, 1990, pp. 665-667.

[90] Nitta, J., et al., "Gate control of spin-orbit interaction in an inverted $In_{0.53}Ga_{0.47}As/In_{0.52}Al_{0.48}As$ heterostructure," *Phys. Rev. Lett.*, Vol. 78, No. 7, 1997, pp. 1335-1338.

[91] Hu, C.-M., et. al., "Zero-field splitting in an inverted $In_{0.53}Ga_{0.47}As$/ $In_{0.52}Al_{0.48}As$ heterostructure: band nonparabolicity influence and the subband dependence," *Phys. Rev. B*, Vol. 60, No. 11, 1999, pp. 7736-7739.

[92] Grundler, D., "Large Rashba splitting in InAs quantum wells due to electron wave function penetration into the barrier layers," *Phys. Rev. Lett.*, Vol. 84, No. 26, 2000, pp. 6074-6077.

[93] Schäpers, Th., J. Knobbe, and V.A. Guzenko, "Effect of Rashba spin-orbit coupling on magnetotransport in InGaAs/InP quantum wire structures," *Phys. Rev. B*, Vol. 69, No. 23, 2004, pp. 235323/1-5.

[94] Egues, J.C., G. Burkard, and D. Loss, "Datta-Das transistor with enhanced spin control," *Appl. Phys. Lett.*, Vol. 82, No. 16, 2003, pp. 2658-2660.

[95] Mal'shukov, A.G., et al., "Spin-current generation and detection in the presence of an ac gate," *Phys. Rev. B*, Vol. 68, No. 23, 2003, pp. 233307/1-4.

[96] Łusakowski, A., J. Wróbel, and T. Dietl, "Effect of bulk inversion asymmetry on the Data-Das transistor," *Phys. Rev. B*, Vol. 68, No. 8, 2003, pp. 081201(R)/1-4.

[97] Cahay, M. and S. Bandyopadhyay, "Conductance modulation of spin interferometers," *Phys. Rev. B*, Vol. 68, No. 11, 2003, pp. 115316/1-5.

[98] Wang, X.F., "Spin transport of electrons through quantum wires with a spatially modulated Rashba spin-orbit interaction," *Phys. Rev. B*, Vol. 69, No. 3, 2004, pp. 035302/1-9.

[99] Meier, G. and T. Matsuyama, "Magnetic electrodes for spin-polarized injection into InAs," *Appl. Phys. Lett.*, Vol. 76, No. 10, 2000, pp. 1315-1317.

[100] Bandyopadhyay, S. and M. Cahay, "Alternate spintronic analog of the electro-optic modulator," *Appl. Phys. Lett.*, Vol. 85, No. 10, 2004, pp. 1814-1816.

[101] Schliemann, J., J.C. Egues, and D. Loss, "Nonballistic spin-field-effect transistor," *Phys. Rev. Lett.*, Vol. 90, No. 14, 2003, pp. 146801/1-4.

[102] Hall, K.C., et al., "Nonmagnetic semiconductor spin transistor," *Appl. Phys. Lett.*, Vol. 83, No. 14, 2003, pp. 2937-2939.

[103] Bandyopadhyay, S. and M. Cahay, "Reexamination of some spintronic field-effect device concepts," *Appl. Phys. Lett.*, Vol. 85, No. 8, 2004, pp. 1433-1435.

[104] Shen, S.-Q., Z.-J. Li, and Z. Ma, "Controllable quantum spin precession by Aharonov-Casher phase in a conducting ring," *Appl. Phys. Lett.*, Vol. 84, No. 6, 2004, pp. 996-998.

[105] Molnár, B., P. Vasilopoulos, and F.M. Peeters, Spin-dependent transmission through a chain of rings: influence of a periodically modulated spin-orbit interaction strength or ring radius," *Appl. Phys. Lett.*, Vol. 85, No. 4, 2004, pp. 612-614.

[106] Zülicke, U, "Spin interferometry with electrons in nanostructures: a road to spintronic devices," *Appl. Phys. Lett.*, Vol. 85, No. 13, 2004, pp. 2616-2618.

[107] Ohno, H., et. al., "Electric-field control of ferromagnetism," *Nature*, Vol. 408, No. 6815, 2000, pp. 944-946.

[108] Fabian, J., I. Žutić, and S. Das Sarma, "Magnetic bipolar transistor," *Appl. Phys. Lett.*, Vol. 84, No. 1, 2004, pp. 85-87.

[109] Fabian, J. and I. Žutić, "Spin-polarized current amplification and spin injection in magnetic bipolar transistors," *Phys. Rev. B*, Vol. 69, No. 11, 2004, pp. 115314/1-13.

[110] Dennis, C.L., et al., "High current gain silicon-based spin transistor," *J. Phys. D*, Vol. 36, No. 2, 2003, pp. 81-87.

[111] Steinmuller, S.J., et al., "Highly efficient spin filtering of ballistic electrons," *Phys. Rev. B*, Vol. 69, No. 15, 2004, pp. 153309/1-4.

[112] Bauer, G.E.W., et al., "Spin-torque transistor," *Appl. Phys. Lett.*, Vol. 82, No. 22, 2003, pp. 3928-3930.

[113] Rudolph, J., et al., "Laser threshold reduction in a spintronic device," *Appl. Phys. Lett.*, Vol. 82, No. 25, 2003, pp. 4516-4518.

[114] Shelykh, I., et al., "Semiconductor microcavity as a spin-dependent optoelectronic device," *Phys. Rev. B*, Vol. 70, No. 3, 2004, pp. 035320/1-5.

[115] Žutić, I., J. Fabian, and S. Das Sarma, "Proposal for a spin-polarized solar battery," *Appl. Phys. Lett.*, Vol. 79, No. 10, 2001, pp. 1558-1560.

[116] Ganichev, S.D., et al., "Conversion of spin into directed electric current in quantum wells," *Phys. Rev. Lett.*, Vol. 86, No. 19, 2001, pp. 4358-4361.

[117] Ganichev, S.D., et al., "Spin-galvanic effect," Nature, Vol. 417, No. 6885, 2002, pp. 153-156.

[118] Ganichev, S.D. and W. Prettl, "Spin photocurrents in quantum wells," *J. Phys.: Condens. Matter*, Vol. 15, No. 20, 2003, pp. R935-R983.

[119] Cho, Y., et al., "Terabit inch^{-2} ferroelectric data storage using scanning nonlinear dielectric microscopy nanodomain engineering system," *Nanotechnology*, Vol. 14, No. 6, 2003, pp. 637-642.

[120] Galindo, A. and M.A. Martín-Delgado, "Information and computation: classical and quantum aspects," *Rev. Mod. Phys.*, Vol. 74, No. 2, 2002, pp. 347-423.

[121] Burkard, G. and D. Loss, "Electron spins in quantum dots as qubits for quantum information processing," in *Semiconductor Spintronics and Quantum Computation*, D.D. Awschalom, D. Loss, and N. Samarth (Eds.), Berlin: Springer, 2002.

[122] Tejada, J., et al., "Magnetic qubits as hardware for quantum computers," *Nanotechnology*, Vol. 12, No. 2, 2001, pp. 181-186.

[123] Kane, B.E., "A silicon-based nuclear spin quantum computer," *Nature*, Vol. 393, No. 6681, 1998, pp. 133-137.

[124] Kane, B.E., "Silicon-based quantum computation," *Fortschr. Phys.*, Vol. 48, No. 9-11, 2000, pp. 1023-1041.

[125] O'Brien, J.L., et al., "Towards the fabrication of phosphorus qubits for a silicon quantum computer," *Phys. Rev. B*, Vol. 64, No. 16, 2001, pp. 161401(R)/1-4.

[126] McKinnon, R.P., et al., "Nanofabrication processes for single-ion implantation of silicon quantum computer devices," *Smart Mater. Struct.*, Vol. 11, No. 5, 2002, pp. 735-740.

[127] Stoneham, A.M., A.J. Fisher, and P.T. Greenland, "Optically driven silicon-based quantum gates with potential for high-temperature operation," *J. Phys.: Condens. Matter*, Vol. 15, No. 27, 2003, pp. L447-L451.

[128] Berman, G.P., et al., "Solid-state quantum computer based on scanning tunneling microscopy," *Phys. Rev. Lett.*, Vol. 87, No. 9, 2001, pp. 097902/1-3.

[129] Khitun, A., R. Ostroumov, and K.L. Wang, "Spin-wave utilization in a quantum computer," *Phys. Rev. A*, Vol. 64, No. 6, 2001, pp. 062304/1-5.

[130] Popescu, A.E. and R. Ionicioiu, "All-electrical quantum computation with mobile spin qubits," Phys. Rev. B, Vol. 69, No. 24, 2004, pp. 245422/1-11.

[131] Loss, D., and D.P. DiVincenzo, "Quantum computation with quantum dots," *Phys. Rev. A*, Vol. 57, No. 1, 1998, pp. 120-122.

[132] Burkard, G., D. Loss, and D.P. DiVincenzo, "Coupled quantum dots as quantum gates," *Phys. Rev. B*, Vol. 59, No. 3, 1999, pp. 2070-2078.

[133] Leuenberger, M.N. and D. Loss, "Spintronic and quantum computing: switching mechanisms for qubits," *Physica E*, Vol. 10, No. 1-3, 2001, pp. 452-457.

[134] Kato, Y., et al., "Gigahertz electron spin manipulation using voltage-controlled g-tensor modulation," *Science*, Vol. 299, No. 5610, 2003, pp. 1201-1204.

[135] Bandyopadhyay, S., "Self-assembled nanoelectronic quantum computer based on the Rashba effect in quantum dots," *Phys. Rev. B*, Vol. 61, No. 20, 2000, pp. 13813-13820.

[136] Recher, P., E.V. Sukhorukov, and D. Loss, "Quantum dot as spin filter and spin memory," *Phys. Rev. Lett.*, Vol. 85, No. 9, 2000, pp. 1962-1965.

[137] Harneit, W., et al., "Architectures for a spin quantum computer based on endohedral fullerenes," *Phys. Stat. Sol. (b)*, Vol. 233, No. 3, 2002, pp. 453-461.

[138] Harneit, W., "Fullerene-based electron-spin quantum computer," *Phys. Rev. A*, Vol. 65, No. 3, 2002, pp. 032322/1-6.

[139] Ardavan, A., et al., "Nanoscale solid-state quantum computing," *Phil. Trans. R. Soc. Lond. A*, Vol. 361, No. 1808, 2003, pp. 1473-1485.

[140] Wu, N.-J., N. Shibata, and Y. Amemiya, "Boltzmann machine neuron device using quantum-coupled single electrons," *Appl. Phys. Lett.*, Vol. 72, No. 24, 1998, pp. 3214-3216.

Chapter 5

Electronic Devices Based on Nanostructures

This chapter deals with nanosized electronic devices. These are intensively studied in order to improve or even to replace the existing silicon electronic devices, which are still the building blocks of very complex integrated circuits incorporated in computers or advanced communication systems. Nanosized electronic devices are expected to increase the integration degree of logical devices on a single chip, and to reduce the dissipated power and the power consumption. They are also assumed to be faster than the actual electron devices and hence are believed to lead to an unprecedented increase in computing speed and the performance of communication systems; the cutoff frequency of nanosized electronic devices is very often a few THz. There are two main tendencies toward nanosize electronics: 1) to downscale the existing Si CMOS transistors to the nanoscale, where the carrier transport is no longer diffusive but ballistic, and 2) to find new innovative devices, such as ballistic devices, resonant tunneling diodes, or single-electron devices working at room temperature. New logical circuit configurations and new memory configurations are studied and fabricated using these new types of devices.

5.1 NANOSCALE FET TRANSISTORS

5.1.1 Downscaling the MOSFET Dimensions up to Few nm

The FET (field-effect transistor) is a transistor whose behavior is controlled by an electrode called gate, capacitively coupled to the active region of the device. The gate is separated from this active semiconductor region called channel by an insulator or a depletion region. The two other terminals of the FET, named source and drain, respectively, terminate the channel. The gate simply modifies the resistance of the channel and thus the carrier flow between source and drain. Therefore, a FET is a genuine switch.

There are many transistor types belonging to the FET family, but in what follows we will first analyze the most illustrious member of this family: the MOSFET (metal-oxide-semiconductor FET). The name MOSFET suggests that the metallic gate is separated from the active region by an oxide that plays the role of the insulator. In a typical example, a doped Si active region is isolated from a metallic gate by a SiO_2 layer. The isolator could be also Si_3N_4 or high-permittivity dielectrics, as encountered in Chapter 3 in the case of CNTFETs. The MOSFETs were originally fabricated with a *p*-channel (PMOS), but the subsequent *n*-channel transistors (NMOS) were found to switch faster than PMOS. Both types of MOSFETs can be combined in the so-called CMOS (complementary MOSFET) transistor, which is a very low-power consuming transistor that preserves the high switching rate of the NMOS.

The MOSFET transistor is the simplest and most effective electronic device, quite easy to fabricate in comparison with other active devices such as bipolar transistors. Due to its simplicity the CMOS was selected as a key element in integrated circuits, which imposed the scaling down of its dimensions up to unprecedented values. The gate length of MOSFETs presently used in commercial microprocessors is 50–70 nm. MOSFETs with a 15 nm gate length have been already demonstrated in the laboratory, MOSFET gates that reach 9 nm being expected in the next ten years. The scaling down of MOSFET dimensions increases the density of transistors and thus the complexity and the functionality of integrated circuits (ICs); the transistor density attains 10^7 transistors on a chip in very-large scale integrated circuits (VLSI) while in ultra-larges scale integration (ULSI) there are more than 10^9 transistors on a chip. The semiconductor technology is so impressive and so cheap that in 2002 the number of CMOS transistors included only in the DRAMs was higher than the number of rice grains produced in a year, the price of one grain of rice being equal to that of 100 transistors [1].

MOSFETs with nanosized gate lengths are the most spread nanoelectronic devices; they prove the Moore law [2], which states that at every 1.5 years since 1970 the transistor count per integrated circuit chip, such as a microprocessor, is doubling. Another version of the Moore law asserts that the dimensions of CMOS have been reduced with 13% per year, which implies an increased speed of logical devices. In particular, for microprocessors this means an increase of the clock speed with 30% per year. As a consequence, the cost per function, for example, the cost per a bit of DRAM, is decreasing with 30% per year due to the reduction of CMOS dimensions, due to an increased chip and wafer size, and an improved technology. The question is how much longer will the Moore law be valid. The problem is that if the gate length decreases new physical phenomena appear at the nanoscale, which impede the function of MOSFET when the gate length is only a few nm. New MOSFET configurations suitable for the nanoscale are needed, and they will be presented in the following.

The MOSFET transistor function can be understood by analyzing first a simple configuration called MOS capacitor. As shown in Figure 5.1, the MOS capacitor consists of a top metal gate and a substrate, which is a doped semiconductor (usually *p*-Si), separated from the gate through an insulating layer (commonly SiO_2). When a negative gate voltage V_G is applied, the resulting electric field confines the holes at the interface between the semiconductor and the insulator. On the contrary, the holes are repealed when V_G is positive, creating a depletion region. The width of the depletion region is given by

$$x_{dep} = (2\varepsilon_s \mid \psi_s \mid eN_s)^{1/2}, \qquad (5.1)$$

where ε_s is the permittivity of the substrate, $\mid \psi_s \mid$ is the surface potential, which measures the bending of the energy bands of Si due to the applied voltage, and N_s is the substrate concentration.

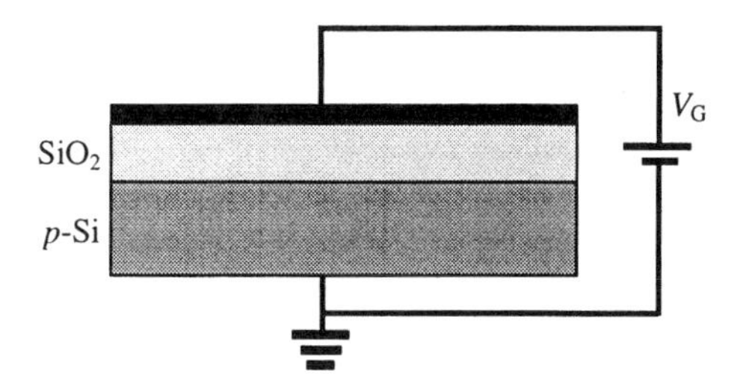

Figure 5.1 The MOS capacitor.

The bending of the conduction energy band of Si induces the transfer of minority carriers (electrons in *p*-Si) at the top of the depleted layer, such that at a certain gate voltage termed threshold voltage V_{th} they equal the concentration of holes in the substrate. Beyond this gate voltage an inversion layer, called also channel, is created beneath the insulator, in this inversion layer the density of electrons prevailing over the hole density. This means that the carrier depletion attains its maximum value when

$$\psi_s = 2\phi_F \qquad (5.2)$$

where $\phi_F = \pm(E_i - E_F)/e = (k_B T/e)\ln(N_s/n_i)$ is the Fermi potential with E_i the intrinsic carrier energy, n_i the intrinsic carrier concentration, E_F the Fermi energy, and the $\pm$ sign corresponds to the p-type or n-type substrate, respectively. Thus, the maximum depletion layer width is

$$x_{\text{dep,max}} = [4\varepsilon_s k_B T \ln(N_s / n_i)/(e^2 N_s)]^{1/2} \tag{5.3}$$

and the charge per unit area in the depletion layer is given by

$$q_{\text{dep}} = \mp e N_s x_{\text{dep}} , \tag{5.4}$$

where the minus sign is assigned to a p-type substrate and the plus sign corresponds to an n-type substrate. The threshold voltage can be computed from the relation

$$V_{\text{th}} = V_{\text{fb}} + 2\phi_F - q_{\text{dep}} / C_{\text{ox}} , \tag{5.5}$$

where V_{fb} is the flat band voltage, which can be viewed as a gate voltage that does not bend the energy bands in the substrate, and C_{ox} is the oxide (insulator) capacitance per unit area. V_{th} has typical values in the range 0.2–0.4 V.

The charge in the inversion layer is given by

$$q_n = -C_{\text{ox}} (V_G - V_{\text{th}}) , \tag{5.6}$$

which is a very important relation for MOSFETs since it says that the charge in the channel and implicitly the current is directly proportional with the gate voltage. If the carriers in the channel are electrons we have an NMOS transistor, while we deal with a PMOS transistor if the holes form the channel.

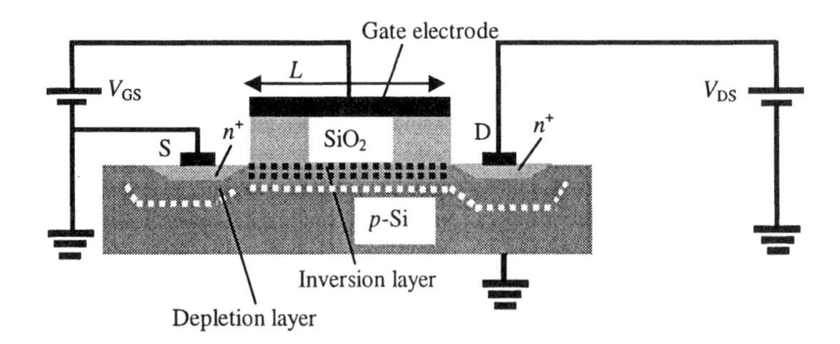

Figure 5.2 The MOSFET transistor (*After*: [1]).

The MOSFET, represented in Figure 5.2, is formed from two diodes called source and drain that encompass the MOS capacitor, the voltages between source S and drain D, and between gate and source being denoted by V_{DS} and V_{GS}, respectively. In the configuration described in Figure 5.2 electrons supplied by the n^+ source, which is forward biased, form the channel. A forward bias corresponds to a gate voltage higher than V_{th}, for which the generated charge density in the channel follows the relation (5.6). If the drain terminal is forward biased with V_{DS}, the electrons in the channel flow towards the drain, the electron distribution in the channel being uniform at small V_{DS}. In this case, the channel with length L and width W is modeled as a resistor, the Ohm's law predicting that, for $V_{\mathrm{GS}} >> V_{\mathrm{DS}}$,

$$I_{\mathrm{DS}} = \mu_n C_{\mathrm{ox}} W (V_{\mathrm{GS}} - V_{\mathrm{th}}) V_{\mathrm{DS}} / L, \tag{5.7}$$

where μ_n is the electron mobility in the channel. As follows from (5.7), the drain current I_{DS} is a linear function of V_{DS}. This is, however, no longer valid if V_{DS} is further increased such that a nonuniform distribution of electrons in the channel is created. Assuming a gradual nonuniform electron distribution in the channel for $V_{\mathrm{GS}} - V_{\mathrm{th}} > V_{\mathrm{DS}}$, such that the density of electrons decreases linearly from a maximum value in the source to a minimum value near the drain, we obtain the behavior of the MOSFET in the so-called triode region:

$$I_{\mathrm{DS}} = \mu_n C_{\mathrm{ox}} W [(V_{\mathrm{GS}} - V_{\mathrm{th}}) V_{\mathrm{DS}} - m V_{\mathrm{DS}}^2 / 2] / L. \tag{5.8}$$

Here $m = 1 + (1 / C_{\mathrm{ox}}) / (\varepsilon_{\mathrm{s}} q N_{\mathrm{s}} / 4 | \phi_{\mathrm{F}} |)^{1/2}$ is a parameter called body effect coefficient, which has typical values 1.1–1.4. In the triode region relation (5.6), which describes the charge in the channel, becomes

$$q_n = -C_{\mathrm{ox}} (V_{\mathrm{GS}} - V_{\mathrm{th}} - mV), \tag{5.9}$$

the parameter V ranging from 0 to V_{DS}.

If V_{DS} is further increased, we reach a regime where there are no charges near the drain, so that the channel is pinched off and the current becomes independent of V_{DS}. The MOSFET works now in the saturation regime, which is onset when $V_{\mathrm{DS}} = V = (V_{\mathrm{GS}} - V_{\mathrm{th}}) / m = V_{\mathrm{DSat}}$. The saturation current, obtained from (5.8), is given by

$$I_{\mathrm{DS}} = \mu_n C_{\mathrm{ox}} W (V_{\mathrm{GS}} - V_{\mathrm{th}})^2 / 2mL. \tag{5.10}$$

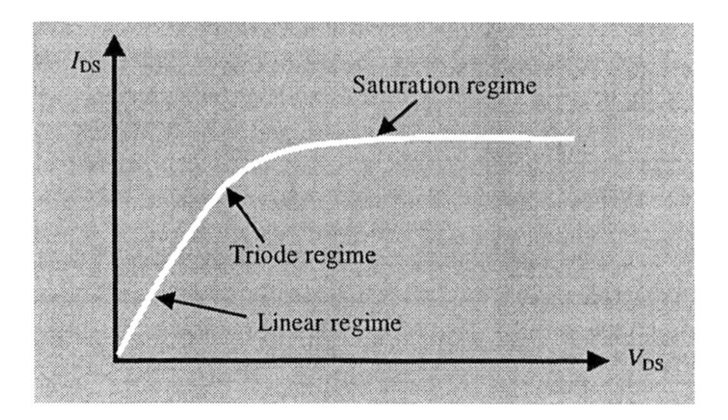

Figure 5.3 The working regimes of a MOSFET.

The three working regimes of the MOSFET are visualized in Figure 5.3. These regimes are predicted by the classical model of MOSFET presented above, which is valid only for long gate lengths.

Other important parameters for any FET are the transconductance or gain,

$$g_{m} = \partial I_{DS} / \partial V_{G} , \tag{5.11}$$

and the channel conductance

$$g_{DS} = \partial I_{DS} / \partial V_{DS} . \tag{5.12}$$

As soon as L becomes smaller the classical MOSFET model is no longer valid. For example, for MOSFETs with small gate lengths I_{DS} in the saturation regime depends also on V_{DS}, causing a reduction of the gate effective length. Also, in a nanosized MOSFET the substrate and the source are not at the same ground potential and therefore V_{th} depends on the ground potential. Moreover, there are also subthreshold effects, which manifest themselves through a non-abrupt transition between the ON state, characterized by $V_{GS} > V_{th}$, and the OFF state, for which $V_{GS} < V_{th}$; this nonabrupt transition is called weak inversion.

The reduction in the horizontal and vertical dimensions of a MOSFET with the same factor, an operation called scaling, is the recipe to obtain nanoscale FETs and complex integrated circuits, such as microprocessors or DRAMs. There are some alternatives to scale down the MOSFETs, each of them with advantages and drawbacks. The scaling issues and the associated effects are reviewed in [1], [3,4].

Table 5.1

MOSFET Scaling (*From*: [1, 3])

MOSFET parameter	*Constant voltage scaling*	*Generalized scaling*
Dimensions	$1/p$	$1/p$
Electric field	p	Q
Doping concentration	p	Pq
Voltage	1	q/p
Current	p	q^2/p
Capacitance	$1/p$	$1/p$
Delay	$1/p^2$	$1/pq$
Power density	p^3	q^3
Power dissipation	p	q^3/p^2

If the voltage scales with the same amount as the MOSFET dimensions, the scaling method is termed constant-field scaling, but this scaling method is not commonly used since it is difficult to work with too small voltages. Another scaling method is to maintain the voltage unchanged and to scale down only the MOSFET dimensions, a case in which the scaling method is termed constant-voltage scaling. The generalized method of scaling is actually a combination of these two methods, with the advantage that the voltage is not overly decreased. The parameter values for the constant-voltage and generalized scaling methods are summarized in Table 5.1, under the hypothesis that long channel operation is valid. A scaled value is obtained by multiplying the corresponding parameter with the coefficient indicated in the table, where p, q are natural numbers.

On the other hand, when the values of various geometrical or electrical MOSFET parameters are scaled down, new physical effects appear which tend to reduce the performances of the MOSFET. Therefore, a simple scaling down of MOSFET parameters will not produce a smaller transistor with better performances unless the scale reduction is accompanied by a new model that takes into account some basic phenomena, described briefly in the following.

From Table 5.1 it follows that the power density, which is equal to the power dissipation per unit area (i.e., the consumed power), dramatically increases ($\propto p^3$ or q^3) when we scale down the MOSFET dimensions and can exceed 150 W in current microprocessors. This is a major limitation in the scaling of MOSFET and is, unfortunately, accompanied by many others.

Another severe limitation, which manifests itself in the flow of current when the transistor is OFF, that is, when $V_{GS} = 0$, occurs due to the subthreshold leakage current given by

$$I_{\mathrm{DS,leak}} = I_0 \exp[e(V_{\mathrm{G}} - V_{\mathrm{th}}) / mk_{\mathrm{B}}T] . \tag{5.13}$$

Here the coefficient m is defined via the subthreshold slope $S = 2.3mk_{\mathrm{B}}T / e = 2.3k_{\mathrm{B}}T(1 + C_{\mathrm{D}} / C_{\mathrm{ox}})$, which is the gate voltage necessary to change I_{DS} with one decade; typical values are 80–100 mV per decade. It is clear that S, and therefore also m, cannot be scaled down except by lowering the temperature. Moreover, the leakage current becomes excessive if we change the threshold voltage too much. As a remedy, the biasing of the substrate or two threshold voltages is used to partially alleviate the lack of scaling of the leakage current.

Another observation that follows from Table 5.1 is that the electric field along the channel increases when we downscale the MOSFET dimensions. Thus, the electrons in the channel reach high kinetic energies and are able to tunnel across the barrier between the silicon and oxide, some of them being even trapped in the oxide. These hot electrons produce an undesired increase of V_{th}, m, and S, a problem that can be, at least partially, overcome by using a light-doped drain.

Short-channel effects occur when the charges in the channel are distributed to all MOSFET terminals due to smaller MOSFET dimensions. When the gate length in the downscaled transistor is much shorter than the source and drains junction depth and the depletion width, it is no longer capable to control entirely the field along the channel and, as an expected consequence, the drain voltage will influence the channel voltage near the source. This unwanted effect is called drain-induced barrier lowering (DIBL) and is an important limitation for down-scaling. DIBL is alleviated partially by shallow junctions or by nonuniform channel implant.

Another major limitation encountered in scaling down is velocity saturation. The velocity of the carriers is proportional to the electric field in the channel, but this proportionality ceases beyond a certain electric field threshold, producing an abrupt decrease of the mobility. Because the electric field increases in short channels, the carrier velocity in these channels is expected to reach saturation; the direct consequence is that the current does no longer increase with the drain voltage and is independent of L.

When the channel width becomes very narrow, the depletion region beneath the gate extends along the channel width direction, increasing the threshold voltage. Higher substrate doping alleviates this effect.

Another thorny problem is related to doping fluctuations in very small channels, where the amount of dopants is small and becomes random in number and location. As a result, there is a large variation of the threshold voltage among devices.

Due to these, and many other, drawbacks not mentioned here (but mentioned in the above-indicated references) the downscaling techniques for MOSFET cannot be used beyond certain dimensions. Therefore, new configurations are experimented now in order to reach the nanometer scale.

A first example of a nanometer transistor is the silicon-on-insulator MOSFET (SOI MOSFET), represented in Figure 5.4, which shows an optimal *S*, no body effects, and very low capacitance of the junctions. The concept of the SOI MOSFET is based on the introduction of a SiO_2 layer, termed buried SiO_2, between the active region of the device and the substrate. If the depletion width beneath the channel is larger than the thickness of the active silicon layer, the SOI MOSFET is termed fully depleted silicon-on-insulator MOSFET, or FDSOI-MOSFET. On the contrary, when the silicon layer is thicker than the depletion width the silicon-on-insulator MOSFET is called partially depleted silicon-on-insulator MOSFET, or PDSOI-MOSFET. This is the most used SOI MOSFET transistor.

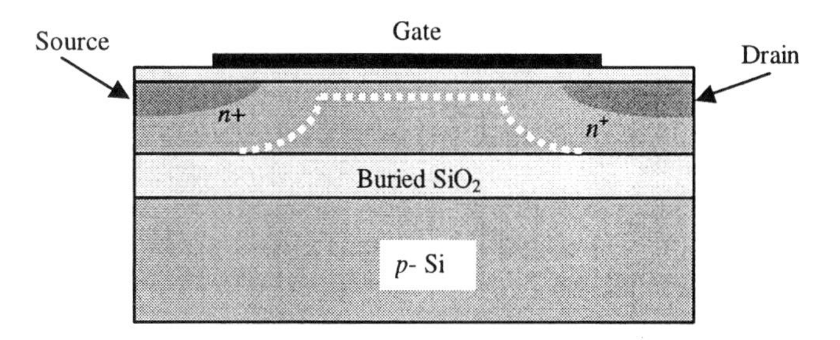

Figure 5.4 The SOI MOSFET (*After*: [1]).

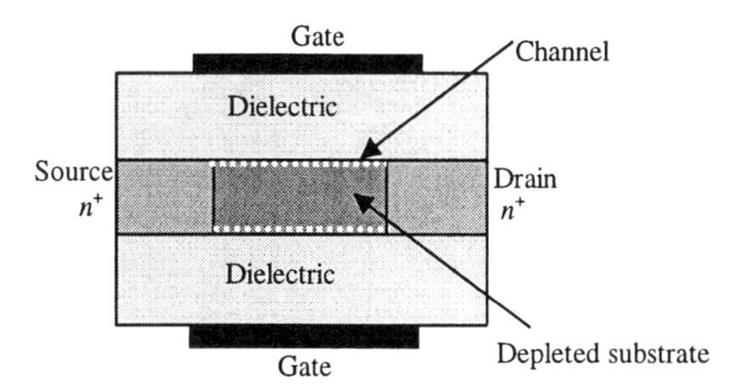

Figure 5.5 The double-gate FET (*After*: [1]).

Another promising configuration, which is represented in Figure 5.5, is the double-gate MOSFET (DGFET). It is a key configuration for downscaling below 10 nm and it is now considered the basic configuration for nanosize MOSFETs. The DGFET has a channel sandwiched between two gates, which controls more effectively the potential in the channel. Therefore, the short-channel effects are no longer dependent on doping but are only a manifestation of the geometry of the transistor, and the DIBL effects no longer exist. Due to this fact, the substrate does not need to be doped or is only lightly doped, which implies that the DGFET has an *S* value of around 60 mV/decade and an excellent drive current.

The DGFET is a nonplanar transistor and is encountered in three basic configurations, as displayed in Figure 5.6. The DGFETs of type II and III are vertical structures, with the channel perpendicular to the substrate, and, hence, more difficult to fabricate, but with the advantage that the density of transistors on cm^2 is much higher than in the case of type I DGFET.

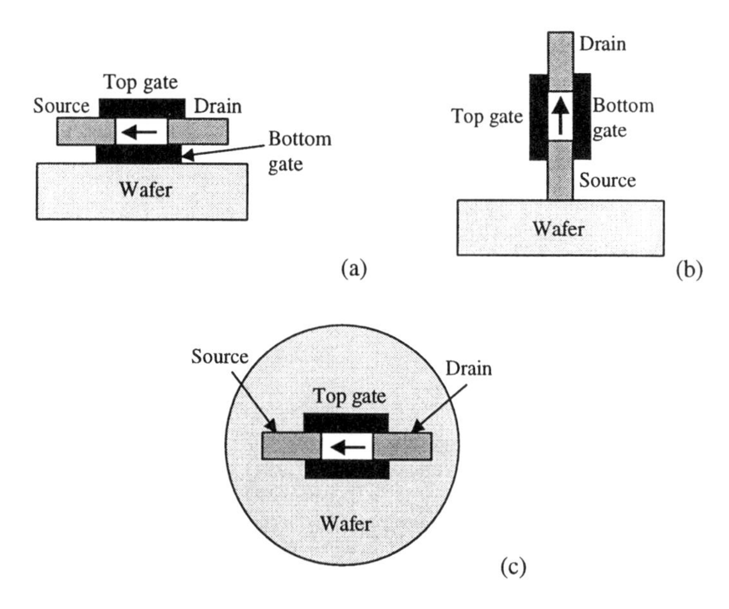

Figure 5.6 The three types of DGFETs: (a) side view of the type I, (b) side view of the type II, and (c) top view of the type III (*After*: [1]).

The short-channel effects are also reduced by downscaling the thickness of the gate dielectric. For gate lengths of 30–40 nm used in nowadays microprocessors, the SiO_2 thickness is about 1.5 nm. As a result, the leakage current is quite large because the electrons in the channel tunnel through the SiO_2 layer. Moreover, the density of the leakage current increases exponentially with the gate dielectric thickness and therefore a further reduction of the SiO_2 thickness below 1 nm is not admissible. Hence, new types of gate dielectric materials with higher dielectric constant, such as Al_2O_3, ZrO_2, and Ta_2O_5, are envisaged. They can have thicker dimensions than the SiO_2 when the transistor is scaled down to nanosized gate lengths.

5.1.2 The Ballistic FET

In this section we describe MOSFET transistors working in the ballistic regime, which represents the ultimate boundary of downscaling where the channel length is in the range 30–80 nm (see Figure 5.7). In this regime, the carriers travel the distance between source and drain without any scattering event.

In the ballistic transport regime the channel length of nanosized MOSFETs is comparable to the mean-free-path of carriers, and thus, the charges in the channel behave like waves. In ballistic MOSFETs the classical model described in the previous section is no longer valid and must be replaced by a quantum treatment. In the ballistic regime the saturation velocity effect is no longer valid, since the carriers travel along the channel without scattering and so their average velocity can exceed the saturation velocity and can attain large values. This phenomenon, called velocity overshoot, enhances the speed of the device. As we will see later, the cutoff frequency of some ballistic transistors is situated in the THz range. The main configurations envisaged for ballistic MOSFETs are the SOI MOSFETs and especially the DGFET.

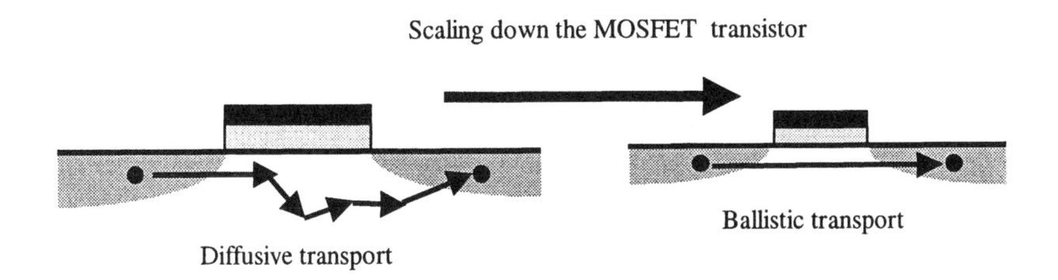

Figure 5.7 By scaling down the MOSFET transistor the diffusive transport transforms into ballistic transport.

Several quantum mechanical methods able to model ballistic MOSFETs are reviewed in [4]. We will present a model proposed in [5] because it utilizes explicitly the transmission and reflection coefficients of electrons. These quantum mechanical parameters are also used in the Landauer formula, which governs the physical phenomena at the nanoscale. Although a 1D treatment of the energy profile of a MOSFET instead of a more complicated 2D picture is used in the model represented in Figure 5.8, the most important new physical characteristics are preserved.

In this model the I_{DS} current is expressed as

$$I_{DS} = TI^{D} - T'I^{S}, \tag{5.14}$$

where I^{D} is the incident carrier flux directed from source to drain with T the corresponding transmission coefficient and I^{S} is the carrier flux originating from drain and directed to the source with the transmission coefficient T'. When there is no bias, the two terms in (5.14) are equal and there is no net current in the device. In the presence of a bias, the second term in (5.14) is cancelled and a net current flows through the device from source to drain. Assuming that the charge density in the channel near the source is described by (5.6), we obtain the charge at the Fermi level via the following integral:

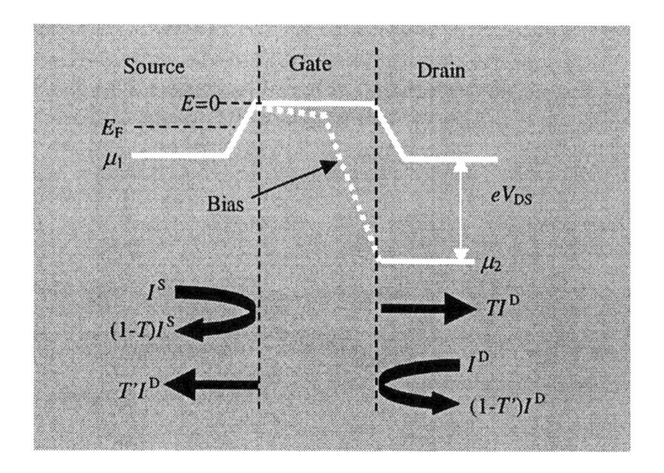

Figure 5.8 The model of a MOSFET taking into account the transmission and reflection (*After*: [5]).

$$q_{E_F} = (1/2\pi^2)\int_{-\infty}^{+\infty} dk_z \int_{-\infty}^{+\infty} dk_x f(E) = mE_{th}K/\pi\hbar^2, \tag{5.15}$$

where $f(E)$ is the Fermi distribution, $K = \ln(1 + E_F / E_{th})$, and $E_{th} = k_B T$ is the thermal energy at temperature *T*. In order to avoid confusion between temperature and the transmission coefficient, both commonly denoted by *T*, the temperature will be replaced whenever necessary with E_{th} / k_B. The expression (5.15), valid for nondegenerate systems, for which $q_{E_F} < 10^{12}$ cm^{-2}, $K << 1$, can be approximated by

$$q_{E_F} = mE_{th} \exp(E_F / E_{th}) / \pi\hbar^2 . \quad (5.16)$$

Under these circumstances, the current passing through the transistor in the presence of a drain voltage is

$$I_{DS}[TI^S]_{eq}[1 - \exp(-eV_{DS} / E_{th})] \cong I_{DS}[TI^S]_{eq} eV_{DS} / E_{th} , \quad (5.17)$$

where the subscript indicates the equilibrium value. From (5.17) it follows that the channel conductance equals

$$g_{DS} = e[TI^S]_{eq} / E_{th} . \quad (5.18)$$

Supposing now that in a nanometric FET working in the ballistic regime $T = 1$, we obtain a contact resistance

$$R_c = E_{th} / eI_{eq}^S , \quad (5.19)$$

which cannot be overcome by any downscaling method or contact technology. The incident carrier flux in a channel with width *W* is then

$$I_{eq}^S = (eWq_{E_F} / 2)v_{th} , \quad (5.20)$$

where v_{th} is the thermal electron velocity, while the minimum source contact resistance is given by

$$R_c W = (325\Omega\mu m)[E_{th} / k_B(300K)][10^{12} / cm^2 / q_{E_F}][10^7 cms^{-1} / v_{th}] . \quad (5.21)$$

The minimum source contact resistance at room temperature, for which $q_{E_F} = 10^{12}$ cm^{-2}, is $R_c W / 2 =$ 160 Ω μm; this lower limit cannot be surpassed. For a nanosized transistor with a width of 10 nm this minimum resistance is very close to the 12.9 kΩ Landauer contact resistance, so that (5.21) can be considered the quantum limit of the channel resistance.

By equating the number of conducting modes with the number of half-waves encompassed into the width we can obtain the minimum contact resistance in the degenerate limit, which is more specific for nanosized transistors with heavily doped channels. The result is

$$R_c W = \pi\hbar\lambda_F / 2e^2 \cong 50\,\Omega\mu\text{m}(10^{13}\text{cm}^{-2} / q_{E_F})^{1/2} \tag{5.22}$$

with λ_F the Fermi wavelength.

Because at low bias voltages the transmission coefficient is given by

$$T = L_{fp} /(L + L_{fp}) = 1/(L / L_{fp} + 1)\,, \tag{5.23}$$

where $L_{fp} = E_{th} / eF_S$ is the mean free-path with F_S the applied field in the source region, the total resistance of the channel is

$$R_{DS} = R_c / T = R_c (L / L_{fp} + 1)\,. \tag{5.24}$$

The saturation drain current,

$$I_{DSat} = TI^S\,, \tag{5.25}$$

depends directly on the transmission, its maximum value in the ballistic regime being

$$I_{DSat,max} = I^S / W = (315\mu\text{A} / \mu\text{m})(m_0 / m)(q_{E_F} / 10^{13}\text{cm}^{-2})^{3/2}\,. \tag{5.26}$$

For $q_{E_F} = 10^{13}\text{cm}^{-2}$, we get $I_{DSat,max} = 1.26$ mA/μm, which is comparable to a high driving current encountered in current VLSI digital devices. However, I^S, given by the expression $I^S = I^S_{eq}(2L_{fp})/(2L + L_{fp})$, is dependent on the bias. The maximum applied field in present nanosized transistors is 10^4 V/cm, which corresponds to a transmission coefficient of about 0.5. This value can increase in DGFET configurations, and can reach 0.8 at an applied field of 6×10^4 V/cm.

Another interesting approach for a nanometer MOSFET description is the analytic model proposed by Natori [6]. The electrical potential energy in the ballistic MOSFET along the channel direction, chosen as the *x* direction, is shown in Figure 5.9. It is clear from this figure that beyond an energy threshold the carriers travel to the drain via thermionic emission, but below this threshold the carriers are transmitted to the drain via tunneling, which is an unwanted effect.

The role of the gate is to lower the height of the potential barrier along the x direction and to allow electrons to cross the channel; the smoothed part of the potential barrier is termed valley. Thus, the height of the channel is modulated by the gate voltage, as shown in Figure 5.9, the drain current increasing when the barrier height is reduced by applying a higher gate voltage.

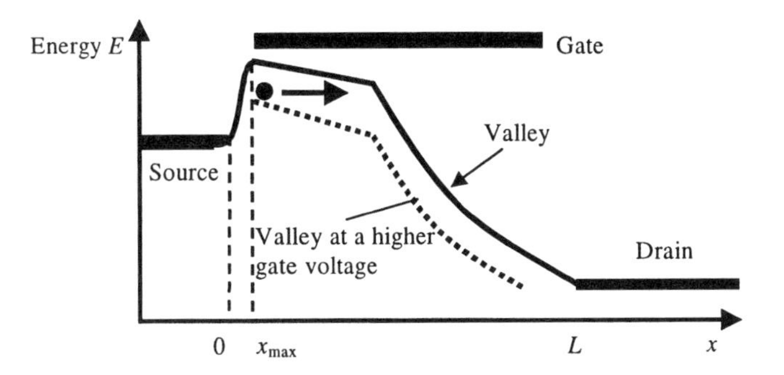

Figure 5.9 The energy distribution along the channel.

In the analytic model in [6] it is also assumed that the carriers are confined along the channel width direction (y direction) by a square barrier potential profile and that carrier confinement takes place also along the channel depth direction (the z direction) due to a triangular potential well in which the electronic states are quantized (see Figure 5.10). The electronic states in the channel are determined by solving the Schrödinger equation

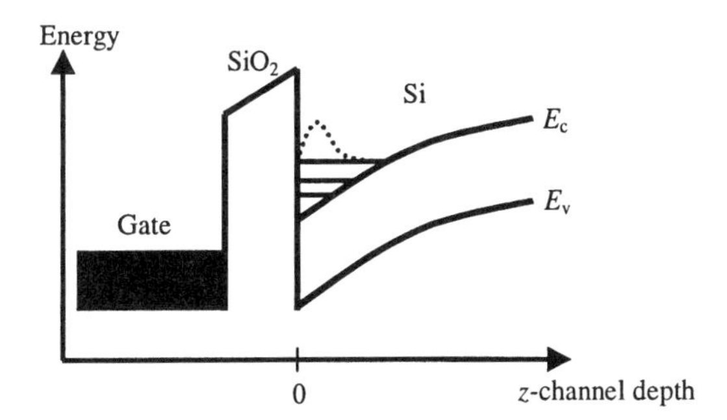

Figure 5.10 The energy distribution along the channel thickness (*After*: [6]).

$$\left(-\frac{\hbar^2}{2m_x}\frac{\partial^2}{\partial x^2}-\frac{\hbar^2}{2m_y}\frac{\partial^2}{\partial y^2}-\frac{\hbar^2}{2m_z}\frac{\partial^2}{\partial z^2}+U(x,z)\right)\psi(x,y.z)=E\psi(x,y,z) \tag{5.27}$$

with the boundary condition $\psi(x,0,z)=\psi(x,W,z)=0$. If the potential varies slower along the x direction than along z, the solution of the Schrödinger (5.27) in the WKB (Wentzel-Kramers-Brillouin) approach has the form

$$\psi(x,y,z)=A[p(x)]^{-1/2}\exp[i/\hbar\int_0^x p(x)dx)](2/W)^{1/2}\sin(n_y\pi y/W)\phi_{n_z}(x,z). \tag{5.28}$$

Here A is a constant, $\phi_{n_z}(x,y,z)$ is the solution of the equation

$$\left(-\frac{\hbar^2}{2m_z}\frac{\partial^2}{\partial z^2}+U(x,z)\right)\phi_{n_z}(x,z)=E_{n_z}(x)\phi_{n_z}(x,z)\,, \tag{5.29}$$

and

$$p(x)=(2m_x)^{1/2}[E-(\hbar^2/2m_y)(n_y\pi/W)^2-E_{n_z}(x)]^{1/2}\,, \tag{5.30}$$

where E_{n_z} is the x-dependent energy level of the n_z node in the z direction, and (n_y,n_z) are quantum numbers assigned to the motion along y and z, respectively.

Relation (5.28) models the wavefunction in the channel as composed of 1D wavefunctions of subchannels or subbands, each of them described by a pair (n_y,n_z). At $x=x_{\max}$ the wavefunction transforms into a plane wave

$$\psi(x,y,z)=A_1(\hbar k_m)^{1/2}\exp(ik_m x)(2/W)^{1/2}\sin(n_y\pi y/W)\varphi_{n_z}(x,z)\,, \tag{5.31}$$

with A_1 a constant and

$$k_m=(2m_x)^{1/2}[E-(\hbar^2/2m_y)(n_y\pi/W)^2-E_{n_z}(x_{\max})]^{1/2}/\hbar\,. \tag{5.32}$$

The drain-source current I_{DS}, calculated with the Landauer formula, is a sum of 1D currents, each corresponding to a certain subchannel. Each subchannel current is proportional to the number of carriers per unit time, the transmission of electrons in the subchannel and the probability that the destination of the carrier is empty. In turn, the number of carriers flowing in the subchannel is proportional to the subchannel density of states, denoted by $D(E)$, and the probability of state occupancy by the carrier, given by the Fermi-Dirac distribution $f(E,\phi_F) = \{1+\exp[(E-\phi_F)/E_{th}]\}^{-1}$. Denoting the source Fermi level at the source subchannel as ϕ_{FS} and as ϕ_{FD} at the drain subchannel, we have

$$I_{DS} = e \sum_{valley} \sum_{n_y} \sum_{n_z} \int \{D^+(E) v_g^+ f(\phi_{FS}, E)[1 - f(\phi_{FD}, E)] - D^-(E) v_g^- f(\phi_{FD}, E) \times [1 - f(\phi_{FS}, E)]\} T(E) dE \tag{5.33}$$

where the superscripts + and – correspond to propagation of charges from source to drain and vice versa, $v_g = p(x)/m_x$ is the group velocity of carriers, and $D(E) = m_x / \pi\hbar p(x)$. Because in ballistic transport $T(E) = 1$ and $\phi_{FD} = \phi_{FS} - eV_{DS}$, equation (5.33) becomes

$$I_{DS} = \frac{W 2^{1/2}}{\pi^2 \hbar^2} E_{th}^{3/2} \sum_{valley} (m_y)^{1/2} \left[\sum_{n_z} F_{1/2}\left(\frac{\phi_{FS} - E_{n_z}(x_{max})}{E_{th}} \right) - \sum_{n_z} F_{1/2}\left(\frac{\phi_{FS} - E_{n_z}(x_{max}) - eV_{DS}}{E_{th}} \right) \right] \tag{5.34}$$

where

$$F_{1/2}(u) = \int_0^\infty y^{1/2} dy / [1 + \exp(y - u)] \tag{5.35}$$

is the Fermi-Dirac integral and $v_g^+ = v_g^- = m_x / \pi\hbar p(x)$. Under the same assumptions the charge in the channel is given by

$$q_n = (eE_{th}/2\pi\hbar^2) \sum_{valley} \sum_{n_z} (m_x m_y)^{1/2} \ln\left\{ \left[1 + \exp\left(\frac{\phi_{FS} - E_{n_z}(x_{max})}{E_{th}} \right) \right] \times \left[1 + \exp\left(\frac{\phi_{FS} - E_{n_z}(x_{max}) - eV_{DS}}{E_{th}} \right) \right] \right\}. \tag{5.36}$$

A comparison with experimental results of a 70 nm MOSFET revealed that the measured $I_{DS} - V_{DS}$ characteristics have reproduced (5.34) in the saturation region, but discrepancies were found in the linear region due to carrier scattering [7].

When only one n_z subband is occupied and when high-level contributions in the sum in (5.34) are neglected, I_{DS} has a simplified expression given by

$$I_{DS} = WI_0[F_{1/2}(u) - F_{1/2}(u - v_d)], \tag{5.37}$$

where $I_0 = eM_v(2m_t)^{1/2}E_{th}^{3/2}/(\pi\hbar)^2$, with $M_v = 2.5$ and m_t the transverse electron mass, $u = \ln\{[(1+\exp v_d)^2 + 4\exp v_d(\exp\rho - 1)]^{1/2} - (1+\exp v_d)\} - \ln 2$, $v_d = eV_{DS}/E_{th}$, and $\rho = 2\pi\hbar^2 C(V_G - V_{th})/(eE_{th}m_t M_v)$, with C the MOS junction capacitance. In this case the charge expressed by (5.36) transforms into

$$q_0 = C(V_{GS} - V_{th}) = 2\pi\hbar^2 C(V_G - V_{th})/eE_{th}m_t M_v, \tag{5.38}$$

where C is independent of V_G. From (5.37) it follows that the ballistic current is independent of the channel length, but depends on the channel width, and that the saturation current becomes independent of the drain voltage when V_{DS} is large and turns in this case into

$$I_{DSat} = WI_0 F_{1/2}[\ln(\exp q_0 - 1)]. \tag{5.39}$$

In the degenerated case, around x_{max}, when the energy separation in the channel is larger than E_{th} the saturation current is given by

$$\begin{aligned} I_{DSat} &= 8\hbar W \mid q_0 \mid^{3/2} / 3m_t(e\pi M_v)^{1/2} \\ &= 8\hbar W[C(V_G - V_{th})]^{3/2} / 3m_t(e\pi M_v)^{1/2}, \end{aligned} \tag{5.40}$$

and the transconductance is

$$g_m = 4\hbar WC[C(V_G - V_{th})]^{1/2} / m_t(e\pi M_v)^{1/2}. \tag{5.41}$$

Other ballistic devices such as *p*-MOSFET, SOI MOSFET, DGFET, and HEMT (high electron mobility transistor) can be modeled in the same way, and similar expressions can be obtained. The power of the Natori method is that it offers quite simple formulas for any FET device.

The HEMT, which is a key component of millimeter-wave and submillimeter-wave devices, is a FET transistor with a 2DEG (two-dimensional electron gas) channel formed in a heterostructure made of AIII–BV semiconductors or nitrides at the interface between a doped wider-bandgap semiconductor and an undoped semiconductor. A typical example of such a heterostructure is *n*-AlGaAs/GaAs. Since the Fermi level in the wider-bandgap semiconductor is higher than the Fermi level in the narrower bandgap semiconductor, the electrons from the doped semiconductor move towards the undoped one, leaving behind them positively charged donors. These donors bend the energy bands of the heterostructure, confining the electrons in the well that forms at the interface, as illustrated in Figure 5.11(a). The separation of the positive and negative charges at the interface and the particular position of the Fermi energy in the well (higher than the first resonance level) confer to the electrons situated in the well special properties. They are free to move in the transverse plane to the heterostructure with a low rate of scattering events, being well separated from donor impurities; hence, electrons attain very high mobilities.

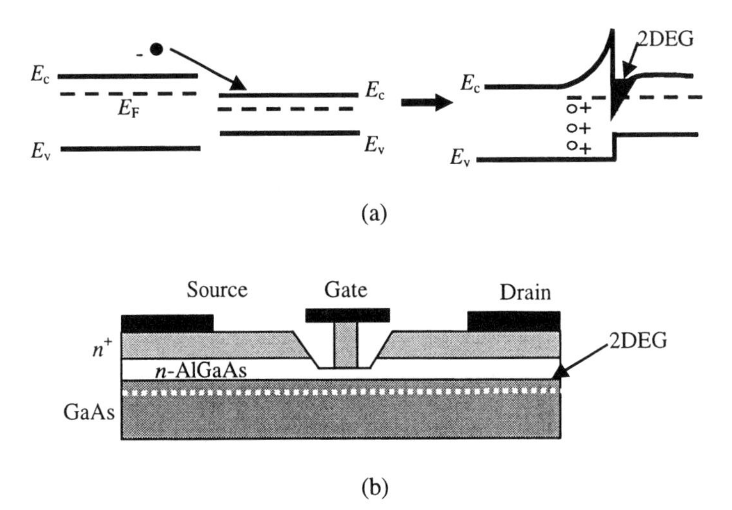

Figure 5.11 (a) The 2DEG formation and (b) the HEMT transistor.

The confined electrons at the heterostructure interface form a 2DEG with a carrier density of $2\times10^{11}-2\times10^{12}\,/\text{cm}^2$ and a mobility exceeding $10^7\,\text{cm}^2\,/\text{Vs}$. This sheet of conductive electrons can play the role of the channel in a FET transistor if other ingredients are added to the heterostructure: a source and a drain electrode, doped n^+ areas beneath them, and a gate electrode which depletes the electrons in the 2DEG layer when a negative voltage is applied. The resulting structure, depicted in Figure 5.11(b), is an HEMT transistor, which is analogous to a MOSFET in which the SiO_2 layer at the SiO_2/Si interface plays the role of the wider semiconductor. The inversion layer in the MOSFET is a 2DEG, as in AlGaAs/GaAs heterostructures, but the SiO_2/Si interface is of much lower quality than in AIII–BV compounds and thus many properties of 2DEG are not evident in normal conditions in MOSFETs.

The theory of the ballistic HEMT was developed first in [6], where analytic forms of the saturation current and channel charges were found. The ballistic transport of the HEMT was studied in [8] using a generalization of the analytic model presented above [9]. This theory is based on the assumption that the channel is situated at the top of the barrier between the source and the drain since the voltage gate lowers its height to modulate the density of electrons passing through it. The ballistic FET transistor behaves classically in channel and quantum mechanically when tunneling is taken into account, the tunneling phenomenon degrading always the performances of MOSFETs and HEMTs. In the Datta-Lundstrom model of the HEMT, when only the top of the source-drain barrier is considered and the transport in the channel is treated with a classical model, the carrier distribution function at the top of the barrier consists of two-equilibrium halves due to the source and drain injection of electrons, as can be seen in Figure 5.12(a); the charge at the top of the barrier is completely determined by the height of the potential barrier and the Fermi level of the source and drain. As the drain voltage increases the carriers located at the drain are impeded to flow because the barrier height is higher towards the channel and thus mainly the electrons from the source flow into the channel. The circuit model of the HEMT, depicted in Figure 5.12(b), is based on three capacitors C_G, C_S, and C_D, which represent the effect of the gate, source, and drain terminal at the top of the barrier, respectively.

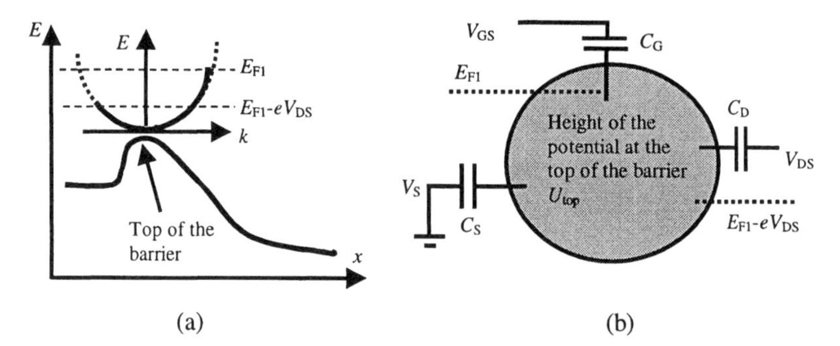

Figure 5.12 (a) The energy behavior for the ballistic HEMT in the Datta-Lundstrom approach, and (b) the Datta-Lundstrom circuit model (*After*: [9]).

In this model the top of the potential between source and drain is given by

$$U_{top} = -e[(C_G V_G / C_{tot}) + (C_D V_D / C_{tot}) + (C_S V_S / C_{tot})] + e^2 n_{mob} / C_{tot}, \quad (5.42)$$

where $C_{tot} = C_G + C_S + C_D$ and $C_G >> C_S, C_D$. The mobile charge in the channel, n_{mob}, is estimated using (5.42) from the known bottom of the band at the top of the barrier and by filling the $\pm k$ states according to the source and drain Fermi levels, respectively. U_{top} is computed iteratively adding mobile charges until a convergence is attained and I_{DS} is computed from the known populations in the two halves of the energy dispersion diagram above the top of the barrier. In [8] a 30 nm InP based InAlAs/InGaAs HEMT was numerically simulated considering it as an intrinsic ballistic transistor with an added extrinsic source-drain resistor. It was found that the HEMT behaves ballistically below a 50 nm gate length.

However, contrarily to the common sense, Shur has demonstrated that the mobility in short channel HEMTs, where the carriers move ballistically, is much lower than the mobility in long channel HEMTs [10]. The effective mobility is expressed as

$$1/\mu_{eff} = 1/\mu_{ball} + 1/\mu_0 \tag{5.43}$$

where μ_0 is the mobility in the long-channel regime dominated by collision effects and μ_{ball} is the quantity assigned to the ballistic movement of carriers. The ballistic mobility is given by $\mu_{ball} = 2eL/\pi m v_{th}$ in the nondegenerate case, while the corresponding expression in the degenerate case is obtained by replacing the thermal velocity $v_{th} = (8E_{th}/\pi m)^{1/2}$ with the Fermi velocity v_F. Relation (5.43) indicates a large decrease of mobility with a decrease in gate length, supported also by experiments; for example, in GaAs the 10, 000 cm^2/Vs mobility assigned to a gate length of 10 μm decreases to 3000 cm^2/Vs for a 150 nm gate length. This significant decrease becomes even more accentuated at low temperatures.

When the drain voltage is low, the drain current can be approximated with [8]

$$I_{DS} = Wq_i(0)\mu_{ball}V_{DS}/L, \tag{5.44}$$

where $q_i(0)$ is the electron distribution at the source. In the opposite case, when the drain voltage is high [8],

$$I_{DSat} = Wq_i(0)\overline{v_{thd}} = Wq_i(0)[v_{thd}F_{1/2}(\zeta)/F_0(\zeta)], \tag{5.45}$$

where $v_{thd} = (2E_{th}/\pi m)^{1/2}$ is the unidirectional thermal velocity and $\zeta = (E_F - E_{S_0})/E_{th}$, with E_{S_0} the first electron subband at the source. Since in a long channel HEMT $I_{Dsat} = Wq_i(0)v_{sat}$ if $L \to 0$, the saturation velocity v_{sat} can be replaced by $v_{thd}F_{1/2}(\zeta)/F_0(\zeta)$ in the conventional formulas of HEMT, in particular in the $I_{DS} - V_{DS}$ characteristic. Thus, the extended concept of mobility in the ballistic regime allows the use of conventional FET equations to evaluate the ballistic behavior.

Ballistic transport at GHz frequency in ungated HEMT structures was recently experimentally confirmed [11]. The ungated HEMT is a HEMT where the current is injected in the 2DEG via an ohmic contact, flows laterally through the 2DEG and exits through another ohmic contact. The ballistic transport was achieved in a different manner compared to the above situation, where the length of the device is shorter than the mean-free-path: in the ungated HEMT the ballistic limit is reached when the frequency of the electric field ω is higher than the scattering frequency, that is, if $\omega\tau_{scat} > 1$, with τ_{scat} the scattering time.

As shown in Figure 5.13, the transition from the ballistic regime, in which $\omega\tau_{scat} > 1$, to the diffusive regime, for which $\omega\tau_{scat} < 1$, can be evidenced by measuring the dependence on frequency of the dynamical impedance of the 2DEG, $Z_{2DEG}(\omega) = R + i\omega L = m/ne^2\tau_{scat} + i\omega m/ne^2$. In fact, the Drude conductivity formula yields $Z^{-1}{}_{2DEG}(\omega) = \sigma(\omega) = ne^2\tau_{scat}/m(1+i\omega\tau_{scat})$, where n is the electron density.

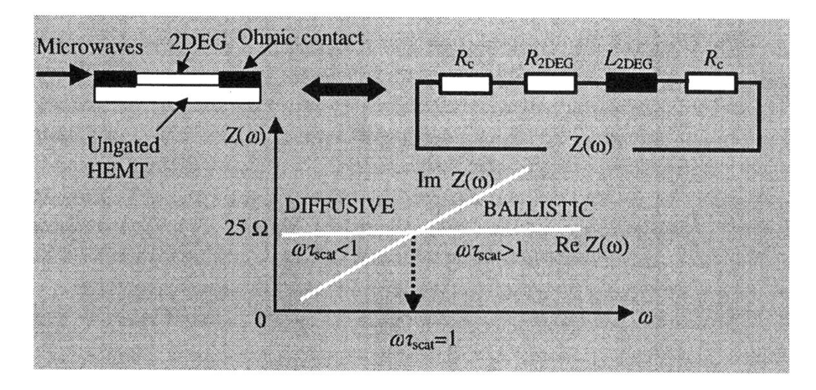

Figure 5.13 The diffusive and ballistic regime evidenced in an ungated HEMT (*After*: [11]).

Measurements of the microwave impedance performed at 4 K on modulation doped GaAs/AlGaAs devices with dimensions of $500\ \mu m \times 500\ \mu m$ indicate, as illustrated in Figure 5.13, that the crossover between the diffusive and ballistic regimes, expected to occur when $\omega = 1/\tau_{scat}$, takes place when $\mathrm{Re}\, Z_{2DEG}(\omega) = \mathrm{Im}\, Z_{2DEG}(\omega)$ if the contact resistance is low compared with the resistance of the 2DEG. This equality was found to hold at the frequency of 2 GHz for a 2DEG resistance of about 24 Ω. When the contact resistance is comparable with that of the 2DEG the real and imaginary parts of Z_{2DEG} are not equal at the frequency at which $\omega = 1/\tau_{scat}$ since the real part of the impedance includes an additional term R_c due to the contact resistance.

In an HEMT with a nanosized gate length, the transport in the 2DEG is ballistic in the sense that there are no electron-phonon or electron-impurity collisions, but the mean free path for interelectronic collisions (determined by the average distance between two electrons) may be smaller than both the gate length and the mean free path of electrons. For example, this is the case in an AlGaAs/InGaAs HEMT at 77 K, where the transport is ballistic when the gate length is around 30 nm, but the average distance between electrons is only 10 nm [12]. As a consequence, the electrons collide and the 2DEG with a large number of collisions, modeled as a plasma wave, generates current instabilities, and thus, oscillations, at THz frequencies. The equation that describes this 2DEG plasma with significant electron-electron interaction is analogous to the shallow water hydrodynamic equation known as the Euler equation. If the surface electron density is given by [12]

$$n_{\mathrm{s}} = C_{\mathrm{G}} V / e \,, \tag{5.46}$$

where C_{G} is the gate capacitance per unit area, and $V = V_{\mathrm{GC}}(x) - V_{\mathrm{th}}$, with V_{GC} the local gate channel voltage, the 2DEG plasma is described by the Euler equation

$$\partial v / \partial x + v \partial v / \partial x = (-e / m) \partial V / \partial x \tag{5.47}$$

accompanied by the continuity equation

$$\partial V / \partial t + \partial (Vv) / \partial x = 0 \,. \tag{5.48}$$

In (5.47) and (5.48) $v(x,t)$ is the local electron velocity and $\partial V / \partial x$ is the electric field along the channel direction. For a constant voltage V_0 between the gate and the channel the electron plasma dispersion law is $k = \pm \omega / s$ where the plasma wave velocity is given by

$$s = (e V_0 / m)^{1/2} \,. \tag{5.49}$$

If the source current between the source and drain is constant and if there is a constant voltage source between the source and the gate, which corresponds to the boundary conditions of an ac short circuit at the source and an ac open circuit at the drain, the entire HEMT channel is in fact a plasma resonator with the fundamental plasma frequency

$$\omega_{\mathrm{p}} = \pi s / 2L \,. \tag{5.50}$$

In most devices based on electron plasma effects the device length must satisfy the inequality $L < L_{cr} = s\mu m / e$, where μ is the low-field mobility. This relation is mathematically similar, up to a constant, to the condition $\omega\tau_{scat} >> 1$, which assures the occurrence of ballistic transport. For a gate voltage of 0.5 V the parameter L_{cr} is 300 nm in GaAs, 100 nm in Si, and 50 nm in GaN [13]. The operation frequency of the devices based on 2DEG plasma must be higher than $1/\tau_{scat}$, which corresponds to 8 THz in Si, 3.5 THz in GaAs, and 10 THz in GaN. Using the above principles, THz oscillators, detectors, mixers, and multipliers were demonstrated [13], but the most interesting application is THz generation. The plasma waves travel back and forth in the channel and increase in amplitude after successive passages in the channel cavity, THz oscillations leaving the device through source and drain. This behavior is similar to that of electromagnetic waves in a resonant cavity. The condition of observing plasma instability is found by considering small ac perturbations for the electron velocity v and applied voltage V. More precisely, if in (5.47) $v = v_0 + v_1 \exp(-i\omega t)$ and $V = V_0 + V_1 \exp(-i\omega t)$ the obtained frequency $\omega = \omega_1 + i\omega_2$ has real and imaginary parts given by

$$\omega_1 = p\pi | s^2 - v_0^2 |^2 / 2Ls , \tag{5.51a}$$

$$\omega_2 = | s^2 - v_0^2 | \ln | (s + v_0)/(s - v_0) | / 2Ls , \tag{5.51b}$$

respectively, where p is an odd integer. The plasma instability is obtained when

$$s > v_0 > 0 , \tag{5.52}$$

and the lowest fundamental frequency plasma mode is determined as

$$f_1 = | s^2 - v_0^2 |^2 / 4Ls . \tag{5.53}$$

Other conditions of plasma wave instability and growth are found in [12].

The predicted THz oscillations were recently experimentally demonstrated in the first manufactured Dyakonov-Shur transistor [14], which is a lattice-matched InGaAs/InAlAs HEMT grown by MBE with a gate length of 60 nm and a drain-source separation of 1.3 μm. The HEMT has a high InGaAs channel mobility and a high carrier density. The required boundary conditions for plasma generation (zero impedance at the gate-source end of the channel, and an infinite impedance at the gate-drain end of the channel), were implemented by shortcutting the gate with the source and by selecting the saturation regime via the applied drain voltage. The HEMT emitted electromagnetic fields in the 0.4–1 THz range depending on the drain voltage, which initiates the plasma emission beyond the threshold of 0.2 V. The detected power was only in the nanowatt range because the transistor emitted through its leads and not via an antenna. Besides, the transistor was placed in a waveguide at 4.2 K and it seems that no impedance match was used. The parameter $V_0 = 80 \pm 30$ mV and the electron density, of $4\times10^{11}\,\mathrm{cm}^{-2}$, were determined from the measured $I-V$ characteristic. Because the parameter s that determines the plasma wave velocity was found to be $s = 5.8 \pm 1.2\times10^5$ m/s, while $v_0 = 4.9 \pm 1.2\times10^5$ m/s, the plasma instability criterion (5.52) is satisfied. If we assume that at V_{DS} = 0 V the gate has an effective length $L_{eff} = L + 2d$, with d the thickness of the wide band barrier layer and if we take into account that L = 60 nm and d = 17 nm, it follows that $L_{eff} = 94$ nm, which gives a plasma frequency of $f_1 = 1$ THz if introduced in (5.53) together with the above determined velocities. By increasing the bias, the gate length decreases and hence the channel length, which is expressed through the relation

$$L_{eff}(V_{DS}) = L_{eff}(0) - (V_{DS} - V_{DSat}) / F_{eff} \tag{5.54}$$

with F_{eff} the effective electric field, reduces. Relation (5.54), which demonstrates the linear dependence of $1/f_1$ on the source-drain voltage, was confirmed by the measured THz emission of the HEMT.

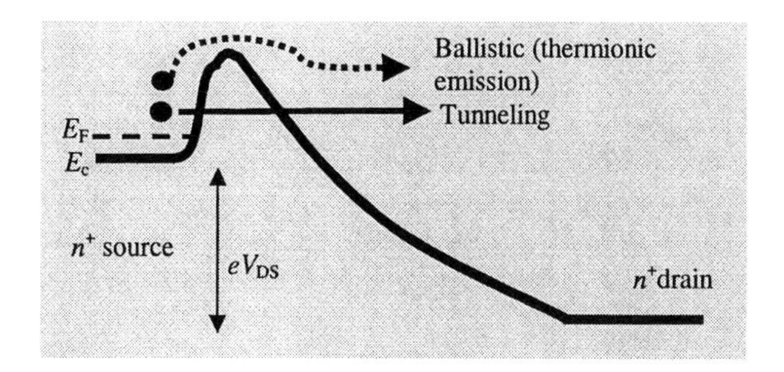

Figure 5.14 SIT potential band diagram (*After*: [15]).

Another ballistic FET is the static induction transistor (SIT), which is a short channel FET in which the current flows vertically from source to drain; although SIT is a vertical FET, a horizontal SIT was reported in [15]. The current in the channel is controlled by the height of an electrostatically induced potential. The SIT in the reverse gate operation regime is similar to a short channel FET, while, when operating under a forward gate bias, the SIT is analogs to a bipolar transistor with an extremely thin base structure, which produces an extremely high cutoff frequency. The potential band diagram of an SIT is illustrated in Figure 5.14.

The drain current is controlled by the potential barrier ϕ_b. The electrons flow over this barrier whose height depends on V_{GS}, V_{DS}, and on the built-in potential. The dependence of ϕ_b on V_{DS} is responsible for the triode-like characteristic of SIT, which lacks a saturation region; the saturation regime is an imprint of most FETs. At large source-drain distances (i.e., for a thick barrier) the drift-diffusion transport is important and produces electron scattering, but if this distance decreases beyond the mean free path value (that is, for sufficiently thin barriers) the transport becomes ballistic and the transport at short distances occurs via tunneling.

In ballistic MOSFETs or HEMTs the gate potential modulates the barrier height, which in turn controls the current flow, while the drain potential has only a small influence on the current. The barrier is screened from the drain potential by the channel. On the contrary, in a ballistic SIT the barrier screening is absent, which is an advantage for very short transit times, the SIT being considered as the FET with the shortest possible electron path. The ballistic SIT is thus among the fastest electron devices, its transit time being given by

$$\tau_t = (1/2\pi f_t) = 2L(m/2eV_{DS}), \tag{5.55}$$

where L is the channel length.

For a drain voltage of 1 V, the ballistic regime, in which electrons tunnel through the potential barrier, is encountered for channel lengths shorter than 20 nm, which correspond to transit times of tens of fs, such that the cutoff frequency of the SIT is beyond 1 THz. For a 10 nm channel the transit time is estimated to be 20 fs and the corresponding cutoff frequency is 8 THz. This can occur only when the gate-source and gate-drain capacitances, C_{GS} and C_{GD}, are very small. The SIT with a 10 nm channel length was fabricated using a special epitaxy technique named molecular layer epitaxy (MLE) [15]. SIT is also known as the FET with the smallest series resistance R_s, and with minimal values for C_{GS} and C_{GD}. Since the parasitic elements are minimized, the ballistic SIT has the impressive status of a THz transistor.

The SIT source-drain current is calculated using the Landauer formula

$$I_{DS} = (4\pi e E_{th} / h^3) \int_{E_c}^{\infty} T(E) m_t(E_x) \ln\left(\frac{1 + \exp[(E_x - E_F)/E_{th}]}{\exp[(E_x - E_F)/E_{th}]} \right) dE_x \tag{5.56}$$

where m_t is the transverse effective electron mass, $T(E)$ is the transmission coefficient, and a parabolic dispersion relation is assumed in the transverse plane (y, z), which is perpendicular to the source-drain direction $-x$. Formula (5.56) can be simplified in some cases. For example, when the electron energy exceeds the Fermi level, and for $E_x > \phi_b$, the transmission is almost 1 and the ballistic current across a thick potential barrier becomes

$$I_{DS,ball} = (4\pi e m_t E_{th}^2 / h^3)\exp[(-\phi_b + E_c)/E_{th}]. \quad (5.57)$$

Another approximation of (5.56) is valid when all carriers are well below the Fermi level and tunnel through the barrier. For a triangular barrier the transmission is calculated at $E_x = E_F$ using the WKB approximation to finally obtain the drain current in the tunneling regime, $I_{DS,tun}$. The ratio between the tunneling and ballistic drain current densities is given by

$$I_{DS,tun} / I_{DS,ball} \cong (E_F - E_c)^2 \frac{\exp\{-8\pi w[2m_t(\phi_b - E_c)]^{1/2}/3h\}}{2E_{th}^2 \exp[(-\phi_b + E_c)/E_{th}]} \quad (5.58)$$

where w is the width of the triangular barrier at $E_F = E_c$ (see Figure 5.15). Equation (5.58) tells us that the tunneling current prevails at low temperatures, while at high temperatures $I_{DS,ball} > I_{DS,tun}$. The maximum width of the potential barrier where tunneling is the prevailing transport mechanism is given by

$$w_{max} \approx \frac{3h}{8\pi[2m(\phi_b - E_F)]^{1/2}} \left[\frac{\phi_b - E_c}{E_{th}} - \ln\left(\frac{2E_{th}^2}{(E_F - E_c)^2} \right) \right]. \quad (5.59)$$

In GaAs, where $\phi_b - E_c$ = 0.85 eV and $E_F - E_c$ = 0.15 eV, the maximum tunneling width is 25 nm at room temperature and increases to 90 nm at 77 K.

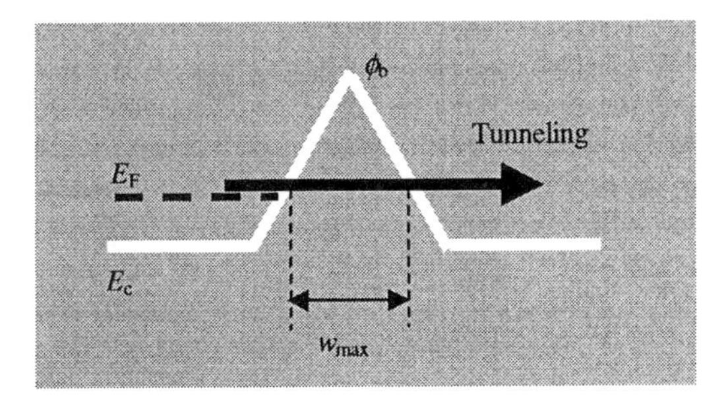

Figure 5.15 SIT carrier tunneling.

5.2 MESOSCOPIC DEVICES AT ROOM TEMPERATURE

The mesoscopic devices, based mainly on AIII–BV semiconductors, are key components for future ultrafast devices. The carrier transport in any mesoscopic device is ballistic and occurs when the mean free path of carriers is longer than the dimensions of the device. The mesoscopic scale refers to devices with dimensions that range between the atomic scale and the macroscopic scale. The dimensions of mesoscopic devices depend strongly on the temperature at which they are designed to work. For example, at very low temperatures (around 0.1 K) the mean free path of electrons is 160 μm in GaAs/AlGaAs heterostructures [16]. Thus, ballistic mesoscopic devices working at low temperatures can be realized using standard lithographical techniques.

An impressive experimental and theoretical work emerged that proves the amazing properties of ballistic quantum devices. Tunneling devices, electron lenses, prisms, beamsplitters, waveguides, quantum interference devices, and even quantum computing devices were designed and experimented. A recent review on ballistic electron devices can be found in [17].

When the operating temperature of mesoscopic devices increases towards room temperature, their dimensions shrink from the microscale to the nanoscale. In AIII–BV semiconductors the mean-free path of carriers at room temperature is in the range 100–200 nm. The most common used heterostructure in electronics, GaAs/AlGaAs, is not suitable for fabricating ballistic devices due to its large depletion length, which is comparable with the mean free path of carriers. Thus, the room-temperature mesoscopic devices use InGaAs/InAlAs or InGaAs/InP heterostructures, where a high sheet carrier density is formed by reducing the depletion length.

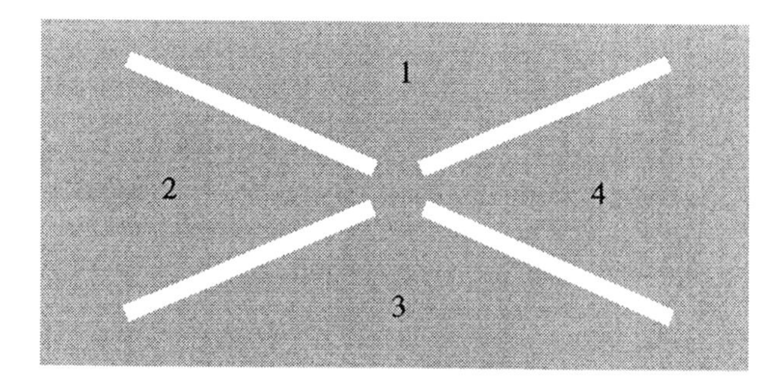

Figure 5.16 The first ballistic device working at room temperature; white areas are insulating lines (*After*: [16]).

According to [16], the first room-temperature ballistic device was a 260 nm square four-terminal structure made on an AlGaAs/InGaAs/GaAs modulation doped heterostructure. The four insulating lines, which are represented in white in Figure 5.16, had a width around 60 nm and were fabricated using the FIB technique, which was described in Chapter 1 [18].

The four-terminal device in Figure 5.16 has very interesting properties, investigated by injecting a current through terminals 1 and 2, and detecting it through the other two terminals. The bend resistance, which was measured experimentally, is defined as $R_{12,34} = V_{43} / I_{12}$. This resistance is positive in the diffusive transport regime, when $V_{43} < 0$ for electrons that are transported diffusively at contact 2 for $I_{12} < 0$, and negative in the case of a ballistic regime, when $V_{43} > 0$ for electrons that are transported ballistically at contact 3 instead of contact 2 for $I_{12} < 0$. The negative sign of the bend resistance $R_{12,34}$ is preserved even at room temperature although its value reduces by a factor of 7. At room temperature the mean free path in this device is 140 nm, a value that is only a bit shorter than the junction length, which is 260 nm, and thus still assures the ballistic character at this high temperature.

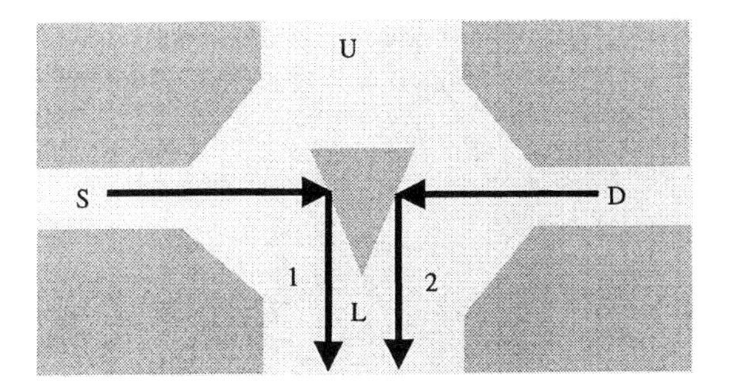

Figure 5.17 The ballistic rectifier (*After*: [19]).

One of the most studied mesoscopic devices at room temperature is the ballistic rectifier [19] where, in contrast to macroscopic electron devices, the angular parameter of electron trajectories dictates the transmission of electrons between terminals, and where an applied voltage is able to control the propagation angle of every ballistic electron. The ballistic rectifier is presented in Figure 5.17. It consists of two narrow channels termed source (S) and drain (D) and two wide channels termed upper (U) and lower (L). At the crossing point a triangular reflector is placed. Darker areas in Figure 5.19 are etched areas, which play the role of insulators (reflectors) for electron propagation. The first version of the ballistic rectifier was designed to work at 4 K, to prove its working principles, and was made using a GaAs/AlGaAs heterostructure with a 2DEG located at 40 nm beneath the wafer surface.

The current is applied between drain and source and the output voltage is collected via L and U. When a current is injected between source and drain electrons propagate towards the lower lead L since ballistic electrons follow the path 1. On the contrary, when the current is injected in the opposite direction, from drain to source, the ballistic electrons follow the path 2, but in both situations the detected voltage between L and U is the same, and is always negative. The equation describing this rectifying mechanism is

$$V_{LU}(I_{SD}) - V_{LU}(-I_{SD}) = 0\,. \tag{5.60}$$

An ideal ballistic rectifier behavior based on (5.60) is displayed in Figure 5.18.

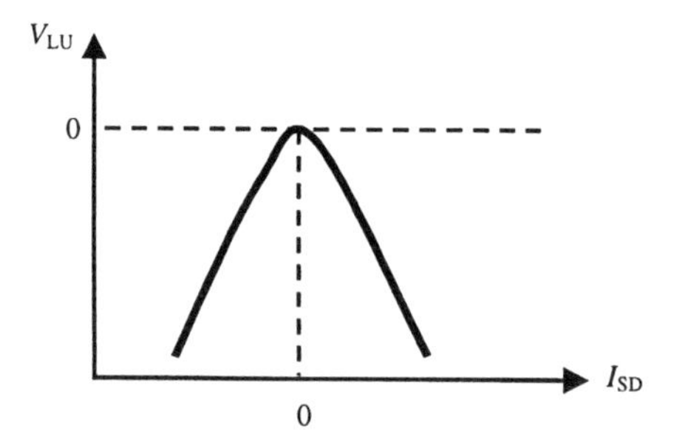

Figure 5.18 Typical current-voltage characteristic of an ideal ballistic rectifier (*After*: [19]).

By defining $R_{SD,LU} = V_{LU} / I_{DS}$ the ideal rectifier can be also characterized by a relation similar to (5.60):

$$R_{SD,LU}(I_{SD}) + R_{SD,LU}(-I_{SD}) = 0\,. \tag{5.61}$$

The ballistic rectifier working at 4 K is not a nanodevice since the mean free path of the carriers is about 6 μm. However, the corresponding room-temperature ballistic rectifier based on an InGaAs/InP heterostructure becomes a nanosized device when the carrier mean free path is only 140 nm. The room-temperature ballistic rectifier is thus 40 times smaller than that at 4 K. This device was tested to work as a millimeter-wave detector up to 50 GHz, but it seems to work at much higher frequencies because, due to the lateral configuration of the contacts, the parasitic capacitances are very small. Moreover, an artificial nanomaterial, which consists of a 2D periodic arrangement of ballistic rectifiers, was demonstrated to work as a millimeter-wave detector, but the cutoff frequency of this nanomaterial could be in the THz domain [20].

The ballistic rectifier has a quadratic response to the input signal, which also allows the generation of second harmonics without producing spurious harmonics. Let us denote by θ the ejection angle of electrons emitted from S, by θ_0 the minimum ejection angle from S to D, and by $v > 0$ the drift velocity. At $V_{SD} = 0$ the velocity components of electrons in the source channel are v_x and v_y, respectively, and therefore $\theta = \arctan(v_y / v_x)$ at $I_{SD} = 0$. Then, the input-output characteristic of the ballistic rectifier is

$$V_{LU} = -(3h / 4e^2 N_{LU})(\sin\theta_e - \sin\theta_0) | I_{SD} |, \quad (5.62)$$

where $\theta_e = \theta_0 + \arcsin(v \sin\theta_0 / v_F)$ and N_{LU} is the number of occupied modes in the upper-lower channel, the corresponding number of modes in the source-drain channel being denoted by N_{SD}. For small $| I_{SD} |$ values (5.62) becomes an

$$V_{LU} = -(3\pi\hbar^2 / 4e^3 E_F N_{LU} N_{SD}) I_{SD}^2, \quad (5.63)$$

expression that illustrates the quadratic character of the input-output characteristic.

An important category of mesoscopic devices working at room temperature is constituted from Y-branch devices, which include T-branch devices as a particular case [21]. A Y-branch and a T-branch device is represented in Figure 5.19.

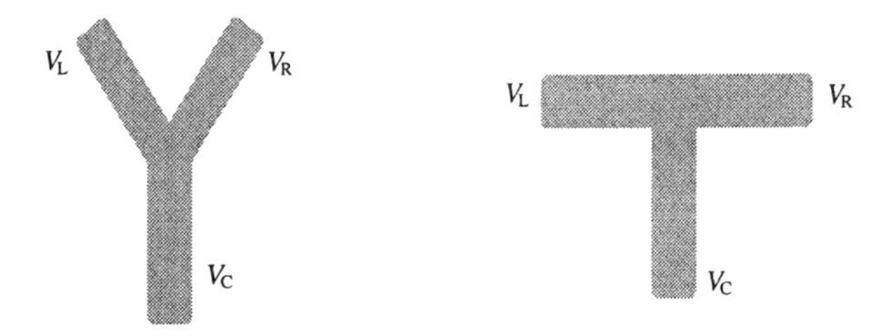

Figure 5.19 Schematic representation of the Y- and T-branch mesoscopic devices.

The three-terminal ballistic devices in Figure 5.19 show several different functionalities for the nanoelectronic applications at room temperature [22]. When the applied voltages on the left and right branches of the Y-branch junction follow the rule

$$\begin{cases} V_L = V \\ V_R = -V \end{cases} \tag{5.64}$$

the output voltage of the central branch, V_C, is always negative and the device demonstrates a rectification effect.

For large V values there is a parabolic dependence between V_C and V that can be found using a simple model. If the three branches are modeled by three saddle-point contacts, as shown in Figure 5.20, the electrostatic potential is given by

$$V(x, y) = V_0 - (1/2)m\omega_x^2 x^2 + (1/2)m\omega_y^2 y^2, \tag{5.65}$$

where V_0 is the potential at the saddle point, x is the carrier transport direction, while y is the corresponding transverse direction. The frequencies ω_x and ω_y express the potential curvature.

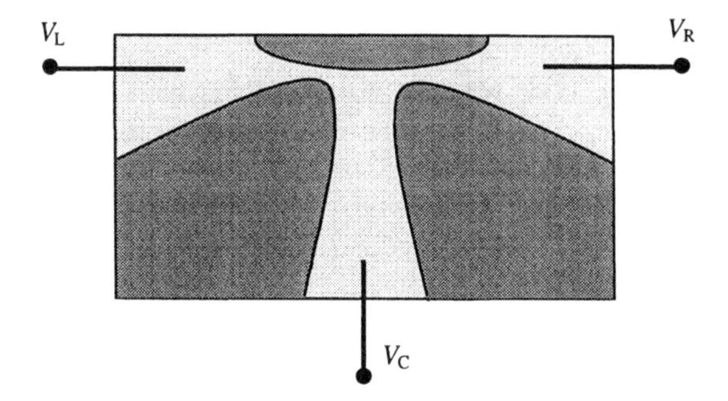

Figure 5.20. The three-terminal ballistic devices consisting of three saddle-point contacts (*After*: [23]).

The device in Figure 5.22 is called the three-terminal ballistic junction (TBJ). The central voltage of the TBJ is

$$V_C = (2\mu_F - \hbar\omega_{max})/2e - [(2\mu_F - \hbar\omega_{max})^2/4e^2 + V^2]^{1/2} < 0, \tag{5.66}$$

where μ_F is the electrochemical potential at the Fermi level and $\hbar\omega_{y(C,L)} = \hbar\omega_{y(R,C)} = \hbar\omega_{max}$. When $|V| << \mu_F / e$, relation (5.66) becomes

$$V_C = -(e / 2\mu_F)V^2 . \tag{5.67}$$

Another property of TBJ is frequency multiplication. If we apply a harmonic excitation between the left and right contact, such that $V_0 = A\cos(\omega_0 t)$, then the output voltage of the central branch is given by

$$V_C = A_1 + A_2 \cos(2\omega_0 t) + A_4 \cos(4\omega_0 t) + ... \tag{5.68}$$

The TBJ also performs basic logical operations. V_C is positive only when both left and right voltages are positive and in this case the logical value 1 can be assigned to it; otherwise it is in the logical state 0. So, the TBJ performs naturally the AND operation with 0 and 1 input states encoded in the negative and positive voltages, respectively, applied on the left and right branches. An OR gate can be implemented by assigning the logical value 1 to negative voltages. TBJs can also act as diodes and transistors.

The room-temperature function of the TBJ rectifier was experimentally proved in [23] using a modulation doped GaInAs/InP heterostructure. The same type of heterostructure was used to demonstrate room-temperature operation of the logic devices implemented with TBJ [24]. A complicated logical operation, namely a half-adder based on a ballistic Y junction, was recently demonstrated to work at room temperature using a modulation doped GaAs/AlGaAs hetero-structure with a 2DEG located at 50 nm beneath the top surface of the device [25].

Ballistic devices such as TBJ based on AlInAs/InGaAs are designed for future THz data processing according to the Monte Carlo simulations made in [26]. In such heterostructures the mean free path of electrons is higher than 100 nm at room temperature and the mobility reaches 14,000 $cm^2V^{-1}s^{-1}$. The room-temperature THz ballistic nanodevices, such as the ballistic channel, the TBJ-based THz multiplexer/demultiplexer device, and the ballistic rectifier, are compatible with the HEMT technology and constitute the building blocks of future terabit data systems.

A last category of ballistic devices with potential applications at room temperature is the quantum interference transistors [27]. A schematic representation of the T-shaped interference transistor, which includes its geometric dimensions, is shown in Figure 5.23. If L_2 and L_3 are less than the mean free path the reflection of electrons from the gate arm produces interference. When L_1 is also small compared to the mean free path, the interference patterns extend over the entire drain-source channel and influence the transmission along the channel and thus the conductance of the device. By tuning the gate voltage, the interference pattern is changed, and thus the transmission and the conductance in the channel can be controlled. This interference-effect-based FET will work at the room temperature when all the lengths indicated in Figure 5.21 will be around a few tens of nm.

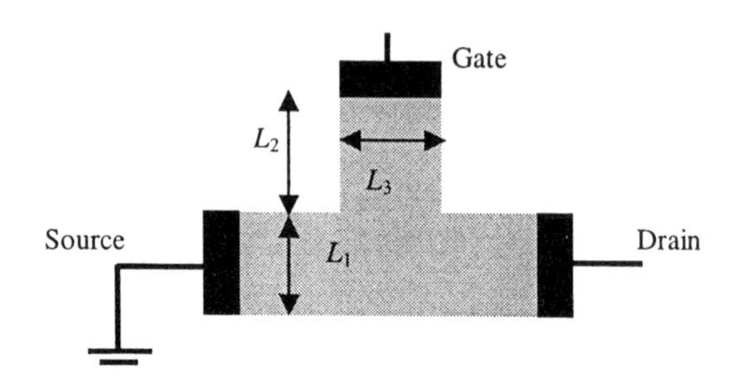

Figure 5.21 Schematic representation of the T-shaped ballistic interference transistor.

5.3 RESONANT TUNNELING DEVICES AND CIRCUITS

The resonant tunneling diode (RTD) is based on the resonant tunneling effect described in Chapter 1. The carriers in the RTD have a short transit time, of less than 0.1 ps, and, additionally, the RTD has low parasitic capacitances and large current densities, which exceed 10^5 A/cm^2. The RTD-based oscillators have produced the highest frequency ever attained with an electronic device, of 0.71 THz, and an output power of 0.3 μW [28]. The RTD is among the fastest semiconductor switching devices with a slew-rate of 300 mV/ps. Due to these properties the RTD is a crucial element in ultrafast circuits, such as binary or multiple-valued logic gates, analog-digital converters, DRAMs, and flip-flops [29].

As shown in Figure 5.22, the RTD is a vertical nanodevice, which consists, in principle, from a sequence of three nanosized undoped layers, for example, AlAs/GaAs/AlAs or AlSb/InAs/AlSb, where the first and the third layers play the role of barriers and the middle layer is the well. The $I-V$ characteristic of an RTD is given by [30]

$$I = (SemE_{\mathrm{th}} / 2\pi^2\hbar^3)\int_0^{\infty} T(E,V)\ln\left(\frac{1+\exp(E_{\mathrm{F}}-E)/E_{\mathrm{th}}}{1+\exp(E_{\mathrm{res}}-E-eV)/E_{\mathrm{th}}}\right)dE, \tag{5.69}$$

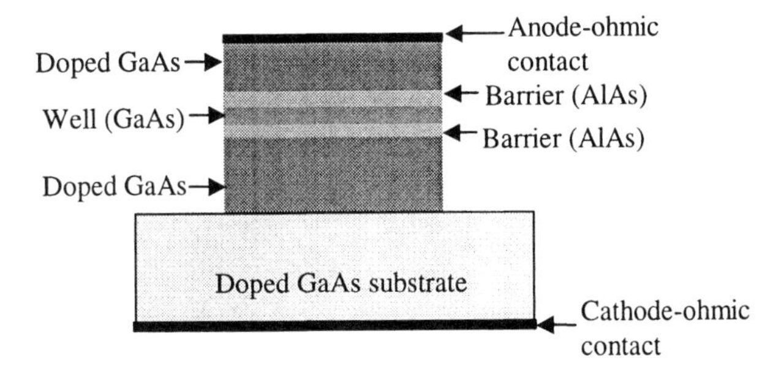

Figure 5.22 The RTD device.

where S is the diode area and E_{res} is the first resonant level in the well. For a symmetric RTD the transmission coefficient can be approximated with a Lorentzian function

$$T(E,V) = \frac{(\Gamma/2)^2}{[E-(E_{res}-eV/2)]^2+(\Gamma/2)^2}, \tag{5.70}$$

with Γ the resonance width. Formulas (5.69) and (5.70) assume that the barriers have equal widths and that half of the bias voltage drops between the emitter and the middle of the well. When Γ is small the transmission coefficient is negligible, apart from the case when the energy E is near resonance (i.e., when $E \cong E_{res} - eV/2$). In this case the integral in (5.69) can be easily computed and the final result is

$$I = (SemE_{th}\Gamma/4\pi^2\hbar^3)\ln\left(\frac{1+\exp[(E_F-E_{res}+eV/2)/E_{th}]}{1+\exp[(E_F-E_{res}-eV/2)/E_{th}]}\right) \times\{\pi/2+\tan^{-1}[2(E_{res}-eV/2)/\Gamma]\}. \tag{5.71}$$

Expression (5.71) is the analytical formula of the current in an RTD and was obtained using too many simplified assumptions (for example, the form of $T(E)$ in (5.70)). Therefore, in practice the current is expressed through an approximate formula that contains some constants, which must be determined by comparing the theoretical and experimental data. Relation (5.71) is replaced in general by

$$j_1(V) = A\ln\frac{1+\exp[(B-C+n_1V)e/E_{th}]}{1+\exp[(B-C-n_1V)e/E_{th}]}\times\{\pi/2+\tan^{-1}[2(C-n_1V)/D]\} \tag{5.72}$$

where $j_1 = I / S$ is the current density. Formula (5.72), which was derived around the resonance, is able to reproduce the peak current of the diode and a region of the negative resistance, but not the valley current. The valley current is introduced by a new current density component $j_2(V)$, given by

$$j_2(V) = H[\exp(n_2 eV / E_{\text{th}}) - 1]. \tag{5.73}$$

Thus the total $j - V$ dependence of the RTD is expressed through

$$j(V) = j_1(V) + j_2(V), \tag{5.74}$$

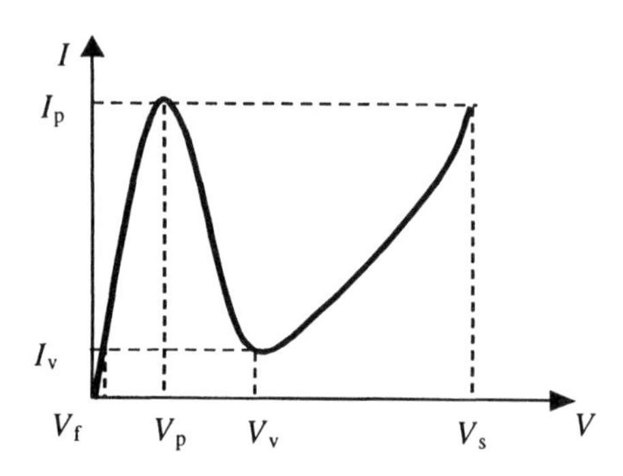

Figure 5.23 Typical *I-V* dependence of the RTD.

which includes a series of fitting parameters: n_1, n_2, *A*, *B*, *C*, *D*, *E*, *H* that are determined from experimental data. The procedure of determining the fitting parameters is described in detail in [31]. The typical $I - V$ dependence of the RTD looks like Figure 5.23. This $I - V$ dependence can be characterized as follows: when the voltage increases from zero the current increases sharply due to tunneling, up to its peak value I_p, which is attained at the voltage V_p. For larger biases the current decreases rapidly, up to its valley value I_v obtained at the voltage V_v. This valley region of the RTD is distinguished by a negative differential resistance, which demonstrates the capability of the device to oscillate. Beyond (I_v, V_v) the RTD displays the behavior of a usual diode.

Several basic RTD parameters can be determined from the $I - V$ dependence. Among them, the first and the second positive differential resistances are given by

$$R_{p1} = V_p / I_p, \tag{5.75a}$$

$$R_{p2} = (V_v - V_p)/(I_p - I_v), \tag{5.75b}$$

the negative differential resistance is defined as

$$R_n = (V_p - V_v)/(I_p - I_v), \tag{5.76}$$

the peak-to-valley voltage ratio is

$$PVVR = V_p / V_v, \tag{5.77}$$

and the peak-to-valley current ratio is given by

$$PVCR = I_p / I_v. \tag{5.78}$$

Some room-temperature values of *PVCR* for several RTD fabricated from various heterostructures are summarized in Table 5.2.

Table 5.2

Peak-to-Valley Current Ratio of Some RTD at Room Temperature

Heterostructure	*PVCR*
AlSb/InAs/AlSb	3.3
AlAs/GaAs/AlAs	1.4
AlAs/InGaAs/AlAs	12

The cutoff frequency of the RTD was calculated in [32] using an equivalent circuit consisting, as shown in Figure 5.24, from the negative conductance $G = -1/R_n < 0$ of the diode determined from the negative resistance region, and a negative inductance defined as

$$L = \tau_N / G, \tag{5.79}$$

where τ_N is the lifetime of the *N*th quasibound state in the well through which the current flows. For a symmetric RTD the first quasibound state through which the tunneling current flows resonantly has a lifetime of

$$\tau_1 \cong \hbar / \Gamma_1. \tag{5.80}$$

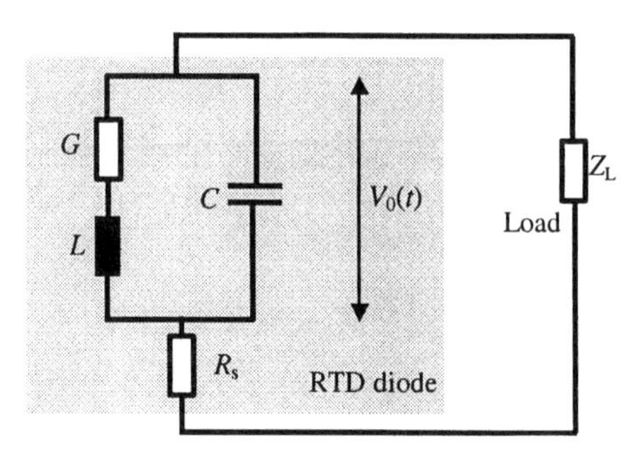

Figure 5.24 The equivalent circuit of an RTD connected to a load.

The other two elements of the equivalent circuit are the series resistance R_s and the total capacitance

$$C = C_q + C_g = -G / R_{esc} + A / [(L_w / \varepsilon_w) + (2L_b / \varepsilon_b) + (L_d / \varepsilon_d)], \quad (5.81)$$

which is composed of a quantum capacitance part C_q and a geometrical capacitance C_g [29]. In (5.81) R_{esc} is the electron escape rate from the quantum well to the collector; L_w, L_b, and L_d are the widths of the quantum well, barrier, and depletion region, respectively, with ε_w, ε_b, and ε_d the corresponding dielectric constants.

The oscillation frequency of the RTD is then

$$f_{RLC} = \frac{1}{2\pi}\left\{\left[\frac{1}{LC}\left(1 - \frac{C}{2LG^2}\right)\right]\left[1 - \left(1 - \frac{1 + 1/GR_s}{(C/2LG^2 - 1)^2}\right)^{1/2}\right]\right\}^{1/2}, \quad (5.82)$$

which becomes in the case $L \to 0$

$$f_{RC} = (2\pi C)^{-1}(-G / R_s - G^2)^{1/2}. \quad (5.83)$$

For the best RTDs made from AlSb/InAs/AlSb typical values of the equivalent circuit elements are $G = -20$ mS up to $G = -60$ mS, C = 2–3 fF, and $R_s = 1\Omega$, which generate oscillation frequencies greater than 1 THz. The oscillations are produced only if

$$G^{-1} + R_s + R_l < 0\,, \tag{5.84}$$

where R_l is the real part of the load impedance Z_l (see Figure 5.24).

Because the RTD is able to oscillate at such high frequencies, it is used for submillimeter-wave generation and detection. The switching time of these rapidly oscillating RTDs, which is about $\approx 5(C/G)$, is very short and is estimated at a few ps. The switching time is 1.7 ps, for example, in the case of the RTD implemented with an AlSb/InAs/AlSb heterostructure [33]. Such short switching times promote RTD as a strong candidate for high-speed logic gates or high-speed circuits. Two important reviews [34,35] present the state-of-the-art of RTD logic devices for more than a decade. The characteristic N-shaped $I-V$ curve leads to an obvious identification of the two stable states of the RTD located in the two positive resistance regions, one before and the other after the negative region, as shown in Figure 5.23. These two stable states are the prerequisite of logic operations and memory elements, as demonstrated by recent logic configurations that use the controlled quenching of series-connected RTDs [31]. The RTD quenching or switching is realized when the applied voltage lowers the resonant state in the RTD below the emitter conduction band minimum. In this situation, resonant tunneling is suppressed and the logic state is assigned as high-logic 1. Before quenching, the assigned logic state is low-logic 0. In Figure 5.25 these two logic states are marked on an ideal RTD $I-V$ characteristic, together with the symbol of the RTD used for logic applications.

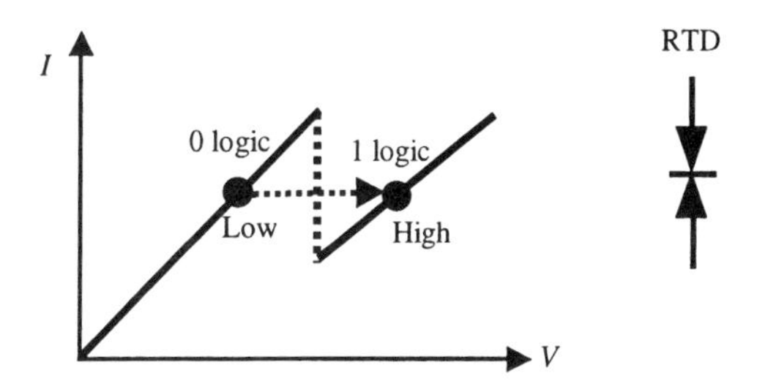

Figure 5.25 The logic states of a RTD (*After*: [31]).

If more RTDs are connected in series and if each of them has a different peak current, the RTDs are quenched one by one in a certain sequence. The first quenched RTD is that with the lowest peak current, followed by the RTD with the second lowest peak current, and so on. By changing the peak currents through external currents, the sequence of quenching is modified and thus various logic functions and, finally, programmable logic gates can be implemented.

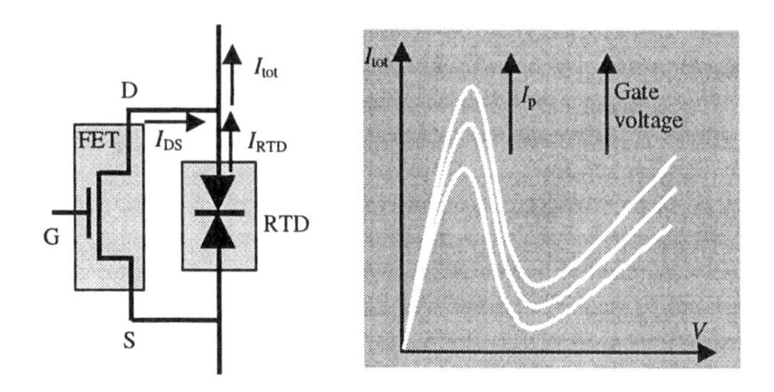

Figure 5.26 The basic cell of a programmable logic-gate-based RTD (*After*: [31]).

The logic circuit consists of a series of RTD diodes with different peak currents, each of them having an FET in parallel, which controls the RTD behavior and determines the sequence of RTD quenching. The logic circuit has a common supply voltage V, which is clocked by a signal with maximum amplitude V_{max}. The value of V_{max} is selected such that when the bias increases from 0 toward its maximum value only one RTD is quenched and thus assigned to a high-logic state. The role of the FET is to change the quenching order of the series of RTDs through the variations of the peak current of the RTD. This can be done by changing the gate voltage of the FET, since the total current passing through the RTD-FET circuit is the sum between the RTD current and the source-drain current of the FET, which depends on the gate voltage. The working principle of the logic circuit, which constitutes the basic unit of a programmable logic gate, is illustrated in Figure 5.26. Because only the FET current is changed via the gate voltage V_G, while the RTD current remains unchanged, the current control equation, which expresses the change in peak voltage of an RTD in parallel with an FET as a function of the FET gate voltage, is given by

$$I_{tot}(V_G) = I_{DS}(V_G) + I_{RTD}. \tag{5.85}$$

Both RTD and FET are integrated together in the frame of a common technology, for example, the InP technology, in which RTDs with high performances and HEMT transistors are implemented via MBE growth on an InP substrate. The programmable gate that uses a chain of RTDs is represented in Figure 5.27.

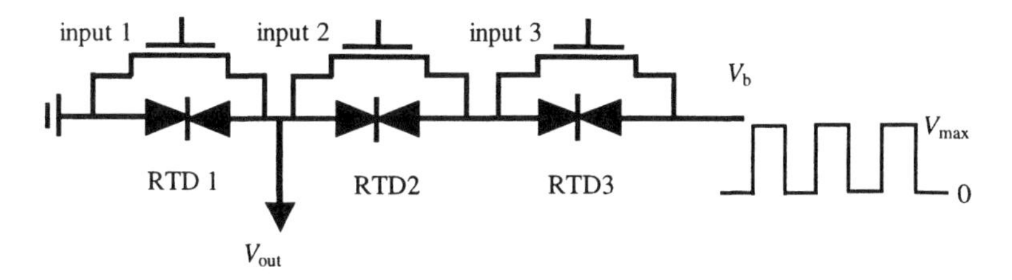

Figure 5.27 The programmable logic gate based RTD; the three RTDs have different areas and so different peak currents (*After*: [31]).

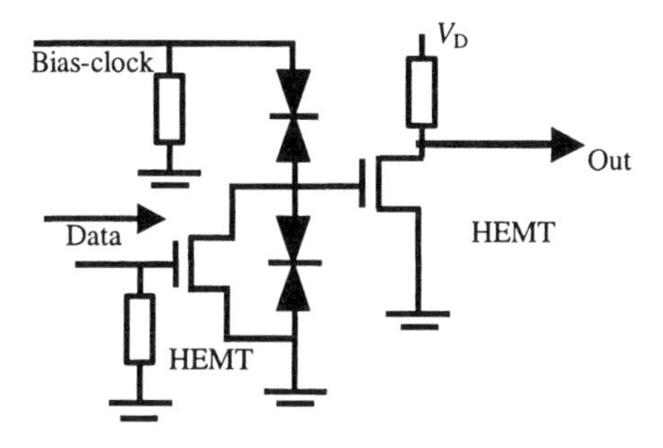

Figure 5.28 The MOBILE circuit (*After*: [36]).

Another category of logic circuits based on the same principle (i.e., on the sequential and controlled RTD quenching) is the MOBILE (monostable-bistable transition logic element) [36]. The MOBILE is composed of two RTDs with different peak values connected in series and subject to a clocked bias, as shown in Figure 5.28. When the bias voltage V_D increases, the output voltage, which is the voltage of the common node of the RTDs, changes to one of the two stable states: logic 0 or logic 1, depending on which RTD is quenched first. Although the MOBILE configuration is very simple, it has a lot of applications, including programmable logic gates, SRAM, flip-flop circuits, or adders, all showing significant potential for high-speed operations (see the references in [31]). So, the MOBILE is a highly functional logic gate able to perform impressive high-speed logical operations. The MOBILE operation at 35 Gb/s was demonstrated in [36] using the InP technology. The HEMTs had a 0.7 μm gate width and a measured cutoff frequency of 50 GHz. The RTD with 1.6 nm AlAs barriers and a 1.8 nm

InAs subwell was located at the center of an InGaAs well. Moreover, the power dissipation was very low, of about 1.4 mW, and was almost constant for a wide range of data rates in the interval 20–40 GB/s.

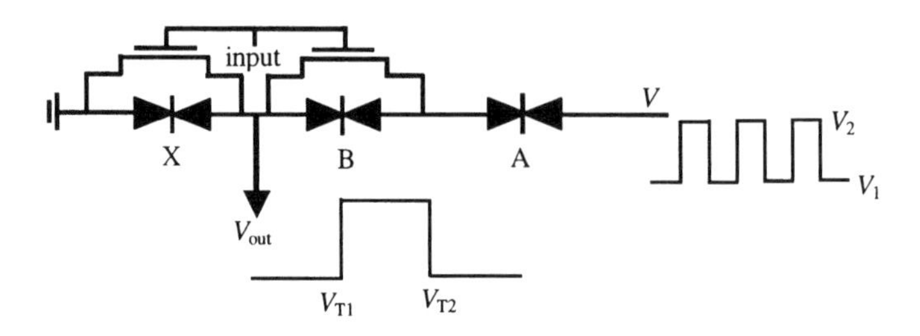

Figure 5.29 The MVL literal gate (*After*: [38]).

More sophisticated devices, including programmable AND/NOR gates and parallel adders in the threshold logic, can be created with the help of the MOBILE circuit [37]. The MOBILE configuration was also used to implement multiple-valued logic (MVL) elements, such as ternary inverters, or literal gates. The MVL literal gate [38] is displayed in Figure 5.29.

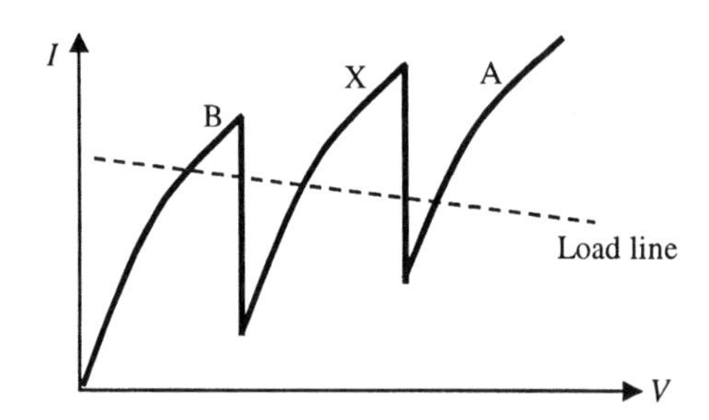

Figure 5.30 The *I-V* characteristic of the literal MOBILE (*After*: [38]).

Let us assume that the peaks of the RTDs B and X increase with different slopes when the input of the MVL literal gate in Figure 5.31 increases. Then, if the clock has the minimum value V_1 and the maximum value V_2 , for an input voltage lower than the threshold voltage V_{T_1} that corresponds to the smallest value of the peak current of B (i.e., for $V_{in} < V_{T_1}$), B is switching off when the clock attains its maximum value. Then X is in the ON state and the output is low. For an input voltage in the region $V_{T_1} < V_{in} < V_{T_2}$, where V_{T_2} is the voltage for which the peak current of X is smallest, the RTD X is switched off and the output is high. Obviously, when $V_{T_2} < V_{in}$ the RTD A is off and the output is low. The $I-V$ characteristic of the literal gate is shown in Figure 5.30.

The multipeaked current-voltage dependence in Figure 5.30, with a sawtooth form, is the key shape in MVL logic. It has three stable states, which can be used to implement multilevel memory cells, or MVL logic circuits. Initially, the MVL sawtooth was implemented with identical series diodes [34]. The MOBILE MVL uses nonidentical RTDs, with peak currents controlled by HEMTs. This configuration offer a huge flexibility to the MVL logic implementation.

The combination of FETs with RTD is useful for the implementation of binary or MVL programmable gates and also in the implementation of memories, such as SRAM [39]. By combining a low current density RTD with a heterostructure FET, both integrated on an InP substrate, it is possible to obtain a 50 nW standby power, which is three orders of magnitude lower than that of memory cells based on AIII–BV semiconductor compounds. Moreover, a tri-state memory cell with 100 nW dissipated power can be implemented by integrating vertically several RTDs. The SRAM cell contains two RTDs in latch configuration, which provides two voltage levels at a point called storage node. Each RTD is connected to the gate of a heterostructure FET, which reads and writes the bits. The SRAM cell is presented schematically in Figure 5.31.

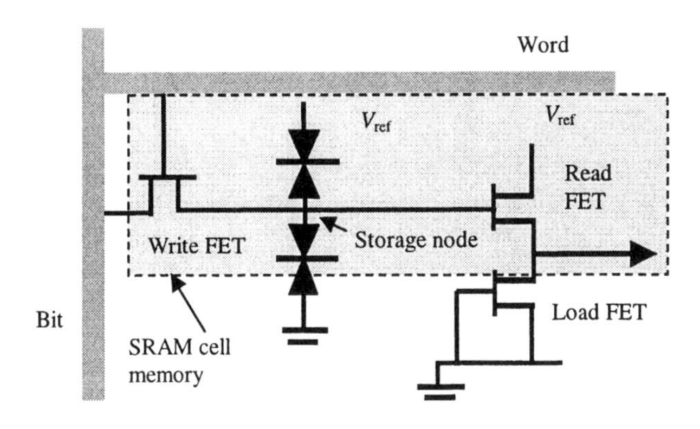

Figure 5.31 The SRAM cell memory (*After*: [39]).

5.4 SINGLE ELECTRON TRANSISTORS AND RELATED DEVICES

Single electron devices and, in particular, the single electron transistor (SET) are based on the effects produced when single electrons are injected and extracted from a nanosized quantum structure, such as a nanocluster or a quantum dot, both termed generically island. Therefore, the rudimentary structure of a single electron device is represented by a charge injector (drain), a nanosized island, and a charge collector (source); a gate voltage applied on a metallic electrode controls the number of charges in the island. The charge injector and collector are often metal tunnel junctions that consist of point contact structures. The main physical effect related to single electron transfer from injector to island is the Coulomb blockade, which consists of the creation of a gap in the energy spectrum of the island that is localized symmetrically around the Fermi energy. The Coulomb blockade effect is explained in Chapter 1. The energy gap around the Fermi level is produced by charge rearrangement inside the island, and becomes significant when the associated potential change is greater than the thermal energy E_{th}. As a result, the electron tunneling is stopped until the charging energy is compensated by an applied bias. This behavior anticipates the $I-V$ characteristic of a single electron device, which has the form of a staircase that consists of a series of plateaus. The plateaus are caused by the Coulomb blockade effect that induces abrupt jumps in current due to rapid charging effects. The behavior of the single electron device, which is a weakly coupled metallic island connected to two metallic electrodes, can be understood from the equivalent circuit depicted in Figure 5.32.

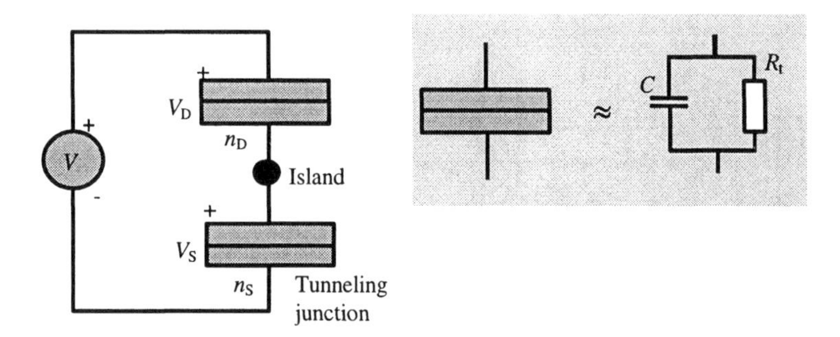

Figure 5.32 The equivalent circuit model of a metallic island weakly coupled to two metallic electrodes on which a bias is applied.

In Figure 5.32 the island is a metallic nanocluster weakly coupled (via a thin insulator) to two thin metallic electrodes. The ensemble composed of the thin insulator and the metallic electrode is a tunneling junction, which injects and extracts charges from the island. This tunneling junction can be modeled as a parallel configuration formed by a tunneling resistance R_t and a capacitance C. The voltage drop on the two tunneling junctions is denoted by V_D and V_S, and the respective capacitances of the equivalent circuits by C_D and C_S, the subscripts originating from drain and source, respectively. In the following, we assume that the thin tunneling barrier between the metal electrodes and the island is very high (i.e., the tunneling resistance is independent of the voltage applied on the junctions), and that the tunneling is sequential, the electrons arriving in the island being immediately relaxed via carrier-carrier scattering. This model of the metallic island connected to two tunneling junctions is detailed in [40]. The energy of the equivalent circuit depicted in Figure 5.32 is given by:

$$E(n_S, n_D) = (C_S C_D V^2 + Q^2)/2C_{tot} + eV(C_S n_D + C_D n_S)/C_{tot} \tag{5.86}$$

where $C_{tot} = C_S + C_D$ and $Q = Q_D - Q_S = -(n_D - n_S)e = -ne$, with n_D and n_S the number of electrons that tunnel through the drain and source junction, respectively. The change in the energy of the metallic island when the carriers are tunneling through the drain junction is

$$\Delta E_D = E(n_S, n_D) - E(n_S, n_D \pm 1) = (e/C_{tot})[-(e/2) \pm (en - VC_S)], \tag{5.87}$$

while the change in energy when the carriers tunnel through the source junction is expressed by

$$\Delta E_S = E(n_S, n_D) - E(n_S \pm 1, n_D) = (e/C_{tot})[-(e/2) \mp (en + VC_D)]. \tag{5.88}$$

For a neutrally charged metallic island, for which $n = 0$, the change in the energy of the island when one electron enters or leaves it,

$$\Delta E_{S,D} = -e^2/2C_{tot} \mp eVC_{D,S}/C_{tot} > 0, \tag{5.89}$$

can only be equal to zero when a threshold bias voltage is applied, which enables electron tunneling between source, island, and drain. If $C_S = C_D = C = C_{tot}/2$ this threshold bias voltage is given by

$$V_{th} = |V| = e/C_{tot}. \tag{5.90}$$

Below the threshold voltage value of e/C_{tot} the tunneling is suppressed and thus $I = 0$. This electron transport regime is called Coulomb blockade and is characterized by a low conductance around the origin of the $I-V$ curve. The Coulomb blockade regime of the ensemble source-island-drain is exemplified in Figure 5.33(a). When no bias is applied, or for a bias below the threshold voltage, a Coulomb energy gap e^2/C_{tot} around the Fermi level opens in the energy band diagram and suppresses tunneling between contacts. Above the threshold voltage, tunneling from source to drain through the island is allowed, and thus, the Coulomb blockade is overcome, as shown in Figure 5.33(b). If C_{tot} is large enough the Coulomb blockade effect is strongly attenuated, and ultimately vanishes, because the threshold voltage becomes very low.

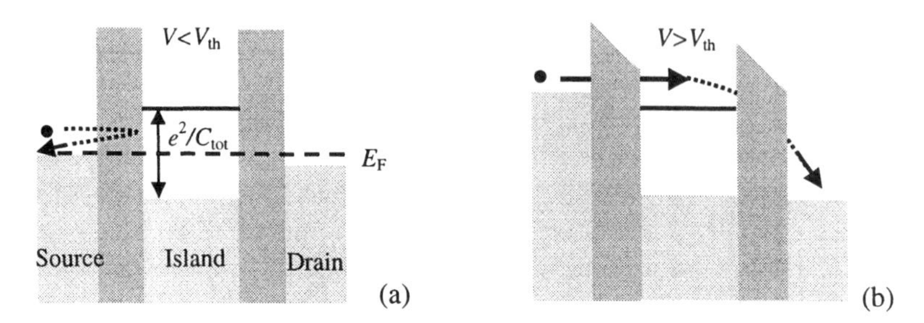

Figure 5.33 (a) The Coulomb blockade regime, and (b) overcoming the Coulomb blockade by applying a high enough bias voltage.

If $V > e/2C$ an electron has already tunneled into the island, for which $n = 1$, and the Fermi energy increases by e^2/C_{tot}, a new gap forming around the Fermi level; tunneling of an extra electron from the island to the drain is now forbidden unless the threshold voltage is increased to $V > 3e/2C$. Between these two threshold values no electron flows through the structure until the electron in the island tunnels into the drain, so that the island returns to the $n = 0$ state and the Fermi level in the island is lowered and another electron can tunnel the structure; this cycle is repeated several times.

If the tunneling resistance in the source junction is much greater than that in the drain junction (i.e., if $R_{\text{t}} = R_{\text{S,t}} >> R_{\text{D,t}}$), but the corresponding capacitances are equal, the current through the ensemble source-island-drain is controlled by the voltage $V_{\text{D}} = V/2 + ne/C_{\text{tot}}$ that drops across the drain junction. The voltage across the drain jumps with e/C_{tot} each time the threshold voltage for the drain junction is attained for increasing n values. Then, the jumps in current are given by

$$\Delta I = e/C_{\text{tot}}R_{\text{t}}, \tag{5.91}$$

and the $I-V$ characteristic of the ensemble source-island-drain takes the specific staircase form represented in Figure 5.34, which reflects the charging effects in the island. This striking $I-V$ shape, which is a macroscopic behavior of quantum phenomena occurs only when the Coulomb charging energy prevails over the thermal energy and when the fluctuations in the number of electrons in the island is small enough to allow the localization of a charge in it. This last condition is fulfilled when

$$R_t >> h/e^2 = 25.8\ \text{k}\Omega\,. \tag{5.92}$$

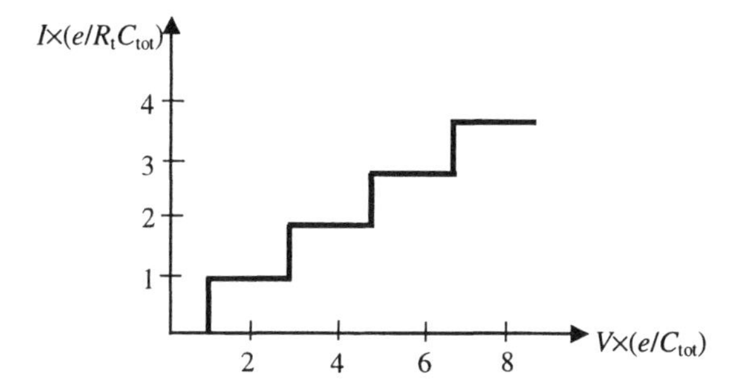

Figure 5.34 The *I-V* characteristic of the single electron device.

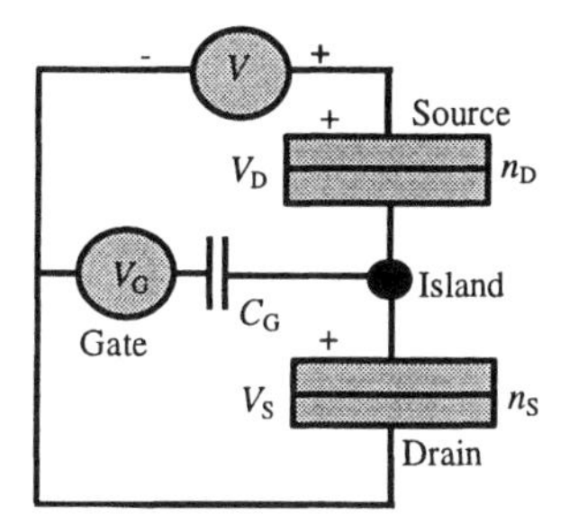

Figure 5.35 The equivalent circuit of the SET transistor.

The single electron transistor (SET) is very similar to the single electron device discussed above, except for an additional gate contact on the island; the equivalent circuit of the SET is displayed in Figure 5.35.

The total capacitance of the SET is $C_{tot} = C_S + C_D + C_G$ and its total energy is given by a similar formula to (5.86), which includes the energy of the gate capacitor. When tunneling events happen in the source junction the change in energy is now expressed as

$$\Delta E_S = (e / C_{tot}) / \{-e/2 \mp [en - Q_p + (C_G + C_D)V - C_G V_G]\}, \quad (5.93)$$

while the change in energy for tunnel events across the drain junction is given by

$$\Delta E_D = (e / C_{tot}) / \{-e/2 \pm [en - Q_p - C_S V - C_G V_G]\}, \quad (5.94)$$

where $Q = -ne + Q_p$, with Q_p a parasitic charge originating in the different work-functions and random charges trapped near the junction. The gate voltage is used to tune the charges in the island and to shift the Coulomb blockade regions for $n \neq 0$. The conditions for tunneling into and out from the island are given by

$$-e/2 \mp [en + (C_G + C_D)V - C_G V_G'] > 0, \quad (5.95a)$$

$$-e/2 \pm [en - C_S V - C_G V_G'] > 0, \quad (5.95b)$$

where $V_G' = V_G + Q_p / C_G$.

Equations (5.93–5.95) form a family of straight lines in the (V, V_G) plane, which intersect each other, as shown in Figure 5.36(a). They form the stability plot of the SET, the shaded areas indicating Coulomb blockade regions. Inside these regions, which are stable regions called Coulomb diamonds, the number of electrons is fixed. It is worth noting that the Coulomb diamond shape is dependent only on the gate and junction capacitances. On the contrary, outside Coulomb diamonds, the number of electrons is fluctuating in a certain range. Above and below Coulomb diamonds, the number of electrons varies between two consecutive integers. In these regions, the sequential tunneling through the island is allowed at a finite source-drain voltage V_{DS}. The maximum blockage region is determined by the condition that $C_G V_G'$ is an integer multiple of the charge (i.e., it is equal to *pe*, with $p \in Z$), while tunneling is allowed if this product has half integer values of electron charge. If we tune the gate voltage at a fixed V_{DS}, the drain current shows a multipeaked structure, which indicates the onset of Coulomb blockade and sequential tunneling regions, as shown in Figure 5.36(b). The current flows only when the number of electrons in the gate is a half integer value, the gate voltage that separates two consecutive peaks being $\Delta V_G = e / C_G$. This behavior is an illustration of charge quantization. In macroscopic conductors the charge quantization is not evident because the electron wavefunctions extend over large distances. On the contrary, in nanosized islands the electrons are localized and their wavefunction is confined in a very small region, a condition that favors the manifestation of charge quantization.

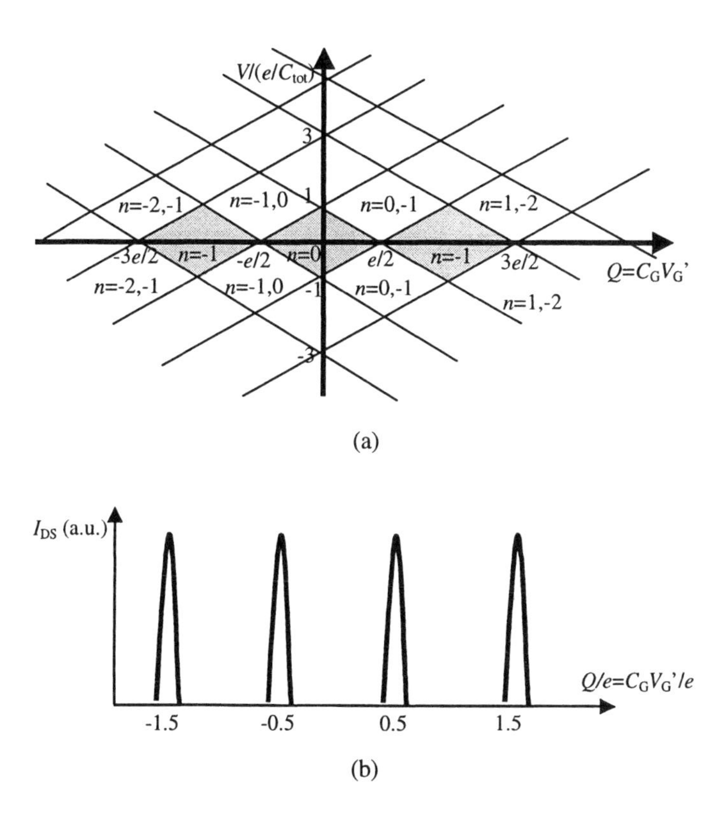

Figure 5.36 (a) The stability diagram of the SET, and (b) the dependence of the drain-source current on the gate voltage.

Single charge devices are modeled with more elaborated methods than the equivalent circuit used above, which include the orthodox theory of SET and Hamiltonian theories. A review of these methods can be found in [40] and in the seminal review paper [41]. However, the equivalent circuit model used above originates from the orthodox SET theory and preserves the main physical phenomena and the behavior of SET and is therefore extensively used due to its simplicity.

As we have pointed out above, the occurrence of the Coulomb blockade is determined by the condition that the charging energy $E_{ch} = e^2/(2C_{tot})$ is greater than the thermal energy. If the thermal energy prevails the electrons fluctuate in the island, including in the Coulomb blockade regions, and the SET behavior is destroyed. If we consider the island as a sphere with a diameter d embedded in a dielectric host with permittivity ε, then $C_{tot} = 2\pi\varepsilon_0\varepsilon d$ and we have room-temperature SET effect (for $E_{th} = k_B T = 26$ meV and $\varepsilon = 4$) only when $d < 12$ nm. More precisely, for $E_{ch} = 260$ meV, which is 10 times larger than the thermal energy at room temperature, we need a Si island with 1.3 nm in diameter and with $C_{tot} = 0.3$ aF. This very low capacitance value is very difficult to obtain technologically. Thus, due to the many technological impediments, much larger islands were used over some decades and the first SETs worked only at a few K. Nowadays, however, some SETs work at room temperature due to advanced nanotechnology processes [42].

The island can be made from any conducting material, and therefore the first SETs that worked around 1 aK were implemented using large island from metal or superconductor materials. However, the main target is to implement SETs using semiconductors, such as Si and GaAs. The silicon SET is the subject of intensive researches with the main aim of room-temperature functionality of the device and the circuits based on it. This aim cannot be easily achieved since an island with a size of 10 nm is no larger than a few tens of times the Si lattice constant [42] and, moreover, a 3D confinement must be achieved.

The straightforward way to form Si nanosized islands is to use nanolithographical methods. In principle, a quantum wire is first formed and is then further confined to form the island. One method to decrease the island size is the SIMOX. In this method, a 2D Si-on-insulator (SOI) layer realized via implanted oxygen (SIMOX) plays the role of the quantum wire. Further spatial confinement is achieved by electron beam lithography dry etching, as shown in Figure 5.37 [43]. The trench height is about 30 nm, allowing an E_{ch} of about 2 meV, which implies that the Si SET is able to work at a temperature which does not exceed a few K.

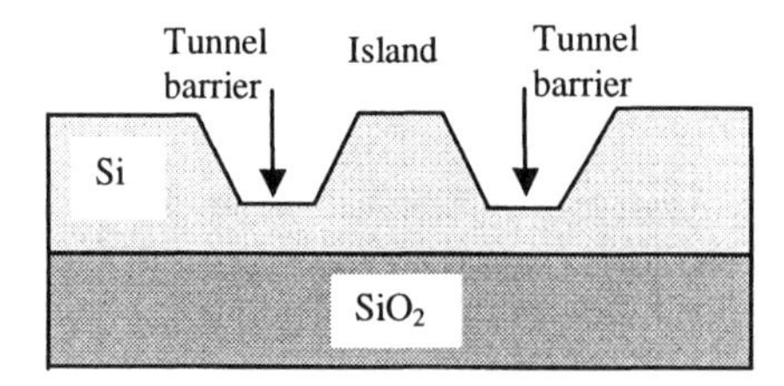

Figure 5.37 Schematic representation of the SIMOX SET (*After*: [42]).

A room-temperature SET was made using a fabrication technique named pattern-dependent oxidation (PADOX) [42]. PADOX consists of the thermal oxidation of a nanosized Si wire connected to two wide Si layers. The thermal oxidation stress is huge in the nanosized wire and induces an additional confinement necessary for the creation of the island. The PADOX SET is a layered structure made from a Si substrate that plays the role of a back gate, a buried SiO_2 layer, and a thin Si layer. The 20–30 nm wide and 100 nm long Si wire is formed in this SOI layer via electron-beam lithography and dry etching. In a subsequent fabrication step the Si wire is thermally oxidized in a dry oxygen medium, as shown in Figure 5.38. The measurement reveals the formation of an island with a diameter of 7 nm, $C_{tot} = 1.5\,\text{aF}$, and a confinement energy of $E_{ch} = 50\,\text{meV} \cong 2E_{th}$. The dimension of the island is below the lithographical limit and is realized via the giant compressive stress, of up to 20 000 atm, which occurs in the quantum wire due to the thermal oxidation process.

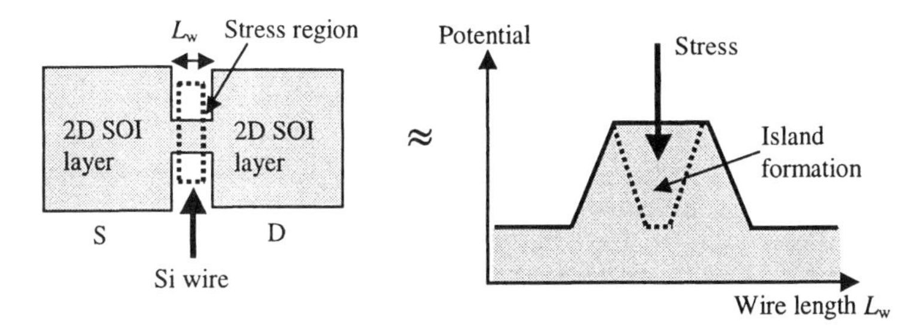

Figure 5.38 The PADOX SET and its energy band diagram (*After*: [42]).

An improved version of PADOX, the vertical PADOX (V-PADOX), is based on the thermal oxidation of a thin Si wire with a nonuniform thickness, for example, with a trench in the middle of the wire. In this way, two twin Si wires are formed. Their edges are unaffected by the oxidation due to the build-up stress, while the central part of the wire around the trench (where the two twin wires are formed) is fully oxidized. In consequence, twin islands form after oxidation, a configuration that is very useful for logical gates. Multiple islands can be created in the same way, allowing the implementation of various logical operations based on SETs. Many other methods of fabrication of room-temperature SETs are based on variants of the PADOX method, but all of them have as a common point the island formation via thermal oxidation.

Another way to fabricate room-temperature SETs is to employ the STM as a nanofabrication tool. For example, nanooxidation of Ti is used to create such a

SET [44]. The fabrication procedure consists in the growth of a thin Ti layer, with a thickness of 3 nm, over a SiO_2/n-Si substrate, followed by the oxidation of the Ti surface by anodization using an STM tip as a cathode. In this way, a layer of TiO_x is formed, the island being produced by surrounding a 30 nm × 30 nm Ti area with TiO_x regions that act as barriers.

Other materials besides Si can be used for SETs. We have shown in Chapter 3 that carbon nanotube-based room-temperature SETs have $E_{ch} = 40\ \text{meV}$. Also, SETs based on AIII–BV semiconductors have been implemented, but they work at temperatures lower than the room temperature.

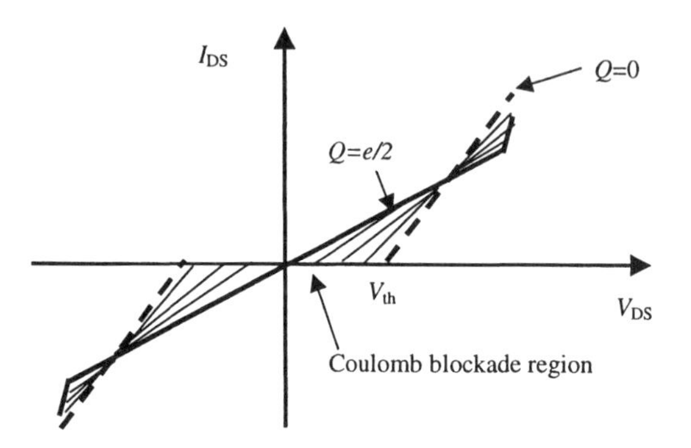

Figure 5.39 The I_{DS}-V_{DS} characteristic of the SET.

The applications of SETs are numerous and are focused on logical circuits and memories. However, a straightforward application of the SET is as a very sensitive electrometer. The working principle of this electrometer is based on the extreme sensitivity of the drain current to changes in the gate voltage. Sub-single electron charge variations lead to measurable changes in the drain current, as can be seen from the output $I_{DS} - V_{DS}$ characteristic of the SET displayed in Figure 5.39 at various values of the $Q = C_G V_G$' parameters (at various gate voltages).

The $I_{DS} - V_{DS}$ characteristic beyond the threshold blockade voltage V_{th} is linear and is given by

$$I_{DS} \cong [V_{DS} - \text{sign}\,(V_{DS}) \times (e / 2C_{tot})] / (R_S + R_D). \tag{5.96}$$

The sensitivity of the electrometer is very high at high operating frequencies where the $1/f$ noise is suppressed. For example, the charge sensitivity at 1 MHz attains $1.2 \times 10^{-5}\, e / \text{Hz}^{1/2}$, which corresponds to an energy sensitivity of $41\hbar$ [45].

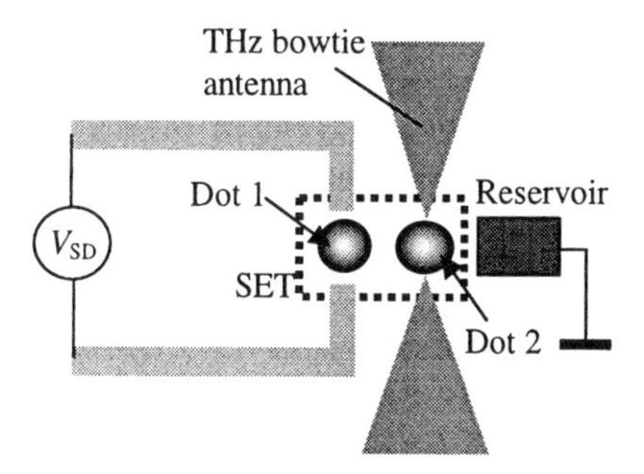

Figure 5.40 Schematic diagram of the SET THz detector (*After*: [46]).

Other analog applications of the SET include dc current standards, standards of temperature and resistance, and detection of infrared radiation. In this respect, a SET THz detector was recently demonstrated based on single-photon detection. The SET works at 50 mK and consists of two parallel GaAs/AlGaAs quantum dots, as shown in the schematic representation in Figure 5.40 [46]. One of the dots is coupled to a dipole antenna illuminated by the THz radiation, which generates an electron-hole plasma. As a result of the photomultiplication effect of 10^8–10^{12} electrons/photons the conductance of the other dot is shifted via electron tunneling. This SET THz detector has a sensitivity of 0.1 photons/0.1 mm^2. The noise equivalent power (NEP) is 10^{-17} W(Hz)$^{-1/2}$, three orders of magnitude superior to the NEP of any THz bolometer, which is in addition much bulkier.

The SET is also a very sensitive sensor of displacement [47,48], especially when it is coupled with a NEMS (nanoelectromechanical system) resonator, which (see Chapter 2) plays the role of a harmonic oscillator. The position of a resonator with a resonant frequency $f_0 = \omega_0 / 2\pi$, at thermal equilibrium with the ambient at a temperature T_a, fluctuates with the rms amplitude $\delta x_{rms} = (k_B T_a / m\omega_0^2)^{1/2}$. According to this formula, the fluctuations decrease by simply reducing the ambient temperature. On the other hand, if the oscillator is treated as a quantum mechanical system it has an intrinsic fluctuation $\delta x_q = (\hbar / m\omega_0^2)^{1/2}$ at the temperature $T_a << T_q = \hbar\omega_0 / k_B$. In a single crystal GaAs nanomechanical oscillator, which is 3 μm long, 250 nm wide and 200 nm thick, the resonant frequency is about 1 GHz [47]. The inequality $T_a << T_q$ is thus fulfilled in this resonator when it operates at an ambient temperature of 10 mK, the temperature at which the SET-based displacement sensor attains the quantum mechanical limit, with important applications in the area of weak forces detection. The corresponding displacement, of 10^{-14} m, is approximately 1,000 times the diameter of the hydrogen atom. The displacement technique based on SET is founded on the use of the SET as an ultrasensitive electrometer. By placing a metallic electrode on the nanomechanical resonator, it becomes capacitively coupled to the gate of the SET, as shown in Figure 5.41. The resonator is biased at the voltage V. The charge between the gate and the resonator is then $Q = VC$, where C is the capacitance between the gate and the nanoresonator. The vibration of the beam produces a change in the charge $\Delta Q = V\Delta C$, which modifies the I_{DS} current. The SET is working in a heterodyne configuration, where it acts as a mixer, because in such a configuration the drain current noise is minimized and the sensitivity of the entire measurement system is increased.

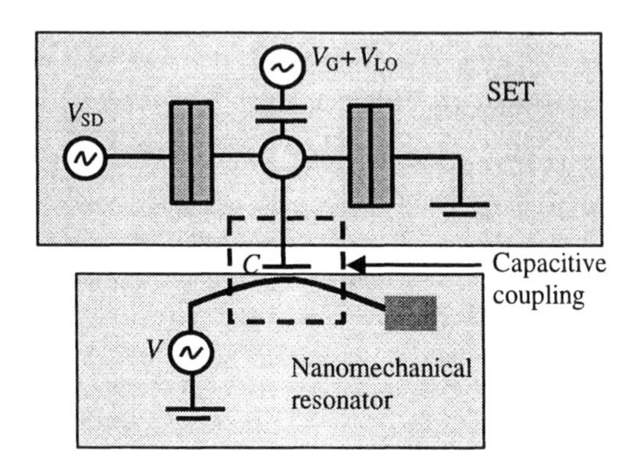

Figure 5.41 Schematic diagram of the displacement sensor based on SET and an NEMS device (*After*: [47]).

The digital applications of SETs, such as memories and logical devices, are very appealing because SETs are very small and consume relatively little power. In most cases, the SET memory is a floating-node-type memory [42], which is composed from two distinct parts: the electron storage part and the charge sensing part. The electron storage task can be performed in various ways, but the most encountered methods are: 1) the electron trap, which consists of two or more tunnel junctions and a capacitor, and 2) storage capacitors serially connected to a transistor (MOSFET or SET).

In an electron trap, as that displayed in Figure 5.44, the potential V of the node displays a hysteretic $V - V_G$ characteristic, which is a typical behavior for any memory. The tunneling of electrons, which are coming in or are going out of the memory node, is controlled by the Coulomb blockade regime, being forbidden in the voltage interval $-Q_{cr} / C < V < +Q_{cr} / C$, where

$$Q_{cr} = [eC/(C + C_G)]\{[1 + (1 - 1/N)C_G / C]/2\} \tag{5.97}$$

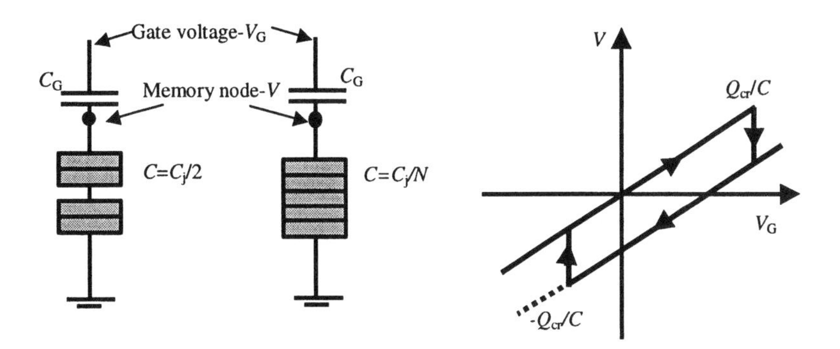

Figure 5.42 SET floating-node memory having as storage element an electron trap with two or N junctions (*After*: [42]).

is the critical charge, N is the number of tunneling junctions, and C is a capacitance that depends on N and the capacitance of a single junction C_j. Thus, the gate voltage V_G controls the state of the memory node V, and so controls its bistable states.

The charge-sensing device connected to the memory node can be a MOSFET, a SET, or a multiple-junction SET. The drain current senses the charges located in the memory node, and the content of the memory is read as a current. Giga-totera-bit SET storage is reviewed in [49].

The SET logic schemes and solutions are well developed and studied for the advantages offered by SET in the implementation of various logical functions, either single-valued or multiple-valued. The logic devices based on SET can be implemented in two ways. A first category is the family of logical devices that includes inverters, NAND, and XOR gates, where the traditional CMOS transistors are replaced by SETs. The second category of logical circuits, which represents advanced logical devices that have no CMOS counterpart, uses elementary charges as bits.

By defining the noninverting SET and inverting SET voltage gains as the rising and falling slopes of the $V_{DS} - V_G$ characteristics as

$$g_{ni} = C_G /(C_G + C_j)\,, \tag{5.98a}$$

$$g_i = C_G / C_j, \tag{5.98b}$$

we see that the inverting gain is greater than 1 if the gate capacitance is larger than the tunnel junction capacitance. So, the SET can amplify the voltage and thus an entire family of logic circuits based on inverters can be built, similar to CMOS logic [50].

An example of a SET inverter is represented in Figure 5.43. The logic NAND and NOR gates can be implemented by slightly changing the inverter circuit. Note that two inverters form the memory cell of a static RAM and three inverters form a ring oscillator. The output voltage of the inverter is high when the input is low and it is low when the input is high. The inverter in Figure 5.46 functions by applying a bias voltage and by tuning the two gate voltages, such that when the input is low SET1 is conducting and SET2 is in a Coulomb blockade state. While the inverter in Figure 5.43 works at 100 mK it is possible to implement SET logic at much higher temperatures (30–77 K) using PADOX [51]. In this case, two islands are integrated in an area of 0.02 μm^2 together with gate electrodes able to control separately their potential. By using PADOX each island is embedded in one branch of a T-shaped silicon wire grown on an SOI substrate. The double-island single-electron device is modeled as two SETs connected in parallel and each transistor can be switched on and off depending on the gate voltage, the entire device acting like an inverter.

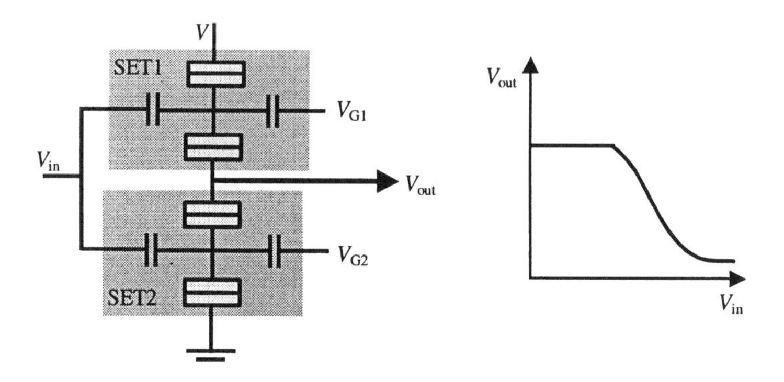

Figure 5.43 The SET inverter and its voltage characteristic (*After*: [50]).

SET implementation of multiple-valued logic is very promising since the $I_{DS} - V_G$ characteristic of the SET, represented in Figure 5.36(b), shows a multitude of peaks that occur periodically as a function of the gate voltage. Based on this property, a universal literal gate and a quantizer were implemented and tested at 27 K [52].

The logic devices that go beyond the analogies with CMOS binary or MVL logic are quantum cellular automata formed from regular arrays of quantum dots linked by tunneling junctions, where an external electron injected in the array induces a change in the charge pattern distribution, and the final pattern corresponds to the computation result performed by the charges themselves.

REFERENCES

[1] Van der Spiegel, J., "Advances in microelectronics – From microscale to nanoscale devices," in *Introduction to Nanoscale Science and Technology*, pp.217-259, M. Di Ventra, S. Evoy, J.R. Heflin, Jr. (eds.), Dordrecht: Kluwer Academic Publishers, 2004.

[2] Moore, G.E., "Lithography and the future of the Moore Law," *Proc. SPIE*, Vol. 2437, 1995, pp. 2-17.

[3] Wong, H.-S.P., et al., "Nanoscale CMOS," *Proc. IEEE*, Vol. 87, No. 4, 1999, pp. 537-570.

[4] Khanna, V.K., "Physics of carrier-transport mechanisms and ultra-small scale phenomena for theoretical modeling of nanometer MOS transistors from diffusive to ballistic regimes of operation," *Phys. Rep.*, Vol. 398, No. 2, 2004, pp. 67-131.

[5] Datta, S., F. Assad, and M.S. Lundstrom, "The silicon MOSFET from a transmission viewpoint," *Superlattices and Microstructures*, Vol. 23, No. 3/4, 1998, pp. 771-780.

[6] Natori, K., "Ballistic metal-oxide-semiconductor field effect transistor," *J. Appl. Phys.*, Vol. 76, No. 8, 1994, pp. 4879-4890.

[7] Natori, K., "Ballistic MOSFET reproduces current-voltage characteristics of an experimental device," *IEEE Electron Device Letters*, Vol. 23, No. 11, 2002, pp. 655-657.

[8] Wang, J. and M. Lundstrom, "Ballistic transport in high electron mobility transistors," *IEEE Trans. Electron Devices*, Vol. 50, No. 7, 2003, pp. 1604-1609.

[9] Rahman, A., et al., "Theory of ballistic nanotransistors," *IEEE Trans. Electron Devices*, Vol. 50, No. 9, 2003, pp. 1853-1864.

[10] Shur, M.S., "Low ballistic mobility in submicron HEMTs," *IEEE Electron Device Letters*, Vol. 23, No. 9, 2002, pp. 511-513.

[11] Kang, S., et al., "Ballistic transport at GHz frequencies in ungated HEMT structures," *Solid-State Electronics*, Vol. 48, No. 10-11, 2004, pp. 2013-2017.

[12] Dyakonov, M.I. and M.S. Shur, "Shallow water analogy for ballistic field effect transistor," *Phys. Rev. Lett.*, Vol. 71, No. 15, 1993, pp. 2465-2468.

[13] Dyakonov, M.I. and M.S. Shur, "Plasma wave electronics: novel terahertz devices using two dimensional electron fluid," *IEEE Trans. Electron Devices*, Vol. 43, No. 10, 1996, pp. 1640-1645.

[14] Knapp, W., et al., "Terahertz emission by plasma waves in 60 nm gate high electron mobility transistors," *Appl. Phys. Lett.*, Vol. 84, No. 13, 2004, pp. 2331-2333.

[15] Nishizawa, J.-I., P. Płotka and T. Kurabayashi, "Ballistic and tunneling GaAs static induction transistors: nano-devices for THz electronics," *IEEE Trans. Electron Devices*, Vol. 49, No. 7, 2002, pp. 1102-1111.

[16] Song, A.M., "Room-temperature ballistic nanodevices," in *Encyclopedia of Nanoscience and Nanotechnology*, pp. 371-389, H. S. Nalwa (Ed.), Stevenson Ranch, CA: American Scientific Publishers, 2004.

[17] Dragoman, D. and M. Dragoman, *Quantum-Classical Analogies*, Berlin: Springer, 2004, pp. 9-62.

[18] Hirayama, Y. and S. Tarucha, "High temperature ballistic transport observed in AlGaAs/InGaAs/GaAs small four-terminal structures," *Appl. Phys. Lett.*, Vol. 63, No. 17, 1993, pp. 2366-2368.

[19] Song, A.M., "Electron ratchet effect in semiconductor devices and artificial materials with broken centrosymmetry," *Appl. Phys. A*, Vol. 75, No. 4, 2002, pp. 229-235.

[20] Song, A.M., et al., "Room-temperature and 50 GHz operation of functional nanomaterial," *Appl. Phys. Lett.*, Vol. 79, No. 9, 2001, pp. 1357-1359.

[21] Goodnick, S.M. and J. Bird, "Quantum-effect and single-electron devices," *IEEE Trans. Nanotechnology*, Vol. 2, No. 4, 2003, pp. 368-385.

[22] Xu, H.Q., "Diode and transistor behaviors of three-terminal ballistic junctions," *Appl. Phys. Lett.*, Vol. 80, No. 5, 2002, pp. 853-855.

[23] Shorubalko, I., et al., "Nonlinear operation of GaInAs/InP-based three-terminal ballistic junctions," *Appl. Phys. Lett.*, Vol. 79, No.9, 2001, pp. 1384-1386.

[24] Xu, H.Q., et al., "Novel nanoelectronic triodes and logic devices with TBJs," *IEEE Electron Device Letters*, Vol. 25, No. 4, 2004, pp. 164-166.

[25] Reitzenstein, S., L. Worschech, and A. Forchel, "Room temperature operation of an in-plane half-adder based on ballistic Y-junction," *IEEE Electron Device Letters*, Vol. 25, No. 7, 2004, pp. 462-464.

[26] Mateos, J., et al., "Ballistic nanodevices for terahertz data processing: Monte Carlo simulations," *Nanotechnology*, Vol. 14, No. 2, 2003, pp. 117-122.

[27] Mitin, V.V., V.A. Kochelap, and M.A. Strocio, *Quantum Heterostructures*, Cambridge: Cambridge University Press, 1999, pp. 389-393.

[28] Brown, E.R., et al., "Oscillations up to 712 GHz in InAs/AlSb resonant-tunneling diodes," *Appl. Phys. Lett.*, Vol. 58, No. 20, 1991, pp. 2291-2293.

[29] Liu, Q., et al., "Unified AC model for the resonant tunneling diode," *IEEE Trans. Electron Devices*, Vol. 51, No. 5, 2004, pp. 653-657.

[30] Schulman, J.N., H.J. De Los Santos, and D.H. Chow, "Physics-based RTD current-voltage equation," *IEEE Electron Device Letters*, Vol. 17, No. 5, 1996, pp. 220-222.

[31] Chen, K.J. and G. Niu, "Logic synthesis and circuit modeling of a programmable logic gate based on controlled quenching of series-connected negative differential resistance devices," *IEEE J. Solid-State Circuits*, Vol. 38, No. 2, 2003, pp. 312-318.

[32] Brown, E.R., C.D. Parker, and T.C.L.G. Sollner, "Effect of quasibound-state lifetime on the oscillation power of resonant tunneling diodes," *Appl. Phys. Lett.*, Vol. 54, No. 10, 1989, pp. 934-936.

[33] Özbay, E., et al., "1.7-ps, microwave, integrated-circuit-compatible InAs/AlSb resonant tunneling diodes," *IEEE Electron Device Letters*, Vol. 14, No. 8, 1993, pp. 400-402.

[34] Capasso, F., et al., "Quantum functional devices: resonant-tunneling transistors, circuits with reduced complexity, and multiple-valued logic," *IEEE Trans. Electron Devices*, Vol. 36, No. 10, 1989, pp. 2065-2082.

[35] Mazumder, P., et al., "Digital circuit applications of resonant tunneling devices," *Proc. IEEE*, Vol. 86, No. 4, 1998, pp. 664-686.

[36] Maezawa, K., et al., "High-speed and low-power operation of a resonant tunneling logic gate MOBILE," *IEEE Trans. Electron Device Letters*, Vol. 19, No. 3, 1998, pp. 80-82.

[37] Pacha, C., et al., "Threshold logic circuit design of the parallel adders using resonant tunneling devices," *IEEE Trans on Very Large Scale Integration (VLSI) Systems*, Vol. 8, No. 5, 2000, pp. 558-572.

[38] Waho, T., K.J. Chen, and M. Yamamoto, "Resonant-tunneling diode and HEMT logic circuits with multiple thresholds and multilevel output," *IEEE J. Solid-State Circuits*, Vol. 33, No. 2, 1998, pp. 268-274.

[39] van der Wagt, J.P.A., A.C. Seabaugh and E.A. Beam III, "RTD/HFET low standby power SRAM gain cell," *IEEE Electron Device Letters*, Vol. 19, No. 1, 1998, pp. 7-9.

[40] Ferry, D.F. and M. Goodnick, *Transport in Nanostructures*, Cambridge: Cambridge University Press, 1997, pp. 230-264.

[41] Likharev, K.K., "Single-electron devices and their applications," *Proc. IEEE*, Vol. 87, No. 4, 1999, pp. 606-632.

[42] Takahashi, Y., et al., "Silicon single-electron devices," *J. Phys.: Condens. Matter*, Vol. 14, No. 39, 2002, pp. R995-R1033.

[43] Ali, D. and H. Ahmet, "Coulomb blockade in silicon tunnel junction device," *Appl. Phys. Lett.*, Vol. 64, No. 16, 1994, pp. 2119-2121.

[44] Matsumoto, K., et al., "Room temperature operation of a single electron transistor made by the scanning tunneling microscope nanooxidation process for the TiO_x/TiO system," *Appl. Phys. Lett.*, Vol. 68, No. 1, 1996, pp. 34-36.

[45] Schoelkopf, R.J., et al., "The radio-frequency single-electron transistor (RF-SET): a fast and ultrasensitive electrometer," *Science*, Vol. 280, No. 5367, 1998, pp. 1238-1242.

[46] Astafiev, O., S. Komiyama, and T. Kutsuwa, "Double quantum dots as a high sensitive submillimeter-wave detector," *Appl. Phys. Lett.*, Vol. 79, No. 8, 2001, pp. 1199-1201.

[47] Knobel, R.G. and A.N. Cleland, "Nanometer-scale displacement sensing using single electron transistor," *Nature*, Vol. 424, No. 6946, 2003, pp. 291-293.

[48] LaHaye, M.D., et al., "Approaching the quantum limit of a nanomechanical resonator," *Science*, Vol. 304, No. 5667, 2004, pp. 74-77.

[49] Yano, K., et al., "Single-electron memory for giga-to-terabit storage," *Proc. IEEE*, Vol. 87, No. 4, 1999, pp. 633-651.

[50] Heij, C.P., P. Hadley, and J.E. Mooij, "Single-electron inverter," *Appl. Phys. Lett.*, Vol. 78, No. 8, 2001, pp. 1140-1142.

[51] Fujiwara, A., et al., "Double-island single-electron devices – a useful unit device for single-electron logic LSI's," *IEEE Trans. Electron Devices*, Vol. 46, No. 5, 1999, pp. 954-959.

[52] Inokawa, H., A. Fujiwara, and Y. Takahashi, "A multiple-valued logic and memory with combined single-electron and metal-oxide-semiconductor transistors," *IEEE Trans. Electron Devices*, Vol. 50, No. 2, 2003, pp. 462-470.

Chapter 6

Optoelectronic Devices Based on Semiconductor Nanostructures

This chapter focuses on the main optoelectronic devices based on nanostructures. The chapter starts with optoelectronic devices that exploit the linear and nonlinear optical properties of nanowires, nanofilms, and nanoparticles, and presents afterwards optoelectronic devices realized via various nanotechnology methods, such as optical antennas and optoelectronic devices based on metamaterials. The chapter ends with optical cascade devices, which probably constitute the most complex devices implemented via nanostructures.

We do not treat here the emerging area of optoelectronic devices based on semiconductor quantum wires or quantum dots, since this issue is extensively treated in a series of monographs [1,2]. The reader who is not familiar with nanoscale optoelectronic devices realized with AIII–BV semiconductors should peruse these monographs because the experimental realization of optoelectronic devices, and especially the fabrication of lasers based on semiconductor quantum wires and quantum dots, constitutes one of the first proofs of the amazing properties of nanostructures and their immediate application in computers and advanced communication systems. The self-assembly techniques described in Chapter 1 were recently employed to extend the types of quantum dots used in optoelectronic devices. In this respect, InGaN quantum dots grown using self-assembly techniques were used as UV lasers [3], and self-assembly of InAs and Si/Ge quantum dots on structured surfaces such as GaAs (001), Si (001), or Si (111) were used for advanced optoelectronic devices [4]. The self-assembled quantum dots are compatible with AIII–BV technologies and show similar optical properties as standard quantum dots made from AIII–BV semiconductors. The self-assembled quantum dots have very large confinement energies, which are higher than the thermal energy at room temperature, and show a broadband gain and large tunability spectra, as well as reduced temperature sensitivity and narrow linewidth. These properties promote them as strong competitors for the existing devices dedicated to optical communications in the 1.3–1.5 μm band. A comprehensive review about self-assembled quantum dot lasers for 1.5 μm applications, widely tunable quantum dot lasers, quantum dot infrared photodetectors, and quantum dot memories can be found in [5].

6.1 OPTOELECTRONIC DEVICES BASED ON NANOWIRES

6.1.1 Optoelectronic Devices Based on Carbon Nanotubes

In Chapter 3 we presented the electrical properties of carbon nanotubes (CNTs) and, based on them, we have investigated a wealth of CNT electronic devices. Here we examine the optical properties of CNTs and the optoelectronic devices based on them.

The linear light absorption and emission in semiconducting SWCNTs (single-walled CNTs) is determined by dipole selection rules and depolarization effects [6]. The depolarization effect indicates that the emitted (or absorbed) light is mainly polarized along the tube axis, the light polarization perpendicular to the semiconducting CNT axis being negligible. The dipole selection rules allow only optical transitions between the valence and conduction bands of semiconducting CNTs within the same subband when the excitation light is polarized along the nanotube axis. The different absorption bands in the SWCNT occur due to the transitions between pairs of van Hove singularities, which appear as spikes in the density of states (see Chapter 3). The van Hove singularities are proportional to $(E^2 - E_0^2)^{-1/2}$ at the energy minima and maxima ($\pm E_0$) of the SWCNT dispersion relation. The energy separation between the highest valence band singularity and the lowest conduction band singularity is given by [7]:

$$E_{11}^{\mathrm{s}}(d) = 2a_{\mathrm{C-C}}\gamma_0 / d \tag{6.1a}$$

$$E_{11}^{\mathrm{m}}(d) = 6a_{\mathrm{C-C}}\gamma_0 / d \, , \tag{6.1b}$$

where the superscript indicates the metallic or semiconducting nanotube character, d is the tube diameter, the other constants having the same notation and signification as in Chapter 3. Defining by $p = 1,2,3,...$ the order of the π valence band and by $p' = 1,2,3,...$ the order of the π^* conduction band, both located symmetrically around the Fermi level, the optical transitions occur according to the

$$\delta p = p - p' = 0 \tag{6.2a}$$

selection rule in the case of parallel light polarization and at

$$\delta p = p - p' = \pm 1 \tag{6.2b}$$

for perpendicular light polarization. The last selection rule is, however, suppressed due to depolarization effects. When we deal with an SWCNT film made from metallic and semiconducting nanotubes the optical transitions occur at the following energies:

$$E_{11}^{s}(d), 2E_{11}^{s}(d), E_{11}^{m}(d), 4E_{11}^{s}(d)...E_{22}^{s}... \tag{6.3}$$

The first and the second optical van Howe transitions are indicated in Figure 6.1.

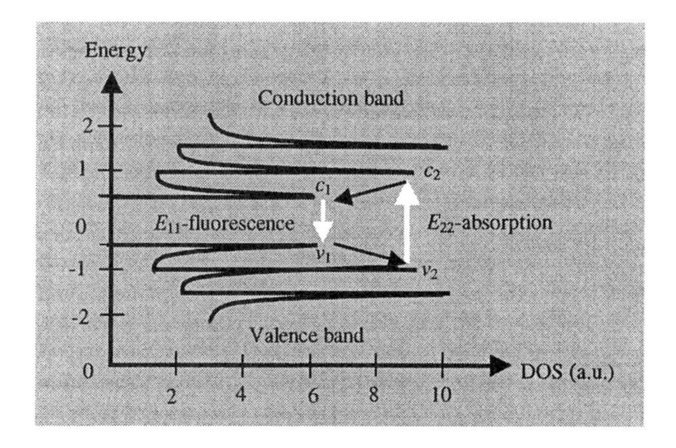

Figure 6.1 Optical transitions of a SWCNT where the solid lines are assigned for the radiative transitions while the dashed lines are assigned for nonradiative relaxation (*After*: [8]).

As shown in Figure 6.1, the energies of the second van Hove peaks establish the value of the optical excitation energy $v_2 \rightarrow c_2$ as $E_{22} = hc/\lambda_{22} = hc\tilde{v}_{22}$, while the location of the first van Hove peaks determine, the emission energy $c_1 \rightarrow v_1$ as $E_{11} = hc/\lambda_{11} = hc\tilde{v}_{11}$. The values E_{11} and E_{22} are determined experimentally from fluorescence and absorption spectra, respectively. On the other hand, the optical transitions $\tilde{v}_{11}$ and $\tilde{v}_{22}$ can be computed using relations similar to (6.1); both depend on the tube diameter, which is related to the tube chirality parameters *n* and *m* (see Chapter 3). The assignment of (*n*, *m*) parameters to interband transitions allows the geometrical characterization of the SWCNTs' optical absorption spectra [8,9].

In general, there is no simple way to compute the optical properties of the CNTs. First principles are used to compute the absorption and a good fit with the experimental results is obtained [10]. However, if we consider the exciton effects in CNTs, modeled using a description of electron-hole pair propagating on a cylindrical surface, the absorption spectrum $\alpha(\omega)$ is given by [11]

$$\alpha(\omega) \propto \sum_{\alpha} |\psi_n(0,0)|^2 / \{(E_n + E_g)[(E_n + E_g - \hbar\omega)^2 + (\hbar\Gamma)^2]\}, \tag{6.4}$$

where $\psi_n(x, y)$ is the n-th exciton state with energy E_n, $\hbar\omega$ is the photon energy of the light excitation, $\hbar\Gamma$ is a phenomenological line width of about 0.05 eV, and $E_g = E_{exc} - E_{bind}$ is the bandgap of the CNT, with E_{bind} the exciton binding energy. E_{bind} has the same dependence on the nanotube diameter and chiral angle as the excitation energy E_{exc}.

Formula (6.4) indicates that the CNT absorption spectrum is a sum of Lorentzian curves and that the absorption depends on the nanotube diameter via E_g. Therefore, all methods for tuning the bandgap of semiconducting SWCNTs described in Chapter 3 are directly applicable, in principle, to tune the optical absorption. The optical absorption spectra in a wide band, from UV to near infrared, can be used to obtain quantitative information on the SWCNT structure.

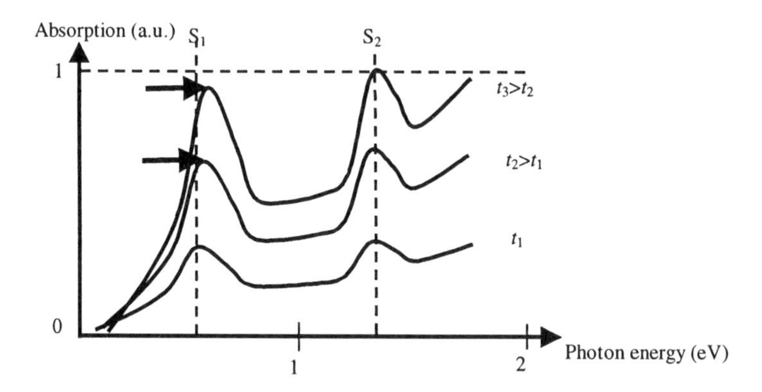

Figure 6.2 The optical absorption spectra of a SWCNT film with different thicknesses (*After*: [12]).

The optical absorption of semiconducting SWCNTs with a diameter in the 1–2 nm range shows mainly two peaks, S_1 and S_2, which correspond to the lowest and the second interband electronic transitions. These peaks, shown in Figure 6.2, are located at excitation energies of 0.7 eV and 1.2 eV, respectively.

This peak sequence can be identified in SWCNT thin films with different thicknesses. Engineering of absorption peaks is a major issue for versatile optoelectronic devices. Doping of the SWCNT film, high-pressure treatment, or chemical exposure was used to change the absorption peaks in a controllable manner [12]. As shown in Figure 6.2, the S_1 peak shows a blue shift of about 5.5 meV/μm when the thickness t of the film increases, as demonstrated in [12] based on measurements of the absorption in SWCNT films with an average tube diameter of 1.31 nm and with thicknesses ranging from 0.2 μm up to 2.5 μm. The absorption peak shift is due to the axial stretch of SWCNTs that originates in the pressure exerted on them when the film with a certain thickness is fabricated. For the SWCNT films in [12] it was estimated that each SWCNT supports a tensile stress in the range of 0.1–0.5 GPa as the film thickness increases with 1 μm.

Also, heating the film at 400 °C for one hour tunes the S_1 absorption band from 0.7 eV to 0.735 eV. The 35 meV red shift is caused in this case by the metal particles present in the SWCNT film, which aggregate and thus produce an axial stress in the film.

The CNT absorption spectrum can be engineered by changing also the average diameter of nanotubes in the thin film. When a multilayer SWCNT film consists of layers with different average diameters of nanotubes, d_1 and d_2, different intrinsic stress is introduced in each layer. For example, if two different SWCNT films with the same thickness but with the S_1 absorption peak located at 0.7 eV and 0.75 eV, respectively, are layered, the S_1 peak of the resulting structure is located at 0.74 eV, as shown in Figure 6.3 [12]. Moreover, since the S_1 absorption peak around 0.7 eV is quite close to the Fermi level, it is possible to shift the Fermi level in order to cancel this transition. An optical switch could thus be implemented using only a gate potential, which shifts the Fermi level and turns on and off the S_1 absorption.

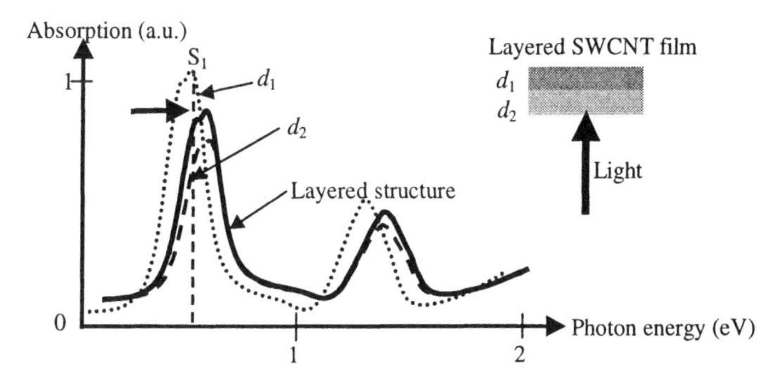

Figure 6.3 The optical absorption spectra of a SWCNT film with different diameter (*After*: [12]).

Absorption engineering has important applications. Since the absorption peaks of the SWCNT are located in the near-IR, the SWCNTs can be used for IR detection. In particular, large areas of SWCNT arrays containing subarrays with certain tube diameters are envisaged as tunable IR detectors. Because, in principle, the optical radiation with energy greater than the CNT bandgap is absorbed, increasing the conductivity of a certain subarray, broad wavelength response, high-temperature operation, and low noise level due to the reduced electron-phonon interaction in 1D quantum systems are expected from such IR detectors [13]. So, highly densely packed (10^{10} cm^{-2}) large areas of ordered and aligned SWCNT or MWCNT (multiwalled carbon nanotube) arrays can cover a broad spectral range, from 1 μm to 10 μm. The IR detectors based on CNTs exceed in performances the QWIR (quantum well infrared detectors) because the CNT has the broadest IR bandwidth and is able to absorb all linearly polarized optical power directed on it. A schematic representation of an IR detector consisting of an aligned CNT array embedded in an Al matrix is displayed in Figure 6.4.

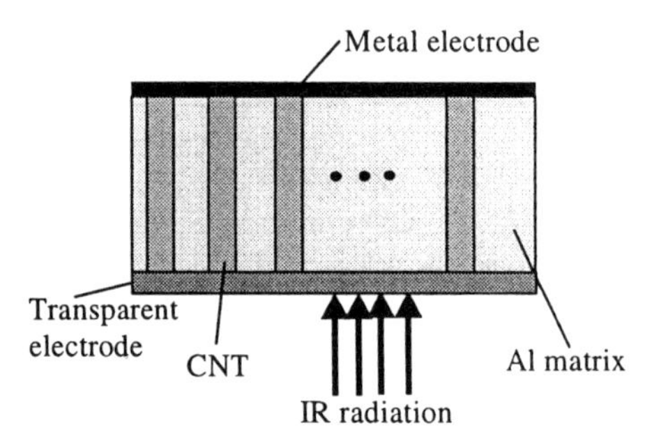

Figure 6.4 CNT IR detector sub-array (*After*: [13]).

The first experimental evidence of the IR detection of CNTs was recently reported [14]. Unlike in the configuration in Figure 6.4, MWCNTs were randomly deposited between two interdigital copper electrodes grown on a Si/SiO_2 substrate, as shown in Figure 6.5. The copper electrode widths and the separation between the fingers were both 1.5 μm, the effective area of the IR photodetector being of 60 μm×60 μm. The $I-V$ dependence of the MWCNT photodetector was insensitive to the IR radiation generated by an IR lamp, but the application of a dc between the two electrodes enhanced the IR sensitivity of the photodetector.

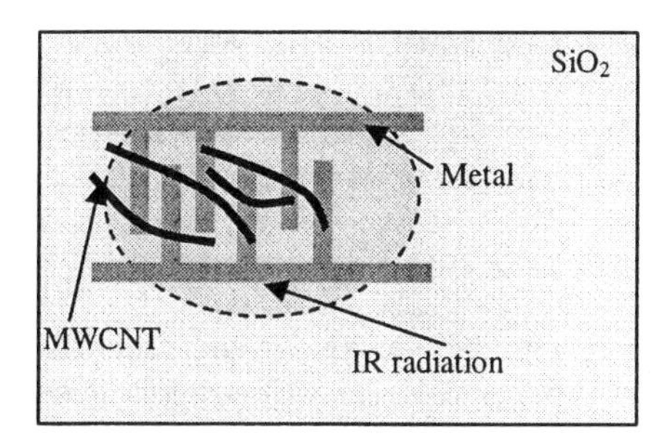

Figure 6.5 MWCNT IR interdigital detector.

MWCNT-based detectors are receptive to the mid-IR spectrum [13]. The sensitivity of these detectors, expressed via the changes in the $I-V$ characteristic, was tested by placing the lamp at different heights above the IR detector. The response time was 400 ms, while the sensitivity increased 10^3 times when the distance between the lamp and the substrate was about 25 cm. It was observed that photocurrents are still generated at room temperature under low applied voltages, around 1 V.

A similar result was demonstrated by illuminating with a lamp in the near-infrared (NIR) wavelength range thin films with 0.2–0.5 μm thicknesses made from randomly oriented SWCNTs with diameters in the interval 0.8–1.1 nm [15]. The film, with an area of 300 μm × 200 μm, was deposited onto a patterned gold substrate with a 200 μm wide insulating gap between the electrodes. The obtained photocurrent response for bias voltages up to 5 V was linearly dependent on the light intensity, which varied in the range 0.1–1.2 mW/mm^2. A tungsten lamp and a filter designed to expose the SWCNT film in the region of maximum absorption (0.8–1.7 μm), which corresponds to the $c_1 \rightarrow v_1$ transition, was used in the experiment. Turning on and off the NIR source the current jumps from a low dark current up to 3 A/cm^2.

The photocurrents in CNT *p-n* junctions show unusual characteristics due to the special features of semiconducting CNTs, which are the only semiconductors where all bands have a direct gap and hence, band-to-band transitions are facilitated in all directions [16]. Moreover, in CNT semiconductors the defect densities are low and the unwanted nonradiative transitions are reduced. The photodetectors based on *p-n* CNT junctions are expected to have a low sensitivity at temperature variations.

In [16] a *p-n* junction in a (17,0) SWCNT was simulated using a tight-binding Hamiltonian under the assumption that the illuminated region consists of parallel rings of 17 carbon atoms with a distance between them alternating between 0.07

nm and 0.14 nm and that the doping is 5×10^{-4} electrons/C atom. The circumferential angular momentum, which is conserved when the light is polarized parallel to the tube axis and which is denoted by an integer J, can be used to label each of the 17 bands as 1, 2, 3, and so on. By defining the photoresponse as $I_{ph}/e\Phi$, where I_{ph} is the photocurrent and Φ the photon flux, the simulations made using 128 illuminated carbon rings have shown that the photoresponse is composed from several sharp peaks ranging from IR up to UV. This broadband photoresponse, schematically represented in Figure 6.6, is explained by the fact that all bands in SWCNT have direct transitions. The existence of the photoresponse at energies below the bandgap is attributed to photon-assisted tunneling. Another distinct feature of the photoresponse is that it depends on the length of the CNT, showing an oscillatory behavior at various photon excitation energies.

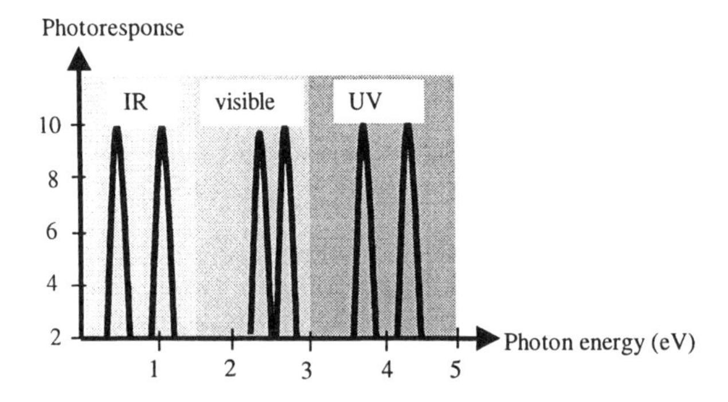

Figure 6.6 The photoresponse of the SWCNT *p-n* junction (*After*: [16]).

Besides the above-mentioned dependence on the film thickness and the tube diameter, the absorption peak of CNT thin films can be tuned by an additional parameter: the variation of the incident optical power induces a significant variation of the absorption value [17,18]. This property was used with impressive results in the passive mode-locking techniques of lasers, with CNT thin film as saturable absorbers. Passive mode-locked lasers, which generate transform-limited optical pulses of ps or sub-ps duration, are among the finest available optical pulse sources. In a mode-locked laser there are many longitudinal modes that oscillate simultaneously. If these oscillations are mode-locked, that is, if a device is able to synchronize the phases of these modes and to produce a constant phase difference between them, the interfering lasing modes produce uniformly spaced pulses.

The key device that performs mode locking, and thus allows the formation of optical pulses and cancels the spurious CW lasing modes is a saturable absorber.

The saturable absorber is an optical device with a nonlinear response, which changes its transparency as a function of the incident optical power in a certain wavelength bandwidth. If the incident optical power is small, the saturable absorber attenuates the light, but the attenuation decreases sharply when the incident power increases beyond a certain threshold. This behavior is typical for SWCNT films with a thickness of less than 1 μm, which consist of SWCNTs with diameters of 1.2 nm and 1.3 nm, with associated absorption peaks at 1.55 μm and 1.68 μm, respectively. The optical bandwidth around 1.55 μm is one of the most important bandwidths in optical communication systems. In this bandwidth the optical absorption of SWCNT thin films decreases with 10% when the light intensity increases from 0.1 kW/cm^2 to 10^4 kW/cm^2, as shown in Figure 6.7.

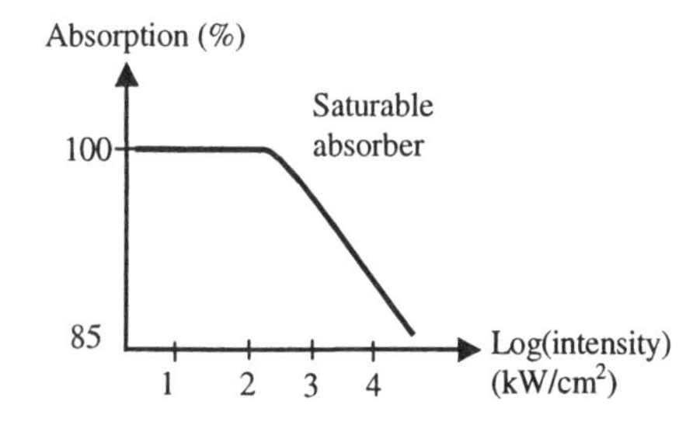

Figure 6.7 The absorption dependence on optical intensity of a saturable absorber (*After*: [18]).

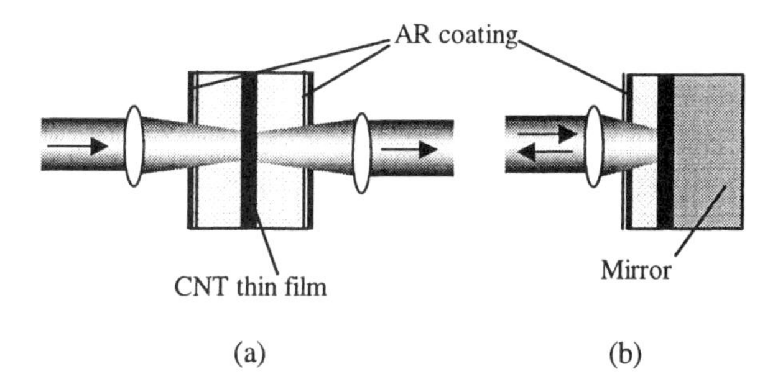

Figure 6.8 (a) Transmission mode and (b) reflection mode of a SAINT (*After*: [18]).

The optical bandwidth of CNT saturable absorbers can be tuned by changing the thickness or the diameter of the CNT. This fact, which is of major importance in optoelectronic applications, was tested experimentally in a ring cavity laser at

1.55 μm. The saturable absorber incorporating nanotubes (SAINT) can be used in two basic designs, as shown in Figure 6.8, depending on the laser configuration. In the transmission mode (T-SAINT), illustrated in Figure 6.8(a), the CNT film is positioned between two quartz substrates, which act as anti-reflection coatings. In the reflection mode (R-SAINT), depicted in Figure 6.8(b), the CNT thin film is placed over a highly reflective mirror and is coated with the same anti-reflection coating as in the T-SAINT configuration. The CNT film can be also placed between two optical fibers covered with ferule in a transmission configuration.

We must emphasize that SAINTs have the same performances as industrial saturable absorbers and in some respects (for example, the ultrafast response) are even better. SAINT has demonstrated ultrafast recovery time of less than 1 ps. It is polarization insensitive, mechanically robust, and less expensive than conventional saturable absorbers based on quantum wells. The ultrafast saturable absorption in CNT can be used for noise reduction, pulse shaping, and in mode-locking configurations.

MWCNT films have other important applications in optoelectronics. It was recently demonstrated that such a film, consisting of randomly oriented metallic MWCNTs, acts as an optical antenna, receiving and emitting electromagnetic radiation at optical wavelengths [19]. The antenna effect is possible when the MWCNT length is comparable with the wavelength of the electromagnetic waves that excites it. Therefore, metallic MWCNTs with a diameter of 50 nm and a length that varies in the 200–1400 nm range are expected to act as antennas in the visible electromagnetic spectrum. The optical antenna effects that can be evidenced in an optical experiment are a drastic reduction of the antenna response when the electric field of the incoming radiation is polarized perpendicular on the dipole axis, and a maximum response when the antenna length is a multiple of $\lambda/2$, with λ the radiation wavelength.

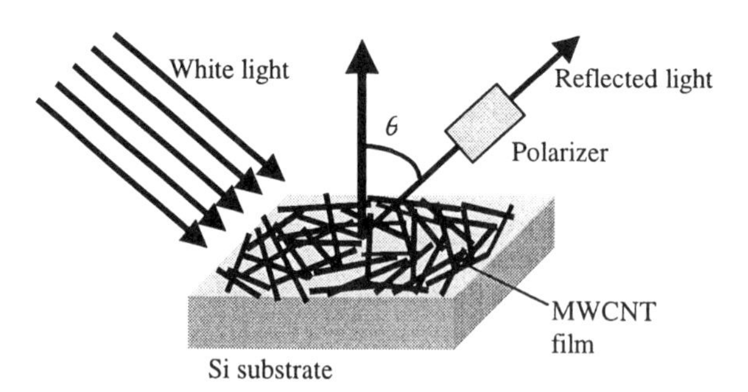

Figure 6.9 MWCNT film optical antenna.

The first antenna effect was tested by illuminating the MWCNT film with unpolarized white light with an electric field $\boldsymbol{E}_{\text{inc}}$, the reflected light being analyzed in a specular direction with a polarizer, as shown in Figure 6.9. The experiment shows that, as the polarizer observation angle θ is varied, the intensity of the reflected light becomes maximum when the polarizer is parallel to the direction $6 = 0$, while, when $6 = \pi / 2$, there is no reflected light and the response is zero. The scattered light intensity follows the law

$$I_{\text{MWCNT}}(\theta) = |\boldsymbol{E}_{\text{inc}}|^2 \cos^2 \theta . \tag{6.5}$$

The scattered optical field from the MWCNT film is maximum when

$$L = m(\lambda / 2) f(6, n) , \tag{6.6}$$

where L is the average length of the MWCNT, m is an integer, and $f(\theta, n)$ is the radiation pattern. $f(\theta, n)$ is 1 in the case of a single dipole and when the average distance between antennas, d_{a}, is either comparable with or much larger than the wavelength λ. In MWCNT thin films we have $d_a << \lambda$, case in which

$$f(\theta, n) = (n^2 - \sin^2 \theta)^2 , \tag{6.7}$$

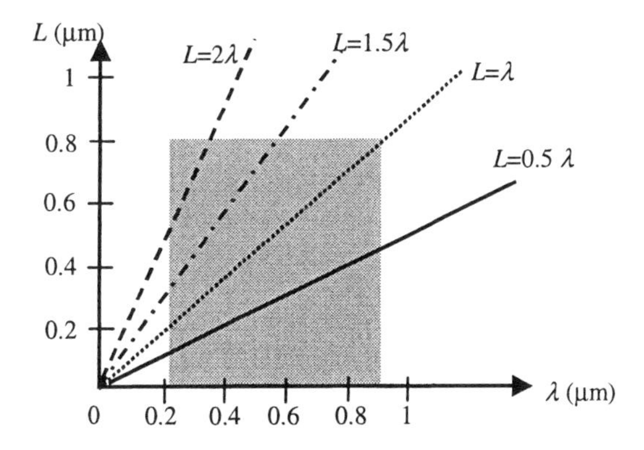

Figure 6.10 The dependence of the average length of MWCNT optical antenna on the wavelength in the case of maximum reflection light detection measurement (*After*: [19]).

where n is the number of nanotubes in the array. The experiments proved that formulas (6.6) and (6.7) are valid by measuring the antenna array response for various lengths and array geometries. The experimental results are represented schematically in Figure 6.10, the shadowed area indicating the region in which experimental results fit the straight lines described by (6.6) for $f = 1$. These results can be used to implement THz and IR devices such as modulators and demodulators.

The optical properties of aligned CNT arrays play an important role in optoelectronic devices and in metamaterial fabrication. The key problem in this case is the computation of the effective electrical permittivity ε_{eff} as a function of wavelength, the ratio between the internal and external radii of the carbon nanotube ρ, and the ratio between the lattice constant and the nanotube external radius R. The CNT films with nanotubes perpendicular and parallel to the surface are called β-aligned and α-aligned, respectively. The α-aligned films are birefringent due to different ε_{eff} values for light polarization along (s-polarization) and normal to (p-polarization) the nanotube axis. There are different methods of calculating the effective permittivity and the optical absorption of aligned CNT arrays arranged in the form of a square or hexagon using the finite-element method, the effective-medium approximation, the gradient approximation, and the series expansion. A brief review of these methods accompanied by many relevant references and a comparison of their accuracy can be found in [20]. We will consider a periodic array of multishell CNTs with an inner radius r and an outer radius R, which are arranged in a square array with a lattice constant $a = 2qR$ [21], where q is a constant. A homogeneous and isotropic insulating medium with the dielectric constant ε_0 surrounds the CNT array. The planar graphite material is highly anisotropic and therefore its dielectric function is a tensor rather than a scalar. In the case of graphite, a dielectric tensor

$$\hat{\varepsilon}(\omega) = \varepsilon_{\perp}(\omega)(\widehat{\boldsymbol{\theta}}\widehat{\boldsymbol{\theta}} + \hat{z}\hat{z}) + \varepsilon_{\parallel}(\omega)\hat{\boldsymbol{r}}\hat{\boldsymbol{r}} \tag{6.8}$$

can be assigned to each point inside the CNT and outside the inner core, where $\hat{\boldsymbol{r}}$, $\hat{z}$, and $\widehat{\boldsymbol{\theta}}$ are unit vectors (versors), $\varepsilon_{\perp}(\omega)$ is the dielectric function perpendicular to the crystallographic c-axis of the nanotube, while $\varepsilon_{\parallel}(\omega)$ is the dielectric function parallel to the nanotube c-axis. When the incoming electromagnetic field is perpendicular on the CNT array (i.e., when $k_y = k_z = 0$, with $\boldsymbol{k}$ the wavevector) the radiation is s-polarized and the electric field is parallel to the CNT at any point. Because in this case the electric field is not modified by the interfaces, the effective dielectric function is a sum of the constituents (CNTs and the insulating medium). On the contrary, the *p*-polarized electric field is seriously perturbed by interfaces, the effective permittivity in this case being given by

$$(\varepsilon_{\text{eff}} - \varepsilon_0)\boldsymbol{E} = f(\hat{\varepsilon} - \varepsilon_0)\boldsymbol{E}_{\text{in}}, \tag{6.9}$$

where

$$\boldsymbol{E} = f\boldsymbol{E}_{\text{in}} + (1-f)\boldsymbol{E}_{\text{out}} \tag{6.10}$$

is the average electric field of the composite medium, f is the filling fraction coefficient, which is an average measure of the quantity of nanotubes in the free space, and $\boldsymbol{E}_{\text{in}}$ and $\boldsymbol{E}_{\text{out}}$ are the average electric fields inside and outside the CNTs, respectively.

When $f \to 0$, the case that corresponds to a plain or hollow CNT cylinder,

$$(\hat{\varepsilon} - \varepsilon_0)\boldsymbol{E}_{\text{in}} = \varepsilon_0 \alpha \boldsymbol{E}, \tag{6.11}$$

and the effective dielectric constant becomes

$$\varepsilon_{\text{eff}} = \varepsilon_0(1 + f\alpha), \tag{6.12}$$

where α is the in-plane polarizability per unit volume.

If $f \ll 1$, a case that corresponds to a sparse distribution of nanotubes, the influence of the multipolar contribution to the optical response is reduced, $\boldsymbol{E} \cong \boldsymbol{E}_{\text{out}}$, and the effective permittivity is given by

$$\varepsilon_{\text{eff}} = \varepsilon_0[1 + \alpha f/(1 - \alpha f L)], \tag{6.13}$$

where $L = 1/2$. The energy loss function $\text{Im}[-\varepsilon_{\text{eff}}^{-1}(\omega)]$ becomes, in this case, $\varepsilon_{\text{eff}}^{-1} = \varepsilon_0^{-1}[1 - \alpha f/(1 + \alpha f L)]$, where the polarizability is

$$\alpha = \frac{2}{(1-\rho^2)} \times \frac{(\varepsilon_{\parallel}\Delta - \varepsilon_0)(\varepsilon_{\parallel}\Delta + \varepsilon_0) - (\varepsilon_{\parallel}\Delta - \varepsilon_0)(\varepsilon_{\parallel}\Delta + \varepsilon_0)\rho^{2\Delta}}{(\varepsilon_{\parallel}\Delta + \varepsilon_0)(\varepsilon_{\parallel}\Delta + \varepsilon_0) - (\varepsilon_{\parallel}\Delta - \varepsilon_0)(\varepsilon_{\parallel}\Delta - \varepsilon_0)\rho^{2\Delta}}, \tag{6.14}$$

with $\rho = r/R$ and $\Delta = (\varepsilon_{\perp}/\varepsilon_{\parallel})^{1/2}$. Equation (6.13) is a generalized Maxwell-Garnett effective dielectric function for a CNT array. In the case of the nonsparse regime the electromagnetic interaction between CNTs must be included in the model via the Bloch waves method. Using the above formulas for $\varepsilon_0 = 1$ and a typical value for the outer radius R = 5 nm, as well as the tabulated values of the real and imaginary parts of $\varepsilon_{\parallel}(\omega)$ and $\varepsilon_{\perp}(\omega)$ for graphite, the effective dielectric permittivity can be computed for different ρ values in the 1–8 eV range. The experiments and simulations performed with this model agree well up in the 0.5–6 eV range, as was recently demonstrated by an experimental study in which the optical absorption of a vertically aligned SWCNT array grown on a polished quartz substrate was measured for *s*-polarized and *p*-polarized incident light, respectively [22]. SWCNTs with 2 nm in diameter were used in the experiment and the optical field polarization was modified by varying the illumination angle θ_s of the CNT array placed on a rotating substrate holder; $\theta_s = 0$ means normal to the substrate. By modifying the photon energy in a very wide wavelength bandwidth encompassing UV, visible and IR using a spectrometer, two major peaks at 4.5 eV and 5.25 eV were recovered. These peaks are also predicted by the Maxwell-Garnett approach and correspond to the maxima in $\mathrm{Im}[\varepsilon_\perp(\omega)]$ and $\mathrm{Im}[-\varepsilon_\parallel(\omega)]$ of the graphite material.

In the *s*-polarization case it was found that the absorption spectrum is independent on θ_s, while a strong dependence on θ_s was observed in the *p*-polarization case. The peaks S_1, S_2 in the absorption spectra were recovered in the *s*-polarization case but they were more pronounced for *p*-polarization. A maximum of absorbance located at 5.25 eV was observed in the *s*-polarization case, which was not shifted when θ_s was changed. On the contrary, for *p*-polarization the absorbance maximum was initially identical to that of *s*-polarization, but a change of θ_s in the $0–\pi/4$ range produced a red shift from 5.25 eV to 4.25 eV.

Aligned arrays of CNTs have very important applications in the emerging area of photonic bandgap (PBG) materials. PBGs are 2D or 3D periodic structures, which stop the propagation of light in prescribed wavelength bandwidths and around certain spatial directions. In this respect, a self-assembly technique was used to grow a large area of well-aligned metallic CNTs on metallic (Ni) dots [23]. In the first step, a suspension of polystyrene nanospheres was applied onto the clean surface of a Si wafer. Then, the Si wafer was immersed in water, resulting a monolayer with a very large area of highly ordered nanospheres. By draining the water the monolayer is deposited again on the Si surface, the highly ordered nanosphere monolayer being subsequently used as a mask to deposit a metallic catalyst, such as Ni. The final result, obtained after chemical removal of the polystyrene particles in toluene, is a honeycomb pattern of Ni dots. The Ni dot honeycomb structure is the basic cell of the CNT array, which contains many such cells. The metallic CNT bundles are grown via PECVD over the Ni dots and the array formed in this way is the PBG. An elementary honeycomb CNT cell is displayed in Figure 6.11. The PBG structure consisting of a multitude of elementary cells reflects the light in a large bandwidth centered around 0.5 μm. The central frequency, as well as the bandwidth, can be tuned by choosing a certain CNT diameter and height.

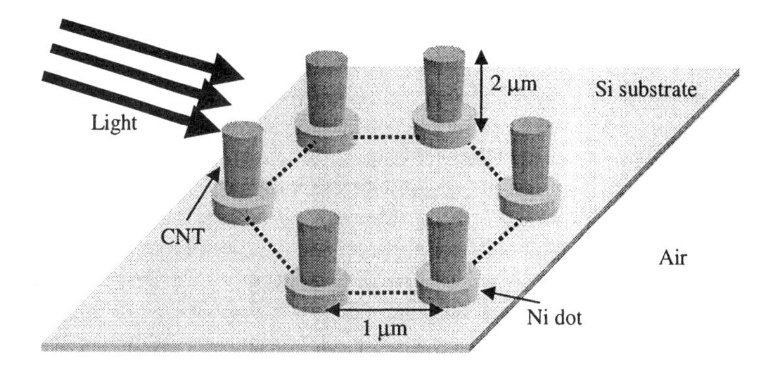

Figure 6.11 Basic cell of a CNT-based PBG.

The semiconducting CNTs also display unusual and very strong nonlinear optical properties. The third order susceptibility $\chi^3(\omega)$, responsible for quadratic electro-optical effects, was theoretically estimated in [24] and experimentally demonstrated in [25]. A straightforward model based on the π-band electron energy dispersion, which shows a parabolic dependence on the wavenumber k (see Chapter 3), allows the calculation of $\chi^3_{\mathrm{QEO}}(\omega) = -\mathrm{Re}\,\chi^{(3)}(-\omega;0,0,\omega)$ and $\chi^3_{\mathrm{EA}}(\omega) = -\mathrm{Im}\,\chi^{(3)}(-\omega;0,0,\omega)$, which are the relevant nonlinear parameters for the quadratic electro-optic effect and electro-absorption, respectively. The calculation of these two parameters is easy, but very laborious and will not be reproduced here; it can be found in [24]. The electro-optic effect refers to the refractive index change due to an applied electric field E_0:

$$n(\omega) = n_0 + \Delta n_{\mathrm{CNT}}(\omega, E_0) = n_0 + (2\pi / n_0)\chi^{(3)}_{\mathrm{QEO}}(\omega) E_0^2 . \tag{6.15}$$

In this respect, the electro-optic effect in SWCNT is characterized by $\chi^3_{\mathrm{QEO}}(\omega)$, while $\chi^3_{\mathrm{EA}}(\omega)$ represents its losses. The susceptibility $\chi^3_{\mathrm{QEO}}(\omega)$ is negative in the long-wavelength region and its magnitude increases with frequency. $|\chi^3_{\mathrm{QEO}}(\omega)|$ is increasing with the SWCNT diameter. In the vicinity of the fundamental absorption edge the susceptibility $\chi^3_{\mathrm{QEO}}(\omega)$ changes sign from negative to positive, and attains a maximum value of 5×10^{-4} e.s.u. at a photon energy located just below the bandgap. The electro-absorption, characterized by $\chi^3_{\mathrm{EA}}(\omega)$, is positive and increases monotonically with the frequency and the nanotube diameter, but in the wavelength region located just below the bandgap the electro-absorption decreases rapidly and changes sign, becoming negative with a resonant peak, which favors the transmission of light. The fact that near the bandgap

$\chi^3_{\rm QEO}(\omega)$ switches between a positive 5×10^{-5} e.s.u. value and a negative value of 10^{-3} e.s.u. for a very slight variation of the laser wavelength is of considerable importance for optical switching devices. Moreover, near bandgap the electro-absorption losses can be minimized, while $\chi^3_{\rm QEO}(\omega)$ can be resonantly enhanced to its highest value. Thus, the electro-optical effect can be maximized while the corresponding losses are very small. The optical nonlinear properties are strongly dependent on the SWCNT bandgap, and hence can be tuned by changing the nanotube diameter. Ultrafast optical switching with SWCNT, in less than 1 ps, is, however, quite difficult to obtain at 1.55 μm, which is an important optical communication wavelength, but it was accomplished using a polyimide-SWCNT composite thin film with a thickness of about 20 μm [25]. The pump-probe method was used to test the switching and nonlinear properties of the film with the help of a 150 fs laser with a wavelength of 1.55 μm. The measured exciton decay time, which represents the relaxation time due to the direct transition of electrons from the conduction to the valence band over the bandgap of 0.57 eV, was less than 0.8 ps. The measured nonlinear optical susceptibility $\chi^3(\omega)$, of 10^{-9}–10^{-10} e.s.u., was very near that indicated by the theory in [24]. A similar susceptibility value was obtained for a film of MWCNTs using a pump-probe configuration in the 0.72–0.78 μm optical range, but the relaxation time, of 2 ps [26], was longer than for polyimide-SWCNT films.

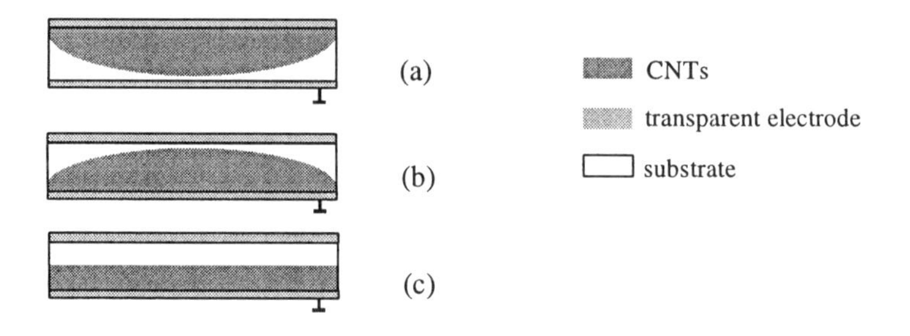

Figure 6.12 (a), (b) Zoom lens configurations fabricated from CNTs grown on curved substrates. (c) Variable phase shifter made from CNTs grown on a planar substrate (*After*: [27]).

Based on the very large electro-optic effect experienced by SWCNTs and described by (6.15), a recent proposal showed that a CNT array grown on a curved substrate can act as a zoom lens, with a large focal length variation controlled by a dc voltage applied on the CNT [27]. The possible zoom lens configurations are displayed in Figures 6.12(a) and 6.12(b). The zoom lens can be either convergent or divergent depending on the semiconductor substrate and/or the shape of the template (convex in Figure 6.12(a), or concave in Figure 6.12(b)). In certain conditions the same lens can act either as a divergent or a convergent zoom lens, depending on the voltage applied on the CNT. Such a lens has never been constructed before and this unusual behavior is due to the rapid change of the sign of Δn when the optical excitation is tuned near the bandgap. Based on the same principle, a CNT array deposited on a planar substrate works as a variable phase shifter for optical signals, as displayed in Figure 6.12(c).

The equivalent focal length f of the zoom lens illustrated in Figure 6.12(a), which consists of a divergent lens made of the substrate material with a refractive index n_s and a convergent CNT lens, is defined as

$$\frac{1}{f} = \frac{n_{CNT}(E_0) - n_s}{R} = \frac{n_0 + \Delta n E_0^2 - n_s}{R}, \tag{6.16}$$

where $\Delta n = 2.25 \times 10^{-8}$, E_0 is measured in V/cm, and R = 46 mm is the radius of curvature of the substrate. The equivalent focal length f of the zoom lens in Figure 6.12(b), which is formed from a divergent CNT lens and a convergent substrate lens, is given by a similar equation to (6.16), with $-R$ instead of R. A refractive index variation of $\Delta n_{CNT} = \Delta n E_0^2 = 0.9$, one of the largest obtainable variations due to the electro-optic effect, requires a dc applied field of about 6.3×10^3 V/cm. This value is far below the electric field breakdown in CNT, which is higher than 10^5 V/cm. The giant variation of the refractive index, $\Delta n_{CNT} = 0.9$, is valid for a (13,0) CNT, for which n_0 = 3.2, excited at an energy of about 0.99 E_g, with E_g = 0.85 eV. The incoming radiation wavelength exciting this CNT array should be about λ = 1.45 μm, inside the optical communication range (1.3–1.55 μm). It is essential to mention that the absorption coefficient of the CNTs is practically zero at this excitation energy.

Equation (6.16) shows that f can be positive (as for a convergent lens) or negative (as for a divergent lens), depending on the sign of the difference between the refractive indices of the CNT and the substrate. For a Si_3N_4 substrate, with n_s = 2, the equivalent focal length is positive, and increases from 24 mm to 46 mm when the applied electric field is tuned between 0 and 6.3×10^3 V/cm, as shown in Figure 6.13(a). If the average height of the CNT array is 1 μm, the above variation of the electric field corresponds to tuning the applied bias from 0 to 630 mV. The focal length of this zoom lens is modified with about 100%, a huge change that is not encountered in usual zoom lenses.

A divergent lens, with a negative equivalent focal length, is achieved using the convex template in Figure 6.12(a) when the substrate has a higher refractive index than the CNT array. This is the case for a silicon substrate with an index of refraction of n_s = 3.5. But when the CNT refractive index increases with 0.9 due to the applied voltage, it becomes higher than n_s, and so the zoom lens changes from divergent to convergent for an increasing bias applied on the CNT. The corresponding variation of the focal length, which varies between –92 mm up to $-\infty$, and then from $+\infty$ up to 115 mm for a dc electric field that increases from 0 to 6.3×10^3 V/cm, is represented in Figure 6.13(b).

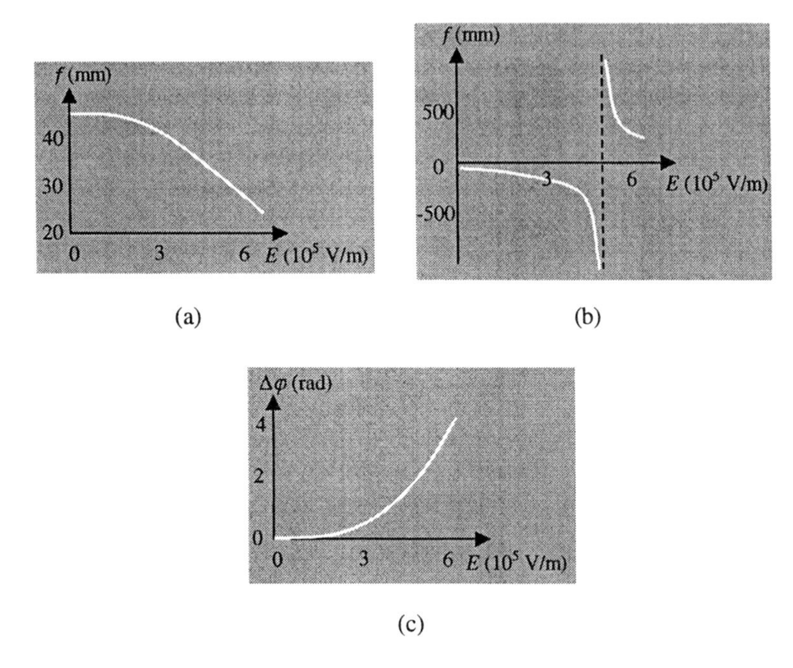

Figure 6.13 Electric field dependence of the focal length of (a) a convergent CNT zoom lens grown on a Si_3N_4 substrate and (b) an initially divergent CNT zoom lens grown on a Si substrate. (c) Electric field dependence of the variable part of the phase shift for a CNT phase shifter (*After*: [27]).

For a variable delay line or a variable phase shifter we can use the configuration presented in Figure 6.12(c), where a uniform semiconductor CNT array is grown over a planar substrate, for example, Si. If both the CNT array and the Si substrate have the same thickness, of t = 1 μm, the tunable part $\Delta\varphi(E_0) = 2\pi\Delta n_{\mathrm{CNT}}(E_0)t/\lambda$ of the phase difference $\varphi = 2\pi[n_{\mathrm{CNT}}(E_0) + n_s]t/\lambda$ $= \varphi_0 + \Delta\varphi(E_0)$ that is introduced by the planar configuration varies with the electric field as depicted in Figure 6.13(c). The phase shift can thus change with about 4 radians, a huge value that can hardly be obtained with any other continuously tunable phase shifter.

CNTs are able to focus not only light but also electron beams. The experimental demonstration of MWCNTs as focusers for electron beams requires a perfect alignment of their axes along the propagation direction of the electron beam, which must have an energy around 200 KeV. This energy corresponds to a wavelength of $\lambda = 2.5 \times 10^{-12}$ m, which is smaller than the MWCNT inner diameter [28]. The Fresnel diffraction of electron waves at the edges is employed to focus the incoming electron beam in a smaller spot. A curved edge is able to focus an electron beam onto a point on the concave transparent side of the edge, enhancing the beam intensity. The Fresnel diffraction is strongest when a nanosized hole in an opaque solid is illuminated, a task almost impossible to be performed unless the natural nanosized hole represents the nanosized empty cylinder cores of MWCNTs with diameters of a few nm and wall thicknesses of tens of nm. An electron lens based on a bent MWCNT array with properly oriented nanotubes is represented schematically in Figure 6.14.

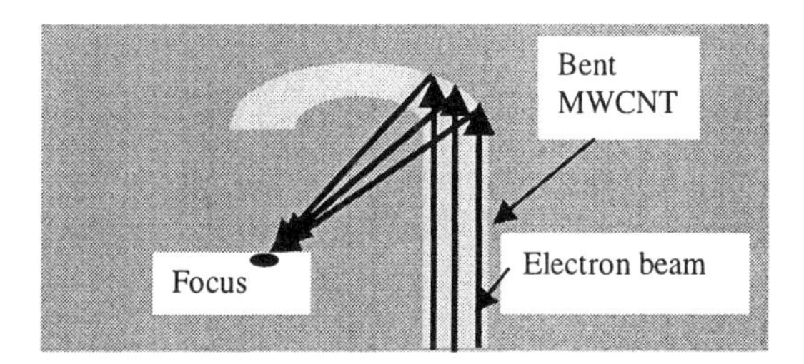

Figure 6.14 Electron focusing through MWCNTs.

Giant optical rectification in CNT films was theoretically predicted [29,30] and experimentally demonstrated [30]. This could have important implications in THz electronics, since optical rectification is one of the most often used physical effects to efficiently generate and detect THz signals [31]. Optical rectification refers to the difference frequency generation or detection, when two interacting waves that excite a material have very close or even identical frequencies. Optical rectification occurs in centrosymmetric crystals due to the quadrupole and magnetodipole mechanisms of optical nonlinearity and is more pronounced in materials with strong spatial dispersion, when the size of the molecules is comparable to the excitation wavelength. This is the case of CNT films with 3–4 μm in thickness that contain randomly arranged nanotubes deposited on a 25 mm × 25 mm Si substrate, which is transparent for THz radiation. If such an unbiased CNT film is excited by a nanosecond pulsed Nd:YAG laser, the measured ratio of the detected dc voltage to the laser power is 500 mV/MW (or 0.5 μV/W) at the wavelength of 1064 nm, and 650 mV/MW (or 0.65μV/W) at 532 nm.

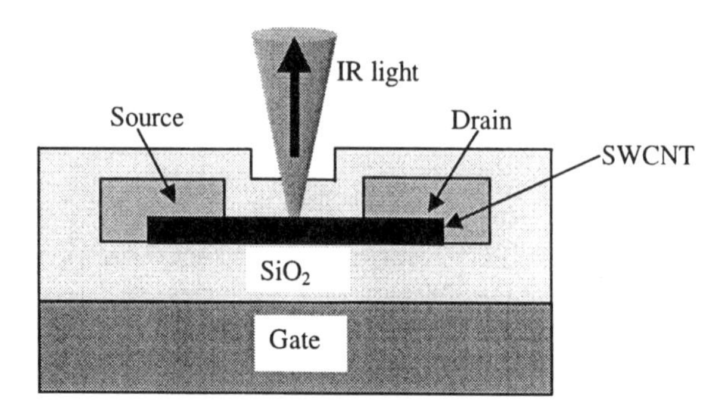

Figure 6.15 Light emission from a SWCNT ambipolar transistor (*After*: [34]).

Light emission from CNTs is another hot topic with serious implications in optical communication or quantum computing. Electrically induced optical emission at 1.5 μm from a CNTFET (CNT field-effect transistor) was experimentally demonstrated in [32]. The CNTFET, which is a typical SWCNT FET transistor in which a SWCNT is placed over a doped substrate that plays the role of a back gate, shows an ambipolar carrier transport. The ambipolar character of electrical transport is conferred by covering the SWCNT with a 100 nm thick SiO_2 layer annealed at elevated temperatures; *p*-type transport occurs when the gate voltage V_G is negative, while *n*-type transport takes place for $V_G > 0$ [33]. However, a hysteresis behavior in the $I_{DS} - V_G$ characteristics of this CNTFET transistor is observed, because the high fields (10^7 V/cm) around the transistor caused by the relative high biases of 40–80 V required for ambipolar transport act as an additional gate, which injects carriers in the gate insulator and thus modifies the configuration of electrostatic fields in the CNTFET. The drain and source contacts are realized by electron-beam lithography followed by the evaporation of 30 nm of palladium. This transistor has two thin Schottky barriers at the source and drain contacts, which allow the thermal-assisted tunneling of electrons and holes through the barriers. Thus, electrons and holes are injected simultaneously in the intrinsic semiconducting SWCNT, a case in which the gate voltage has a value between the drain and source voltages, the gate-induced electric fields having opposite signs at the source and drain contacts, respectively. The ideal situation is when $V_{DS}/2 = V_G$ and the source is grounded. The simultaneous injection of electrons and holes in the SWCNT with a 1.5 nm diameter and a length in the range of 50–100 μm produces the emission of light (see Figure 6.15) in the wavelength range of 1.5–2 μm [34]. This IR luminescence is due to the recombination of the electrons and holes injected at the source and drain contacts. It was found that the IR luminescence is concentrated in a small region of the SWCNT and that the position of this region along the nanotube can be controlled with the help of the gate voltage. In this way, the intensity and the shape of the emitted IR spot can be monitored. This phenomenon cannot be encountered in 3D semiconducting optoelectronic devices.

It is worth noting that simultaneous fluorescence and Raman scattering experiments from SWCNTs [35] have demonstrated that the fluorescence intensity of a CNT does not fluctuate at moderate excitation intensities of less than 70 kW/cm^2. This is in deep contrast with the fluorescence of individual semiconductor quantum dots or that of most molecules, which show emission intermittency, blinking, or bleaching emission at any excitation wavelength. This property will allow the realization of photon sources with a very narrow linewidth. However, at high excitation intensities the fluorescence of the SWCNT starts to fluctuate due to nonradiative transitions.

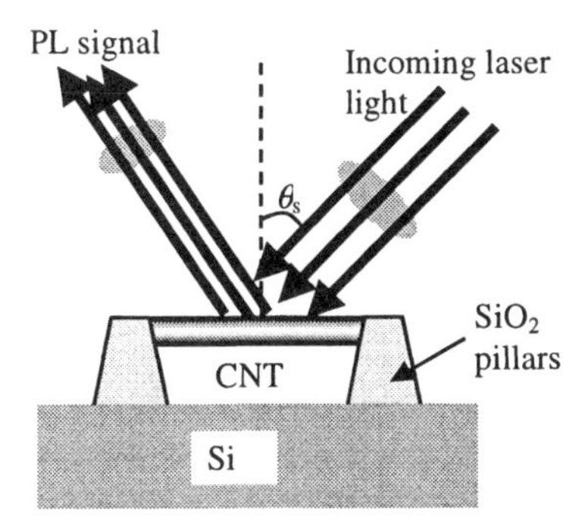

Figure 6.16 Light emission from a suspended CNT on nanopillars (*After*: [36]).

The photoluminescence (PL) of single SWCNTs was studied in air at room temperature using single SWCNTs grown on SiO_2 pillars, as depicted in Figure 6.16 [36]. In this case, there is a unique one-to-one relationship between the optical absorption resonances and the emission peaks. Resonances are about 30 nm wide and are linearly varying in the 0.75–1.1 eV range depending on the nanotube diameter and laser excitation. The experiments were made with a tunable

Ti:sapphire laser (725–837 nm) with a 2 μm spot size focused on the nanotube suspended on the two pillars, which are distanced at 1 μm. A strong polarization dependence of the PL was observed, similar to that noticed in absorption. The emission is maximum at an illumination angle $\theta_s = 0$, for which the electrical component of the field is parallel to the CNT axis, the emission being strongly reduced when the illumination angle is larger than 40º. The emission depends on the angle as $[\cos^2 \theta_s]^{0.7}$.

6.1.2 Optoelectronic Devices Based on Semiconducting, Metallic Nanowires and Nanostructured Surfaces

This section starts with the analysis of optoelectronic devices based on metallic or semiconducting nanowires other than CNTs, and ends with some optical properties of nanostructured surfaces.

Semiconducting nanowires have important applications in optoelectronics, such as in LEDs, nanolasers, or photodetectors for broadband optical wavelength detection. These new devices could be also important in chemical or biological analysis. It is expected that these nanoscale optoelectronic devices will be organized into building blocks (i.e., as an array of nanolasers), to enhance their performances. At present, the composition of CNTs that form an array cannot be perfectly controlled, the array being a mixture of semiconductor and metallic nanotubes, successful efforts are beginning to emerge regarding the separation of metallic and semiconducting CNTs; the realization of a large CNT array with prescribed chemical and geometrical properties is not hopeless [37,38]. On the contrary, an array of nanowires can be easily chemically synthesized in single crystals with prescribed chemical composition, diameter, length, and doping; for a recent review on semiconducting nanowires see [39].

Maybe the most astonishing applications of semiconducting nanowires are as optically or electrically pumped nanolasers, which can work at room temperature [40–42]. There are several types of semiconducting nanowires, each of them lasing at a different wavelength, as displayed in Table 6.1.

Table 6.1

Nanowire Lasing Wavelengths (*From*: [40])

Nanowire	*Lasing wavelength* (eV)
ZnO	3.26
GaN	3.31
CdS	2.43

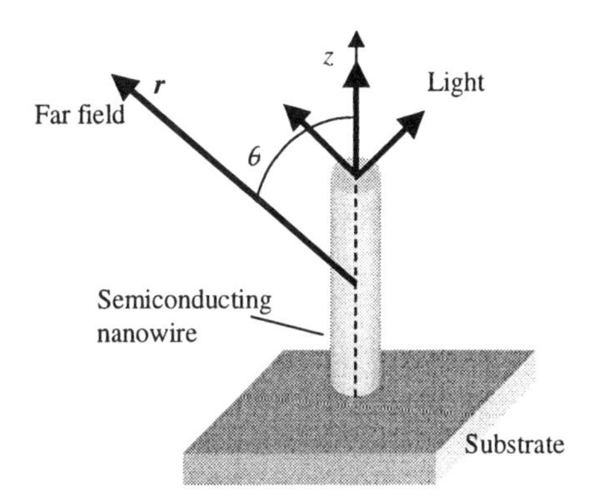

Figure 6.17 Schematic representation of the nanolaser.

The ZnO, GaN, or CdS nanowires have typical diameters in the 20–200 nm range and have lengths in the 2–40 μm interval. The nanowires positioned perpendicular on the substrate emit along their axis, that is, in the vertical direction, denoted as the *z*-axis in Figure 6.17, due to the strong confinement of guided modes that propagate along this axis. The strong confinement factor of these nanolasers is due to the very high difference between the refractive indices of the semiconducting nanowire and the air. It is believed that the top and the bottom facets of the nanowire act as two reflectors and form a Fabry-Perot-like cavity [43].

Although in nanowires few radially dependent components of both transverse electrical modes TE_{0m} and transverse magnetic modes TM_{0m}, as well as radially dependent hybrid modes such as HE_{nm} and EH_{nm}, propagate, only the lowest-guided modes, HE_{11}, TE_{01}, and TM_{01}, are considered for reflectivity calculation. The reflectivity is strongly dependent on the guiding modes, lasing wavelength, and nanowire radius. For example, in nanowires with a radius R = 60 nm, HE_{11} is the only propagating mode for energies smaller than 3.55 eV, this fact being confirmed also by experiments. For thicker nanolasers, however, other guided modes must be also taken into account, especially for ZnO and GaN nanowires.

The reflection coefficient of the top facet increases with increasing frequency, the corresponding reflection coefficient of the electric field, $r = (n-1)/(n+1)$, being similar to that encountered for normal incidence on an infinite dielectric air-interface. The mode with the highest confinement is HE_{11}. On the contrary, the bottom facet reflectivity does not display a monotonic increase with the frequency, as for the top facet, but still HE_{11} has a reflectivity similar to that for infinite interfaces. The TE_{01} mode has also a high reflectivity and could dominate in the range $1 < \omega R / c < 2.5$, where c is the speed of light. The threshold gain of the nanolaser and the quality factor of the Fabry-Perot-like cavity are, respectively,

$$g_{\text{th}} = -(1/L)\ln|r_1 r_2| \tag{6.17}$$

$$Q = -Lk_z / \ln|r_1 r_2| \tag{6.18}$$

where r_1 and r_2 are the amplitude reflection coefficients of the top and bottom facets of the nanolaser, respectively, and L is the nanowire length. The physical mechanism responsible for gain at room temperature seems to be the electron-hole plasma developed inside the semiconducting nanowire. For the HE_{11} mode and $\omega R/c = 1$, the estimated value of the threshold gain is $g_{\text{th}} = 4000\,\text{cm}^{-1}$ and that of the quality factor is $Q = 60$ if $r_1 = 0.3$, $r_2 = 0.05$, $L = 10$ μm, and $R = 60$ nm [43]. On the contrary, for the TE_{01} mode at $\omega R/c = 1.5$ and for $r_1 = 0.71$ and $r_2 = 0.2$ we get $g_{\text{th}} = 1800\,\text{cm}^{-1}$ and $Q = 217$. These calculated values are in quite good agreement with those reported experimentally for single GaN nanowire lasers optically pumped by a laser with a wavelength of 310 nm [41], except that in the experiments the nanowires are thicker (have diameters greater than 300 nm) and longer (L is at least 40 μm) compared with the values used in the example above. Experimental data provide typical values of $g_{\text{th}} = 400\text{–}1000\,\text{cm}^{-1}$ and $Q = 500\text{–}1500$ for different nanowire geometries. Experiments also confirm the main features of the nanolasers: 1) the lasing is produced in the modes guided by the nanowires, and 2) the light is emitted along the nanowire axis. The modes supported by nanowires with radii in the 10–100 nm range are HE_{11}, TE_{01}, and TM_{10}. The TE_{01} mode has only one transverse electric field component, but the other two modes have electric fields with both longitudinal and transversal components. Therefore, it is necessary to determine both the longitudinal and transverse gains and the accompanying confinement factors, which are very large for nanowires and exceed well the corresponding values in semiconductor heterostructure lasers [44].

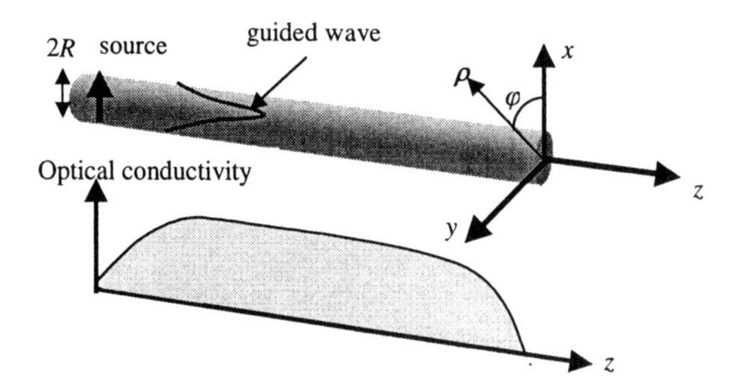

Figure 6.18 The guiding mode lasing model in a nanolaser (After: [44]).

As shown in Figure 6.18, the nanolaser can be seen as a dielectric cylindrical waveguide, the lasing action being modeled through an optical conductivity that links locally the induced current density $\boldsymbol{j}$ and the optical field $\boldsymbol{E}$. The optical conductivity, defined as $\boldsymbol{j} = \boldsymbol{\sigma E}$, is given by

$$\sigma = \begin{pmatrix} \sigma_{xx} & 0 & 0 \\ 0 & \sigma_{yy} & 0 \\ 0 & 0 & \sigma_{zz} \end{pmatrix} = \begin{pmatrix} \sigma_{\perp} & 0 & 0 \\ 0 & \sigma_{\perp} & 0 \\ 0 & 0 & \sigma_{\parallel} \end{pmatrix}. \quad (6.19)$$

Let us consider in the following discussion that the optical axis of the crystal, which is the *c*-axis in GaN, is at the same time the *z*-axis of the nanolaser and that the nanolaser is placed in air, for which $\varepsilon = 1$ and $\sigma = 0$. If only one mode propagates in the positive z direction of the dielectric waveguide with complex electric and magnetic fields $\boldsymbol{E}(\boldsymbol{\rho})$ and $\boldsymbol{H}(\boldsymbol{\rho})$, respectively, which depend only on the transverse coordinate $\boldsymbol{\rho}$ that is a vector in the *x*-*y* plane (see Figure 6.18), the electric field at a certain position $\boldsymbol{r}$ is $\boldsymbol{E}(\boldsymbol{r}) = A(z)\boldsymbol{E}(\boldsymbol{\rho})\exp[-i\omega t + ik_z z] + c.c.$ The magnetic field $\boldsymbol{H}(\boldsymbol{r})$ has a similar form. By defining the power as $I(z) = |A(z)|^2 I_0$, where

$$I_0 = (c/4\pi)\int[\boldsymbol{E}(\boldsymbol{\rho})\times \boldsymbol{H}^*(\boldsymbol{\rho}) + \boldsymbol{E}*(\boldsymbol{\rho})\times \boldsymbol{H}(\boldsymbol{\rho})]\hat{z}d\boldsymbol{\rho}, \quad (6.20)$$

the power change along the *z*-axis is given by

$$I(z+\Delta z) - I(z) = -dz\int \boldsymbol{E}(\boldsymbol{r})\boldsymbol{j}(\boldsymbol{r})d\boldsymbol{\rho}, \quad (6.21)$$

and the modal gain of the nanolaser is defined as

$$dI/dz = g_{\mathrm{m}}I \quad (6.22)$$

where $g_{\mathrm{m}} = g_{\parallel} + g_{\perp}$. The longitudinal and transverse total gains, $g_{\parallel}$ and $g_{\perp}$, are given, respectively, by

$$g_{\parallel} = -(2/I_0)\int \sigma_{\parallel} |E_{\parallel}|^2 \, d\boldsymbol{\rho} = \Gamma_{\parallel} g_{\parallel}^0, \quad (6.23a)$$

$$g_{\perp} = -(2/I_0)\int \sigma_{\perp} |E_{\perp}|^2 \, d\boldsymbol{\rho} = \Gamma_{\perp} g_{\perp}^0, \quad (6.23b)$$

depend on the optical conductivity, guided mode, and frequency. The coefficients $\Gamma_{\parallel}$ and $\Gamma_{\perp}$ in (6.23a) and (6.23b) are the longitudinal and transverse confinement factors while

$$g^0_{\parallel,\perp} = -(4\pi\sigma_{\parallel,\perp}) / c(\varepsilon)^{1/2} \tag{6.24}$$

is the material gain. In [44] a numerical solution was adopted to determine the confinement factors for various modes, which are then compared with the analytical values in (6.23a) and (6.23b). It was found that, while the longitudinal confinement factor $\Gamma_\parallel$ increases with $\omega R / c$, the transverse confinement factor $\Gamma_\perp$ decreases with the same parameter, $\omega R / c$. Unlike in semiconductor heterostructure lasers, the confinement factors can be greater than 1 even for higher guiding modes, such as HE_{21}, where $\Gamma_\perp + \Gamma_\parallel = 1.4$. This fact is due to the waveguide modes, which guide light in nanowires with a slower group velocity than a plane wave and thus gain more energy.

The optical conductivity and the modal gain $g = \Gamma_\parallel g^0_\parallel + \Gamma_\perp g^0_\perp$ can be computed numerically for several values of the nanowire radius and for different modes using the band structure of GaN [44]. For example, in a nanowire with R = 60 nm HE_{11} is the only propagating mode and the highest value of its modal gain is 1500 cm^{-1} for $\Delta E \cong E - E_g - \Delta_1 - \Delta_2 = 30$ meV, where Δ_1 and Δ_2 are known dispersion parameters. For a nanowire with a radius R = 75 nm the TE_{01} mode attains a maximum modal gain value of 2000 cm^{-1} at ΔE = 24 meV, while HE_{11} shows a slightly lower value of the gain, the corresponding value for the TM_{01} mode being very low. The situation is almost the same for an increased radius of 105 nm, with the exception that the modal gain for the TE_{01} mode increases slightly and attains 2200 cm^{-1}.

The intensity of the far-field radiation of nanolasers was numerically computed for various modes and frequencies in [45]. The TE_{01} and TM_{01} modes do not emit at $6 = 0$ (the significance of this angle is displayed in Figure 6.17), the intensity maxima for $\omega R / c$ between 1 and 2 being observed for θ in the 30°–50° interval; thus, the nanolasers are very divergent light sources. On the contrary, the maximum light intensity is at $6 = 0$ for the HE_{11} mode at $\omega R / c = 1$, independent of polarization.

Using the catalytic growth technique many semiconductor nanowires, including those from the IV–IV group materials (Si, Ge, SiGe), the III-V binary group (GaN, GaP, GaAs, InP, InAs), the III–V ternary group ($Ga_{1-x}In_xP$), and the II-VI group (ZnS, ZnSe, CdS, Cd), can be fabricated with the prescribed composition and geometry [39]. The exploitation of this wealth of nanowires in optoelectronics is, however, at the beginning. Among the first applications is the LED depicted in Figure 6.19, consisting of crossed *p*- and *n*-type nanowires that emit light at the crossing point. The crossed nanowires form a *p-n* diode, which emits light when it is forward biased. Such a diode, which emits at 820 nm, was made using InP nanowires. Because different nanowires emit light at different wavelengths, a broad optical spectrum (300–800 nm) that extends from UV up to

NIR can be covered using similar multiple-wavelength emission devices, as indicated in Figure 6.20. In these configurations various nanowires of *n*-type, such as GaN, CdS, or CdSe, cross a *p*-type Si nanowire, which is used as a hole injector for the other nanowires. The UV and NIR spectrum is covered by the emission in bandwidths around 365 nm, 510 nm, and 700 nm, respectively. As will be seen later, we can also realize waveguides and photodetectors using nanowires, so that an optical spectrum analyzer can be implemented at the nanoscale and can be used as a lab-on-a-chip for chemical or biological analysis. The nanowire LED can be also integrated with a FET, which is also made from a crossed wire configuration and which electrically drives the LED, to obtain a nanoscale optical switch.

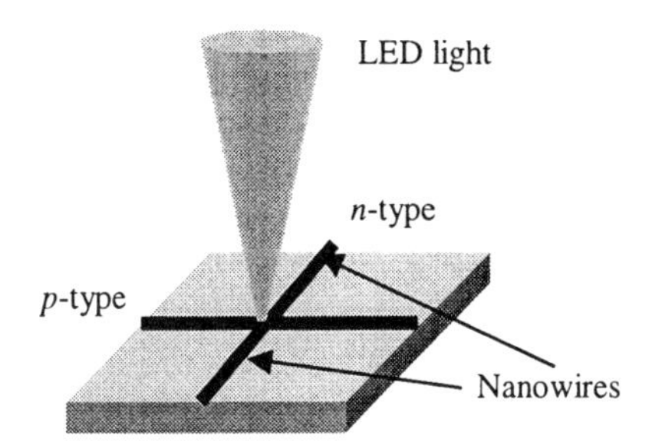

Figure 6.19 The crossed nanowire LED (*After*: [39]).

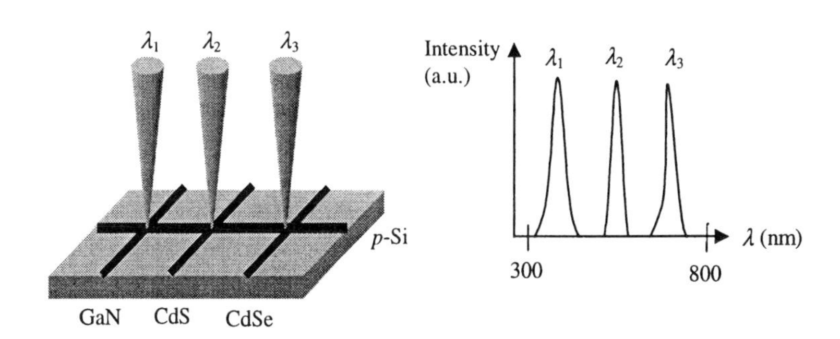

Figure 6.20 The crossed nanowire LED array (*After*: [39]).

An electrically pumped laser can be fabricated using the same principle of hole injection from a *p*-Si substrate, and adding an additional electrode that is able to pump electrons. The electrodes sandwich a CdS nanowire, as illustrated in Figure 6.21. Such a nanowire was shown to lase at 495 nm and 8 K [42].

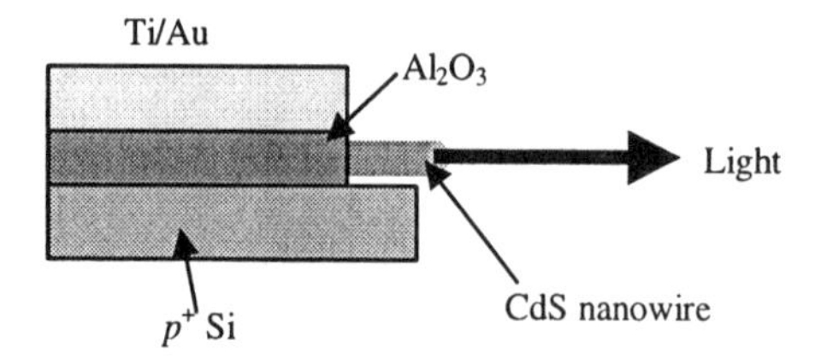

Figure 6.21 Electrically pumped nanowire laser (*After*: [39]).

The nanowire can also act as an optical waveguide, similar to an optical fiber. The nanowire works like a single-mode waveguide if

$$1-(2\pi R/\lambda)(n_1^2-n_0^2)^2<2.4\,, \tag{6.25}$$

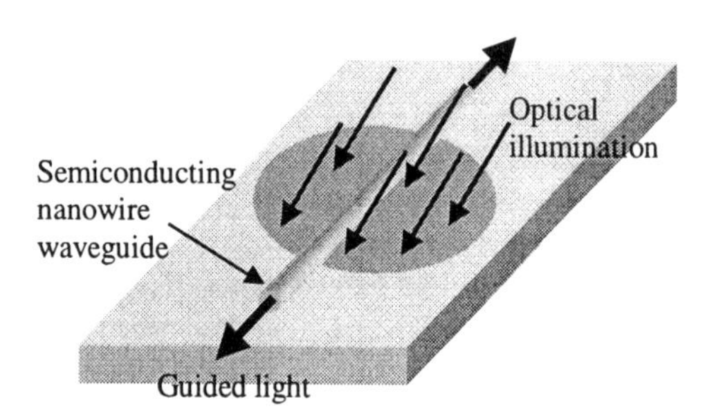

Figure 6.22 Nanowire optical waveguide.

where n_1 and n_0 are the refractive indices of the nanowire and the surrounding medium, respectively. Such a waveguide action of a 50 μm long CdS nanowire, illustrated in Figure 6.22, was demonstrated in [42] using a uniform illumination from a mercury lamp. The loss of the nanowire was 1–2 dB when tested with the help of the scattering optical microscopy method [46]. The losses are due to surface roughness, which can be minimized.

The guided light in the nanowire can be modulated by adding a biased metallic cantilever, which acts as a parallel plate capacitor, above the nanowire; the resulting structure is illustrated in Figure 6.23. The intensity in the nanowire waveguide changes with 30% when the cantilever is bent upon applying a dc voltage V_{dc} in the range of 0–20 V.

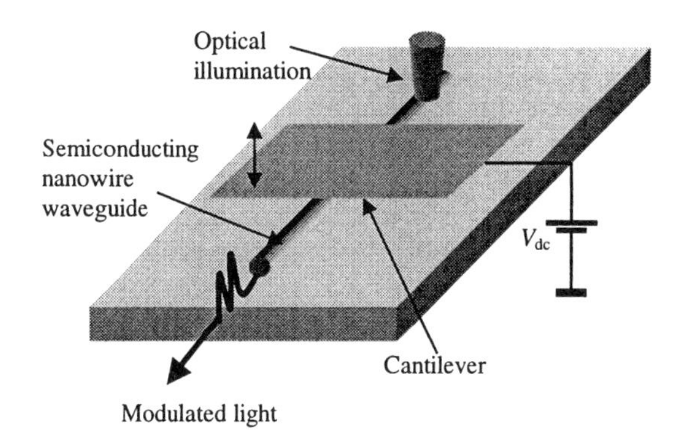

Figure 6.23 Nanowire light modulator (*After*: [46]).

A promising application of nanowires in optoelectronics is the nanowire-grid polarizer [47]. The polarizer is a key element in many optical communication systems. However, the conventional polarizer devices are bulky, require complex fabrication processes, and cannot be easily integrated with other optoelectronic devices. Therefore, a metal wire grid that acts as a polarizer and that can be easily integrated is of great practical significance. Such a device, comparable with existing polarizers in terms of performance, was designed and fabricated for wavelengths in the range of 1–2 μm using nanosized high-aspect ratio dielectric walls as support for metal nanowires. The performance of polarizers is mainly characterized by three parameters: 1) the extinction *Ex*, defined as the ratio T_{TM}/T_{TE} between the transmission of light for the TM and TE polarizations, 2) the total energy efficiency *TEE*, defined as $TEE = T_{TM} + R_{TM} = 1 - loss_{TM}$, where R_{TM} is the reflected power for TM modes, and 3) the transmittance T_{TM} of light for the TM polarization. *TEE* is always less than 1 due to electromagnetic losses, the main goal being to obtain a *TEE* value as close as possible to 1, while maintaining a quite high value for *Ex*.

Taking into account all aspects mentioned above, a nanowire polarizer seems to be a viable solution. The nanowire polarizer consists of a linear array of SiO_2

dielectric walls with a width of 60 nm and a height of 515 nm, one side of each wall being covered with a 24 nm thick gold film. The entire nanowire grid structure is patterned on a bottom antireflection layer, which is grown on a glass substrate. Finally, the structure is coated with a top antireflection coating, as shown in Figure 6.24.

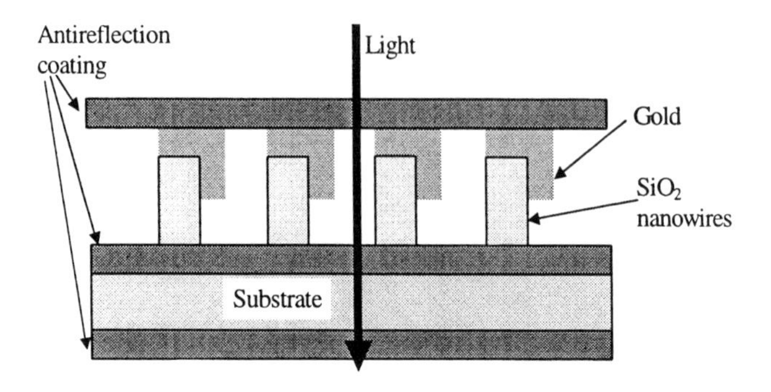

Figure 6.24 The nanowire-grid polarizer (*After*: [47]).

The performances of this polarizer are quite impressive: 0.98 transmittance for the TM polarization, Ex > 40 dB over a 100 nm bandwidth centered on the main optical communication wavelength 1550 nm, and TEE > 98%. The SiO_2 nanowires etched in straight walls were fabricated via the UV nanoimprint lithography technique described in Chapter 1. Another optoelectronic device based on the above nanowire-grid polarizer is a monolithically integrated isolator. The isolator, reported in [48], consists of a polarizer integrated together with a magneto-optic Faraday rotator, and attains a very low loss value (of 0.2 dB) and a high isolation (of 35 dB) around 1500 nm.

Periodic structures like PBG or metamaterials, which display simultaneously a negative permittivity and permeability (or a negative index of refraction), are hot topics of optoelectronics due to their wealth of applications that range from super-lenses, filters, waveguides, up to high-performance semiconductor lasers where the spontaneous emission is suppressed. Recent reviews on these topics are [49-50]. These structures are based on hundreds or thousands of elementary structures such as holes or wires, which are much smaller than the operating visible or IR wavelength and are periodically arranged in a 1D, 2D, or 3D array. So, PBG or metamaterials with negative index of refraction are composed from nanosized structures such as nanowires, nanoparticles, and nanoholes [51]. For example, a CNT array studied in the previous section can constitute a PBG. A nanosized PBG is, for example, a 3D crystal in the form of a woodpile, as that in Figure 6.25, realized by nanofabrication techniques. This PBG works in the near-IR range and it was realized by step-by-step stacking GaAs or InP nanoribbons in a periodic structure with 700 nm periodicity and with an accuracy of 30 nm using wafer-fusion and a laser-alignment method [52]. (The nanoribbons are nanowires with a rectangular cross-section.) This structure reflects almost 99% of the incoming light at normal incidence in the 1–2 μm wavelength range (the reflection coefficient is 23 dB) when the woodpile consists of only 4 layers. By increasing the number of woodpile layers at 8 the reflection increases up to 40 dB, this high reflection being preserved at other incident angles.

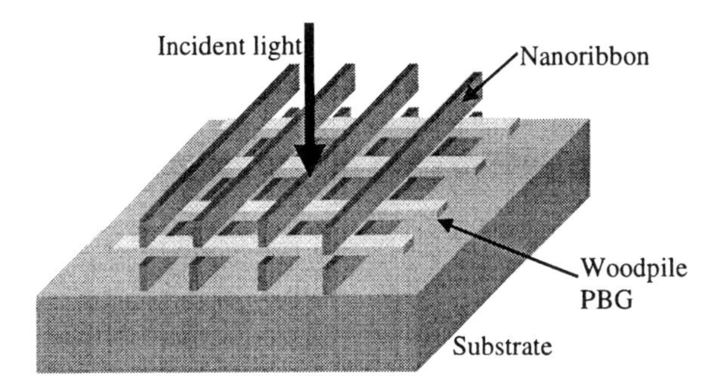

Figure 6.25 PBG in the form of a woodpile.

Simulations have demonstrated that a PBG with Ag nanowires can act as an IR superlens. A superlens is a flat surface of metamaterial, with negative refractive index n_m, which is able to focus a point source located at one side of the metamaterial, with positive refractive index n_1, into a point located on the other side, with positive refractive index n_2, as sketched in Figure 6.26 [50]. The distance u from the object to the metamaterial slab, the distance v from the slab to the image, and the thickness of the metamaterial t are related by $v = |n_2 / n_m| t - |n_m / n_1| u$. In particular, for $n_1 = n_2 = 1$ and $n_m = -1$ this relation becomes $u + v = t$. In PBGs with low thickness this relation cannot be fulfilled, although the negative refraction effect is present, due to the near field regions. These regions produce a confinement of light, which becomes self-collimated. However, in a 2D square lattice consisting of Ag nanowires with a diameter of 60 nm and a lattice constant of 70 nm the relation $u + v = t$ is satisfied for TE fields with a wavelength of 400 nm [53]. The minimum feature size that can be resolved by such an unconventional lens is 100 nm and can be obtained with a slab formed from three PBGs. By varying the PBG dimensions, as well as the diameter of the nanowires, the superlens effect can be extended to the near-IR regime. Submicron imaging with a planar silver lens, which confirms the concept of the superlens, was experimentally demonstrated in optics [54].

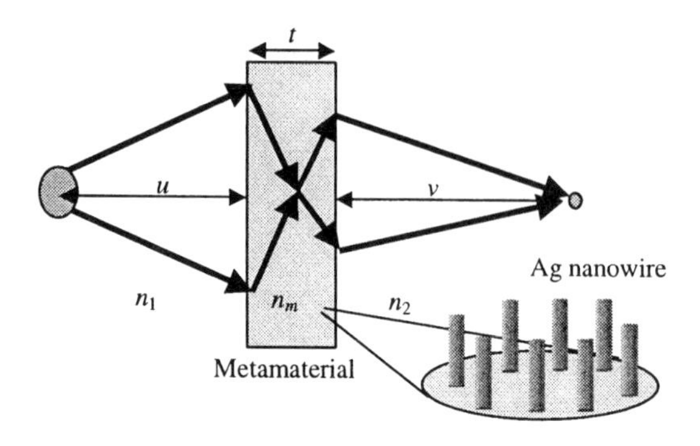

Figure 6.26 The metamaterial superlens.

The nanoribbons are also used for subwavelength photonic integration [55]. In particular, it was experimentally demonstrated that ribbon-shaped nanocrystals of binary oxide are very flexible, and display electrical conductivity and the lasing effect. SnO_2 nanoribbons, for example, are wide bandgap semiconductors with $E_g = 3.6$ eV. Very long SnO_2 nanoribbons with an aspect ratio higher than 1000 are able to guide the PL emitted light in the bandwidth 400–600 nm. Experiments have used SnO_2 nanoribbons with lengths of 400–700 μm, which are sufficiently long to be bent into various geometric shapes, thus creating additional functions such as filtering of light propagating. In addition, the nanoribbons can be easily connected to other photonic components, such as nanoribbon rings, nanoribbon couplers, and junctions between SnO_2 nanoribbons and ZnO nanowires. Hence, nanowire light sources, such as ZnO nanolasers, and nanowire detectors can be linked to form a nanophotonic integrated circuit, for example a Mach-Zehnder interferometer.

Single-mode Si nanowires with subwavelength diameters are also able to act as waveguides of evanescent waves in the 300–600 nm bandwidth. This principle was used to build an optical sensor in the Mach-Zehnder configuration, as shown in Figure 6.27 [56]. An external factor that acts on the sensing area modifies the phase shift, which in turn modifies the detected current of the sensor. The simulation results have shown that the sensitivity of this sensor, defined as the product between the derivative of the phase shift with respect to the refractive index change and the inverse of the effective length of the sensitive area, is 7.5 μm^{-1}, while the sensitivity of a conventional Mach-Zehnder sensor based on planar semiconductor waveguides is only 0.7 μm^{-1}.

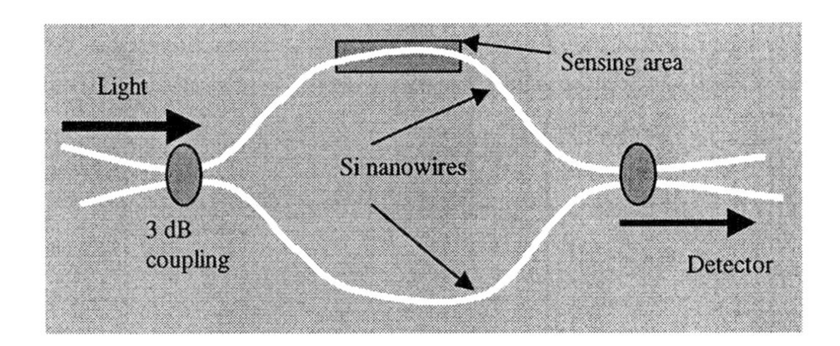

Figure 6.27 Mach-Zehnder sensor based on Si nanowires (*After*: [56]).

A UV photodetector based on a ZnO nanowire, which can be directly connected with nanoribbon waveguides or with a Mach-Zehnder interferometer, is displayed in Figure 6.28. The ZnO nanowire has a diameter in the range of 50–300 nm and can be dispersed on a structure consisting of a series of gold electrodes in order to form a rudimentary photodetector [57]. In dark conditions, the ZnO nanowires behave like insulators and display a resistance higher than 3.5 MΩcm. When the nanowire is illuminated with an incoherent UV light generated by a UV lamp the resistivity of the nanowire decreases with five orders of magnitude. The photocurrent I_{ph} varies with the incoming optical power P_{opt} as

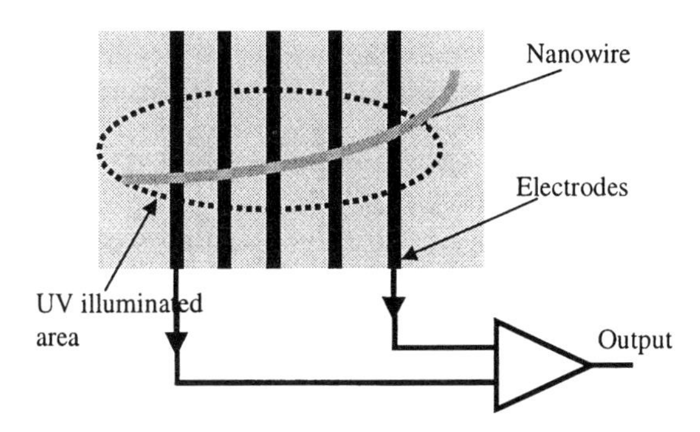

Figure 6.28 The UV nanowire photodetector (*After*: [57]).

$$I_{\mathrm{ph}} \propto P_{\mathrm{opt}}^{0.8}. \tag{6.26}$$

Depending on the optical power, the resistivity of the nanowire photodetector is tuned with 4 to 6 orders of magnitude. Thus, a simple ZnO nanowire has a very high sensitivity to the UV radiation. High wavelength selectivity is also observed, because the nanowire photodetector responds only in a bandwidth centered around 365 nm, but this central wavelength can be tuned by changing the diameter of the nanowire.

Figure 6.29 Columnar nanostructured thin film.

Optical filters and mirrors can be implemented by exploiting, for example, the optical properties of nanostructured thin films. Because the nanostructured thin films are realized using various materials and various deposition methods, their aspect are either columnar, polycrystalline, amorphous, or lacunar [58]. The columnar thin film nanostructures, as those represented in Figure 6.29, are very similar to the nanowire arrays described in this section. Such columnar structures are encountered in Ta_2O_5 if deposited with an evaporation technique or in ZnS grown using similar deposition techniques. Three basic models are used to model the index of refraction of columnar thin films. In the simplest model, which is a linear model, the index of refraction of the thin film n_{f} is related to the index of refraction of the interstices n_{i} and to the index of refraction of the bulk material n_{m} by

$$n_{\mathrm{f}} = pn_{\mathrm{m}} + (1-p)n_{\mathrm{i}}, \tag{6.27}$$

where p is the packing density, which represents the ratio between the volume of the solid part of the nanostructured thin film and the total volume of the film. The packing density p is determined by measuring the index of refraction n_{f} in air and in a humid atmosphere. In another model, called the Maxwell-Garnet model and encountered in connection to the CNT arrays (see relation (6.14)), the refractive index of the nanostructured film is given by

$$n_i^2 = [(1-p)(n_m^2+2) + p(n_i^2+2)n_m^2]/[(1-p)(n_m^2+2) + p(n_i^2+2)] , \qquad (6.28)$$

while in the Bragg-Pippard model the index of refraction is

$$n_i^2 = [(1-p)n_i^4 + (1-p)n_i^2 n_m^2]/[(1-p)n_m^2 + (1+p)n_i^2] . \qquad (6.29)$$

The linear model is used only for materials with low index of refraction, since it gives errors for high refraction index materials. The Bragg-Pippard model describes in many cases better the experimental results than the Maxwell-Garnett model. Multilayers of nanostructured thin films can be grown to act as Bragg filters. The columnar nanostructures are highly anisotropic.

6.2 OPTOELECTRONIC DEVICES BASED ON NANOPARTICLES

6.2.1 Optoelectronic Devices Containing Nanoparticles

The nanoparticles will play an essential role in optoelectronic devices. Nanoparticles are also termed nanocrystals when the bulk material from which the nanoparticle originates is a crystal. This is, in particular, the case of direct semiconductors for which the nanocrystal is also known as a quantum dot [59]. Nanocrystals are made from the most common semiconductors, as described in [59]. Light amplification, enhanced optical nonlinearities, and advanced optical signal processing are examples of optoelectronic applications demonstrated by nanocrystals. However, as will be discussed below, metal nanoparticles have also tremendous applications in innovative optoelectronic devices because a new domain of photonics, called plasmonics, has been developed based on their properties.

As we have already pointed out, we will not treat the issue of quantum dot devices, such as lasers or detectors, because the subject is well covered in the literature; some updated references regarding these issues have been given at the beginning of this chapter. However, we will briefly describe here a new application that involves quantum dots: the development of single photon sources, which have huge applications in quantum communications, quantum optics, and quantum cryptography; for recent reviews see [60,61].

The single-photon source is an optical source that emits a single photon at a time and consists of a direct semiconductor quantum dot coupled to a microresonator. This microresonator is a semiconductor micropillar, which encloses the quantum dot and is sandwiched between two distributed Bragg reflection mirrors

(DBRs), as shown in Figure 6.30. The coupling between the quantum dot and the radiation attains its maximum value when the dot is placed in the antinode of the standing wave associated to the selected cavity mode. Self-assembled InAs/GaAs quantum dot arrays are also used as single-photon sources [62].

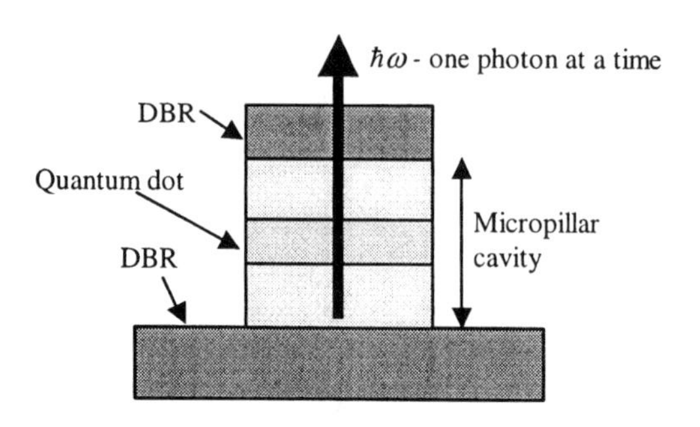

Figure 6.30 Single photon source based on a quantum dot (*After*: [60]).

The most studied nanoparticles nowadays are the Si nanocrystals. Their applications in photonics are investigated thoroughly due to the hope of their integration with silicon VLSI devices, such as microprocessors. Indeed, the Si technology is the most developed and efficient technology worldwide, but it is dedicated exclusively for electronic devices and not for photonic devices. The reason is that bulk Si is a quite poor semiconductor material for light generation and propagation because it displays an indirect bandgap. Therefore, in bulk Si the nonradiative recombination rate is much higher than the radiative rate and, consequently, the majority of electron-hole pairs recombine without emitting a photon. The Auger recombination mechanism and the free-carrier absorption are the main obstacles in the realization of a Si laser, because they prevent the achievement of population inversion. This rampant inefficiency is encountered even at very low temperatures, but, fortunately, it can be overcome if Si is nanostructured. The reason is that Si nanoparticles have a band structure that is completely different from that of bulk Si and, in particular, it favors the drastic reduction of nonradiative effects. Therefore, Si nanoparticles open the possibility of light interconnections inside VLSI devices, which could replace the present metallic connections that are characterized by quite large signal delays in the case of complex VLSI devices that contain more than 10^8 transistors [63].

Si nanocrystals are obtained in various ways and forms. The most encountered assembly of Si nanocrystals is the porous silicon, described briefly in Chapter 1. Porous Si is an assembly of Si nanocrystals of various dimensions, with different shapes and sizes, which are typically of a few nm. Porous Si looks like a sponge with a large number of nanosized pores. As a result of spatial confinement and of the much greater contribution of defects and electronic states at the interfaces, the radiative recombination efficiency increases with several orders of magnitude compared to bulk Si [63]. Therefore, porous Si is an efficient light emitter in the visible range even at room temperature. A review of the PL properties of porous Si can be found in [63,64]. The simple fabrication procedure of porous Si was described in Chapter 1, but Si nanocrystals can be also fabricated with other techniques. In particular, it is possible to produce silicon nanocrystals embedded in a dielectric matrix such as SiO_2 films or substrates. In contrast to porous Si fabrication, the ion-implanted Si nanocrystals and PECVD grown Si nanocrystals are compatible with the standard Si technology [63]. The nonlinear coefficient of the Si nanocrystal is very large, and can attain the value Re $\chi^{(3)} = 1.5\times10^{-9}$ e.s.u., while Im $\chi^{(3)} = 0.6\times10^{-10}$ e.s.u., the absorption in the visible range being moderate. The absolute value of $\chi^{(3)}$, $|\chi^{(3)}| \cong$ Re $\chi^{(3)}$, is with three orders of magnitude larger than in bulk Si, for which Re $\chi^{(3)} = 6\times10^{-12}$ e.s.u. Both the real and the imaginary parts of $\chi^{(3)}$ depend on the nanocrystal size. This significant enhancement in the nonlinear coefficient is due to quantum confinement in Si nanocrystals. In addition, optical gain was reported in ion-implanted Si nanocrystals in quartz substrates [63]. The doping of Si nanocrystals with rare-earth materials, such as Er, paves the way to Si lasers. Many other optoelectronic devices, such as PBG, sensors, microcavities, or light emitting diodes are realized using Si nanocrystals. An updated review of these devices is reported in [63,65].

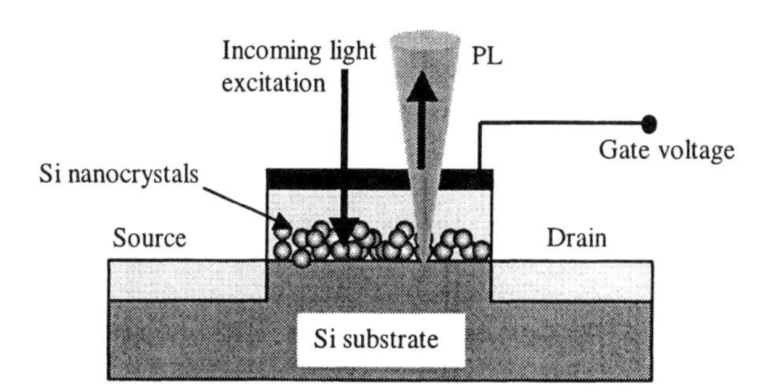

Figure 6.31 Optical memory based on Si nanocrystals (*After*: [66]).

A major achievement of Si nanocrystals is the fabrication of an optical memory [66]. This optical memory based on Si nanocrystals works at room temperature, in contrast to similar AIII-BV quantum dot memories, which work at cryogenic temperatures. Moreover, this optical memory is compatible with present Si technology since it is realized as a ring gate FET transistor on a 300 mm Si wafer on which the Si nanocrystals are implanted beneath the gate, and are separated from the gate electrode by a 40 nm polysilicon semi-transparent and conducting layer. The device is schematically represented in Figure 6.31. The optical memory is in fact a FET in which the PL is controlled by the gate voltage. The PL displays a maximum value at 760 nm when the gate voltage is zero, and decreases progressively as the gate voltage departs from zero in both directions. At +6 V or –6 V the PL is quenched due to the nonradiative Auger recombination produced by the charge excess. The PL quenching is reversible, can be tuned by the gate voltage in Si nanocrystals, and is reproducible even after days of intensive testing [66]. The memory effect is demonstrated by a hysteresis effect in the PL and in the capacitance-voltage $C-V$ curve, which originates in the nonvolatile charge storage in Si nanocrystals. The optical memory can be read or erased either optically or electrically. The programmable logic state of this memory is read optically via the level of the PL emitted by the FET. An optical excitation at 457 nm improves the retention time of the memory, which reaches 100 s. This is the first optical memory based on Si nanoparticles. The electrical nonvolatile memories based on nanoparticles are more advanced and commercially available. In this case, the polysilicon gate is replaced by an array of silicon nanocrystals embedded in the gate oxide.

6.2.2 Plasmonic Optical Devices

Another major application of nanoparticles is the development of a new class of optoelectronic devices, named plasmonic devices. These devices tend to bridge the gap between the microscale, which is specific for the most optoelectronic and photonic devices, and the nanoscale. The diffraction limit forbids the lowering of the lateral dimensions of light waveguides, such as optical fibers, dielectric waveguides, or dielectric PBGs, to values smaller than a few hundred nanometers for visible and infrared wavelengths. Thus, while the dimensions of electronic devices can be lowered to a few nm (see, for example, the nanoscale devices presented in Chapters 3–5), the dimensions of optical field propagating structures and devices cannot be reduced to a few nm due to the diffraction limit beyond which the optical field is no longer confined and hence cannot be guided. However, the diffraction limit can be surpassed using surface plasmons or surface plasmon-polariton surface waves abbreviated further as SPP [67,68]. SPP waves propagate at the interface between a negative refraction index medium and a positive refraction index medium, for example, a metal-dielectric structure. SPP surface modes are produced due to the interaction between the coherent electron oscillations in the metal surface and the interfacial propagating optical field. The SPP are travelling waves confined to the metal-dielectric interface and decay in both media that support them; thus, SPP are evanescent traveling waves. We have shown in Chapter 2 that the diffraction limit is not applicable for evanescent waves. A simple SPP configuration is a nanosized metal film sandwiched between dielectric materials with different permittivities. When an SPP wave is launched, leaky radiation modes are also produced, which form an SPP rainbow jet that can be visualized with a CCD camera. This image confirms the propagation of SPP waves [69].

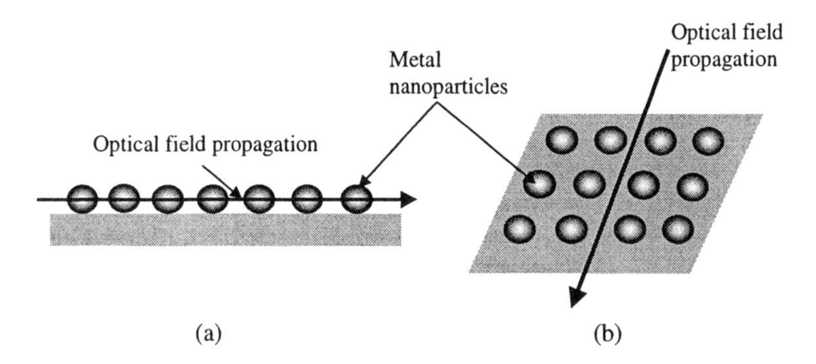

Figure 6.32 Plasmonic optical waveguides: (a) linear chain, and (b) bi-dimensional array (*After*: [68]).

When the resonant frequency of the SPP coincides with the wavelength of the optical field, the light interacts strongly with a noble metal nanoparticle or a chain of nanoparticles and excites collective motions of electrons, the net result being light guiding through such nanosized structures. The interaction of light with nanoparticles with a diameter much smaller than the optical wavelength produces an oscillating polarization of the nanoparticle, which results in an oscillating dipole SPP resonance located in the visible optical spectrum for Au and Ag nanoparticles [70]. With d denoting the spacing between the nanoparticles arranged in either a linear array or a square array, λ the wavelength of the optical field, and λ_{res} the collective dipole resonant frequency of SPP, it is found that if $d \geq \lambda$ the lifetime of SPP oscillations and λ_{res} are due to far-field dipolar interactions, which have a $1/d$ dependence. On the contrary, if $d << \lambda$, λ_{res} and the lifetime of SPP oscillations depend on the high electromagnetic fields confined near the particles, which display a $1/d^3$ dependence. When $\lambda \cong \lambda_{res}$ the optical energy is driven into the nanoparticle, an effect that can be evidenced by a strong enhancement of the scattering cross-section in optical excitation measurements. An application of these interactions between nanoparticles and optical fields is a light waveguide made from a linear chain of nanoparticles, as that represented in

Figure 6.32(a). The Au nanoparticles patterned on a glass substrate have a diameter of 50 nm and are spaced by 25 nm air gaps [70]. Simulations have shown that such a waveguide propagates optical wavelengths with energies in the range 2–2.3 eV. Moreover, the calculations have shown that a 2D metal nanoparticle array patterned on a thin Si membrane acts also as an optical waveguide and propagates light in the optical communication bandwidth 1520–1580 nm with an efficiency of 90%. The 1/e energy attenuation length of this structure, represented schematically in Figure 6.32(b), is greater than 300 μm [68].

The SPP resonance λ_{res} depends on the size and shape of nanoparticles, on the distance d between them, and on the environment of the array of nanoparticles (i.e., substrate, solvents, or adsorbates). Based on this property, plasmonic nanoparticle arrays are utilized as sensors for chemical or biological substances. The chemical modification of nanoparticles induced via SAM techniques enhances the sensing capability of plasmonic nanoparticle arrays because in this way molecular identification elements can be incorporated in the structure. For example, the shift of λ_{res} of individual Ag nanoparticles is used as a sensitive optical sensor for various molecules, with subpicomolar limit of detection [71].

As another application, light scattering can be electrically controlled by covering the Au nanoparticles with an electro-optical material, such as liquid crystal. In this way, the nanoparticle array plasmon resonance λ_{res} is shifted with 50 meV by applying a dc field of 10 kV/cm [72].

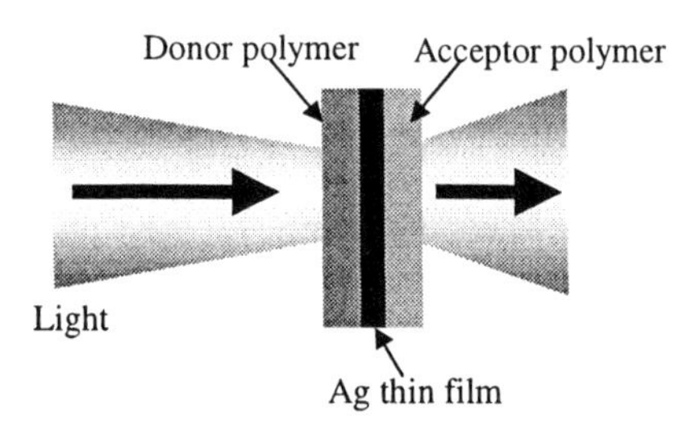

Figure 6.33 The principle of molecular plasmonic light energy transfer (*After*: [73]).

Molecular plasmonic waves were demonstrated by growing on a quartz substrate a molecular donor polymer layer followed by a thin Ag film and an acceptor molecular layer, which is also a polymer but of a different kind from the donor polymer [73]. The resulting structure is illustrated in Figure 6.33. The donor polymer is formed from cromophore molecules and the acceptor polymer is formed from fluorophore molecules. The light is absorbed by the cromophore molecules, is radiated further by the SPP formed in the Ag thin film and is emitted by the fluorophore molecules. The energy transfer has a high efficiency, of more than 70%, over distances of 200 nm, which are 20 times larger than the size of biomolecules. This result enhances the output performances of organic light emitting diodes and could be also used in biosensors.

Similar to nanocrystals, the metal nanoparticles have also enhanced optical nonlinear properties. An Ag nanoparticle system embedded in a SiO_2 glass matrix, for example, shows an ultrafast response of 360 fs and has $\chi^{(3)} = 1.5\times10^{-7}$ e.s.u. [74]. Thus, it can be used as an ultrafast optical switch. The optical nonlinearity of metal nanoparticle composites fabricated by negative ion implantation can be controlled through the dielectric constant of the matrix substrate. For instance, the optical nonlinear coefficient $\chi^{(3)}$ of Cu nanoparticles embedded in $SrTiO_3$ is two times higher than that of Cu nanoparticles embedded in SiO_2 [75]. This is explained by the fact that the local SPP mode, denoted by a positive integer l, must satisfy the relation

$$\varepsilon_\mathrm{p} / \varepsilon_\mathrm{d} = -(l+1)/l , \tag{6.30}$$

where ε_p and ε_d denote the dielectric permittivities of the particles and the embedding dielectric, respectively. The lowest SPP mode ($l=1$) of Cu nanoparticles is located at 2.14 eV for the SiO_2 matrix and at 1.95 eV for the $SrTiO_3$ matrix substrate. The local field enhancement f is the ratio between the applied field and the induced field in the nanoparticle and is given by

$$f = 3\varepsilon_\mathrm{d} / [\varepsilon_\mathrm{p} + 2\varepsilon_\mathrm{d}] , \tag{6.31}$$

while the effective third-order nonlinear coefficient of the nanoparticle system is

$$\chi^{(3)} = pf^2 \mid f \mid^2 \chi_\mathrm{p}^{(3)} , \tag{6.32}$$

with $\chi_\mathrm{p}^{(3)}$ the nonlinear coefficient of the metal nanoparticle. Thus, the effective nonlinear coefficient is proportional to $f^2 \mid f \mid^2$ and is greater when the dielectric permittivity of the matrix is increased.

6.2.3 Random Lasers

The term "random" itself and the fact that random lasers are lasers without cavities are intriguing concepts for photonics. The random laser is an unconventional light source where the feedback mechanism is no longer realized by two mirrors that enclose the amplifying material, but by the disorder-induced multiple light scattering, as evidenced in Figure 6.34. The light is trapped and scattered several times inside the disordered medium until it is amplified and quits the medium in a random direction. The disordered medium plays simultaneously the roles of the amplifying region and of the reflector cavity, the lasing medium being justifiably named mirrorless laser. The random laser is characterized by an incoherent feedback or by a coherent feedback, depending on the scattering process, which can provide, respectively, an intensity feedback or an amplitude feedback [76]. The incoherent feedback is nonresonant, whereas the coherent feedback is resonant. The disordered medium consists of nanoparticles or nanowires, and therefore the random laser is in fact a nanolaser.

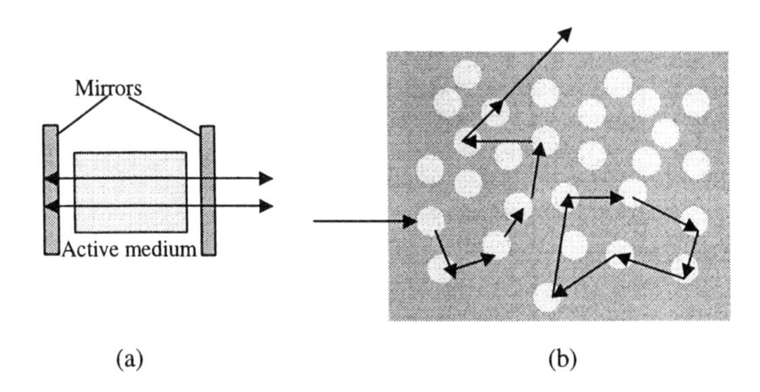

Figure 6.34 Feedback mechanisms in (a) a conventional Fabry-Perot laser, and (b) a random laser.

In Figure 6.34(b) two distinct scattering photon paths have been illustrated. The random, but closed, path, or recurrent scattering, corresponds to a random coherent laser, in which the photons return to the initial scatterer. The random free path that is not closed corresponds to a random incoherent laser, in which the photons are scattered enough times to amplify the incident radiation and are then ejected in a random direction.

The random laser can be characterized by several specific parameters. The scattering mean free-path, denoted by L_s, is the average distance for which the photons travel unscattered between two consecutive scattering events. The transport mean free-path L_t is the average distance of light propagation before its direction changes and becomes random [76]. These two characteristic lengths are related by

$$L_t = L_s / [1 - \langle \cos \theta \rangle] , \tag{6.33}$$

where $\langle \cos \theta \rangle$ is the average of the scattering angle, which has different values for different scattering mechanisms. For example, $\langle \cos \theta \rangle = 0$ in the case of Rayleigh scattering, whereas for the Mie scattering regime $\langle \cos \theta \rangle = 0.5$.

The gain and the amplification of the random laser are specified by another two parameters: the gain length L_g and the amplification length L_a, respectively. The gain length is the length after which the gain increases with e = 2.71, while L_a is the length between the beginning and the ending point of L_g. In a homogeneous medium $L_a = L_g$, but in a diffusive regime, as that specific for random scattering, $L_a = (DL_g / v)^{1/2}$ or

$$L_a = (L_t L_g / 3)^{1/2}, \tag{6.34}$$

where $D = vL_t / 3$ is the diffusion coefficient defined in terms of the transport velocity of light v. Light propagation in a random medium with size L has different regimes, similar to those encountered for charge carriers: 1) ballistic, when $L \le L_t$, 2) diffusive, when $L >> L_t >> \lambda$, and 3) localized, if $kL_s \cong 1$, where k is the wavenumber of light propagation. A review on these issues can be found in [77].

In a conventional laser the gain material amplifies the light due to the stimulated emission effect, and the cavity acts as a positive feedback mechanism, creating a balance between the optical gain and the loss of a resonant mode at a certain gain threshold. This threshold occurs only at a resonance frequency of the cavity because only at resonance the photons are trapped inside the cavity and the gain compensates the propagation losses. The photons are reflected by the cavity mirrors a certain number of times, the time spent by the photons inside a cavity being frequency dependent for conventional lasers. On the contrary, no such dependence exists in random lasers.

If we consider an active medium filled with scatterers, such that the photon mean free path is smaller than the dimensions of the scattering medium but much longer than the optical wavelength, the photon propagation is diffusive and the random laser with an incoherent feedback can be modeled by the equation

$$\partial S(\boldsymbol{r},t) / \partial t = D\nabla^2 S(\boldsymbol{r},t) + (v / L_g) S(\boldsymbol{r},t), \tag{6.35}$$

where $S(\boldsymbol{r},t)$ is the photon energy density. Assuming that the solution of (6.35) is separable and can be expressed as a sum of weighted exponentials (modes) of the form

$$S(\boldsymbol{r},t) = \sum_n p_n \phi_n(\boldsymbol{r}) \exp[-(Dq_n^2 - v / L_g)t], \tag{6.36}$$

and that the boundary conditions require that $\phi_n = 0$ at the edge of the random medium, the character of the solution of (6.35) changes from an exponential decay to an exponential increase when

$$Dq_1^2 - v / L_g = 0. \qquad (6.37)$$

If the scattering medium is formed from small spheres with a diameter $2R$, $q_n = \pi n / R$, and thus $q_1 = \pi / R$, while $q_1 = 1.73(\pi / a)$ for a scattering medium consisting of cubes with sides of length a. Because in both cases the eigenvalue q_1 is proportional with (scatterer dimension)$^{-1}$, (6.37) defines a critical volume

$$V_{cr} \propto R^3 \propto (L_t L_g / 3)^{3/2}. \qquad (6.38)$$

The fulfillment of the above relations is the condition that a photon generates at least another photon before leaving the random medium, and hence triggers a cascade process in which one photon generates two photons, two photons generate four photons, etc. However, the number of photons does not increase indefinitely because the gain saturates and the cascade process stops after a number of steps, the laser exhibiting a finite linewidth. Among the several types of incoherent lasers we mention the powder lasers, where the constituting particles are micron-sized, and the paint lasers, which consists of a methanol solution of dyes and TiO_2 particles with a diameter of 250 nm that play the role of the scattering centers [76]. Although these lasers work at room temperature and are optically pumped, it is rather difficult to fabricate them because it is challenging to prepare such powders or liquid suspensions of nanoparticles.

The random lasers with coherent feedback are produced using ZnO films with nanosized thickness, clusters formed from ZnO nanoparticles with an average diameter of 100 nm, or ZnO nanowires [76]. In this type of laser the photons are mainly localized in the random medium rather than display a diffusive transport as in the case of incoherent random lasers. As an example of a laser with coherent feedback, a sharp UV lasing at 380 nm attributed to Anderson localization of light in the random medium was observed in a ZnO nanocluster pumped at 266 nm by the fourth harmonic of a Nd:YAG laser [78]. The Anderson localization is due to quantum interference in the multiple-scattering regime and is characterized by an exponential spatial decrease of the light amplitude [77]. It is interesting to note that the transition between the coherent and incoherent feedback regimes of random lasers can be studied by varying the amount of scattering in the gain medium [76] using a dye solution containing ZnO nanoparticles. If this mixture of nanoparticles is optically excited at $\lambda = 532$ nm, for a ZnO nanoparticle density of 2.5×10^{11} cm^{-3} an increase of the pump intensity beyond a certain threshold produces a drastic spectral narrowing of the scattered light, and the emission peak increases significantly. This situation corresponds to an incoherent random laser. If the ZnO nanoparticle density is increased at 5×10^{11} cm^{-3} two thresholds appear successively as the pump intensity is increased, which is the signature of the transition between the incoherent and the coherent feedback regimes. Finally, at a ZnO nanoparticle density of 10^{12} cm^{-3} a series of discrete peaks appear as the pump intensity increases, these peaks forming a single linewidth at a certain threshold that is lower than in the case of incoherent feedback; this behavior characterizes a coherent feedback. A comprehensive review of theories dedicated to random lasers, and especially to random lasers with coherent feedback, can be found in [76].

REFERENCES

[1] Dragoman, D. and M. Dragoman, *Advanced Optoelectronic Devices*, Berlin: Springer, 1999.

[2] Grundmann, M. (ed.), *Nano-Optoelectronics*, Berlin: Springer, 2002.

[3] Arakawa, Y., "Progress in GaN-based quantum dots for optoelectronic applications," *IEEE J. Selected Topics Quantum Electron.*, Vol. 8, No. 4, 2002, pp. 823-832.

[4] Patella, F., et al., "Self-assembly of InAs and Si/Ge quantum dots on structured surfaces," *J. Phys.: Condens. Matter*, Vol. 16, No. 17, 2004, pp. S1503-S1534.

[5] Fafard, S., "Quantum-confined optoelectronic systems," in *Introduction to Nanoscale Science and Technology*, M. Di Ventra, S. Evoy, and J.R. Helfin (eds.), Boston: Kluwer Academic Publishers, 2004, pp. 443-483.

[6] Samsonidze, G.G., et al., "Interband optical transitions in left- and right-handed single-wall carbon nanotubes," *Phys. Rev. B*, Vol. 69, No. 20, 2004, pp. 205402/1-11.

[7] Saito, R. and H. Kataura, "Optical Properties and Raman Spectroscopy of Carbon Nanotubes," in *Carbon Nanotubes*, M.S. Dresselhaus, G. Dresselhaus, and Ph. Avouris (eds.), Berlin: Springer, Topics in Applied Physics 80, 2001, pp. 213-247.

[8] Bachilo, S.M., et al., "Structure-assigned optical spectra of single-walled carbon nanotubes," *Science*, Vol. 298, No. 5602, 2002, pp. 2361-2366.

[9] Hagen, A. and T. Hertel, "Quantitative analysis of optical spectra from individual single-wall carbon nanotubes," *Nano Lett.*, Vol. 3, No. 3, 2003, pp. 383-388.

[10] Guo, G.Y., et al., "Linear and nonlinear optical properties of carbon nanotubes from first principles," *Phys. Rev. B*, Vol. 69, No. 20, 2004, pp. 205416/1-11.

[11] Pedersen, T.G., "Exciton effects in carbon nanotubes," *Carbon*, Vol. 42, No. 5-6, 2004, pp. 1007-1010.

[12] Cao, A., et al., "Tailoring the optical excitation energies of single-walled carbon nanotubes," *Appl. Phys. Lett.*, Vol. 85, No. 9, 2004, pp. 1598-1600.

[13] Xu, J.M., "Highly ordered carbon nanotube arrays and IR detection," *Infrared Physics & Technology*, Vol. 42, No. 6, 2001, pp. 485-491.

[14] Liu, L. and Y. Zhang, "Multi-wall carbon nanotubes as a new infrared detected material," *Sensors and Actuators A*, Vol. 116, No. 3, 2004, pp. 394-397.

[15] Levitsky, I.A. and W.B. Euler, "Photoconductivity of single-wall carbon nanotubes under continuous-wave near-infrared illumination," *Appl. Phys. Lett.*, Vol. 83, No. 9, 2003, pp. 1857-1859.

[16] Steward, D.A. and F. Léonard, "Photocurrents in nanotube junctions," *Phys. Rev. Lett.*, Vol. 93, No. 10, 2004, pp. 107401/1-4.

[17] Set, S.Y., et al., "Ultrafast fiber pulsed laser incorporating carbon nanotubes," *IEEE J. Selected Topics Quantum Electron.*, Vol. 10, No. 1, 2004, pp. 137-146.

[18] Set, S.Y., et al., "Laser mode locking using a saturable absorber incorporating carbon nanotubes," *J. Lightwave Technology*, Vol. 22, No. 1, 2004, pp. 51-56.

[19] Wang, Y., et al., "Receiving and transmitting light-like radio waves: antenna effect in arrays of aligned carbon nanotubes," *Appl. Phys. Lett.*, Vol. 85, No. 13, 2004, pp. 2607-2609.

[20] Wu, X., et al., "Optical properties of aligned carbon nanotubes," *Phys. Rev. B*, Vol. 68, No. 19, 2003, pp. 193401/1-4.

[21] García-Vidal, F.J. and J.M. Pitarke, "Optical absorption and energy-loss spectra of aligned carbon nanotubes," *Eur. Phys. J. B*, Vol. 22, No. 2, 2001, pp. 257-265.

[22] Murakami, Y., et al., "Polarization dependence of the optical absorption of single-walled carbon nanotubes," *Phys. Rev. Lett.*, Vol. 94, No. 8, 2005, pp. 087402/1-4.

[23] Kempa, K., et al., "Photonic crystals based on periodic arrays of aligned carbon nanotubes," *Nano Lett.*, Vol. 3, No. 1, 2003, pp. 13-18.

[24] Margulis, V.A., E.A. Gaiduk, and E.N. Zhidkin, "Quadratic electro-optic effects in semiconductor carbon nanotubes," *Phys. Lett. A*, Vol. 258, No. 4-6, 1999, pp. 394-400.

[25] Chen, Y.-C., et al., "Ultrafast optical switching properties of single-wall carbon nanotube polymer composites at 1.55 μm," *Appl. Phys. Lett.*, Vol. 81, No. 6, 2002, pp. 975-977.

[26] Elim, H.I., et al., "Ultrafast absorptive and refractive nonlinearities in multiwalled carbon nanotube films," *Appl. Phys. Lett.*, Vol. 85, No. 10, 2004, pp. 1799-1801.

[27] Dragoman, D. and M. Dragoman, "Carbon nanotube zoom lenses," *IEEE Trans. Nanotechnology*, Vol. 2, No. 1, 2003, pp. 93-96.

[28] Krüger, A., M. Ozawa and F. Banhart, "Carbon nanotubes as elements to focus beams by Fresnel diffraction," *Appl. Phys. Lett.*, Vol. 83, No. 24, 2003, pp. 5056-5058.

[29] Margulis, V.A., E.A. Gaiduk, and E.N. Zhidkin, "Electric-field-induced optical second-harmonic generation and nonlinear optical rectification in semiconducting carbon nanotubes," *Optics Communications*, Vol. 183, No. 1-4, 2000, pp. 317-326.

[30] Mikheev, G.M., et al., "Giant optical rectification effect in nanocarbon films," *Appl. Phys. Lett.*, Vol. 84, No. 24, 2004, pp. 4854-4856.

[31] Dragoman, D. and M. Dragoman, "Terahertz fields and applications," *Prog. Quantum Electron.*, Vol. 28, No. 1, 2004, pp. 1-66.

[32] Misewich, J.A., et al., "Electrically induced optical emission from a carbon nanotube FET," *Science*, Vol. 300, No. 5620, 2003, pp. 783-786.

[33] Martel, R., et al., "Ambipolar electrical transport in semiconducting single-wall carbon nanotubes," *Phys. Rev. Lett.*, Vol. 87, No. 25, 2001, pp. 256805/1-4.

[34] Freitag, M., et al., "Mobile ambipolar domain in carbon-nanotube infrared emitters," *Phys. Rev. Lett.*, Vol. 93, No. 7, 2004, pp. 076803/1-4.

[35] Hartschuh, A., et al., "Simultaneous fluorescence and Raman scattering from single carbon nanotubes," *Science*, Vol. 301, No. 5638, 2003, pp. 1354-1356.

[36] Lefebvre, J., et al., "Photoluminescence from an individual single-walled carbon nanotube," *Phys. Rev. B*, Vol. 69, No. 7, 2004, pp. 075403/1-5.

[37] Maehashi, K., et al., "Chirality selection of single-walled carbon nanotubes by laser resonance chirality selection method," *Appl. Phys. Lett.*, Vol. 85, No. 6, 2004, pp. 858-860.

[38] Melechko, A.V., et al., "Vertically aligned carbon nanofibers and related structures: Controlled synthesis and direct assembly," *J. Appl. Phys.*, Vol. 97, No. 4, 2005, pp. 041301/1-39.

[39] Huang, Y. and C.M. Lieber, "Integrated nanoscale electronics and optoelectronics: Exploring nanoscale science and technology through semiconductor nanowires," *Pure Appl. Chem.*, Vol. 76, No. 12, 2004, pp. 2051-2068.

[40] Johnson, J.C., et al., "Single nanowire lasers," *J. Phys. Chem. B*, Vol. 105, No. 46, 2001, pp. 11387-11390.

[41] Johnson, J.C., et al., "Single gallium nitride nanowire lasers," *Nature Mater.*, Vol. 1, No. 2, 2002, pp.106-110.

[42] Duan, X., Y. Huang, and C. Lieber, "Single-nanowire electrically driven laser," *Nature*, Vol. 421, No. 6921, 2003, pp. 241-244.

[43] Maslov, A.V., and C.Z. Ning, "Reflection of guided modes in a semiconductor nanowire laser," *Appl. Phys. Lett.*, Vol. 83, No. 6, 2003, pp. 1237-1239.

[44] Maslov, A.V. and C.Z. Ning, "Modal gain in a semiconducting nanowire laser with anisotropic bandstructure," *IEEE J. Quantum Electron.*, Vol. 40, No. 10, 2004, pp. 1389-1397.

[45] Maslov, A.V. and C.Z. Ning, "Far-field emission of a semiconductor nanowire laser," *Optics Letters*, Vol. 29, No. 6, 2004, pp. 572-574.

[46] Barrelet, C.J., et al., "Nanowire photonic circuit elements," *Nano Lett.*, Vol. 4, No. 10, 2004, pp. 1981-1985.

[47] Wang, J.J., et al., "High-performance nanowire-grid polarizers," *Optics Letters*, Vol. 30, No. 2, 2005, pp. 195-197.

[48] Wang, J.J., et al., "Monolithically integrated isolators based on a nanowire-grid polarizers," *IEEE Photonics Technology Letters*, Vol. 17, No. 2, 2005, pp. 396-398.

[49] Sakoda, K., *Optical Pproperties of Photonic Crystals*, Berlin: Springer, 2002.

[50] Ramakrishna, S.A., "Physics of negative index metamaterials," *Rep. Prog. Phys.*, Vol. 68, No. 2, 2005, pp. 449-521.

[51] Perrin, M., et al., "Left-handed electromagnetism obtained via nanostructured metamaterials: Comparison with that from microstructured photonic crystals," *J. Opt. A*, Vol. 7, No. 1, 2005, pp. S3-S11.

[52] Noda, S., et al., "Full three-dimensional photonic bandgap crystals at near-infrared wavelengths," *Science*, Vol. 289, No. 5484, 2000, pp. 604-606.

[53] Hu, X. and C.T. Chan, "Photonic crystals with silver nanowires as a near-infrared superlens," *Appl. Phys. Lett.*, Vol. 85, No. 9, 2004, pp. 1520-1522.

[54] Melville, D.O.S., R.J. Blaikie, and C.R. Wolf, "Submicron imaging with a planar silver lens," *Appl. Phys. Lett.*, Vol. 84, No. 22, 2004, pp. 4403-4405.

[55] Law, M., et al., "Nanoribbon waveguides for subwavelength photonics integration," *Science*, Vol. 305, No. 5688, 2004, pp. 1269-1273.

[56] Lou, J., L. Tong, and Z. Ye, "Modelling of silica nanowires for optical sensing," *Optics Express*, Vol. 13, No. 6, 2005, pp. 2135-2140.

[57] Kind, H., et al., "Nanowire ultraviolet photodetectors and optical switches," *Adv. Mater.*, Vol. 14, No. 2, 2002, pp. 158-160.

[58] Flory, F. and L. Escoubas, "Optical properties of nanostructured thin films," *Prog. Quantum Electron.*, Vol. 28, No. 2, 2004, pp. 89-112.

[59] Gaponenko, S.V., *Optical Properties of Semiconductor Crystals*, Cambridge: Cambridge University Press, 1998.

[60] Nakwaski, W. et al., "Single-photon devices in quantum cryptography," *Opto-Electronics Review*, Vol. 11, No. 2, 2003, pp. 127-132.

[61] Lounis, B. and M. Orrit, "Single-photon sources," *Rep. Progr. Phys.*, Vol. 68, No. 5, 2005, pp. 1129-1179.

[62] Shields, A.J., et al., "Self-assembled quantum dots as a source of single photons and photon pairs," *Phys. Stat. Sol. (b)*, Vol. 238, No. 2, 2003, pp. 353-359.

[63] Bettotti, P., et al., "Silicon nanostructures for photonics," *J. Phys.: Condens. Matter*, Vol. 14, No. 35, 2002, pp. 8253-8281.

[64] Dragoman, D. and M. Dragoman, *Optical Characterization of Solids*, Berlin: Springer, 2002, pp. 362-371.

[65] Pavesi, L., S. Gaponenko, and L. de Negro (eds.), *Towards the First Si Laser*, NATO Sciences Series II, Mathematics, Physics and Chemistry, Vol. 93, Dordrecht: Kluwer Academics, 2003.

[66] Walters, R.J., et al., "Silicon optical nanocrystal memory," *Appl. Phys. Lett.*, Vol. 85, No. 13, 2004, pp. 2622-2624.

[67] Maier, S.A., et al., "Plasmonics – A route to nanoscale optical devices," *Adv. Mater.*, Vol. 13, No. 19, 2001, pp. 1501-1505.

[68] Maier, S.A., "Plasmonics – Towards subwavelengths optical devices," *Current Nanoscience*, Vol. 1, No. 1, 2005, pp. 17-22.

[69] Bouhelier, A., and G. P. Wiederrecht, "Surface plasmon rainbow jets," *Optics Letters*, Vol. 30, No. 8, 2005, pp. 884-886.

[70] Maier, S.A., P.G. Kik, and H.A. Atwater, "Optical pulse propagation in metal nanoparticle chain waveguides," *Phys. Rev. B*, Vol. 67, No. 20, 2003, pp. 205402/1-5.

[71] McFarland, A.D. and R.P. Van Duyne, "Single silver nanoparticles as real-time optical sensors with zeptomole sensitivity," *Nano Lett.*, Vol. 3, No. 8, 2003, pp. 1057-1062.

[72] Müller, J., et al., "Electrically controlled light scattering with single metal nanoparticles," *Appl. Phys. Lett.*, Vol. 81, No. 1, 2002, pp. 171-173.

[73] Andrew, P. and W.L. Barnes, "Energy transfer across a metal mediated by surface plasmon polaritons," *Science*, Vol. 306, No. 5698, 2004, pp. 1002-1005.

[74] Inouye, H., et al., "Ultrafast optical switching in a silver nanoparticle system," *Jpn. J. Appl. Phys.*, Vol. 39, Part I, No. 9A, 2000, pp. 5132-5133.

[75] Takeda Y., et al., "Control of optical nonlinearity of metal particle composites fabricated by negative ion implantation," *Thin Solid Films*, Vol. 464-465, No. 1, 2004, pp. 483-486.

[76] Cao, H., "Lasing in random media," *Waves in Random Media*, Vol. 13, No. 3, 2003, pp. R1-R39.

[77] Dragoman, D. and M. Dragoman, *Quantum-Classical Analogies*, Berlin: Springer, 2004, pp. 63-102.

[78] Cao, H., et al, "Microlaser made of disordered media," *Appl. Phys. Lett.*, Vol. 76, No. 21, 2000, pp. 2997-2999.

Chapter 7

Molecular and Biological Nanodevices

Organic materials have become quite widespread in electronic and optoelectronic applications for a variety of reasons that include lower costs, larger flexibility in designing materials with desired parameters, the possibility of fabricating circuit architectures and/or materials with parameters that are either better or different from those of semiconductor materials or heterostructures (see, for example, [1,2] and the references therein), and last, but not least, the possibility to fabricate electronic components such as thin-film transistors and ring oscillators on almost any substrate. The fabrication of such devices on flexible polyetherether ketone film and even on paper has been recently reported [3]. However, despite the number of published reports concerning organic nanoelectronic devices, they have not yet reached a breakthrough in applications, due, in part, to their general lower conductivity and mobility compared to semiconductor devices.

On the other hand, living cells face the same challenges as human technologies in that they must fabricate materials, convert, transmit, and make use of energy, generate motion, process and store information, and so on, based on principles that are sometimes totally different from the nanotechnologies and with goals that do not necessarily match those of humans. For example, the term "cost" certainly has a different meaning and the accuracy of cellular information processing is undoubtedly more important than speed. At the same time, the cell is not optimized for a specific function. Living cells have unique properties such as self-assembling, self-repairing, and self-replicating. Biomolecular responses are, in many cases, almost nondissipative and of a single-molecule precision and specificity, being superior to any present-day nanoelectronic sensors or actuators. An interesting review of the working principles of living cells focused on the lessons we can learn on how to develop systems and materials for nanotechnology can be found in [4]. Maybe the best example of the differences between biomolecules and inorganic matter is that, in contrast to the great majority of inorganic crystals, biomolecular crystals prefer only a few space groups. They have, in general, no inversion center or mirror symmetry, since these symmetry operations are prohibited in chirally pure biomolecular crystals. The biomolecules have a propensity to crystallize in space groups with screw axes, the number of observed structures increasing with the number of screw axes in the group [5].

Although organic or even biological materials differ from semiconductors, their conduction properties can be described similarly in terms of discrete energy levels and allowed or forbidden energy bands and, therefore, the devices based on them do not imply a different physics. From the point of view of nanoelectronics and nanooptoelectronics applications the interest in these materials is mainly focused on single-molecular or few-molecule devices, which are believed to herald a new revolution in miniaturization. On the other hand, biological systems such as molecular motors function on quite different principles than micro or macro physical systems constructed by humans, and thus can inspire us in designing mechanical or electrical devices with unique functionalities. Moreover, in compu-ting applications biological systems with large number of molecules are attracting interest due to the possibility of performing parallel computation in a manner that is different from either quantum or classical computing. Biological information processing, which is the basis of life, should aid us in designing a new type of computing machine. In view of these considerations, the focus of this chapter is on few- or single-molecular electronic and optoelectronic devices, on electronic or optoelectronic devices inspired from biological systems, on organic devices and architectures with no counterpart in semiconductor technology, and on biological computing techniques, which can either mimic classical computation algorithms or establish new classes of algorithms for efficient parallel computation. The chapter ends with a few examples of integration of biodevices and nanodevices.

7.1 CHARACTERIZATION AND MANIPULATION OF MOLECULAR SYSTEMS

The characterization and manipulation of molecular and biological systems are fascinating interdisciplinary research areas requiring in-depth studies of biology, chemistry, and physics. A review of this topic, even a brief one, would not only be lengthy but would require the introduction of a specific vocabulary that would no doubt appear as exotic to most scientists not involved directly in this area of research. To keep the discussion as simple as possible we will not deal with the details of manufacturing molecular systems, which is mostly a chemical rather than a physical process, but will focus on the characterization of their conduction properties and their manipulation at either the single-molecule level or at the level of assembling complex architectures.

7.1.1 Electrical Conduction of Molecules

The concepts that allow the description of conduction properties of semiconductor materials and devices can be modified in order to accommodate molecular devices [6,7]. For example, the filled and empty energy levels in semiconductors, which determine the position of the Fermi level, are replaced in molecular devices by the discrete highest occupied molecular orbital (HOMO) and lowest unoccupied molecular orbital (LUMO), respectively, whereas conjugated linear polymeric systems are used instead of metal interconnects and degenerate contacts. The Fermi level modification in semiconductors by doping is substituted by the change in ionization potential and electron affinity of molecules through chemical substitution, which modifies the energy and geometry of molecular orbitals, whereas bandgap engineering of semiconductors becomes designing molecular orbitals. In this respect, strong σ-bonds act as tunnel barriers, while π-conjugation, which denotes overlap and delocalization of electron orbitals, offers a low barrier path for electron transport. π-conjugated molecules, such as oligo(phenylene-ethynylene)s, polyporphyrins, polyphenylenes, or polythiophenes, have HOMO-LUMO gaps. The molecules are conductive if delocalized molecular orbitals span the molecule around the Fermi level, and are nonconducting if the molecular orbitals around the Fermi level are localized and do not span the molecule. In favorable cases, electrons can travel through LUMOs and holes through HOMOs. Not only electrical, but also mechanical, optical, and other properties, can be modified by synthetic chemistry, that is, by controlling the placement of individual atoms and functional groups on molecules.

Molecular conduction is defined and measured in a metal-molecule-metal geometry, in which electrons from the metal enter the molecule through tunneling. Metal-molecule-metal junctions with reproducible connections to both ends of the molecule are, however, still difficult to fabricate. Electron transmission at a metal-molecule interface is determined by the alignment of the metal Fermi level relative to the HOMO and LUMO levels, barriers analogous to Schottky barrier in semiconductors forming at interface and determining the transport mechanism (tunneling, defect-mediated transport, or thermionic emission). The conduction of molecules is generally described by a coherent resonant tunneling process when the energy of molecular orbitals is in resonance with tunneling electrons, or by a coherent nonresonant tunneling process when the electronic states of the molecules are energetically well separated from the tunneling electrons. In the first case, the rate of electron transport is independent of the molecule length and is dominated by contact scattering, while in the second case the rate of electron transport decreases exponentially with the length of the molecule L [7], the tunnel transparency decaying as $T = T_0 \exp(-\gamma L)$ with $\gamma = 0.4$ Å^{-1} [8]. Because $\gamma = 2$ Å^{-1} in a vacuum gap, it was supposed that the tunneling electrons can be guided by long molecular wires with an effective cross section below 1 nm^2 (smaller than the de Broglie wavelength of an electron). This supertunnel phenomenon can be interpreted as a permanently driven superexchange electron transfer. In the superexchange mechanism, which describes the conduction through a single molecule or through small assemblies of molecules that bridge two reservoirs of charge carriers (metallic leads) when the difference between the bridge energy and the Fermi level is larger than $k_B T$, the bridge mediates the electronic coupling (tunneling) between the leads so that the bridge population is negligible. On the other hand, when the difference between the bridge energy and the Fermi level is smaller than $k_B T$, the bridge states are populated by charge carriers and conduction takes place by diffusive hopping of carriers on the bridge [9].

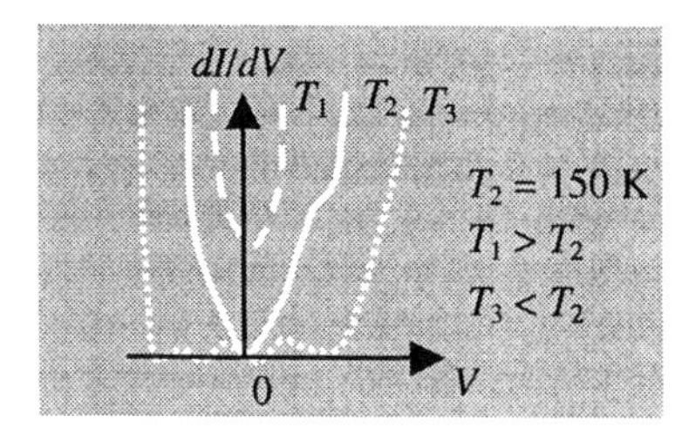

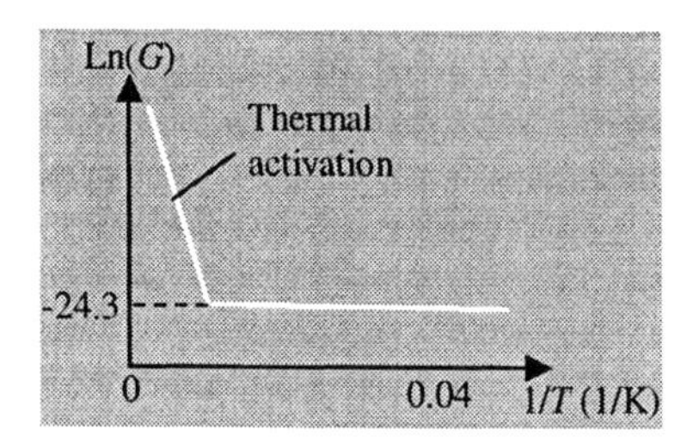

Figure 7.1 Typical temperature dependence of the differential conductance as a function of bias for a metal-1-nitro-2,5-di(phenylethynyl-4'-mercapto) benzene-metal junction (left) and the correspondence Arrhenius plot of zero-bias conductance (right) (*After*: [9]).

The simplest way to adjust the energy alignment across the molecular bridge is to apply a potential on the metallic leads of the metal-molecule-metal configuration. In particular, measurements on the Au-1-nitro-2,5-di(phenylethynyl-4'-mercapto) benzene molecule-Au system in the temperature interval 10–300 K revealed that the transition between the coherent tunneling and the thermally activated incoherent hopping conduction mechanisms is determined by the vibrational temperature of the molecule [9]. As the temperature increases, the current through the junction increases, which suggests the presence of a thermally activated conduction mechanism in parallel to superexchange (in the superexchange mechanism the current is independent of temperature). The differential conductance of a typical junction has a low value region around zero bias at 150 K that widens as the temperature decreases and that increases with temperature above 150 K, as can be seen from Figure 7.1. The barrier for the thermally activated regime, obtained from an Arrhenius plot of zero bias conductance, is about 56 meV, which corresponds to rotational barriers between adjacent phenyl rings in the molecule. As the temperature increases, the molecule changes its configuration from a coplanar ring to one in which the rings fluctuate around the optimized zero value dihedral angles. This conformational change leads to the suppression of coherence and to the encouragement of hopping.

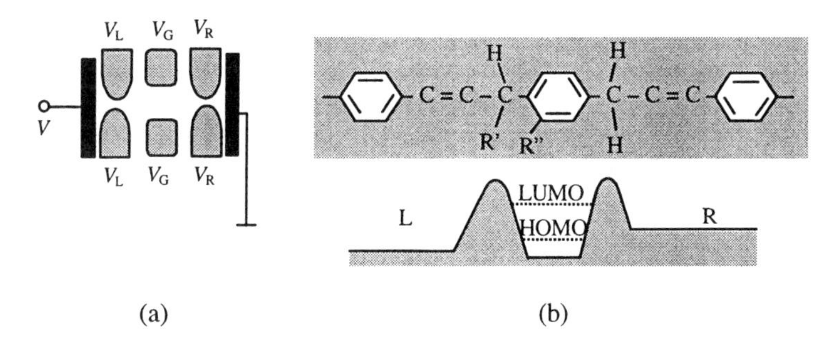

Figure 7.2 (a) Schematic diagram of a quantum dot and (b) the equivalent structure of a molecular wire (*After*: [10]).

To evidence the similarities and differences between semiconductor and molecular conduction, we have schematically represented in Figure 7.2 the equivalent molecular structure of a quantum dot. In a quantum dot defined by gates the height of the energy barriers between the well and the left and right reservoirs is determined by the applied voltages V_L and V_R, and the gate voltage V_G, the chemical potentials of the reservoirs fix the number of electrons in the dot at equilibrium, while the bias voltage V controls the direction of the current. In the molecular wire depicted in Figure 7.2(b), the barriers between the well and electron reservoirs are formed by σ bonds, which are intercalated in the π bonded backbone of the wire, the height of the left barrier and the well energy levels being determined by the functional groups R' and R", respectively [10]; the dot structure that forms in the molecular wire has a scale of a few angstroms rather than tens of nanometers. As in a quantum dot, the HOMO and LUMO energy levels in the well formed in the molecular wire are discrete. The molecule has a rectifier behavior if the barrier structure is asymmetric.

Molecules, unlike quantum dots, are robust at room temperature. On the other hand, high current levels through molecules, especially through biological molecules, can degrade their performances and functionalities; one reason is excessive heat dissipation. A rigorous and straightforward method to calculate the voltage drop, charge transfer and, $I-V$ characteristic of molecular conductors is described in [11]. This method partitions the structure into contacts and the molecular device and combines in a self-consistent way the nonequilibrium

Green's function formalism with standard methods of quantum chemistry. However, the metal-molecule contact results from mixing the end part of the molecular wire orbitals (HOMO, LUMO) with the surface metallic orbitals of contacts, a phenomenon that can be seen as a doping effect without introducing states in the HOMO-LUMO gap. The resistance of the metal-molecule-metal junction depends on the chemical structure of the molecule, its chemical binding to the electrodes and its conformation in the junction [8]. Unlike in electronic circuits, where adding a branch does not change the electronic properties of the others, adding a new branch to a molecule creates a new molecule with a different electronic structure, So, Ohm's and Kirchhoff's laws are inapplicable for intra-molecular circuits, which form with molecular wires and without metallic pads. In particular, the conductance of a molecular circuit composed of a molecule with two branches bonded in parallel and connected to the pads through molecular wires, as depicted in Figure 7.3, is given by

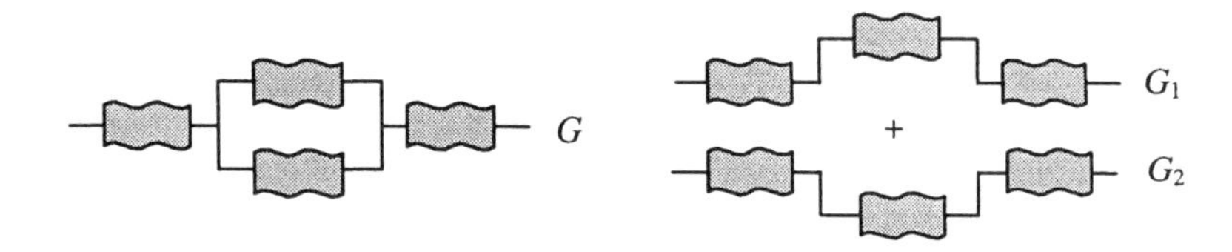

Figure 7.3 The conductance G of a molecule with two branches bonded in parallel and connected to the pads through molecular wires (left) can be expressed in terms of the conductances of the two complete individual branches (right) (*After*: [8]).

$$G = G_1 + G_2 + 2(G_1 G_2)^{1/2}, \tag{7.1}$$

instead of the standard rule of parallel association of conductance $G = G_1 + G_2$, where G_1 and G_2 are the conductances of the complete individual branches in which the metal-molecule-metal junction is decomposed according to Figure 7.3 [8].

The electronic transport properties of a molecular wire are much more dramatically affected if contacted with low-dimensional leads, such as CNTs, than if contacted with bulky electrodes [12]. Calculations in a tight-binding approach show that the 1D conductance of a molecular wire, modeled as a homogeneous linear chain, placed between two semi-infinite armchair (l,l) CNT leads, seen as donor and acceptor electron reservoirs, is dominated by topology properties. It depends on the symmetry of the channel wave function transverse to the transport direction and obeys a universal scaling law $G = G(\Gamma M^{1/2})$ in the multiple-contact configuration, where Γ is the local contact strength and M is the number of hybridization contacts between the CNT and the molecular bridge; M can vary between 1 and $2l$.

The importance of the molecule-electrode contact in determining the $I-V$ characteristic of a molecular device is evidenced also in [13], where it is demonstrated that the shape of the $I-V$ characteristic of a benzene ring placed between two electrodes is determined by the molecule, but the absolute magnitude of the current depends on the nature of atoms at the molecule-electrode contact.

First-principle calculations suggest that a benzene ring connected to two bulk electrodes can operate at very large electric fields without breakdown [14]. In conventional microelectronics, electromigration produces current-induced breakdown, but, up to 5 V, current-induced forces do not significantly affect the current through a 8-Å-long benzene molecule connected to two leads as atomic gold wires break at smaller voltages. At larger voltages, the contact depleted of electrons weakens, the S-metal bond expands acting as an extra barrier for tunneling electrons, and thus induces a significant reduction of the current, allowing for large-bias operation of molecular wires. The benzene-1,4-dithiolate molecule, which is attached to metallic leads through the sulfur end atoms, undergoes dynamic changes in structure, including twists around an axis perpendicular to its plane and breathing oscillations at resonant tunneling via antibonding states. However, the strong σ bonds in carbon-based molecular structures render them more resistant to current induced-forces than, for example, atomic gold wires.

For molecular circuitry self-assembly molecular layers are essential, in particular self-assembled monolayers (SAM), which are complete monolayers of adsorbates in which a nonreactive tail functionality prevents the growth of another layer. In self-assembly, molecules bind selectively to specific transition metals for electrical contact through clip groups such as disulfides, isonitriles, carboxylic acid, or the sulfur headgroup of thiolates; the latter adsorbes to Au{111} surfaces to form SAMs but not to a Si substrate, while some silane groups form SAMs on SiO_2 [7]. The growth of well-ordered, energetically favorable alkanethiol SAMs, for example, is driven by the strong covalent bond between Au and the linking S group that binds the molecule to the surface, and by the van der Waals interactions between adjacent hydrocarbon tails that induce close packing of hydrocarbon chains. Alkanethiolate SAMs are stable but dynamic, the adsorbed molecules being continuously exchanged with other thiol species in solutions, especially at SAM defect sites. The electrical properties of self-assembled molecular wires depend dramatically on their orientation, packing, and order [15]. The competing forces caused by electric field, intermolecular interactions, substrate-molecule chemisorption, and tip-molecule physisorption influence the $I-V$ characteristics measured with scanning tunneling spectroscopic techniques.

7.1.2 Measurement Techniques for Molecular Conduction

There are various methods used to fabricate molecule-metal junctions and to measure the conductivity of molecules, which include, but are not limited to, scanning probe microscopy, the manufacture of nanopores and mercury-drop junctions, the methods of crossed wires, nanoparticle bridges, and mechanically controllable break junctions [7].

The scanning probe microscopy method (see Chapter 2) allows the analysis of a few or even a single molecule using different techniques: 1) STM, in which the topography of the SAM is measured by scanning a tip over a SAM in the constant-tunneling-current regime or by measuring the tunneling current between a single molecule in the SAM and the STM tip, 2) the conductive probe AFM (CP-AFM) technique, in which a metal-coated AFM tip is in direct contact with the SAM under a controlled load, the resulting tip-SAM contact area enclosing about 75 molecules (this method is suitable for measuring the electric properties of thin films), and 3) nanoparticle coupled CP-AFM, in which a bifunctional molecule M with thiol groups on each end, inserted in an insulating SAM, anchors Au nanoparticles on the tip of the AFM to the surface via covalent bonds (as shown in Figure 7.4(a)), the metal-molecule-nanoparticle structure being able to address single molecules. However, if the Au nanoparticle is in contact with more than one molecule the current increases since each molecule contributes independently and equally to the current.

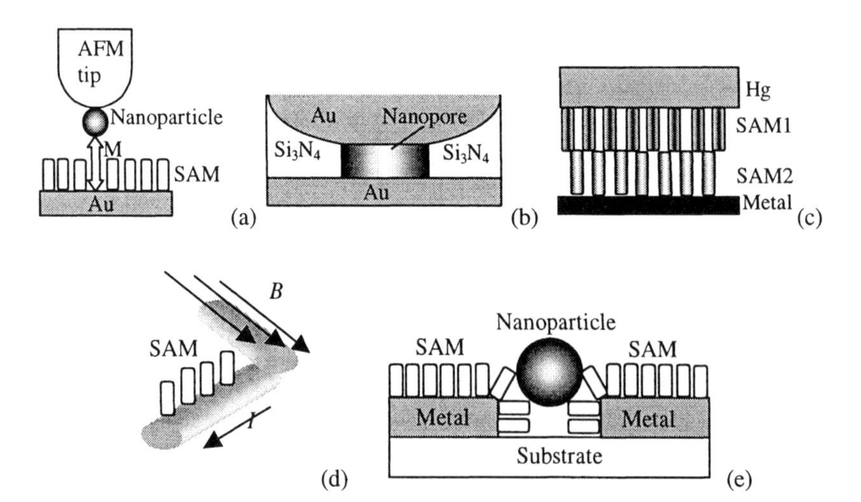

Figure 7.4 Different methods to measure molecular conductivity: (a) nanoparticle coupled CP-AFM, (b) nanopore, (c) mercury drop junction, (d) crossed wires, and (e) bridging molecules (*After*: [7]).

The method of nanopore fabrication is suitable for measuring the conduction through a small number (up to thousands) of organic molecules. The nanopore, as illustrated in Figure 7.4(b), is a SAM of conjugated molecules placed between two electrodes. It is fabricated by evaporating an Au contact on top of a 30–50 nm wide aperture in a suspended silicon nitride membrane, after which a SAM forms in the device immersed in a solution of molecules. Finally, a bottom Au electrode evaporates onto the sample and seals it.

In the mercury-drop junction method a mercury drop is used as an electrode on which a thiol-based SAM develops, the final junction being created by forming a mechanical contact between the SAM supported on the suspended Hg drop, denoted as SAM1 in Figure 7.4(c), and a SAM on a solid surface, labeled SAM2. The resulting structure is Hg-SAM-SAM-metal; variants include Hg-SAM-SAM-Hg and Hg-SAM-semiconductor.

When the crossed wires method is used for measuring the conductivity, the metal-molecule-metal junction forms between a wire functionalized with a SAM and another perpendicular wire, as can be seen in Figure 7.4(d). The wire spacing is controlled by the Lorentz force that appears in the presence of an applied magnetic field B due to the dc current I in the wire perpendicular to the magnetic field.

Metallic nanoparticles can also bridge the gap between electrodes functionalized by SAM of desired molecules, forming a metal-molecule-nanoparticle-molecule-metal structure, as shown in Figure 7.4(e). The nanoparticle becomes trapped over the gap if an ac electric field is applied across the contacts.

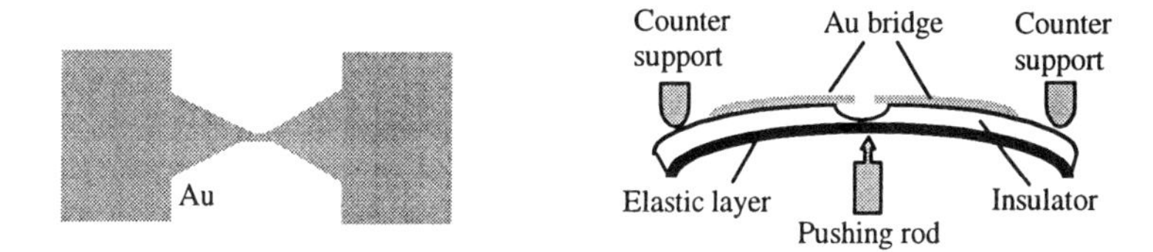

Figure 7.5 Top view (left) and side view (right) of the mechanically controllable break-junction technique (*After*: [17]).

In the mechanically controllable break junction method, a notched metal wire (generally Au) is attached to a flexible substrate, which is bent with a three-point bending mechanism, as shown in Figure 7.5, until the notch fractures and a gap with a controllable separation is produced. This gap acts as an adjustable tunneling junction [16]. Before breaking, one or a few atomwide chains of metals form, after breaking sharp metallic tips being available for molecule attachment. After fracture a molecule is deposited on the two open-end electrodes and forms a SAM on each electrode, the electrode separation being adjusted until a single molecule bridges the gap. The spacing between electrodes can be controlled up to the incredible precision of 10 pm [17]. This method is difficult to adapt for long and flexible molecules since multiple contacts between other molecules in the junction can develop, producing unwanted parallel currents. Moreover, this method is not suitable for the fabrication of molecular transistors since it is difficult to place gate electrodes on the structure.

As demonstrated by recent experiments, in mechanically controllable break-junctions the contact geometry and hence the contact resistance and/or the nature of charge transport between the molecule and electrodes can be varied [18]. The bridge can be broken repeatedly and controllable by applying, for instance, a high enough voltage between the Au pads, but the contact is regenerated with the constriction atoms arranged differently when the pushing rod of the bending mechanism is withdrawn.

Molecules that have a thiol group at the end can be contacted with electrodes, as described in [19]. This method of fabricating metal-molecule-metal junctions approaching the single molecule limit consists of depositing the source and drain electrodes on different faces of a square tip with a diameter in the 20–30 nm range. The gate electrode is first deposited on the top face of the tip, then the drain electrode is deposited on one face, and the tip is dipped into a solution of dithiol molecules until a SAM is formed on the drain. The source electrode is subsequently deposited on the opposite side of the tip until the desired conductance is detected. At low temperatures the conductance of the molecular junction varies in steps, which are periodic in the source-drain voltage, indicating resonant tunneling through coupled electronic and vibration levels. The position of the steps can be tuned by a gate voltage. The device conductance, especially for longer molecules, diminishes when the junctions are exposed to air.

7.1.3 Engineering Electrical Properties of Molecules

Unlike semiconductor materials, the electrical properties of molecular systems can be quite easily engineered. One of the most promising molecules in nanoelectronics is DNA (deoxyribonucleic acid). Each strand of DNA is about 2 nm wide and consists of a linear chain of four possible bases on a backbone of alternating phosphate ions and sugar molecules. The bases are called adenine (A), cytosine (C), guanine (G), and thymine (T). A unit consisting of a sugar molecule, a phosphate, and a base is called a nucleotide. Two single strands of DNA hybridize if the sequences of their bases are complementary, that is, every A base opposes a T base and every G base opposes a C base; this property confers DNA molecular recognition ability. Complementary bases in different strands bind through hydrogen bonds, the spacing between neighboring base pairs in a DNA double helix being 0.34 nm. If a double-stranded DNA is heated above the melting temperature, which depends on the ion concentration of the ambient and the G-C content in the sequence, the two strands separate into single strands. The DNA can be produced in large amounts and at a high quality and low price, as, for example, by polymerase chain reaction. It is also possible to generate any possible base sequence and to copy specific sequences at high speed.

There is no general agreement on the conductive behavior of DNA molecules. Insulating, semiconducting, highly conductive, and even proximity-induced superconducting behaviors have been reported in various surrounding conditions, with various DNA sequences and lengths (see the references in [20,21]). Experiments suggest that over short distances the conduction mechanism is hole hopping from one G-C base pair to the next over distances of up to 40 base pairs. In [22] an insulating behavior of double-stranded DNA molecules between nanofabricated electrodes was reported in either single molecules or small bundles of DNA, irrespective of the base-pair sequence, the substrate (mica or SiO_2), the electrode material (gold or platinum), the electrostatic doping fields, or the length between contacts (40–500 nm). The lower bound for the resistance of a DNA molecule was found to be 10 TΩ. On the contrary, for shorter double-stranded (G+C)-rich DNA of about 10.4 nm, placed in either vacuum or air, large-bandgap semiconducting behavior was reported in [23] down to cryogenic temperatures. It is believed that electron transfer through DNA is based on the overlap between π orbitals in adjacent base pairs and that irregular base-pair sequences tend to localize charge carriers and hence reduce the electron transfer rate.

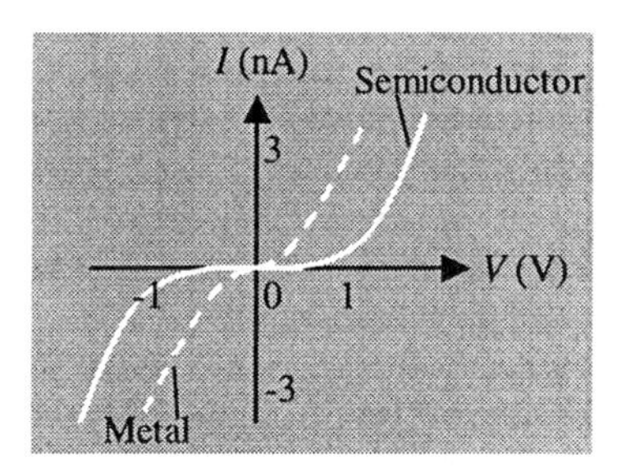

Figure 7.6 Typical *I-V* characteristic for a semiconducting (solid line) and a metallic (dashed line) DNA molecule (*After*: [24]).

Besides the advantage of self-assembly and molecular recognition ability, most important for DNA applications in nanoelectronics is the possibility of engineering DNA such as to modify its conductivity. For example, if the imino proton of each base pair in the phage λ-DNA is replaced by a metal ion the semiconducting DNA with a bandgap of a few hundred meV at room temperature changes to a metallic-like conductor [24]. The metal ions in [24] are Zn^{2+} spaced by about 4 Å in a 15 μm long λ-DNA molecule with sticky ends that bind to electrodes spaced by 10 μm. λ-DNA is a molecule produced in bacteria with a length of 48,000 bases and a known base pair sequence with no characteristic pattern. Typical $I-V$ characteristics for metallic and semiconducting bundles of DNA are represented in Figure 7.6; the semiconducting behavior is evidenced by the plateau in the $I-V$ characteristic around zero bias, which indicates the presence of a conductance gap.

The possibility of engineering the conductivity of DNA opens the prospect of creating all-DNA junction devices. Molecular *p-n* junctions can be created since (G+C)-rich DNA shows *p*-type conduction, while (A+T)-rich DNA shows *n*-type conduction. DNA enriched with (G+C) sequences can be obtained by oxygen hole doping, which increases the conduction with several orders of magnitude [25]. Moreover, the low conductivity of native DNA can be increased by several techniques developed to combine metal and DNA, which range from the inclusion of a few metallic atoms to complete metal coverage of the DNA [20].

Doping with nonbiological molecules such as fullerenes can be used to engineer the electrical properties of other materials, such as carbon nanotubes (CNTs). For example, the conductivity of semiconducting single-walled CNTs placed between semi-infinite Au(111) electrodes can be modified by encapsulating a C_{60} fullerene molecule in them [26]. Tight-binding simulations and the Landauer-Büttiker formalism show that the resulting peapod structure (see Figure 7.7) has an enhanced electrical conductivity due to charge transfer from the nanotube to the C_{60} molecule. The C_{60} encapsulation does not change significantly the geometry of either the nanotube or the fullerene, but the current increases with about 6% for a (17,0) CNT with a diameter of 13.48 Å and a length of 16.33 Å biased with 0.4 V, the difference in current between the pristine and peapod nanotubes remaining roughly constant for larger bias. This difference in current is due to a transfer of about 0.62 electrons from the nanotube to C_{60}, in (17,0) peapod structures holes being the primary carriers.

Figure 7.7 Schematic representation of a peapod structure consisting of a C_{60}-doped CNT (*After*: [26]).

7.1.4 Manipulation of Single Molecules

Another area of rapid progress in molecular engineering is manipulation with an STM tip, which induces chemical reaction steps such as adsorption, re-adsorption, dissociation, diffusion, and bond formation at a single molecule level with atomic scale precision; this topic is reviewed in [27]. STM manipulation of molecules is based on the ability to control applied electric fields, tip-adsorbate interaction forces, or tunneling currents. Tunneling electrons can break molecular bonds when the tip is positioned above the location of the bond at a fixed height. Tunneling electrons are then injected into the molecules, by applying a pulsed bias, when the transferred energy is larger than the specific bond-dissociation energy. Larger organic molecules can thus be selectively broken into smaller building blocks since different bonds have different dissociation energies. The dissociation rate depends on the tunneling current.

To relocate molecules to an assembling place across a surface it is necessary to increase the tip-adsorbate interaction force by approaching the tip towards a target particle and then moving the STM tip to the desired position. This lateral manipulation is completed when the adsorbate is left at the desired position on the surface after the tip retracts to the normal image height. Depending on the nature of the tip-adsorbate interaction the lateral manipulation can be performed in pushing (for repulsive interaction), pulling (for attractive interaction) or sliding modes (when the molecule is bound to the tip or trapped under the tip). In large molecules different tip-molecule interactions can exist at various parts so that lateral manipulation can also be used as a method to study the conformational changes or the mechanical stability of a molecule. Molecular building blocks can also be relocated on a surface by using vertical manipulation, in which single atoms or molecules are transferred between the tip and the substrate, and viceversa, through the application of electrical fields, voltage pulses, or mechanical contact between the tip and the molecule. This type of lateral manipulation is useful if the surface has obstacles that cannot be overcome by lateral manipula-tion. Transferring an atom or molecule to the tip apex functionalizes the tip in the sense that it improves the tip sharpness and hence the image contrast. The vertical manipulation can also join different molecular fragments since it forms a bond between the molecule and the substrate or the tip during transfer and re-deposition processes.

So, STM techniques can be used to construct new molecules. For example, to construct a $C_{12}H_{10}$ molecule at 20 K two C_6H_5 building blocks were first prepared by breaking the C-I bond in a C_6H_5I molecule. The I is transferred to the tip apex by vertical manipulation, increasing the image contrast, while the phenyl C_6H_5

fragments, bonded after dissociation to the step-edge Cu atoms at a Cu(111) step-edge are brought in close proximity by lateral manipulation and then 0.5 V voltage pulses are applied to join the phenyl couple (see [27] and the references therein). These processes are schematically represented in Figure 7.8.

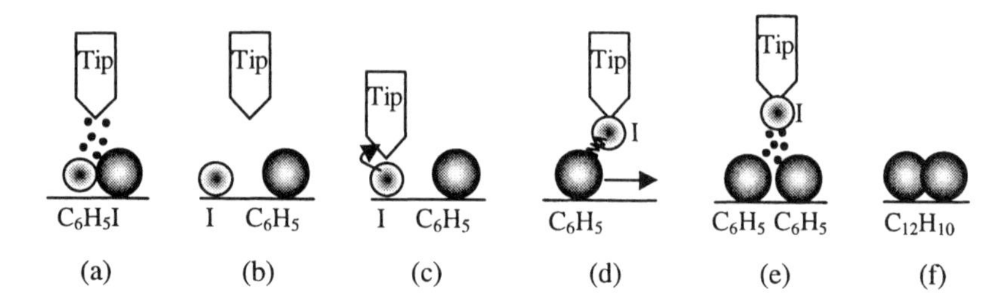

Figure 7.8 Stages of STM fabrication of a $C_{12}H_{10}$ molecule: (a) injection of tunneling electrons on a C_6H_5I molecule (b) breaks the C-I bond, then (c) the I atom is transferred to the STM tip, (d) the functionalized tip pulls the phenyl C_6H_5 molecule into the desired location, where (e) by injecting tunneling electrons (f) two phenyl molecules are bonded (*After*: [27]).

The manipulation of single molecules, such as stretching, injection of functional macromolecules into living cells, or DNA pullout from chromosomes, can also be done with AFM in combination with an optical microscope and a nanomanipulator. The AFM is also a method for measuring the intra- and inter-molecular forces, in particular the binding force of receptor and ligand molecules on the cell surface [28].

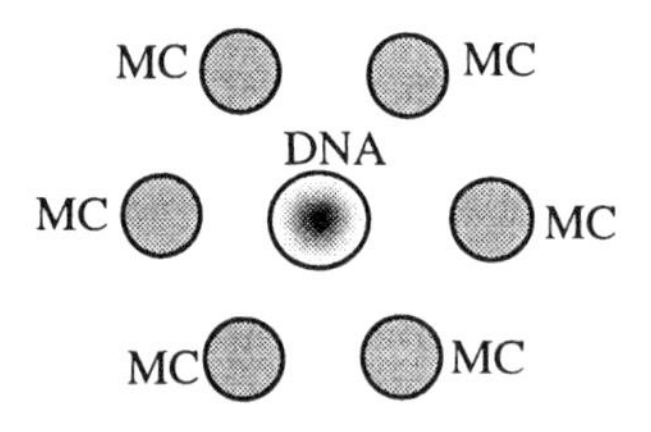

Figure 7.9 DNA manipulation with a magnetic tweezer composed of six microcoils integrated on a platform in a microfluidic channel (*After*: [30]).

Single DNA molecules end labeled with paramagnetic particles with 3 μm diameter can be manipulated with transverse magnetic tweezer systems, consisting of small permanent magnets in solution that can be placed as close as 10 μm from the molecule in the focal plane of the microscope objective [29]. Large and constant forces can thus be applied on the DNA molecule, the molecule being fixed in desired positions by micropipettes while pipette bending indicates the amount of tension. Another more sophisticated magnetic tweezer is a micromachined platform designed for controlling the stretching and rotation of single DNA molecules through magnetic forces produced by arrayed microcoils [30]. The DNA molecule in a fluidic channel is bonded with one end onto a gold surface while the other end is bonded onto a magnetic bead and thus can be manipulated by the magnetic field of hexagonally aligned microcoils, denoted by MC in Figure 7.9. Rotation of the DNA molecule, for example, is accomplished by circular permutation of the currents in the microcoils. The coils, which are 50 μm wide, 20 μm thick, and spaced by 30 μm, generate a 2D magnetic field gradient of 1.25 T/m at the hexagon center for the movement of the magnetic bead, corresponding to a pN force for DNA manipulation. Experiments with λ-DNA confirmed the functionality of the platform.

Electronic nanotweezers can be also used to manipulate DNA molecules since DNA is charged at neutral pH [31]. The manipulation is based on the phenomenon of dielectrophoresis (DEP), which designates the response of the molecule in a dc or an ac electric field gradient that causes a force on any (charge or neutral) polarizable object. To manipulate single molecules micro- or nano-electrodes with a planar or castellated geometry are needed for particles experiencing positive DEP (i.e., particles that move toward higher electric field regions), while quadrupoles are employed for particles with negative DEP, which move toward smaller electric field regions.

7.2 MECHANICAL DEVICES BASED ON MOLECULAR SYSTEMS

7.2.1 Molecular Motors and Quantum Ratchets

Molecular motors are crucial for life-sustaining motions since they support sustained, efficient, and directional movements of cellular components within cells (allowing the formation and maintenance of complex structures), or of entire cells in space, to distant reaction sites. In a certain sense, the motion of entire organisms can be viewed as an end result of molecular motors. Motorized transport of cells and cellular components is more efficient than diffusion, especially over distances of more than a few microns. There are several types of molecular motors [32], such as linearly moving cytoskeletal motors, rotary motors,and track, laying motors. Cytoskeletal motors can be divided into three families: myosin, kynesin, and dynein motors. These are characterized by a unique and conserved amino acid sequence that binds to the ATP (adenosine triphosphate) molecule and converts the ATP chemical energy into mechanical force. Myosin motors move along actin filaments; they drive muscle contraction and are involved in cell motility, vesicle transport, cytokinesis, and so forth. Kinesin and dynein motors move along microtubules, actuate the directional motility of chromosomes, membranous vesicles, RNA (ribonucleic acid), and so on, and can even power the beating of flagella and cilia. An example of a rotary motor is the F1 ATPase, which spins in mitochondrial membranes generating ATP and is powered by a proton gradient; if ATP is supplied, it can be driven backwards. DNA and RNA manipulating enzymes are examples of track-laying motors since they move along the track that they synthesize.

Most of the molecular motors can be viewed as stepping machines that move along their track through a series of binding and unbinding reactions accompanied by shape changes. The majority of molecular motors have a heads-on-a-stalk configuration in which the head contains track-binding functions and the stalk recognizes and binds to other motors, to cargo, and so on [32]. As shown schematically in Figure 7.10(a), the molecular motor diffuses back and forth under Brownian motion until it binds to a site in the progress direction along the track. Then, as illustrated in Figure 7.10(b), it changes conformation during the stage of strong binding driven by the turnover of ATP, for example. In the last step, shown in Figure 7.10(c), the ATP regenerates the weak binding state and supplies the energy required to detach the motor from the track, making it ready to be reattached to a new site along the progress direction. For example, the energy of one ATP molecule is sufficient for a kinesin motor to walk an 8 nm step along the microtubule protofilament axis. Coordinated and repetitive stepping depends on the interrelation of chemical and mechanical events, in the sense that strain or pulling in the opposite direction of progress tends to stabilize the motor attachment to the track, while pulling in the progress direction tends to detach the motor from the track. Coordinated action of several motors is strain dependent. The exertion and sensing of strain are done using lever arms.

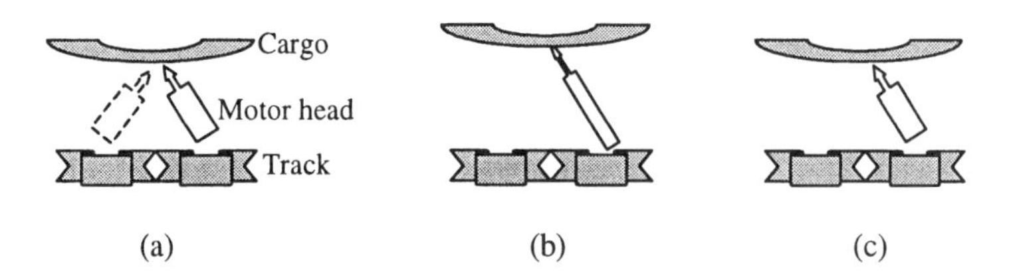

Figure 7.10 Diagram of conformational switching in a molecular motor: (a) Brownian diffusion followed by binding to a site in the progress direction along the track, (b) conformation changes during the stage of strong binding, and (c) detachment of motor from the track during the weak binding state (*After*: [32]).

The duty ratio of a molecular motor is defined as the ratio between the time spent in strong binding states, during which the motor is attached to the track and can exert and support force, and the time spent in weak binding states, in which the motor tends to detach from the track. Kinesin motors that walk along microtubules have a high duty ratio and so are porters, whereas team-working muscle myosin motors are rowers since they have a low duty ratio. In general, rowing motors are not synchronized.

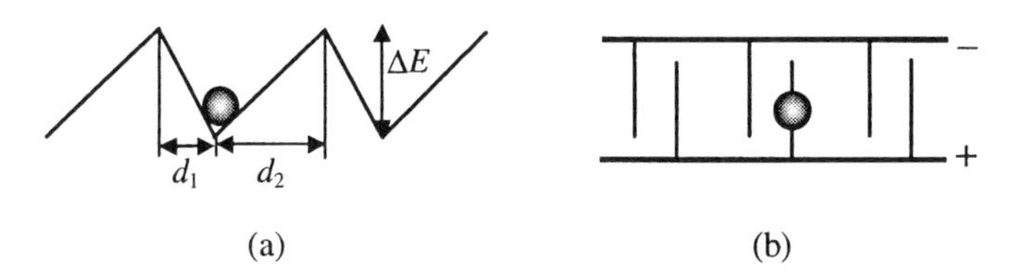

Figure 7.11 (a) A molecule in an asymmetric sawtooth potential that can be implemented (b) by interdigitated electrodes (*After*: [34], [38]).

Molecular motors can be theoretically modeled as thermal (or Brownian) ratchets, the term ratchet describing asymmetric potentials that induce a particle flow when subject to external fluctuations in the absence of any net macroscopic force. More precisely, molecular motors are modeled as Brownian particles in an overdamped environment, moving along asymmetric periodic potentials driven by non-equilibrium fluctuations of either the potential or the fluctuating force [33,34]. For example, in a 1D sawtooth potential as that represented in Figure 7.11(a), in which the valley is surrounded by potential peaks with height ΔE and located at distances d_1 and d_2 at the left and right, respectively (we assume that $d_1 < d_2$), a periodic switching on of the potential to the sawtooth shape and off to a flat potential induces a net motion to the left at small temperatures. The reason is that, when the potential is switched on, the diffusing particle in the flat potential has a higher probability to fall back into the original valley or to a neighboring valley to the left than to a neighboring valley to the right. On the contrary, a fluctuating force that alternates between $+F$ and $-F$ values with zero time average induces jumping events to the right with a probability $\exp[-(\Delta E - Fd_2)/k_B T]$, when the force $F(t) = +F$ points to the right, and jumping events to the left with a probability $\exp[-(\Delta E - Fd_1)/k_B T]$, when $F(t) = -F$. The fluctuating force induces a net motion to the right since $d_1 < d_2$. More complex molecular motor behaviors can be explained by taking into account the interaction between particles [34]. For example, for hard-core repulsion between particles, the direction of particle motion can change its sign several times if the density of particles increases from 0 to 1, the average velocity depending in a complex way on the size of particles close to the maximal particle density.

The drift velocity v developed by a molecular motor, modeled as a diffusive particle moving in a 1D lattice with periodic and asymmetric transition rates, satisfies the relation $|v| \leq 2ND/d$ where D is the diffusion coefficient, d the length of the elementary cell of the lattice, and N the number of different internal states in a full mechanochemical cycle of a motor [35]. The equality is reached when all transitions are unidirectional and have the same magnitude.

The efficiency of molecular motors, whether designed to pull loads, to achieve high velocity, or other tasks, is defined as the ratio between the minimum energy (or power) input needed to accomplish the same task as the engine and the actual energy (or power) input [36]. From this definition it follows that the efficiency cannot exceed unity and that it equals unity only when the task is performed in the energetically most favorable way. This definition confirms that rectified diffusion of Brownian particles can be converted to uniform motion with an efficiency that approaches 100% and that thermally driven motors, in which the temperature is constant in time along the periodic potential but inhomogeneous in space, can move faster than isothermal motors for a given power input.

The efficiency of Brownian motors is increased if the motors are coupled. Elastically coupled particles subjected stochastically to a flat and a periodic asymmetric potential may move even in the absence of thermal diffusion, due to the action of the interparticle springs that restore to their natural length, which is incommensurate with the period of the asymmetric potential [37]. Such a movement is prohibited for isolated particles in the absence of thermal noise. Moreover, the center of mass of the elastically coupled particles moves faster than in the single-particle model, the velocity reaching a maximum at a specific coupling constant of springs, while the efficiency of energy conversion is larger, attaining a peak for a certain load. A vast literature exists on the efficiency of different types of single or coupled molecular motors, and sometimes analytically expressions are found for specific models. The interest in this chapter is, however, limited to the way in which the principles of molecular motors can be taken over and exploited in nanotechnology.

The molecular Brownian ratchets can be replicated in nanotechnology and used to transport, for example, DNA molecules trapped in asymmetric potential wells generated by charging interdigitated electrodes patterned on a Si chip, as shown in Figure 7.11(b) [38]. Net transport of DNA molecules is observed via fluorescence measurements as the asymmetric potential (the bias V applied between the interdigitated electrodes) is switched on and off repeatedly; when the electrodes are discharged the potential is flat and the molecules undergo isotropic Brownian motion. Moreover, since the transport rate of a molecule depends on its diffusion constant D and charge q, it is possible to spatially separate molecules with different sizes that are initially trapped in the same well. The bias V must be switched on for times t_{on} sufficiently long to localize particles in the trapping wells, that is, for $t_{on} \geq d_2^2 k_B T / qVD$, and then must be switched off for times $t_{off} \cong d_1^2 / 2D$ such that the particle moves always to the left. (In t_{off} the particle moves roughly along a distance d_1 and is trapped by the next well if it diffuses to the left, and trapped by the original well when it diffuses to the right, if $d_1 << d_2$.) The probability P that a particle moves one well to the left in a single cycle is

$$P = (1/2)\text{erfc}[(d_{eff}^2 / 4Dt_{off})^{1/2}], \quad (7.2)$$

where d_{eff} is the effective distance traveled by the particle, which takes values between the inner and the outer edge-to-edge distance between electrodes equal to d_1 and $3d_1$, respectively. The steady state flux of particles Φ through the device,

$$\Phi = P/(t_{on} + t_{off}), \quad (7.3)$$

is maximum when $t_{off} = d_{eff}^2 / 2D$. In the experiment in [38] 2 μm wide electrodes placed such that $d_1 = 2$ μm and $d_2 = 18$ μm, and biased with a switching frequency between 0.7 Hz and 8 Hz were used to transport a rhodamine-labeled DNA 50-mer trapped at positive electrodes with a diffusion constant $D(d_1 / d_{eff})^2 \cong 3.5 \times 10^{-8}$ cm²/s; the bias V was switched between 0 V and 1.6 V with a ratio $t_{on} / t_{off} = 1/3$.

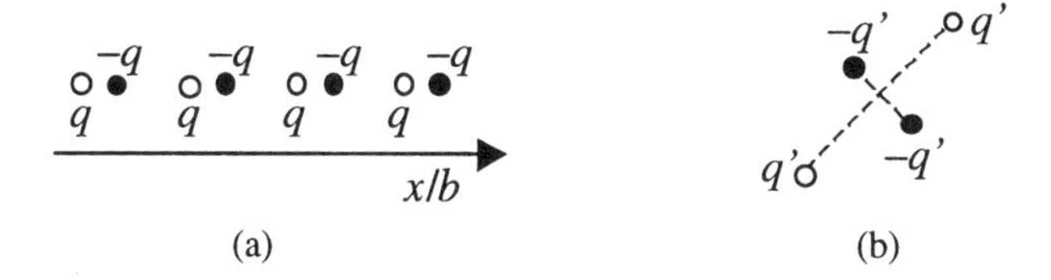

Figure 7.12 Brownian motor based on Coulomb interactions and consisting from (a) a track and (b) a moving entity (*After*: [39]).

Another human-made molecular motor, based on Coulomb interactions and represented in Figure 7.12, was described in [39]. It consists of a track, which is a periodic 1D asymmetric chain of alternating charges q and $-q$, placed respectively at x-coordinates $(0.1+n)b$ and $(0.4+n)b$, with n an integer and b a constant, and a neutral moving entity composed of two charges q' and two charges $-q'$. The moving entity, placed at a distance $0.25b$ above the track, is subjected to a directed translational motion if it rotates around its center in one direction, but remains in its original location if rotated in the opposite direction. This behavior is maintained even for a random rotation of the charge arrangement. The directions of open and blocked translational motion are determined by the spatial arrangement of the charges in the moving entity, and can be switched by charge inversion. In addition to the "forward gear" and "reverse gear" of this motor a "neutral gear" exists, characterized by blocked translational motion in both directions for a specific charge arrangement.

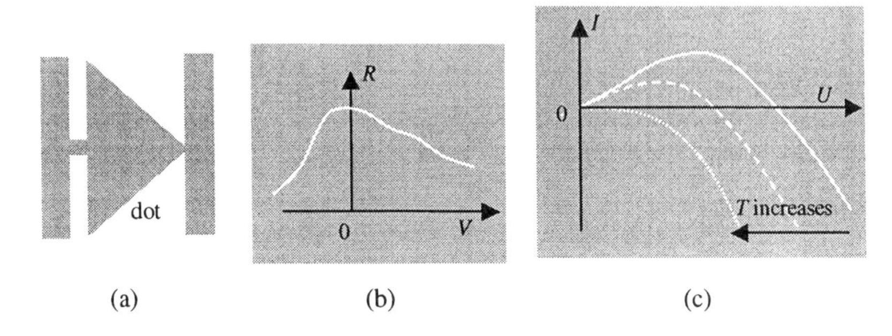

Figure 7.13 (a) Triangular-shaped quantum dot, (b) typical dependence of its differential resistance on a bias, and (c) the dependence of its time-averaged current on the amplitude of a slowly rocking voltage (*After*: [40, 42]).

The principle of operation of molecular thermal ratchets, which can be described by classical mechanics, has inspired researchers to design electronic devices in which electrons behave as quantum ratchets. Namely, the electrons are transported in an asymmetric potential in which quantum effects play a significant role. For example, in the 1μm long ratchet-like triangular-shaped GaAs/AlGaAs quantum dot contacted by two point contacts, as in Figure 7.13(a), the random movement in the asymmetric potential drifts electrons in one direction even when the averaged applied bias is zero, as long as a source of energy exists [40]. The theory of the quantum ratchet [41] models the electron as a quantum Brownian particle, which moves at low temperatures under the influence of quantum noise in a periodic ratchet potential that is tilted alternatively to the right and left by a slowly varying (periodic or random) rocking bias. The quantum nature of electron transport is best evidenced by the dependence of the differential resistance on a dc bias V. As shown in Figure 7.13(b), this dependence is asymmetric around $V = 0$, which suggests quantum interference effects inside the ratchet; the potential, which is not mirror symmetric with respect to the two leads, is not the same when the voltage is reversed such that the transmission coefficient for electrons as well as the current-transmitting channels in the emitter depends on the sign of the voltage.

A quantum theory of the electron, which includes quantum tunneling, predicts that a net current is produced by the ratchet even at T = 0 and that, unlike in classical ratchets, the direction of the current can change at finite temperatures. This prediction was confirmed in [42] for a triangular-shaped ratchet as well as for an array of ratchet cells in the tunneling regime rocked by a square-wave voltage with an amplitude of 1 mV at a frequency of 191 Hz applied in periodic bursts with a repetition frequency of 17 Hz. Measurements were performed at temperatures between 0.4 K and 4 K, the temperature dependence of the current as a function of the rocking amplitude U looking like in Figure 7.13(c). This unusual behavior of the quantum ratchet is caused by the different directions of the two contributions to the rocking-induced current: excitation over the energy barrier of the ratchet and tunneling, which flow on average in different directions, such that the net current direction depends on the temperature, Fermi energy, and the rocking voltage. The barrier is, in this case, the confinement energy for the lowest mode electrons at the narrowest point in the constriction. More precisely, when a square-wave rocking voltage of amplitude U is applied to a quantum ratchet similar to that represented in Figure 7.13(a), the net current, calculated as the time average over one period, is given by $I_{\text{net}} = [I(U) + I(-U)]/2$, or [43]

$$I_{\text{net}} = (e/h)\int_{-\mu}^{\infty} \Delta T(E,U)\Delta f(E,U)dE \ , \tag{7.4}$$

where E is the energy of electrons measured relative to the average of the chemical potentials of the electron reservoirs on the left and right sides $\mu = (\mu_L + \mu_R)/2$, $\Delta T(E,U) = T(E,U) - T(E,-U)$, with $T(E,U)$ the transmission probability of electrons across the structure at bias U, and $\Delta f(E,U) = f_R(E,U) - f_L(E,U)$, with $f_{L,R}(E,U) = 1/\{1 + \exp[(-E \pm eU/2)/E_{\text{th}}]\}$ the Fermi-Dirac distributions in the left and right reservoirs, respectively. $E_{\text{th}} = k_B T$ is the thermal energy at temperature T. At low temperatures ΔT is negative for electrons with energies under the barrier height and positive for electron energies above it and Δf, which is centered around E = 0, has a width that depends on temperature and U. The net current changes sign from positive to negative as the temperature increases.

Recent studies show that adiabatically rocked quantum ratchets, in which the asymmetric potential is tilted periodically and symmetrically, can act as heat pumps [43]. In adiabatically rocked ratchets the symmetric tilting takes place on a much slower scale compared to other time scales, the system behaving as a nonlinear rectifier. As indicated earlier, a net current of particles is produced when the current is averaged over a full rocking period because during each half-cycle the height or shape of the potential deforms in a different manner. However, even when the net particle current is zero, the net energy current is non-vanishing because the average energies transported in opposite directions are different. The heat in the two electron reservoirs changes by $\Delta Q_{\mathrm{L,R}} = E \pm eU/2$ upon transfer of one electron from right to left under a bias U, such that the heat currents that enter the left and right reservoirs when a voltage U is applied across the device are

$$q_{\mathrm{L,R}} = (\mp 2/h)\int_{-\mu}^{\infty}(E \pm eU/2)T(E,U)\Delta f(E,U)dE \tag{7.5}$$

and the net heat flow in the reservoirs, defined as $q_{\mathrm{L,R}}^{\mathrm{net}} = [q_{\mathrm{L,R}}(U) + q_{\mathrm{L,R}}(-U)]/2$ equals $q_{\mathrm{L,R}}^{\mathrm{net}} = \mp\Delta\Theta/2 + \Xi/2$, where $\Delta\Theta = q_{\mathrm{L}}^{\mathrm{net}} - q_{\mathrm{R}}^{\mathrm{net}}$ is the heat pumped from the left to the right side of the ratchet due to the energy sorting property of the device with asymmetric barriers, and $\Xi = q_{\mathrm{L}}^{\mathrm{net}} + q_{\mathrm{R}}^{\mathrm{net}} = (U/2)[|I(U)| + |I(-U)|]$ is the ohmic heat averaged over one cycle. The performance of the ratchet as heat pump is characterized by $\chi(T,U) = \Delta\Theta/\Xi$, which is non-vanishing even when the net current vanishes. The ratchet can cool one reservoir only when $\chi > 1$, a condition that could be met in ratchets with resonant tunneling barriers. For usual quantum ratchets $\chi << 1$.

Electron pumps, which pump electric charges from a reservoir at low electrochemical potential to one at a high electrochemical potential, can also be implemented with molecular wires [10]. A stochastic modulation between two configurations of the dot formed in the molecular wire chemically drives electron pumping. In a dot structure as that depicted in Figure 7.2(b), the height of the left barrier and the energy levels in the dot can be modified if the substituents R' and R' form an active site for the catalytic conversion between a substrate S and a product P. If S and P are electron donating the height of the left barrier and the energy levels in the dot are raised, while electron-withdrawing properties of S and P lower the height of the left barrier and the energy levels in the dot. Electron pumping can be driven by the nonequilibrium fluctuations of the energy levels in the dot determined by the far-from-equilibrium chemical reaction between S and P. The molecular wire must be chemically asymmetric in order to allow chemically driven electron pumping. Suppose now that, starting with a well in equilibrium to the left reservoir, open to it, and isolated from the right by a high barrier, as in Figure 7.14(a), the well energy level increases quasi-adiabatically, and simultaneously the right barrier is lowered and the left barrier is raised, such that in the configuration depicted in Figure 7.14(b) the electrons flow to the right reservoir, which reaches equilibrium with the dot after a waiting time. During the switching process there is no charge transfer between dot and reservoirs, and the local equilibrium within the dot is preserved. Then, the left barrier is lowered while simultaneously decreasing the well energy level and raising the right barrier. After a waiting time the equilibrium is established between the dot and the left reservoir, the system reaching the state represented in Figure 7.14(a). The switching can be either periodic or stochastic. During the cycle the average number of electrons transferred from left to right is $f_L - f_R$, where the Fermi distributions in the two configurations are $f_L = 1/[1+\exp(-E_1/E_{th})]$ and $f_R = 1/[1+\exp(E_2/E_{th})]$, respectively, with the energies E_1 and E_2 measured relative to the reservoir to which the well is open. The energy dissipated per cycle is given by $(E_1+E_2)(f_L - f_R)$ and the efficiency of the electron pumping process is $\Delta\mu/(\Delta\mu+E_1+E_2)$, where $\Delta\mu = \mu_R - \mu_L$ with μ_L and μ_R the chemical potentials of electrons in the left and right reservoirs, respectively. The efficiency can reach unity for large values of $\Delta\mu$; electron pumping can be also achieved in a four-step adiabatic and reversible change of the dot configuration (see the reference in [10]). The quasiadiabatic pumping mechanism is similar to a flashing ratchet, which can model biomolecular pumps that drive ions up an electrochemical gradient.

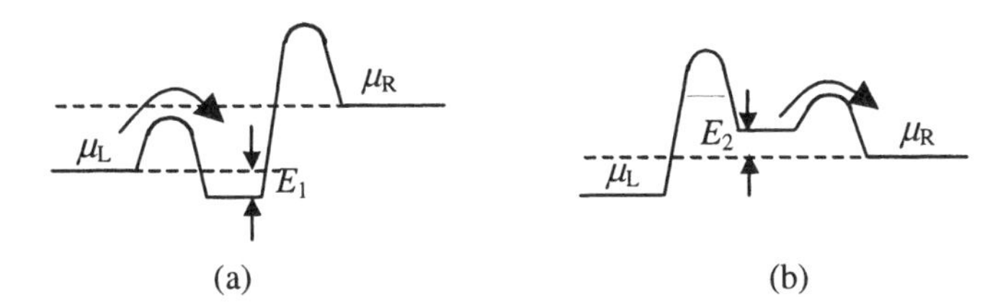

Figure 7.14 (a), (b) The two configurations of a dot formed in a molecular wire that allow electron pumping, by a quasiadiabatic and irreversible switching between them (*After*: [10]).

A closely related phenomenon to the adiabatic mechanism of molecular pumping is the energy transfer from the chemical gradient in a nonequilibrium chemical reaction catalyzed by a membrane enzyme to an electric *RLC* circuit [44]. The oscillating or fluctuating electric field applied on the membrane drives thermodynamically uphill the transport of ions through the membrane, the membrane enzyme having several conformational states, with correspondingly different electrical properties, which interact differently with Na and K ions on opposite sides of the membrane. The transporter (membrane enzyme) is a Brownian ratchet in which the anisotropy is introduced by the different dissociation constants for substrates on the opposite sides of the membrane. Spontaneous amplification in the net polarization of the transporter proteins, and hence conversion of chemical to electrical energy that drives electrical oscillations, occurs if the transport reaction is far from equilibrium (if there is a large concentration gradient between the two sides of the membrane).

In a similar manner, cation-selective asymmetric nanopores in polyethylene terephthalate (PET) rectify the current with the preferred direction of cation flow from the narrow towards the wide opening of the pore and can transport potassium ions against the concentration gradient on opposite sides of the PET membrane in the presence of external field fluctuations. The working principle of the ion pump is again the ratchet mechanism, the internal potential of the nanopore resembling an asymmetric tooth [45]. The nanopore, which can have controlled conical shapes and diameters between 10 nm and 10 μm, and the rectifying $I-V$ characteristic are shown in Figure 7.15. The rectifying behavior is more pronounced for basic pH values for which the pore walls are negatively charged. The current is in this case an ion current and the applied voltage can have a ramp, triangular, or sine shape.

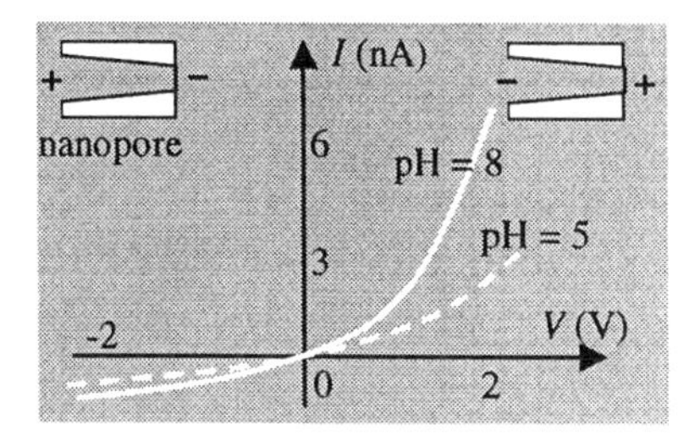

Figure 7.15 Typical rectifying *I-V* characteristic of an asymmetric nanopore, with *I* the ion current; the sign convention for the applied voltage is shown relative to the conical nanopore (*After*: [45]).

Not only linearly moving motors but also rotary motors have been studied in relation to established nanotechnology techniques. For example, in [46] it was predicted by quantum chemistry calculations that triptycene rotors can mount on the Si(100)-2 × 1 surface in vacuum using the –COOH or –OH groups to attach themselves to the surface. The –COOH group creates a flexible coupling to the surface while the –OH coupling has a stiff axle. These groups react with Si dimers on the Si(100) surface forming a Si-O and a Si-H bond. Because the rotational barrier of triptycene is about 0.1 eV when –COOH groups are used for attachment and 0.16 eV when –OH groups are used for the same purpose, the molecule can rotate easily on the Si surface at room temperature. To slow down the rotation and/or to image the triptycene rotor with an STM, the temperature must be lowered or C_{60} molecules must be used to block the molecular rotation. C_{60} molecules are situated in the troughs between Si dimer rows when deposited on the Si(100)-2 × 1 surface at room temperature, but move on the Si dimer rows after annealing at 600 °C; the fullerene molecules can be manipulated with an STM tip. Triptycene rotor chains on the Si surface can be built along the dimer row using the steric interaction between adjacent surface sites, but slippage events between adjacent triptycene molecules are likely to occur at room temperature since the slippage barrier is only 0.4 eV.

A single-molecule rotor that moves on an atomically clean Cu(100) surface was evidenced in [47]. The propeller-shaped hexa-tert-butyl decacyclene molecule is surrounded by like molecules, which form a supramolecular bearing. STM images showed that the single molecule and the bearing, which together have a six-lobed structure, exist in two spatially defined states, one a rotating state and the other an immobilized state, which are laterally separated by 0.26 nm. In the low-symmetry site with respect to the surrounding molecules the molecule rotates (is in a disengaged state), its 29 kJ/mol rotational barrier at room temperature being overcome by the thermal energy, while in the higher-symmetry site it is immobilized (it is in the engaged state), the 117 kJ/mol rotational barrier being higher than the thermal energy. (Note that 1 eV = 1.6×10^{-19} J.) The lateral shuttle motion between the two states is hampered by a translation-energy barrier of 42 kJ/mol, the transition between the two states being possible only through STM manipulation.

Artificial molecular elevators, with a quite complex structure and dimensions of 3.5 nm in diameter and 2.5 nm in height, have been also fabricated [48]. They can develop forces up to 200 pN and can perform nontrivial mechanical movements. The complex and stable structure of the elevator consists of a trifurcated riglike component with three legs interlocked by a platform, each leg having two different notches at different levels. The platform can be raised and lowered between the two levels separated by 0.7 nm using the energy provided by an acid-base reaction.

7.2.2 Molecular Nanoactuators

DNA-based machines are versatile due to the simple molecular recognition chemistry according to which two strands of single-stranded DNA hybridize to form a double-stranded DNA only if their base sequences are complementary. This is the reason why complex structures can be constructed among the combinatorial space of nucleotide sequences that label the DNA strands.

In particular, a nanoactuator DNA structure, which pushes two components apart (DNA tweezers pull them together) powered by DNA hybridization [49], has been experimentally demonstrated. As shown in Figure 7.16, the DNA nanoactuator consists of two DNA strands, A and B, which hybridize to form a loop. The two double-stranded regions of the DNA strand A serve as rigid arms, connected via a flexible hinge formed by the single-stranded DNA region of A. The two rigid arms are pulled towards each other by thermally driven fluctuations of the single-stranded region of B, so that they are separated in the relaxed configuration shown in Figure 7.16(a) by a mean angle smaller than that obtained when the single-stranded region of B is pulled straight. The two arms can be pushed apart into the straightened configuration illustrated in Figure 7.16(b) by a fuel strand F, which has a region that is complementary to the single-stranded region of B, and thus hybridizes with B. The pushing force can reach 37 pN. The nanoactuator is restored to the original configuration by binding the fuel strand F with its complement G through branch migration, which temporarily breaks the base pair bonds by thermal fluctuations as shown in Figure 7.16(c), the resulting waste product of F fully hybridizing with G; the end result of the cycle is depicted in Figure 7.16(d). So, the actuator is cycled between the relaxed and straightened configuration by successively adding F and G. This clocked molecular motor is less susceptible to form dimers compared to molecular tweezers due to its loop structure.

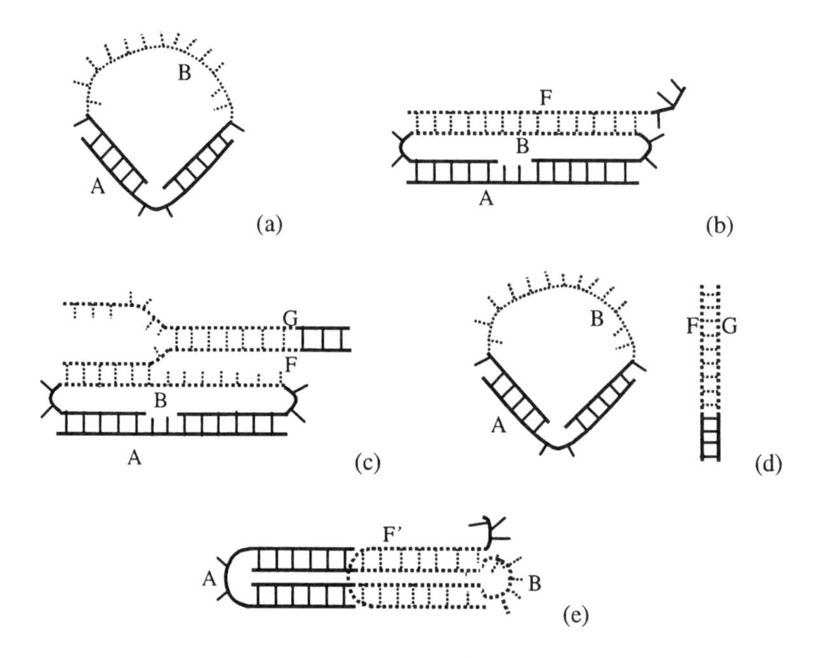

Figure 7.16 The operation of a DNA nanoactuator: the nanoactuator in (a) the relaxed configuration is brought into (b) the straightened configuration upon hybridizing with the fuel strand F, which is then (c) removed from the actuator by its complement G through strand displacement by branch migration, and (d) the actuator returns to its original state accompanied by a double-stranded waste product. (e) The closed configuration (*After*: [49,50]).

Recent studies [50] have, in fact, revealed that the DNA nanoactuator is switchable between three distinct mechanical states: a relaxed, straightened, and closed configuration. The closed configuration, depicted in Figure 7.16(e), is reached after the addition of another fuel strand F', in this configuration the nanoactuator being similar to a molecular tweezer. From the closed configuration the nanoactuator reaches the relaxed configuration upon addition of the complement of F', G', this branch-migration reaction being accompanied by the waste product obtained by hybridizing F' and G', as discussed above. The three-state device is driven in the two rigid states by the hybridization energy and can be operated against an external load. It can be used, for example, to selectively grab or release other molecules, or to hold chemical components in place for following reactions. The selectivity is assured by the fact that two DNA strands hybridize to form the double-stranded helix only if each A base (G base) in one strand binds to a T base (C base) in the other. The DNA nanoactuator can eventually be part of a molecular assembly line.

The remarkable selectivity of molecular reactions is evidenced by the fact that it is possible to detect a single base mismatch between two 12-mer oligonucleotides in the hybridization of two complementary DNA strands. The detection is based on the transduction of DNA hybridization via surface stress changes into a nanomechanical deflection of microfabricated cantilevers, each of them functionalized on one side with a different oligonucleotide base sequence [51]. The cantilevers bend in contact with complementary oligonucleotide solutions due to a difference in surface stress between the functionalized gold and nonfunctionalized Si surface after hybridization of the oligonucleotides in the solution with complementary oligonucleotides immobilized on the cantilever surface. Although large nonspecific responses of individual cantilevers are recorded, the differential deflection, defined as the difference between the deflections of the cantilevers covered by one and by the other oligonucleotide after the corresponding hybridizations, is a true molecular recognition signal. The hybridization of the 12-mer oligonucleotide produces an actuation force of about 300 pN, the differential signal of two 12-mer oligonucleotides that differ through a single base being 10 nm.

The same single base mismatch of a 12-mer oligonucleotide probe can be detected using silicon nanowires on which a single-stranded probe DNA molecule is covalently immobilized [52]. The silicon nanowires are 20 μm long, 50 nm wide, and 60 nm high. When the target DNA attaches to its complementary strand on the Si nanowire the carrier concentration in a *p*-type nanowire increases due to an increase of negative charges introduced by DNA, while the carrier concentration in an *n*-type nanowire decreases. In both cases a change in the conduction is observed compared with the case when no hybridization occurs (when there is a base mismatch in the two strands). The conduction changes with 12% and 46% for *n*-type and *p*-type silicon nanowires, respectively, for 25 pN of DNA in solution. Silicon *p*-type nanowires functionalized with peptide nucleic acid receptors that hybridize with wild type DNA can also be used to distinguish, via conductance measurements, between wild type and mutant DNA samples at concentrations down to tens of fM and so can help identify certain genetic diseases [53]. Other biological molecules can be detected as well in this way.

Even biological interactions can be electrically detected through the variation in conductance of a nanometer-sized gap between two electrodes placed in close proximity with a functionalized surface that concentrates the ligand/receptor complex (for example, biotin/streptavidin or biotin/antibiotin antibodies) between electrodes [54]. The conductance varies with two or three orders of magnitude since the receptor in solution is labeled with gold particles, which are smaller than the gap and become inserted into the gap.

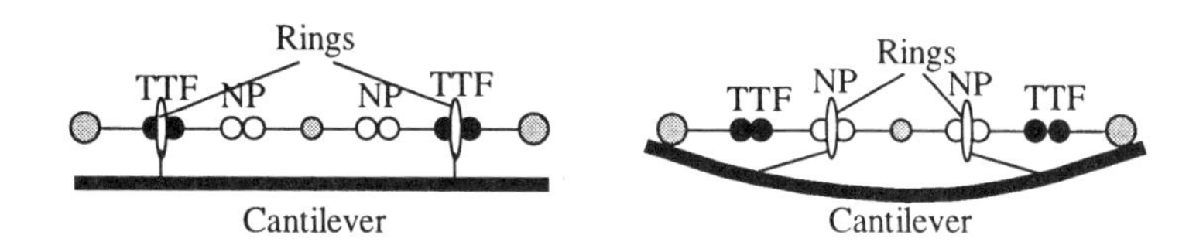

Figure 7.17 Extended (relaxed) and contracted configurations of 3[rotaxane] molecules attached to a gold-coated Si cantilever (*After*: [55]).

An array of microcantilevers coated on one side with a SAM of bistable redox-controllable [3]rotaxane molecules, which imitate the contraction and extension movements of molecular muscles, was used to harness the power of molecular motors [55]. More precisely, [3]rotaxane molecules consist of two cyclobis(paraquat-p-phenylene) $CBPQT^{4+}$ rings that have two stable positions with respect to two tetrathiafulvalene (TTF) and two naphthalene (NP) stations: the $CBPQT^{4+}$ rings can be positioned either around the TTF or around the NP stations. During the so-called power stroke chemical oxidation of the TTF stations drives the $CBPQT^{4+}$ rings to the NP stations due to electrostatic charge-charge repulsion between the rings and the oxidized dicationic form of the TTF stations, while in the diffusive stroke the TTF stations return to their neutral form after reduction and the $CBPQT^{4+}$ rings are restored around the TTF stations by thermal activation (see Figure 7.17). The $CBPQT^{4+}$ rings can be anchored as a SAM to a gold surface deposited on a cantilever by incorporating a disulfide tether in each of them. In this way, a gold-coated Si cantilever is contracted during the oxidation process, in which the inter-ring distance decreases, and relaxes to its original position during the reduction process. The bending of a cantilever with length L

and Young's modulus E during oxidation is characterized by a deflection $w = ML^2 / 2EI$, where M is the moment of the cantilever beam and I the area moment of inertia of the beam's cross section. Experimental data show that cantilevers placed in a transparent fluid cell are bent upwards with 35 nm when an oxidant solution is added and return to the original position in the presence of a reductant solution, the disulfide-tethered [3]rotaxane molecules anchored on an array of cantilevers harnessing the energy of chemical reaction into large-scale mechanical work.

Besides CNTs (see Chapter 2), organic molecules can be used to fabricate nanocantilevers. Microtubules, for example, are ideally suited as cantilevers since they are filamentous assemblies of protein subunits in the form of hollow cylinders with internal diameters of about 15 nm and external diameters of 25 nm. Measurements of the Young's and shear moduli of microtubules have revealed a length-dependent flexural rigidity. The microtubule has a high stiffness on length scales comparable to the cell size but its end can oscillate close to the cellular membrane [56]. If a cantilevered microtubule clamped on one side is loaded with a rate less than 1 pN/s by a second microtubule transported by kinesin motor proteins adsorbed to the surface, a forcemeter able to detect piconewton forces is fabricated [57]. This forcemeter can measure the force necessary to rupture receptor/ligand bonds since the freely swinging end of the cantilevered microtubule bends due to the perpendicular movement of the second microtubule once a link is established through a receptor/ligand pair. After the bond between the receptor and ligand is ruptured the cantilevered microtubule returns to its original position and the second microtubule continues its path. The successive processes are displayed in Figure 7.18. Once the bending is detected by fluorescence microscopy the applied force is determined from the flexural rigidity of microtubules. For example, the streptavidin/biotin bond ruptures at 5 pN.

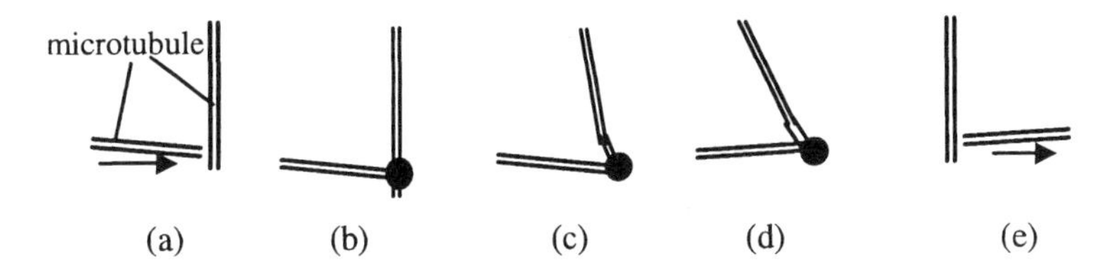

Figure 7.18 The moving microtubule propelled by kinesin motor proteins (a) approaches a vertical cantilevered molecule, (b) contacts it, (c), (d) bends it, and then (e) releases it (*After*: [57]).

7.3 MOLECULAR NANOELECTRONICS

Organic molecules promise electronic devices with miniaturized dimensions as well as the possibility of controlling the molecular design (and, thus, achieving engineered properties and specific functions) in the chemical synthesis stage. In addition, the study of biological systems can eventually inspire nanotechnology to achieve self-replicating and programmable architectures able to manipulate atom-by-atom both nonbiological and biological matter. Last, but not least, organic molecules are synthesized in low-cost, massively parallel processes, which can overcome the technological challenges and the increasing expenses of techniques able to produce Si-based devices with dimensions lower than 100 nm.

The molecular nanotechnology is not mature enough so the research is presently focused on producing hybrid structures in which the functionality of the non-molecular part is enhanced by adding molecules. Recent studies reviewed in [58] suggest that the chemical bond between specific molecules and metal semiconductor bulk materials or quantum particles could be also engineered such that interface effects could further increase the functionality of molecular-based electronic devices.

7.3.1 Molecular Electronic Devices

The first proposal of a molecular device was a D–σ–A rectifier molecule, which consists from an electron donor D with low ionization potential, an electron acceptor A with high electron affinity, and a covalent bridge or spacer σ that acts as a tunneling barrier. In this device rectification occurs by through-bond tunneling, analogous to rectification by an asymmetric multilevel resonant tunneling structure in solid-state physics; the D^+–σ–A^- state is relatively accessible from the neutral ground state D–σ–A, while the D^-–σ–A^+ state has an energy higher with several eV [59].

In this first molecular rectifier the molecule was supposed to be rigid. The difficulty of predicting the geometric and electronic properties of the molecule-metal interface led to the development of molecular rectifiers in which the switching behavior is caused by conformational motions of molecules driven by electric fields, such that the interface structural details are not determinant [60]. The current across a flexible molecule can be modeled by a generalization of the Landauer formula:

$$I = \int_0^\infty dE[f(E) - f(E+eV)]\langle g(E,V)\rangle / e, \quad (7.6)$$

where $f(E)$ is the Fermi-Dirac function, V is the applied potential, and $\langle g(E,V)\rangle$ is the thermal average of the conductance at temperature T over the available conformations α. This average is influenced by the external field via changes in the energy of the conformations, $E_\alpha(V)$. More precisely, for V constant in time and for observation times shorter than the system-dependent equilibration time between different conformations,

$$\langle g(E,V)\rangle = \sum_\alpha g_\alpha(E,V)\exp[-E_\alpha(V)/E_{\text{th}}]/\sum_\alpha \exp[-E_\alpha(V)/E_{\text{th}}] \tag{7.7}$$

in the Born-Oppenheimer approximation, with $g_\alpha(E,V)$ the conductance in the α conformation. Different conformations can have different conductances because the metal-molecule interaction is different, or because the intramolecular interaction changes. Unlike in rigid junctions, where rectification is possible only if $g(E,V)$ is asymmetric with respect to the potential, in flexible junctions another possible rectification mechanism appears: selective population under external bias of conformations with different conductances. For example, rectification can occur if the molecular junction has two conformations with similar energies but sufficiently different conductances g_1, g_2 and different molecular dipoles projections on the interelectrode axis μ_1, μ_2 if $|\mu_1 - \mu_2|V/d > E_{\text{th}}$, with d the distance between the electrodes. An example of a molecule that has different conformations for different applied potentials is shown in Figure 7.19.

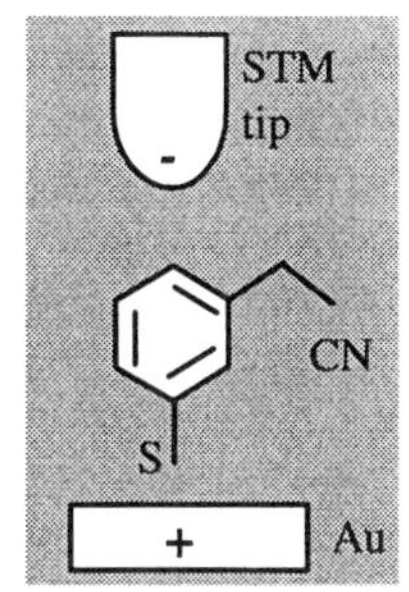

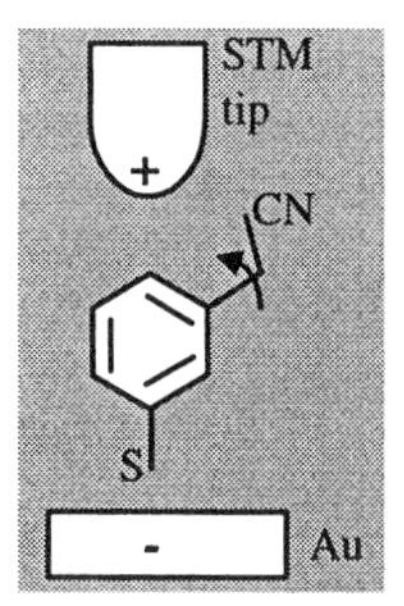

Figure 7.19 Example of a flexible molecular junction, in which the conformation of the molecule changes with the polarity of the applied potential (*After*: [60]).

The possibility of electrically addressing a single conducting molecule was demonstrated in [61]. The polarizable molecule is electrostatically trapped in the narrow spacing (about 4 nm) between stable Pt electrodes, being attracted to the gap between electrodes by an applied electric field, which has a maximum value in the gap. In this controlled deposition of single molecules the negatively charged side of the molecule is attached to the positive electrode while the positive side is attached to the negative electrode, a bridge between the electrodes being created if the gap between them is smaller than the molecule size (see Figure 7.20). To trap only a single molecule it is necessary to connect a high resistor R in series with the electrodes such that after trapping a molecule the main part of the voltage V drops across the resistor and the electric field in the gap becomes sufficiently weak to prevent trapping another molecule. The electrostatic trapping energy $E_{\text{trap}} = (C_p - C_g)V^2/2$, where C_p and C_g are the capacitances between electrodes with and without the trapped molecule, should be larger than the thermal energy $E_{\text{th}} = k_B T$ at room temperature. Experimental results in [61] confirmed room-temperature trapping of 20 nm Pd colloids covered with a monolayer of ligands, the $I-V$ characteristic of the device at 4.2 K showing a Coulomb blockade regime for $V < V_{\text{ch}}$ with V_{ch} the gap voltage and an exponential increase of current for $V > V_{\text{ch}}$ due to the suppression of the effective tunnel barrier by the applied bias. At both 4.2 K and room temperature the $I-V$ characteristic can be fitted by $I = (2\pi E_{\text{th}}/eR_0)\sinh(eV\tau/\hbar)/\sin(2\pi\tau E_{\text{th}}/\hbar)$, where R_0 is the junction resistance at $E_{\text{th}} = 0$ and no bias, and $\tau = L/(2U/m)^{1/2}$ is the tunneling traversal time of an electron with mass m through a barrier of height U and length L. This characteristic is exponential if $E_{\text{ch}}\tau/\hbar$ is comparable with or higher than unity, where $E_{\text{ch}} = eV_{\text{ch}}/2$ is the charging energy in the Coulomb blockade regime; experimental results agreed with $E_{\text{ch}}\tau/\hbar = 0.5$.

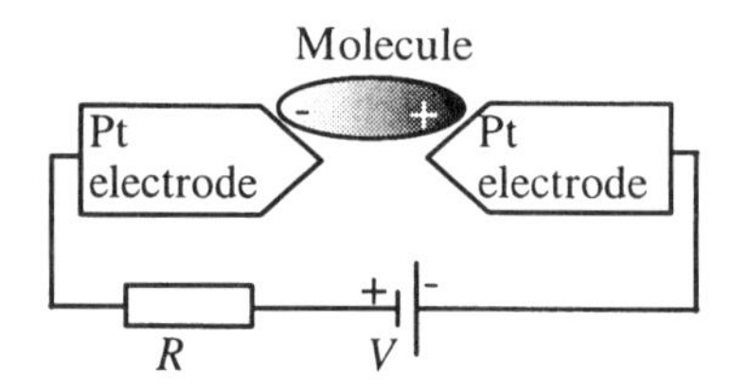

Figure 7.20 Electrostatic trapping of a single molecule between two electrodes (*After*: [61]).

A method to efficiently immobilize DNA molecules in the gaps between electrodes has been developed in [62]; it allows even DNA bundles to span the gap. In this method droplets containing long λ-DNA fibers are applied onto chip substrates with microstructured electrodes. The dried droplet formed after receding meniscus has a large concentration of DNA in the center and a zone with radially extended DNA fibers at the periphery surrounded by the droplet boundaries. Because the DNA interaction with perpendicular electrodes is insignificant and does not affect DNA orientation, while DNA interacts and becomes partially aligned when parallel to electrodes, the droplet must be positioned such that a region with oriented DNA fibers is parallel with an electrode array. In this case the DNA spans the 2 μm gap region between electrodes, adsorbing to the electrodes above and below the gap due to attractive interaction. Aligning of DNA fibers can be observed either by fluorescence, when DNA is labeled with a fluorescence dye, or by SEM imaging if it is labeled with positively charged nanoparticles.

Negative differential conductance is a useful property in electronic applications. This property is predicated in the nonlinear transport through a benzene ring, which is weakly coupled to electrodes and which undergoes relaxation by radiative transitions [63]. The effect is due to the interplay between the charging effects and the specific structure of the molecular orbitals, the negative differential conductance appearing when the molecule couples to the electrodes at the para-position, a case in which a blocking state of the molecule is occupied that cannot decay for symmetry reasons. Coupling at the meta-position generates steps in the $I-V$ characteristic but no negative differential conductance; typical $I-V$ characteristics for the two positions are displayed in Figure 7.21.

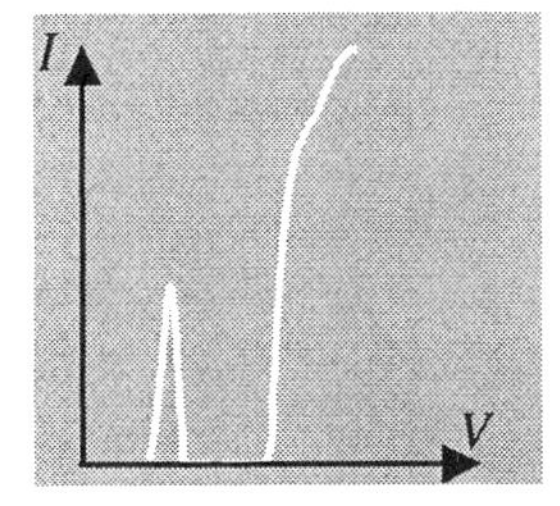

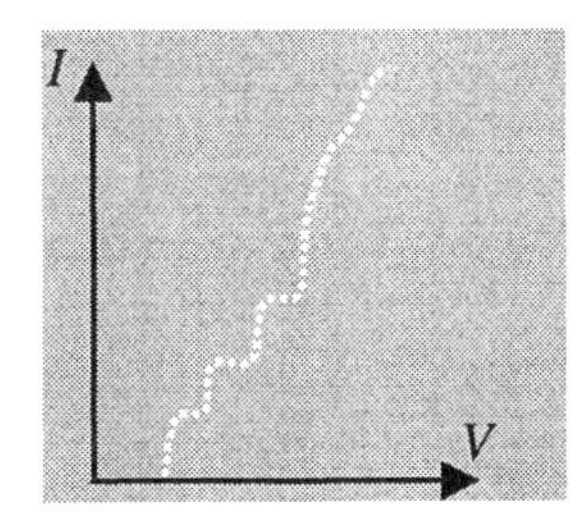

Figure 7.21 Typical *I-V* characteristic for a benzene ring weakly coupled to electrodes in the para position (solid line) and meta position (dashed line) (*After*: [63]).

The prediction of negative differential resistance in molecular devices was confirmed in [64] at room temperature, in circuits containing gold nanowires deposited on SAMs of insulating C_{16} (16-mercaptohexadecanoic acid) molecules, on which electronically active 4-{[2-nitro-4-(phenylethynyl)phenyl]-ethynyl} benzenethiol molecules are adsorbed. The nanowires were fabricated inside the 6 μm long pores of polycarbonate track etched membranes, with 70 nm diameter, by replication techniques. The $I-V$ characteristic of the device showed a standard diode behavior under negative biasing from 0 to –2 V and a negative differential resistance region at positive biasing from 0 to 2 V, with peak currents around 3×10^{-9} A and a peak-to-valley ratio of 1.8 to 2.21. Room-temperature negative differential resistance was also measured through styrene molecules on degenerately doped *n*-type Si(100) surfaces for negative sample bias (positive voltage induces electron-stimulated desorption), or through the organic 2,2,6,6-tetra-methyl-1-piperidinyloxyl molecule for positive sample bias on *p*-type Si(100) surfaces and for negative sample bias on *n*-type Si(100) [65]. Since multiple negative differential resistance regions can be observed in the $I-V$ characteristics recorded by ultrahigh vacuum STM spectroscopy, the most plausible explanation of this phenomenon is a resonant tunneling between the discrete orbitals of the adsorbed molecule and the narrow band of free charge carriers in the degenerately doped Si substrate.

Another molecular system in which reproducible negative differential resistance was observed is a Hg-alkanethiol/arenethiol-Au bilayer molecular junction formed by joining together a drop of Hg coated with teradecanethiol SAM with an oligo(phenylene-ethynylene) SAM on a gold surface [66]. The resulting room-temperature negative differential resistance in the $I-V$ characteristic for positive bias has a peak-to-valley ratio that reaches values up to 4.5 for several devices, the average peak voltage being 0.69 V with a standard deviation of 0.12 V. A typical dependence of the current density j on voltage is represented in Figure 7.22. The high-quality contacts on both sides of the bilayer are assured by strong S-metal bonds between the Au and Hg metal electrodes and thiol termini, while a van der Waals contact develops at the interface between the two SAMs. Negative differential resistance components can play an active role in information processing, or memories.

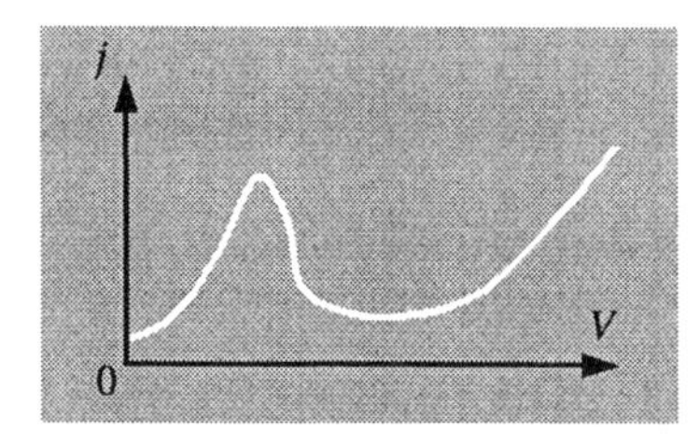

Figure 7.22 Typical current-voltage characteristic of a molecular bilayer (*After*: [66]).

Molecular devices can be also used for switching or as resonant tunneling structures in which tunneling occurs through electronic molecular states. These devices can be much faster than similar electronic devices since energy and electron transfer processes within molecules can occur in less than 1 ps. Moreover, molecular devices are natural multiplexers, since light of a given frequency selectively addresses certain components, and can decrease heat production via efficient fast photo-driven processes. Examples of ultrafast molecular switches based on electron transfer (see [67] and the references therein) include the modification of the electronic state and hence the transmission characteristic of a bridge molecule in a covalently linked donor-bridge-acceptor system, the sequential photo-induced electron transfer in a donor-acceptor array, and the control of the rates of electron transfer reaction through applied electric fields.

The smallest molecule placed between two electrodes that has a bistable behavior and is thus capable of switching has been studied theoretically in [68]. It is a 1,4-benzene-dithiolate (BDT) molecule with two sulfur end groups that bind, respectively, to a gold substrate and to a monatomic gold STM tip. This molecule has two low-energy conformations with different symmetries and conductance values (a high-conducting and a weak-conducting state). Flipping between them occurs in response to the lateral motion of the STM tip for only 3 Å or, for some STM tip positions, spontaneously at finite temperatures, or in response to current pulses through the molecular wire. As a result, the conductance shows a bistable behavior and switching, caused by the different overlap of the π orbitals of the molecule with the orbitals on the Au tip for the two conformations. If the STM tip is modeled by a tetrahedron of Au atoms and the Au(111) substrate by a cluster of three Au atoms, the ON state corresponds to the molecule orientation with the ring facing the tip, while in the OFF state the edge of the ring faces the tip.

An example of an all-carbon molecular switch consists of a molecule of fullerene C_{60} that bridges two (5,5) CNTs, which play the roles of donor and acceptor electron reservoirs [69]. The electronic transport properties of a single C_{60} molecular junction can be manipulated through charge transfer processes controlled by a gate potential, through electromechanical processes when an STM tip compresses the molecule, or through altering the orientation of the molecule with respect to the substrate. In the proposed all-carbon molecular switch, simulations performed in the Landauer formalism using a density functional theory approach combined with a Green function technique predict a three-orders-of-magnitude variation of the conductance if the orientation of C_{60} is changed or if one of the CNTs is rotated around the symmetry axis at a fixed CNT-C_{60} distance. The conductance of the molecular junction is mainly determined by tunneling through the gap between the HOMO and the LUMO of the C_{60}, and its value depends significantly on the structural relaxation of the junction.

A bi-stable and fast switching molecular device was proposed in [70]. It consists of a molecular quantum dot characterized by an intrinsic switching of the current value due to attractive electron correlations that appear when the degeneracy of the molecular level is larger than two. Under these circumstances, simulations have shown that the dot shows current hysteresis as a function of the bias voltage, the tunneling current becoming bistable in some voltage range. Unlike repulsive electron correlations, which cause Coulomb blockade in quantum dots but no switching, attractive electron correlations in molecular quantum dots, modeled by a negative Hubbard U in the molecular eigenstate weakly coupled with left and right leads, can generate an intrinsic non-retarded current switching. Negative U can have a pure chemical origin or can originate in strong electron-phonon (vibronic) interactions, which form polarons in the molecular quantum dot if the tunneling time is comparable to or larger than the characteristic phonon time. Unlike bare electrons, polarons attract each other at short distances, comparable to the interatomic spacing. If at zero temperature $2(\Delta - |U|) < eV < 2\Delta$ and $|U| < 2\Delta / d$, where Δ is the lowest unoccupied molecular level with respect to the chemical potential and d the degeneracy, the nonlinear kinetic equation reduces to $2n = 1 - (1-n)^{d-1}$, where n is the expectation number of electrons on the conducting molecular level. This equation has one root for $d = 2$: $n = 0$, but two roots for $d = 4$: $n = 0$ and $n = 0.38$, or $d = 6$: $n = 0$ and $n = 0.48$, which correspond, respectively, to two stationary states with low and high current. For larger applied biases, for which $eV > 2\Delta$ the kinetic equation has only one solution, $n = 0.5$. The current is related to n as $I = 2nI_0$, with $I_0 = ed\Gamma / 2\hbar$ and Γ the tunneling rate between the dot and the lead. The hysteresis behavior of the current at zero temperature is represented in Figure 7.23 for $d = 4$. When the bias increases from zero the molecular dot remains in the low-current state until $V_2 = 2\Delta / e$, a decrease of voltage from above this threshold value being associated to the high-current state until $V_1 = 2(\Delta - |U|)/e$. The current jump at V_2 disappears for $d >> 1$, when the two solutions correspond to $n = 0$ and $n = 0.5$. The voltage range of the hysteresis loop narrows with increasing temperature.

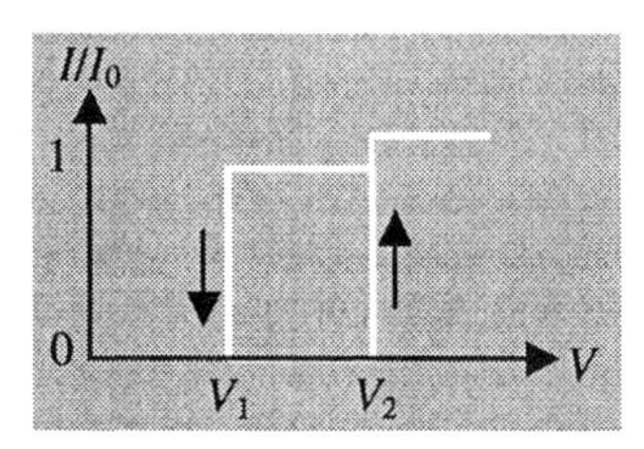

Figure 7.23 Current-voltage hysteresis loop for a molecular quantum dot with attractive electron correlations (*After*: [70]).

An ultrafast THz molecular switch with switching times of the order of 0.1 ps was predicted in [71]. It consists of a molecular complex modeled by two active sites with intramolecular coupling and connected to left and right particle reservoirs. Simulations performed in the tight-binding approximation, which take into account the charge buildup at the molecular site through a nonlinear term, show that an incident electromagnetic pulse with characteristic times in the THz range can trigger switching between different current states of the device. The device can act as tunable detector for THz radiation, since for a suitably applied external bias the electromagnetic pulse turns off the passage of current. The switching is caused by bistable states in the $I-V$ characteristic at high voltages, analogs to the bistable region in an RTD. The bistable states originate from the dynamic coupling between the sites in the molecule.

A molecular SET has been experimentally demonstrated in [72]. The device was fabricated by preparing first the metal Ti/Au electrodes with a gap distance of 20 nm followed by growing a SAM of nitro-amine molecules on top of the electrodes. Au nanoparticles comparable to or larger than the gap distance were finally trapped electrostatically in order to connect the source and drain. The nanoparticles act as quantum dots. The resulting double-junction structure showed single-electron tunneling if the SAM is sufficiently thin to act as a tunnel barrier. The current-voltage characteristic was nonlinear for temperatures below 70 K, showing periodic Coulomb oscillations as a function of the bias applied on a back gate electrode.

A three-terminal molecular device consisting of a single molecule DNA with semiconducting behavior connected with CNT source, drain, and gate electrodes has been assembled and measured in [73] using a triple-probe atomic force microscope (T-AFM); the device and its $I_{DS}-V_{DS}$ characteristic are schematically represented in Figure 7.24. The T-AFM is composed of a conventional AFM with a conductive CNT probe and a nanotweezer with two CNT probes that are prevented from adhering to each other by being vibrated at 100–800 kHz. After connecting the CNT source, drain, and gate electrodes to the single DNA molecule using the T-AFM, a gate bias V_G was applied by fixing the CNT AFM probe onto the single-walled CNT gate electrode. The observed nonlinear room-temperature $I_{DS}-V_{DS}$ characteristic under dry nitrogen atmosphere for $V_G = 0$ and a source-drain distance of 25 nm indicates the semiconducting behavior of the DNA molecule; the nonlinear character is maintained for higher gate voltages but the voltage gap around $V_{DS} = 0$ decreases, the DNA molecule acting as an electronic switch (of the current value) when the gate is biased.

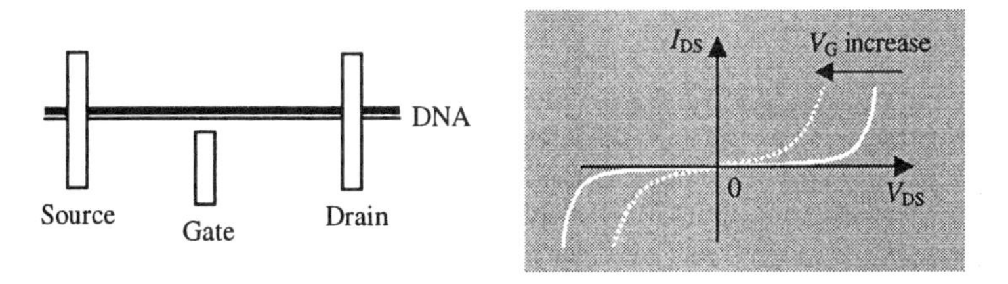

Figure 7.24 A three-terminal DNA device (left) and its *I-V* characteristic (right) (*After*: [73]).

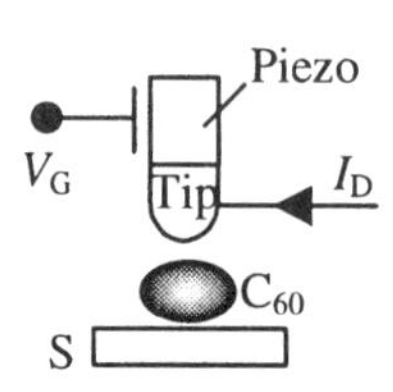

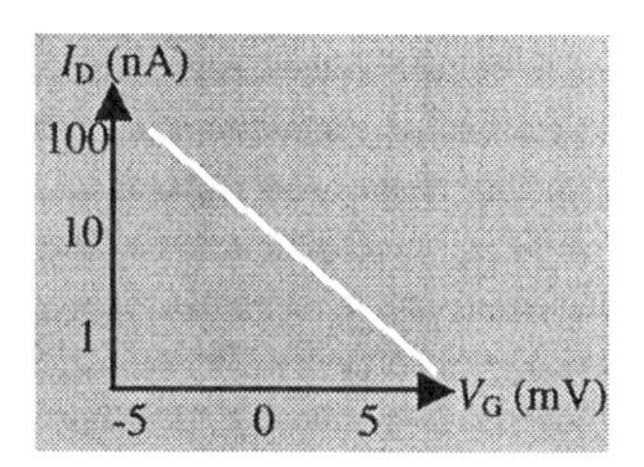

Figure 7.25 Schematic representation of the electromechanical single C_{60} amplifier (left) and its *I-V* characteristic (right) (*After*: [8]).

A shift in the energy of a molecular orbital, which can modulate tunneling through single C_{60} molecules, for example, even in the nonresonant tunneling regime (and hence change the resistance), can be induced by applying a small force of the order of nN with a metallic STM tip [8]. The same principle can be applied to mechanically reduce molecular level degeneracy and to design a three-terminal electromechanical single C_{60} amplifier (see Figure 7.25), which is the tunneling equivalent of the Schockley valve effect.

The ultimate limit of scaling down of electronic devices is a single atom in which electrons hop off from and on to electrical leads. In [74] a single-atom transistor was demonstrated, in which electron transport takes place through definite charge states (the spin-degenerate Co^{2+} and the non-degenerate Co^{3+} states) of a single transition metal atom (Co in this case). The Co ion is bonded to polypyridyl ligands, which are attached to insulating tethers with variable lengths. Because the coupling of the Co ion to electrodes is modified via the length of the insulating tether, it is possible to fabricate single-electron transistors based on the Coulomb blockade effect or devices based on the Kondo effect. In single molecular complexes with longer insulating tether the current below 100 mK is strongly suppressed up to a threshold voltage that depends on the bias voltage V_G, and then increases in steps. This is a characteristic behavior of single-electron transistors with the island consisting of a single Co ion attached to electrodes by organic tunneling barriers. On the contrary, in complexes with shorter insulating tether molecules the conductance initially decreases below $2e^2/h$, which indicates the presence of a tunneling gap between the electrodes, and then increases significantly with increasing voltage. The differential conductance $\partial I/\partial V$ has a peak at $V=0$, which depends logarithmically on temperature between 3 and 20 K, indicating (together with the peak splitting in an applied magnetic field) the presence of the Kondo effect (i.e., the formation of a bound state between the conduction electrons in the leads and a local spin on the island, which is the S = 1/2 spin of the Co^{2+} ion; the bound state increases the conductance at low voltages). The Kondo-assisted tunneling appears for strong coupling between the transition ion and the electrodes. Single-electron transistors that show Coulomb blockade have also been fabricated with single C_{60} [75], C_{70}, and C_{140} fullerene molecules [76]. Excited energy levels were found in these SETs, which appear as a sudden increase in current for a specific value of the gate voltage and are caused by vibrational molecular modes that couple with the motion of tunneling electrodes. These modes are of the bouncing-ball type in C_{60} and C_{70} fullerenes, while an intercage stretch mode is observed in C_{140}.

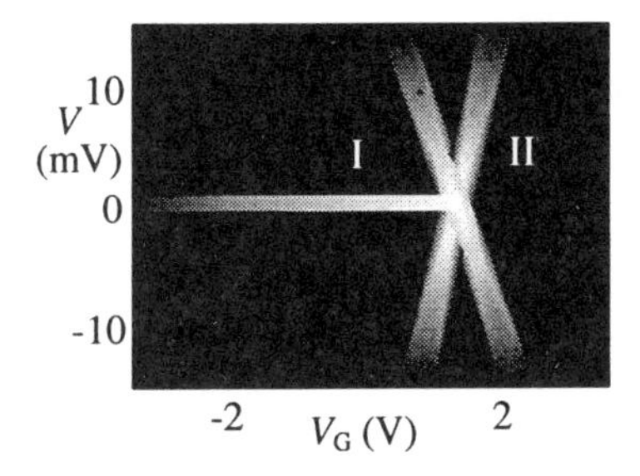

Figure 7.26 Schematic representation of a typical bias dependence of differential conductance in a single-molecule transistor; brighter regions indicate higher values of the differential conductance (*After*: [77]).

The Kondo effect can be observed also in single-molecule transistors, in which the spin and orbital degrees of freedom of molecules that incorporate transition metal atoms are chemically controlled. In particular, in single-molecule transistors where a divanadium molecule ([(N,N',N''-trimethyl-1,4,7-triazacyclononane)$_2$-$V_2(CN)_4(\mu$-$C_4N_4)$]) acts as spin impurity the Kondo resonance can be observed up to 30 K and can be reversibly tuned via the gate voltage, which alters both the charge and the spin state of the molecule [77]. The differential conductance $\partial I / \partial V$ of such transistors has two gap regions, denoted by I and II in Figure 7.26, which are bounded by two broad conductance peaks that slope linearly as a function of the gate voltage V_G. The peaks cross at a gate voltage $V_G = V_{ch}$ for which the conductance gaps vanish, and a sharp zero-bias differential conductance peak reminiscent of the Kondo resonance in quantum dots is observed up to 10 K in the gap region with a smaller V_G (gap region I). The origin of the conductance gap is the finite energy required to add or remove an electron from the molecule due to single-electron charging and the discrete molecular level spacing. The conductance gap varies reversibly and linearly with V_G since a more positive gate voltage stabilizes an additional electron on the divanadium molecule, and disappears at $V_G = V_{ch}$ where two different charge states of the molecule have the same total energy. The sharp zero-bias differential conductance peak is a Kondo resonance, that is, a many-electron effect originating from the exchange interaction between a localized spin and the conduction electrons in the leads, which occurs when the electronic state of the molecule has a non-zero spin S. The Kondo resonance manifests itself in splitting of the position of the conductance peak in an applied magnetic field with $2g\mu_B B/e$, which is twice the Zeeman splitting. The peak height of the Kondo resonance decreases and its width increases in a characteristic manner with increasing temperature, facts observed experimentally. The Kondo resonance appears only in one conductance gap since in that region $S = 1/2$ for the ground state of the positively charged molecule, whereas in the other gap region, which develops for $V_G > V_{ch}$ upon one electron addition, the ground state is neutral and has $S = 0$.

7.3.2 Molecular Architectures for Nanoelectronics

The scaling down of electronic devices from the present-day CMOS electronics, which operates in the Boltzmann electron transport regime, to ballistic and tunneling regimes characteristic for single-electron transfer is complicated from a conceptual point of view. An alternative is to develop new architectural concepts based on electron transport characteristics inside molecules. Thus, in passing to the next level of nanoelectronics, molecular devices that perform specific functions must be interconnected in nanoscale circuits.

A molecular architecture is a method to create circuits containing molecular devices, which are interconnected and are placed in desired locations. Standard semiconductor technologies, such as lithography, are not suitable for molecular electronics, so that other techniques, especially self-assembly, are used to design molecular circuits. A brief review on nanofabrication techniques for molecular devices can be found in [8]. To make efficient use of the smallness of single-molecular devices, such as diodes or transistors, the connecting wires should be also thin and the density of molecular devices must be as high as possible. In order to achieve this, molecular devices and the connecting wires are grown on low-conducting (even insulating) DNA templates (or scaffolds).

Thin metallic wires, with the capability of self-recognition to link the ends of the DNA to the contacts, can be fabricated by DNA metallization [20]. DNA-templated growth of metal structures consists of several steps: first, metal complexes or ions bind to specific sites on the DNA molecule, then these sites become seeds by treating them with a reducing agent, and finally the seeds act as catalysts for further reduction of metal and the subsequent enlargement of the metal clusters. Metal necklaces of separated nanoclusters with diameters between

3 nm and 5 nm and spacings ranging from one to several nm, or continuous nanowires with diameters as small as a few tens of nm can be fabricated depending on the reaction time and the concentrations of the metal solution and the reducing agent. The low-conductance behavior of DNA prevents shortening of nanosize clusters. Even molecular lithography on single DNA molecules can be achieved, with metallization gaps on sites where protection molecules are bound to the DNA at places defined by self-recognition. Moreover, DNA anchor molecules that bind to specific sites can position metal nanoclusters connected to DNA at specific target locations. Most metallization experiments were made with gold, platinum, and palladium, the DNA resistance decreasing up to a few tens of ohms over lengths of a couple of microns. The corresponding conductivities reach values comparable to those of polycrystalline metal films. Metallic room-temperature behavior is obtained for metal coatings wider than 50 nm.

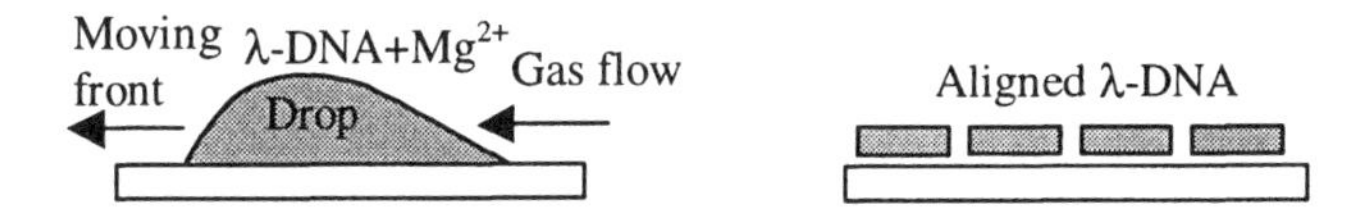

Figure 7.27 Schematic representation of the fluid-flow-assisted molecular combing technique (*After*: [79]).

DNA metallization implies mechanical alignment of DNA molecules, since, due to their high flexibility, they are commonly obtained as random coils. Methods for straightening single DNA molecules include AFM, the use of micro-needles, of magnetic or optical tweezers, while simultaneous manipulation of multiple DNA molecules can be achieved with a moving water-air interface (method known also as molecular combing), with applied sweeping electric fields (method known also as electric poling), or using hydrodynamic flow. Even molecular motors have been proposed for (unfortunately, not very reliable) straightening of DNA molecules (see [78] and the references therein). In particular, molecular combing was proved efficient for fabrication of 1D parallel and 2D crossed metallic nanowire arrays grown on DNA templates [79]. In the fluid-flow-assisted molecular combing technique illustrated in Figure 7.27, unidirectional alignment of DNA molecules is achieved by first placing a drop of λ-DNA solution containing Mg^{2+} onto a glass or mica surface, and by then driving the solution to advance along a given direction using compressed air flow. (The divalent cation Mg^{2+} tunes the adhesion of negatively charged DNA molecules on negatively charged glass and mica surfaces.) Straightened and parallel DNA chains remain behind for an appropriate interaction between the surface and the

DNA molecules, metallization through electroless palladium deposition promptly following DNA stretching. A 2D DNA array can be fabricated in a similar way, by realizing a second alignment along another direction.

A method of fabricating highly ordered assemblies of gold nanoparticles along well-stretched λ-DNA templates has been presented in [80]. In a first step, the gold nanoparticles are surface-functionalized by coating them with a monolayer of oxidized aniline. These functionalized nanoparticles strongly interact with DNA molecules and can assemble in continuous or necklace wires on DNA according to the fabrication method. More precisely, continuous, uniform deposition of functionalized Au nanoparticles on double-stranded DNA is achieved if the nanoparticles are brought into contact with stretched DNA molecules that form highly aligned patterns fixed on the surface, obtained by depositing a droplet of λ-DNA solution on a polycarbonate-coated coverslip and then sucking up the droplet with a pipet; the DNA molecules are stretched by the surface tension. Discontinuous necklace assembles of gold nanoparticles, with larger interparticle spacing, are obtained by first assembling functionalized Au nanoparticles on DNA molecules and then stretching and fixing the DNA molecules. Unlike in the first method, the attachment of functionalized Au nanoparticles to unstretched DNA molecules can occur from any direction and more than one particle can bind to the DNA, surrounding it at close position.

Not only metal clusters but also *p*-type CuS semiconductor nanocrystals can be grown selectively and densely on phage λ-DNA templates in solution or on λ-DNA stretched on a surface [81]. In the first case the sources of Cu and S are, respectively, an equimolar concentration of $CuCl_2$ and a stoichiometric solution of Na_2S, while in the second case S is introduced via H_2S, the selective growth of CuS on DNA being possible because Cu^{II} binds strongly to the phosphate backbone and bases of the DNA. Chains of CuS nanoparticles with 1 nm up to 10 nm in diameter and separated by less than 40 nm can be grown in this way, nanoparticles with larger diameters being obtained when the DNA is first immobilized on a substrate. The number of nucleation sites for CuS increases when bundles of DNA are used; the density of CuS nanoparticles grows with the number of strands in the bundle.

The fabrication method of DNA scaffolds for targeted deposition of other nanoparticles or molecules is described in [82]; both sheet and tube conformations can be created by self-assembling 4×4 DNA tiles. A square aspect ratio 4×4 tile, as that shown in Figure 7.28(a), contains nine DNA strands arranged in four arms. In a modified version of the tiles, in which DNA have sticky ends, the 4×4 tiles can self-assemble in 2D ordered lattices through the sticky-end association, as illustrated in Figure 7.28(b). Another slight modification of the 4×4 tiles produces a curvature of the DNA lattice and eventually a tubular structure similar to a CNT, which flattens if placed on the substrate, forming regular ribbons; the height of the ribbon is twice the height of a layer of the DNA lattice. The distance between adjacent tile centers is about 20 nm, sufficiently large to trap nano-particles or molecules if the loops at the tile center are functionalized with molecular groups that selectively react to the molecule intended to be trapped. The ribbons, on the other hand, have quite regular widths, of 60 nm, and are often longer than 10 μm. They can be used as templates for metallic nanowires. Metallization of DNA nanotubes in solution results in hollow metal tubes with about 40 nm in diameter; the diameter of a metallized double-stranded DNA in solution is 15 nm. Other biomaterials besides DNA can also form templates. For example, self-assembled 2D crystals of a bacterial surface layer protein with a lattice constant of 18 nm have been used to fabricate highly ordered arrays of gold nanoparticles with 5–10 nm in diameter by deposition from preformed colloids [83].

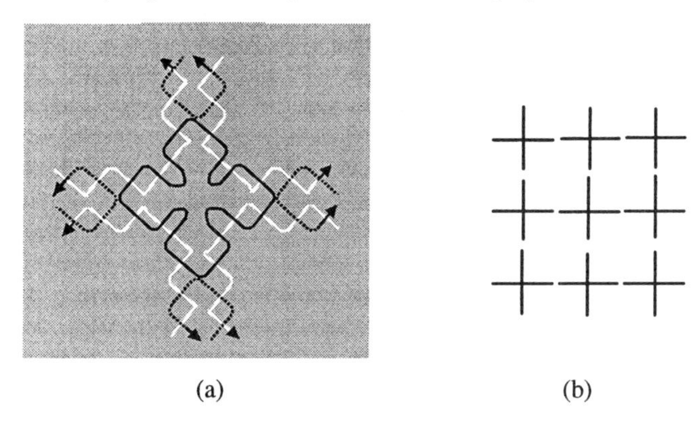

Figure 7.28 (a) The structure of a 4 X 4 DNA tile and (b) a 2D grid self-assembled from 4 X 4 tiles with sticky ends (*After*: [82]).

As an example of alternative nanotechnologies and architectures to the CMOS technology, we mention CNT FET (CNTFET) transistors self-assembled into circuits with the help of DNA molecules as scaffolding [84]. To produce a CNTFET, a semiconducting CNT is first localized at a chosen address on a DNA scaffold molecule via homologous recombination of a protein, for example (the RecA protein from *Escherichia coli* bacteria in [84]), the DNA/CNT assembly is then stretched on a Si wafer, and the DNA is afterwards metallized with the protein doubling as sequence-specific resist to form conductive wires that make electric contact with the CNT.

By fabricating lattices of DNA and then attaching a tag of single-strand DNA to a certain position on the lattice and the complementary single-strand DNA to one end of a CNT, it is possible to attach in a controllable way the end of the CNT to the scaffold. Similarly, different connections on the lattice can be achieved by using multiple pairs of DNA tags. These connections can attach both ends of the CNT to the scaffold or can create a CNTFET by attaching two perpendicular CNTs that cross each other such that the top metallic CNT acts as a gate for the bottom semiconducting CNT channel. In an analogous manner, CNT wires at each side of cavities created by cross-connected nanotubes can be implemented by specifying a DNA tag at the gap between the wires, which attracts a metallic nanosphere that acts as a nucleation site for chemical electroless plating. Thus, CNTFET circuits can be constructed with this technology, the present limitations in DNA self-assembly technology imposing, however, an upper limit to lattice sides of 2 μm. Multiple lattice nodes must be interconnected for larger scale systems. The design tools needed for placement, routing, and electrical simulation of this new self-assembling circuit technology are discussed in [85]; logical molecular circuits can be designed in this way, although DNA-directed manipulation of CNTs is still at an embryonic stage.

Note that, although self-assembled objects from organic molecules with well-defined size and structure can generate a variety of nano- or microscale objects, molecular self-assembly of ill-defined macromolecules can generate tubes with macroscopic dimensions. For example, transparent tubes with millimeter-scale diameters and centimeter-scale lengths can be produced by adding an amphiphilic hyperbranched multiarm copolymer to a selective solvent [86]. These multiwalled tubes, which have almost the same inner diameter and the same distance between neighboring walls, as well as a uniform wall thickness of about 400 nm, can be observed with the naked eye. The tubes are flexible and can be bent in solvent (acetone) such that no-fracture crossed-knot patterns are obtainable, but collapse into ribbon-like objects when taken out of the solvent.

7.4 MOLECULAR-BASED OPTIC AND OPTOELECTRONIC DEVICES

Optical properties of molecular devices can be significantly changed by applying a radiation of another wavelength. For example, in [67] it was shown that optical absorption bands in charge transfer chromophores can be red shifted if a covalently linked donor-acceptor D–A molecule bonded to a chromophore is first irradiated with a 130 fs laser pulse at 416 nm that photogenerates the ion pair D^+–A^- at a fixed distance and orientation relative to the chromophore. The local electric field caused by this process shifts the charge transfer absorption of the chromophore with 15 nm and thus significantly diminishes the chromophore absorption at 530 nm; the donor molecule in this experiment is zinc 5-phenyl-10,15,20-tripentylporphyrin, the acceptor is pyromellitimide, while the chromophore is a 9-aminoperylene-3,4-dicarboximide. A picosecond molecular optical switch can be implemented in this way.

On the other hand, photovoltaic elements can be implemented with molecular systems. As an example, the steady-state photoresponse of an ITO (indium tin oxide)/PM (purple membrane) heterostructure was studied in [87]. PM patches of about 300 nm in diameter are obtained by crystallization of the light-sensitive bacteriorhodopsin (bR) protein within the cellular membrane. These patches could be used as future nano-scale photodetectors. PM has an absorption peak at 568 nm with a half-width of 50 nm, the absorbed photon producing within picoseconds a conformational change of the bR, which induces proton transfer towards the extracellular side of the protein. Non-contact SPM measurements revealed that the PM/air interface becomes negatively charged at illumination with a 635 nm photodiode and that the maximum value and the time evolution of the photovoltage, in particular the turn-on and turn-off time constants, is significantly influenced by the relative humidity of the atmosphere. The maximum photovoltage, of 2.5 V for a 4 μm thick PM multilayer illuminated with a 36 W/m^2 light intensity occurs at a relative humidity of 15%, the effect disappearing as the relative humidity approaches 40%.

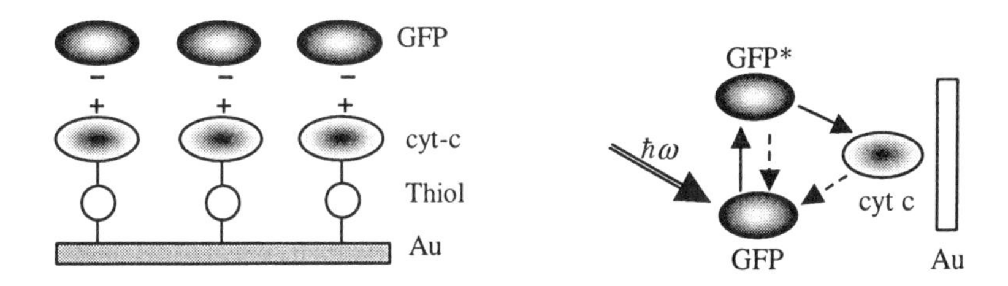

Figure 7.29 Schematic representation of the layered structure of a biophotodiode (left) and its electron transfer mechanism (right) (*After*: [88]).

An artificial molecule that mimics biological photosynthesis and that generates photoinduced current and has rectifying properties has been evidenced in [88]. It consists of a GFP (green fluorescent protein)/cytochrome *c* hetero-protein layer adsorbed onto an Au substrate. A self-assembled layer of cytochrome *c* is first deposited on Au through a thiol (-SH) functional group synthesized on the cytochrome *c* surface, and then the GFP is spontaneously adsorbed by electrostatic attraction. The device shows photoelectrical response, that is, a reproducible photocurrent is measured with an STM under illumination with a 488 nm light wavelength, the intensity of which increases when a forward bias is applied. The mechanism responsible for photocurrent generation is depicted in Figure 7.29: under illumination the electrons of GFP are excited to a higher-energy state, denoted by GFP*, from which some of them return to the ground state by emission of green fluorescence at 510 nm (drawn with dashed line), while others are transferred to the cytochrome *c*, which acts as an electron acceptor. The photocurrent is generated by the photoinduced one-way charge transfer, this mechanism also explaining the rectifying behavior of the device: under illumination at 488 nm the photocurrent generated for a forward bias of 1 V is 10 nA, while its value for a reverse bias of –1 V is –3.5 nA. This asymmetric spectrum is observed by scanning tunneling spectroscopy only above regions covered by GFP.

Another class of optoelectronic devices is the light-driven electric switch. Specific organic molecules can switch their electric behavior upon illumination [17]. For example, photochromic molecules self-assembled on gold switch from conducting to insulating after a switching time of about 20 s when illuminated with visible radiation with a wavelength of 546 nm, the reverse process, expected to appear under illumination with 313 nm radiation, failing to materialize due to the influence of the gold leads attached to the molecule (the reverse process is observed in solution). Photochromic molecules consist of conjugated units linked by a switching element, the covalent bonds of which rearrange upon exposure to light with specific frequencies, such that conjugation throughout the molecule can be switched on and off. At the same time the energy levels of the molecules are modified, the absorption spectrum and even the color of molecules are changing. In particular, in the 1,2-bis[5’-(5”-acetylsulfanylthien-2”-yl)-2’-methylthien-3’-yl] cyclopentene molecule electronic conduction in dark conditions (closed state) takes place via an alternation of single-double carbon bonds extended throughout the molecule, this alternation being broken on the central switching ring upon illumination (in open state) with wavelengths in the range 500–700 nm. The resistance after illumination is with about three orders of magnitude larger than in dark conditions, jumping from MΩ to GΩ values. The failure of switching back from the open to the closed molecular state is explained by quenching of the excited state in the open form under the influence of gold electrodes.

Another light-driven molecular electric switch is the azobenzene molecule, which can switch reversibly between two isomers (a *trans* and a *cis* configuration), with dramatically different conductances, upon photoexcitation [89]. The transformation from the *trans* to the *cis* structure occurs under illumination with a 365 nm wavelength radiation, while a light source with 420 nm wavelength converts the *cis* configuration back to the *trans*. The *trans* configuration has an energy lower with 0.6 eV than the *cis* configuration, and is longer with 1.98 Å than the *cis*, the two configurations being separated by a 1.6 eV barrier. A good contact between the azobenzene molecule and the gold electrodes is achieved by replacing the hydrogen atom with CH_2S and thus fabricating an Au-S-azobenzene-S-Au molecular wire. To account for the change in length of the molecule between the isomer configurations one gold lead should be fixed while the other should be contained within a conducting CNT such that the length change caused by photoexcitation translates into a current change in the CNT.

Single-electron tunneling characteristics of porphyrin-based molecules, which are inserted into the oxide layer of a metal-oxide-semiconductor structure, as illustrated in Figure 7.30, can also be optically switched [90]. The porphyrin molecules act as Coulomb islands, their staircase $I-V$ characteristics, which originate from single-electron tunneling, being reversibly shifted by illumination. In this way, the tunneling current is switched optically rather than electrically (through a gate voltage), the structure acting as an optoelectronic single-electron device. Optically sensitive molecules are attractive as Coulomb islands since they have a uniform structure and appropriate dimensions and because their size and structure can be chemically controlled. The porphyrin molecules (more precisely the tetrakis-3,5di-t-butylphenyl-porphyrin derivative), with a number density of 1.4×10^4 μm^{-2} and hence an average distance of 2.7 nm between molecules (comparable to the size of the molecule), were sandwiched between two SiO_2 layers with thicknesses of 3.1 and 1.9 nm, covered with a sufficiently thin Au electrode, such that light with wavelength 430 nm can penetrate through it. The $I-V$ characteristic of the SiO_2/molecule/SiO_2 multilayer at 5 K, before and after illumination with a light intensity of 7.8 $\mu W/mm^2$, is represented in Figure 7.30. As a result of illumination the Coulomb staircase $I-V$ characteristic was shifted to lower voltages, the threshold voltage $V_{th} = e/2C$ being shifted with about 40 mV from the original 90 mV value. The shifted curve returns to the dark characteristic when the light is turned off, a similar effect being observed for illumination with wavelengths in the range 390–750 nm. In particular, the tunneling current for a fixed bias voltage of, say, 130 mV, can be switched from a low value in dark condition (OFF) to a high value (ON) under illumination. The switching mechanism is not fully understood.

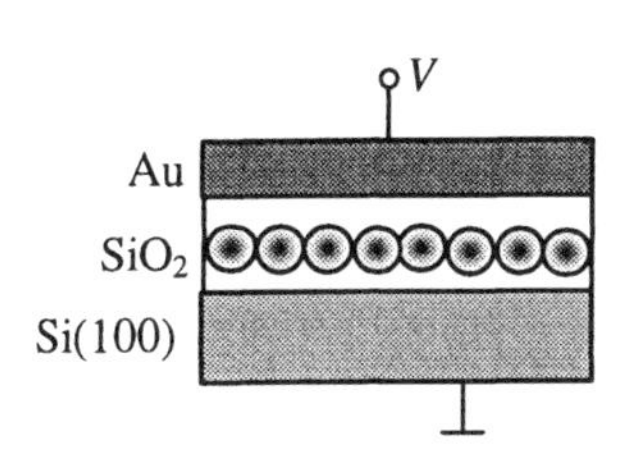

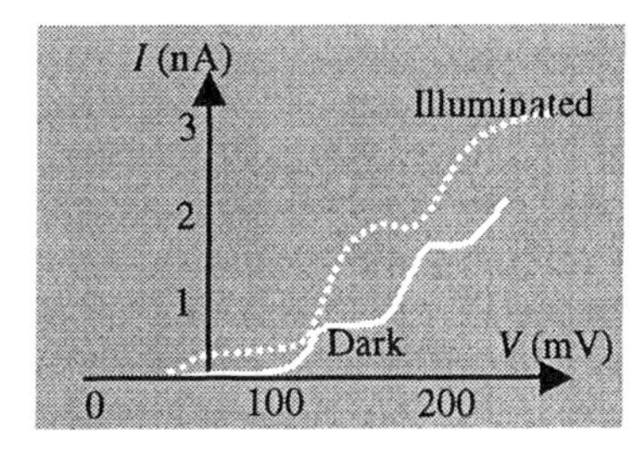

Figure 7.30 Schematic representation of a multilayered SiO_2/molecule/SiO_2 structure formed by inserting porphyrin molecules in the oxide layer of a metal/oxide/semiconductor structure (left) and its *I-V* characteristic (right) (*After*: [90]).

7.5 MOLECULAR COMPUTING DEVICES

Molecular computing devices include both data storage devices and molecular systems that perform logical operations. Most of these devices are designed following concepts developed for classical electronic computers, although it would be reasonable to presume that computing machines could resemble the human brain only if they would adopt structures developed by nature.

Molecular memories are expected to be ultradense and to show low-power consumption since they work with few electrons at the molecular scale. A configurable molecular memory crossbar architecture that connects molecular switches in a 2D grid, as shown in Figure 7.31, was developed in [91]. In a reconfigurable crossbar the wire dimensions can be scaled down to molecular sizes, while scaling up the number of wires, the efficient communication with other circuits being ensured by the small number of communication wires, $2N$, which is required to individually address 2^N nanowires.

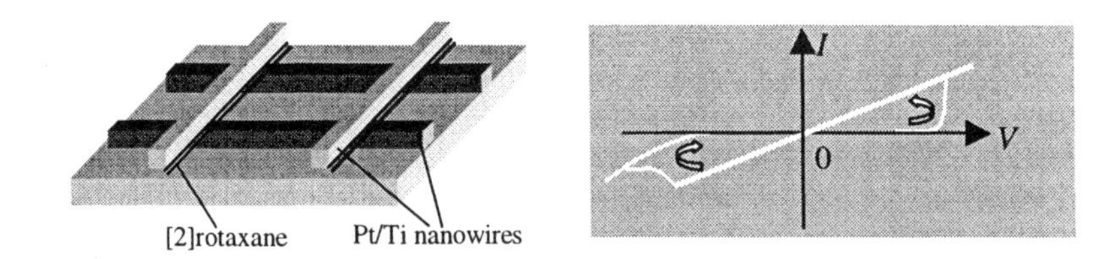

Figure 7.31 Schematic diagram of a configurable crossbar architecture that connects molecular switches in a 2D grid (left) and the reversible switching behavior of [2]rotaxane molecules placed between Pt and Ti electrodes (right) (*After*: [91,92]).

In the configuration presented in Figure 7.31, a monolayer of [2]rotaxane is sandwiched between the top and the bottom arrays of Pt/Ti nanowires such that a Pt/rotaxane/Ti junction is formed at each cross point. This junction, with an active area of 1600 nm^2, which encompasses about 1100 rotaxane molecules, is a nonvolatile and reversible switch that can act as an ultra-high-density memory, with a density of 6.4 Gbits/cm^2. A bit can be written at a cross point by applying a voltage pulse V on the top nanowire and grounding the bottom nanowire in order to reduce the low-voltage resistance of the cross point below 5×10^8 Ω; the duration of the pulse is 0.5 s and the amplitude exceeds 3.5 V (up to 7 V). The [2]rotaxane molecule has an intrinsic mechanical bistability originating from the motion of the cyclophane ring that is mechanically interlocked with a dumbbell component, which can be activated by low-voltage redox (reduction-oxidation) reactions. When a Langmuir-Blodgett [2]rotaxane molecular layer is placed between Pt and Ti electrodes it functions as a current- or voltage-controlled switch, or as tunable resistor over a three-orders-of-magnitude range (from 10^2 to 10^5 Ω), this reversible hysteretic switching behavior (see Figure 7.31) being common also to other molecular species [92]. To read the memory, a bias voltage of 0.5 V is applied to the selective row, all other rows and columns being grounded; the resistance of the bit is determined by measuring the current flowing to ground via the respective column. A positive voltage between 3.5 and 7 V turns on the molecular switch, while a negative voltage between −3.5 and −7 V turns it off. The resistance ratio between the ON and OFF states is at least ten, but decays in time, reaching unity after a few to several hundred cycles. A similar proposal for a molecular memory crossbar architecture can be found in [93]. In a related room-temperature experiment, an ultrahigh-data storage with a density of 10^{13} bits/cm^2 was demonstrated in an *N*,

N-dimethyl-*N'*(3-nitrobenzylidene)-*p*-phenylenedia-mine organic thin film by applying a voltage pulse of 3.5 V for 5 ms between the STM tip and the substrate of the film. The recorded area, with an average size of 0.7 nm, becomes conductive while the unrecorded area is still highly resistive [94].

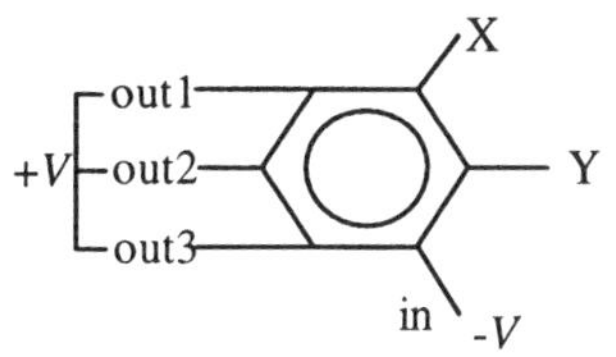

Figure 7.32 A benzene molecule connected to an input electrode *in* and three possible output electrodes (*After*: [95]).

A molecular data storage device inspired from the electron transport characteristics inside molecules has been presented in [95] and is represented schematically in Figure 7.32. When a benzene molecule with two chemical side-groups X and Y is driven by a bias voltage between the input electrode *in* and output electrodes *out*1, *out*2, and *out*3, the probability of an electron to reach one of the possible output states depends on X and Y. The states of single atomic orbitals X and Y can be associated with logical bits in the sense that 0 denotes no orbital and 1 the presence of an orbital at either X or Y. Thus, there are four possible input states (0,0), (0,1) (1,0) or and (1,1) that could be stored. A change of the X and Y groups conjugate or de-conjugate them from the central phenyl ring, controlling the energy levels of the molecule around the HOMO-LUMO gap; the change of X and Y can occur through site-specific electrochemical reactions or through the control of their rotation angle by an STM tip. Calculations in the tight-binding representation show that in the ballistic transport regime with all four electrodes attached, the four different molecular storage states result in four different sets of output currents. The storage states can be determined by measuring the current at only two output electrodes. In the ballistic transport regime the output currents are calculated from the transmission coefficients, by taking into account electron wavefunction propagation through the molecule and the eventual interference effects. On the contrary, in the tunneling regime the propagation paths to the possible output electrodes are independent from each other and only three storage states can be identified. The tunneling transport regime is closer to reality since the molecule is attached to the metal electrodes through polyacetylene chains that act as tunneling wires, and transform the ballistic electron propagation into a tunneling propagation regime by introducing a gap in the transmission coefficient around the Fermi energy. Recent results show that the molecular memory scheme presented in Figure 7.32 can be extended (at least in principle) to perform intra-molecular logic functions, for example, XOR and AND gates that form a half-adder [96].

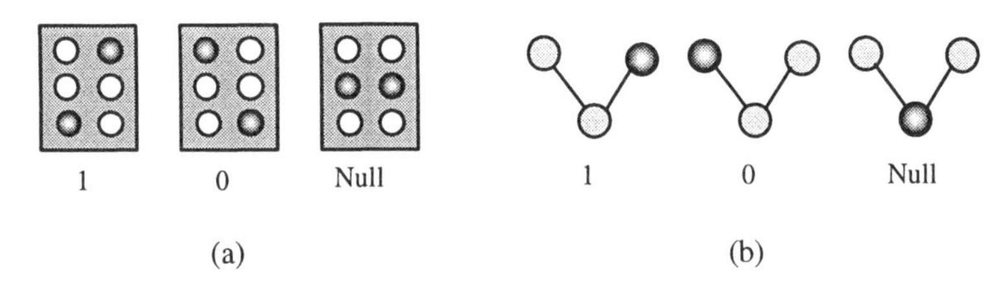

Figure 7.33 (a) In a clocked automata with a cell consisting of an array of six quantum dots the logic values are encoded in the position of the extra charges (dark circles); (b) in clocked molecular automata a half-cell can be encoded in the charge position in a V-shaped molecule (*After*: [98]).

Boolean computers rely on operations with binary logical bits that can have only two values: 0 and 1 logic. In present-day computers these are encoded into the low-current and high-current state of transistors fabricated in the Si-based integrated technology. Alternatively, the binary inputs can be encoded in the low-output and high-output states of molecular switches stimulated by chemical, electrical or optical signals. The implementation of basic logical operations such as AND, NOR, and OR, as well as of more complex functions, is reviewed in [97]. Another proposal for molecular cellular automata uses different redox sites of molecules to represent binary information, and implements the device-device coupling through Coulomb interaction between neighboring molecules. Clocked control of molecular quantum dot cellular automata (QCA) can be achieved with local electric fields [98]. In a clocked QCA cell one must encode not only the 0 and 1 logic states but also a null state, encoding that is easily done in a six-dot array that represents a cell in which the position of two extra mobile charges (electrons or holes) define the desired states, as illustrated in Figure 7.33(a). In molecular QCA the logic states of half-cells are defined by the positions of a charge in a V-shaped molecule, as shown in Figure 7.33(b).

Figure 7.34 A fullerene C_{60} molecule enclosed in a C_{480} nanocapsule can encode logical bit values through its position inside the nanocapsule (*After*: [99]).

Molecular computation can also be performed, at least in principle, with multiwall carbon C_{480} nanocapsules that contain electrically charged C_{60} fullerenes [99]. The logical states are in this case represented by the position of C_{60} inside the nanocapsule, as shown in Figure 7.34. The van der Waals force dominates the interaction between the fullerene molecule and C_{480}, and stabilizes the fullerene molecule at either end of the capsule, but weak covalent interaction, which is proportional to the contact area between C_{60} and C_{480}, and an image charge interaction (if C_{60} is electrically charged), is also present. For electrically charged fullerene molecules, which form when a K atom is encapsulated in C_{60} so that the valence electron of K is transferred via C_{60} shell to the graphitic structure, the information can be written by switching the equilibrium position of the C_{60}^{+} ion through an applied bias of 1.5 V between the end caps of C_{480}; recent studies indicate, however, that the charge transferred to the C_{480} is less than 1 electron per molecule. Destructive reading is done by measuring the current pulse in the connecting wires generated by C_{60}^{+} motion under an applied probing voltage. High-storage densities can be implemented by placing nanocapsules at cross-points between arrays of nanowire electrodes, such that each capsule can be individually addressed, and/or several memory elements can be addressed in parallel. The potential traps of the capsule near its ends are deep enough (about 0.24 eV) to assure stability of stored information well beyond room temperature; the information can be destroyed for $T \geq 3000$ K. Because position switching of C_{60}^{+} is expected to occur in about 10 ps the access rate of the memory is close to 0.1 THz, which imply a data throughput rate of 10 Gbyte/s.

Not only Boolean logical operations can be performed with molecules, but also optical computation. The interest in optical computation is justified by the high speed of operations, which is comparable to light velocity and thus is much higher than electron velocity in solid-state computers, and by the fact that optical signals with closely spaced wavelengths can be transmitted in parallel through a single waveguide since, unlike electrical signals, optical signals with different wavelengths do not interact. In both optical computing and signal processsing, negative light intensities would be of advantage. A unique opportunity to generate negative light intensities is provided by bacteriorhodopsin (bR) films, which can transform its purple state into a stable yellow form M by absorption of a photon in the yellow spectral range [100]. This transformation takes place in about 50 μs, the M state switching back to the bR state in about 200 ns after absorption of a photon in the blue spectral range. Thus, a film prepared with equal concentrations of bR and M can encode negative and positive light intensity values as enhancements in the concentrations of bR and M, respectively, which correspond to excess of blue and yellow light. For example, a difference of 1D Gaussian functions with widths σ_1 and σ_2,

$$\Delta G = G(x,\sigma_1) - G(x,\sigma_2) \tag{7.8}$$

where $G(x,\sigma) = (2\pi\sigma^2)^{1/2}\exp(-x^2/2\sigma^2)$ can be implemented as follows: the Gaussian distribution is mimicked by an out-of-focus image and the subtraction is accomplished by illuminating the film of mixed bR and M states with two yellow and blue images, each having a different Gaussian distribution. The spatial distribution of concentration obtained in this way reflects the difference of Gaussian intensities, which takes both positive and negative values. Because the photoinduced transitions from the bR to the M state and vice versa are accompanied by charge movement of opposite polarity, which can be detected via electrical signals, this implementation of negative light intensity can facilitate the integration of optical with electrical computers. This would be of great practical importance since there is no optical analog of a transistor and the conversion between binary data and levels of optical intensity is achieved with sophisticated optoelectronic devices.

One of the DNA-based logical devices is based on the fact that phosphate bridges in DNA act as tunnel junctions in the Coulomb blockade regime and hydrogen H-bonds have capacitive properties. Internal chemical bonds in DNA can thus be used to implement stable, nanoscale room-temperature logical devices, which are also very fast since they are based on the single electron effect [101]. A schematic representation of such a logic device is displayed in Figure 7.35(a). In this device the P-bond, namely, the phosphorus bridges that connect different DNA units (grains) of a DNA strand, each composed of a sugar and a base, forms a tunnel junction for a net charge. The tunneling can be either coherent or stochastic depending on the coupling to the degrees of freedom of the environment. The capacitive property of the H-bond, which connects complementary bases in different DNA strands, is due to the proton that screens a net charge density on either side of the bond by displacing towards this side. The darkened rectangles in Figure 7.35(a) represent the grains, which have inductive properties due to hopping of additional electrons, while V and V_G are the external and gate voltages, respectively. The fabrication of the structure in Figure 7.35(a), which consists of a main strand and a gate strand, such that the end base of the gate strand is attached to a complementary base in the middle of the main strand, can be done with manipulation techniques. The strands are metal-coated with the exception of the grain in the main strand connected to the gate strand, its neighboring P-bonds, and the connective H-bond. The resulting device, which consists of a grain connected to a voltage source by two tunnel junctions and capacitively coupled to a gate voltage, is in fact a DNA-SET transistor if the inductive properties of the grain are neglected. For incoherent tunneling across identical tunneling junctions and smeared discrete energy levels in the grain, the $I-V$ characteristic has a voltage threshold above which the current raises steeply if there is a gap in the density of energy states in the grain; the bias V_G varies around this threshold voltage. The energy gap arises naturally since the DNA is not conductive, and can be increased using longer sections of DNA that consist of several grains. Boolean logic values 0 and 1 are identified with the low- and high-current states of the DNA-SET.

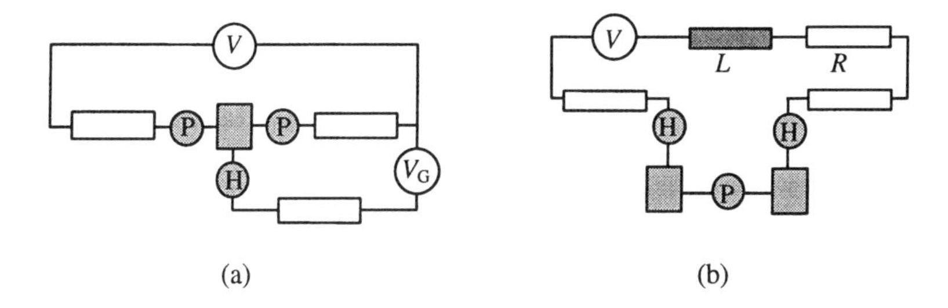

Figure 7.35 (a) DNA-SET logical device based on internal chemical bonds and (b) DNA qubit (*After*: [101]).

A DNA quantum bit, or qubit, can be implemented using a short DNA strand, which contains two sugars linked by a P-bond, connected in series to two metal-coated long strands by two H-bonds, to an external voltage source, a resistance R and an inductance L, as shown in Figure 7.35(b) [101]. The qubit maintains quantum coherence over long time periods if the interaction with the environment is weak.

A multilevel molecular memory for nonvolatile data storage of three bits per cell, which implies eight levels in contrast to two levels in the standard one-bit-per-cell configuration, has been demonstrated in [102]. The eight levels are implemen-ted by electrical charges placed at eight discrete levels in redox active molecules, which are self-assembled on single-crystal nanowire FETs, the reading and writing operations being performed by current sensing and by applying gate voltage pulses, respectively. This memory device retains its charge storage ability up to 600 h. Different memory states are represented by different number of charges stored in molecules interfaced with the channel of an FET in which hot electrons are injected from the channel into the floating gate through a tunneling oxide layer. A schematic representation of the molecular memory FET with two bits per cell is given in Figure 7.36, where S and D are source and drain electrodes. One, two, three, or four charges stored in the redox molecules can be identified with the 00, 01, 10, and 11 logic states; these states can be modified by altering the population of the reduced/oxidized molecules. In a FET with a 10 nm wide and 2 μm long In_2O_3 nanowire channel on which Fe-terpyridine redox molecular wires are absorbed, a storage density of 40 Gbits/cm^2 can be achieved at room temperature for a monolayer deposition of molecules assuming that each cell has eight levels. The memory device can be programmed by applying negative gate pulses of different amplitudes, or by sweeping the gate voltage V_G from 0 to negative values, since different values of the negative gate bias correspond to different conduction levels at $V_G = 0$. The readout is performed with a small electric field applied between source and drain contacts, the value of the current indicating the logical value of the memory. Erasing the stored information can be achieved by sweeping V_G from 0 to a large positive value or by applying positive gate pulses of high amplitudes.

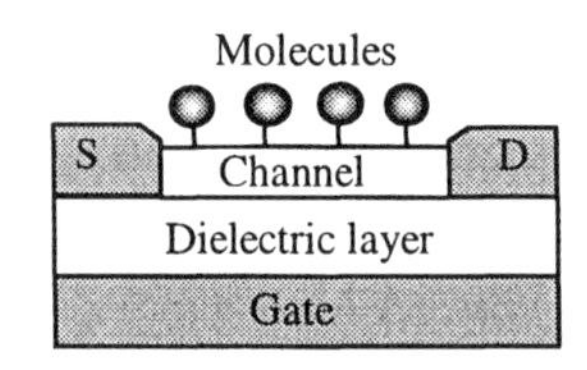

Figure 7.36 Molecular memory field-effect transistor with 2 bits per cell (*After*: [102]).

Apart from mimicking electronic, optical or quantum computation algorithms, specific computing algorithms can be designed for molecular computers. The first successful manipulation of a large set of 20-mer DNA sequences in test tubes to solve a small instance of the Hamiltonian path problem has been reported in [103]. The problem resides in finding a path through a connected graph of nodes such that each node is visited exactly once. In [103] each edge in a graph is encoded as a DNA fragment with two sticky ends that represent the starting and ending nodes of the edge. A sequence of DNA is allocated to each node in the graph such that any edge that starts at a particular node uses this sequence at one end and the complement of the DNA sequence at the ending node. The fragments are then mixed together and form strings of edges representing feasible paths in the graph, all strings other than the Hamiltonian paths (which have as many edges as nodes in the graph) being discarded by biochemical techniques. DNA computing has also been reported in [104], where it was shown that a large class of NP-complete satisfaction problems can be solved by encoding the problem in DNA molecules. (An NP-problem is a problem verifiable in nondeterministic polynomial time; NP-complete problems have the property that any other NP-problem can be translated into these problems.) This and other works on DNA computing (for example, the references in [105–107]) are justified by the potentiality of massive parallelism. DNA Boolean logic functions using separate operator strands and input strands can be implemented using hybridization and ligation, as shown in [105]. Any Boolean variable X is represented by two distinct and noncomplementary DNA strands, x and $\bar{x}$ (the bar denotes negation), the presence of x indicating that X is true, while $\bar{x}$ indicates that X is false. Similarly, if a structure f is assembled after hybridization of the input strands with operator strands and subsequent ligation and denaturing, the value of the Boolean function F is true, whereas F is false if $\bar{f}$ is obtained. Since any Boolean operation can be decomposed in NOR, AND, and OR gates, the DNA implementation of the addition operation $\wedge$ was studied in [105], the function $F = X_1 \wedge X_2$ being true, for example, if a spot on a hybridization plate becomes fluorescent at location 1 and false if fluorescence is obtained at location 2. Since $\overline{F} = \overline{X}_1 \vee \overline{X}_2$, with $\vee$ the OR operation, the DNA implementation of addition is in fact an implementation of OR if the negation of the input and output strands is considered. The method presented in [105] for a two-bit addition requires more than 30 gates and can be competitive with electronic computation only when the DNA parallelism is exploited by partitioning the reaction vessel into separate active volumes, which allows the addition to a specific test tube of only those operator strands actually used in the computation; a separate test tube is used to compute each output bit. Other computer architectures specific for DNA parallel computation and enabled by self-assembly can be found in [108].

In another DNA-based algorithm, the information is encoded in the order of nucleotides (bases). However, 3D structures constructed from DNA molecules could be more efficient in problem solving. For example, using these structures the Hamiltonian path problem can be solved in a constant number of steps irrespective of the size of the graph. In an attempt to enlarge the problems that could be solved with 3D DNA structures to the 3-satisfiability and 3-vertexcolorability problems k-armed branched junction molecules were used to represent the k-degree vertices of the graph [106]. The solution of the problem can be found by the construction of one graph structure in many copies.

An intramolecular computing model, which encodes the input bits in the coupling constants of the Hamiltonian that governs the time evolution of a single molecule prepared in a non-stationary state, has been proposed in [109]. The coupling constants of the Hamiltonian can be modified, for example, by switching between different conformations of the molecule with an STM. In this way reliable logic gates such as AND, XOR or HALF-ADDER can be implemented, the computational result being read by performing a quantum measurement on the molecule, which can be either a measurement of the population of a molecular level after a certain computational time or of the population of an auxiliary quantum system (an ancilla) coupled to the molecule.

Spintronic computing systems and algorithms (see Chapter 4) could also be implemented with molecules since the first step towards them, which is the coherent transfer of electron spins between quantum dots with different radius coupled by a benzene ring, has been recently demonstrated [110]. The instantaneous transfer of spin polarization created by optical pumping from the larger to the smaller quantum dot is believed to be mediated by the π-conjugated molecule.

We have shown that molecular devices can be used for Boolean computation, which uses mathematical algorithms created by man, but it could be pondered upon if biological systems should not be more efficiently used in computational algorithms inspired from biological processes. After all, the living cell can be regarded as the smallest DNA-based computer driven by self-organizing chemical reactions and molecular motors. And indeed, efforts are spent to conceptualize the analysis, model, and design of living cells in engineering terms; it is hoped that a "silicon mimetic" approach or an *in vivo* logic circuit terminology could help us understand the biological functionality [111]. But the cell language is not yet very precisely formulated, although efforts are made to unveil the laws of molecular semiotics [112], which study the signs that mediate computation, measurement and communication on the level of cells and molecules, or to develop a specification language to model, analyze, and verify biological systems seen as hybrid systems comprising a non-trivial mixture of continuous activities and discrete events [113]. Perhaps the ultimate goal of molecular computing, when biology will be able to fully understand and manipulate biological processes, would be to program ourselves in order to perform certain tasks, such as maintaining our youth, becoming top athletes or scientists, and so on. However, there is a long path until we will transform ourselves in molecular machines.

7.6 INTEGRATION OF BIO- AND NANODEVICES

Biology and nanotechnology, although based on quite different concepts, are in many cases sources of inspiration for one another and, moreover, tend to merge and to develop hybrid non-organic/biological systems with unique performances and functionalities. It would be unfair not to briefly mention some of the most remarkable achievements in the domain of integration of bio- and nanodevices; a detailed study of this subject is not within the aims of the present chapter so that only a bird's eye overview is possible. The interested reader can consult the references for more information.

In this chapter we have mainly dealt with molecular devices for engineering applications but the reverse is also true: nanodevices can replace certain biological systems and functions in different prosthetic applications. Perhaps one of the most amazing achievements in this field is the realization of a nanotube stereo-cilia device, called also nanoear due to its similarity to the hair cells in human ears. This device, composed of an array of highly ordered and parallel CNTs, senses molecular motions via the acoustic waves generated by molecular activities [114]. A not less impressive realization is the development of implanta-ble epiretinal and subretinal stimulator devices consisting of microelectrode arrays integrated into flexible and rigid substrates [115], or the fabrication of a 3D structured diaphragm based on the design of the *Ormia ochracea* fly ear, which can be used as a biomimetic directional or differential microphone [116]. The design of high-quality achromatic nanolenses or multielement lens systems can be improved if inspiration is drawn from biology [117]. Biological eyes, whether compound or not, use gradient index lenses implemented by up to a thousand layers of cells with different refractive indexes induced by a different concentration of protein, and are able to change focus maintaining the center of the lens in a fixed position by changing the shape of the lens through muscle contraction. The greatest challenge in bio-inspired nano-optics is finding the appropriate materials, which could turn out to be organic. Biological designs that produce striking optical effects such as strong angle-dependent iridescence, ultra-long-range visibility, or intense UV visibility in certain species of butterflies are in fact photonic structures of reduced dimensions, that were implemented in optoelectronics only very recently. Even 3D photonic crystals have existed in biological designs for millions of years and are used, for example, by *Parides sesostris* to reflect bright green color over a broad angle. These biological 3D photonic crystals can function also as anti-reflection structures over broad angles or frequency ranges [118].

For future biotechnology applications in nanoelectronics it should be important, for example, to interface neurons and living cells to bulk silicon or porous silicon electronic devices and, eventually, to convert the neural activity into optical or electrical signals. Recently, it was shown that biochemical molecules, in particular neurons, can be adsorbed on the surface of porous silicon, and preliminary experiments suggest that bioactivity (the electrical activity of neurons) could be eventually converted into a quantifiable optical signal taking advantage of the visible luminescence of microporous Si with nanometer-sized crystallites [119]. *Aplysia* neurons, for instance, cultured on porous Si preserve normal membrane properties and produce action potentials while forming a neuron-semiconductor contact; these neurons survive at least a week on porous Si substrates.

Nanotechnology can eventually offer tools that interface with biomolecular processes, introduce stimuli, and analyze responses with both spatial and temporal resolution because nanoscale devices have the same size as the biomolecular machines of cells. A revolution in biology may occur, similar to that in electronics, with the advent of nanotechnology. In this respect, arrays of closely spaced vertically aligned carbon nanofibers (with a pitch of 5 μm, diameter of 30 nm, and 7 μm long), which are first modified with either adsorbed or covalently linked plasmid DNA, can be inserted into viable cells, which retain their viability after insertion. This intracellular interface can be used for biochemical manipulation, for *in vivo* diagnostics, or *in vivo* logic devices [120]. *In vivo* computation has been reported using biological processes such as gene expression and protein synthesis, genetic toggle switch as cellular memory unit, or molecular switches based on DNA or RNA/ribozyme (see the references in [107]). The introduction of plasmid DNA into living cells in order to control the cellular response is one approach towards designing biomaterials [121], which can speed the healing processes, deliver drugs or genes, and improve tissue regeneration through incorporation and controlled release of growth factors, biomimetic adhesion sites, DNA, or other biologically active materials. Another approach to design biomaterials is to use microfabrication techniques to create, for example, microchips with multiple reservoirs for drugs that can be released through precise activation (see the references in [121]).

The integration of biological systems with microelectronics requires selective and reversible recognition between the sensor surface and analyte, a single-event sensitivity of detection or signal processing and high levels of compatibility between the biological interface and the artificial device. Kinesin or F1-ATPase molecular motors can be tethered to nanopatterned surfaces so that the self-propelling enzymes can be combined with NEMS in order to fabricate bio-smart interfaces. A simpler alternative is the functionalization of inorganic substrates, structures (quantum dots, for example), or devices [83]. Yet another example of the integration of biological systems with nanodevices is the use of molecular glues and lubricants, such as SAMs, Langmuir-Blodgett monolayers and multilayers, or nanocomposite molecular layers, in order to diminish the gap between mating surfaces in MEMS and high-density storage technologies [122]. These molecular lubricants with a thickness of less than 10 nm could supply a low static friction coefficient and shear modulus, a reduced surface energy, and tailored dynamic behavior, while providing the ability to self-repair and increasing the shelf life and the ability to sustain temperature and humidity variations.

The possibility of integrating biodevices with nanodevices will soon lead to the development of artificial biology. Actually, artificial mechanical red blood cells with a diameter of about 1 μm and equipped with pressure, chemical, and thermal sensors as well as with an onboard nanocomputer have already been designed. The amount of oxygen per unit volume expected to be provided by this cell is 236 times larger than that delivered by a natural red cell. Nerve cell membranes are imitated by artificial ion-based nanomachines consisting of membrane-forming molecules, which are chemically tethered to a gold surface, ion channels within the membrane, a reservoir space between the membrane and the surface that stores ions, and receptors attached to the membrane that recognize specific molecules. These and other nanodevices, which have applications in nanomedicine, are reviewed in [107], while a 3D ionic polymer-metal composite artificial muscle made with ion-exchange membranes, which changes its conformation and generates forces associated with small displacements in response to electric biases of a few volts, is described in [123]. These achievements are quite remarkable for the incipient stage of integration of biology and nanotechnology, and undoubtedly they will be followed by more spectacular successes from which both medicine and technology will benefit.

REFERENCES

[1] Dalton, L.R. "Rational design of organic electro-optic materials," *J. Phys.: Condens. Matter*, Vol. 15, No. 20, 2003, pp. R897-R934.

[2] Preezant, Y., Y. Roichman, and N. Tessler, "Amorphous organic devices—degenerate semiconductors," *J. Phys.: Condens. Matter*, Vol. 14, No. 42, 2002, pp. 9913-9924.

[3] Eder, F., et al., "Organic electronics on paper," *Appl. Phys. Lett.*, Vol. 84, No. 14, 2004, pp. 2673-2675.

[4] Ball, P., "Natural strategies for the molecular engineer," *Nanotechnology*, Vol. 13, No. 5, 2002, pp. R15-R28.

[5] Kitaev, Y.E., et al., "Why biomolecules prefer only a few crystal structures," *Phys. Rev. E*, Vol. 67, No. 1, 2003, pp. 011907/1-8.

[6] Reed, M.A., "Molecular-scale electronics," *Proc. IEEE*, Vol. 87, No. 4, 1999, pp. 652-658.

[7] Mantooth, B.A. and P.S. Weiss, "Fabrication, assembly, and characterization of molecular electronic components," *Proc. IEEE*, Vol. 91, No. 11, 2003, pp. 1785-1802.

[8] Joachim, C., J.K. Gimzewski, and A. Aviram, "Electronics using hybrid-molecular and mono-molecular devices," *Nature*, Vol. 408, No. 6812, 2000, pp. 541-548.

[9] Selzer, Y., et al., "Temperature effects on conduction through a molecular junction," *Nanotechnology*, Vol. 15, No. 10, 2004, pp. S483-S488.

[10] Astumian, R.D. and I. Derényi, "Towards a chemically driven molecular electron pump," *Phys. Rev. Lett.*, Vol. 86, No. 17, 2001, pp. 3859-3862.

[11] Damle, P.S., A.W. Ghosh, and S. Datta, "Unified description of molecular conduction: from molecules to metallic wires," *Phys. Rev. B*, Vol. 64, No. 20, 2001, pp. 201403/1-4.

[12] Cuniberti, G., G. Fagas, and K. Richter, "Fingerprints of mesoscopic leads in the conductance of a molecular wire," *Chemical Physics*, Vol. 281, No. 2-3, 2002, pp. 465-476.

[13] Pantelides, S.T., et al., "Molecular electronics by the numbers," *IEEE Trans. Nanotechnology*, Vol. 1, No. 1, 2002, pp. 86-90.

[14] Di Ventra, M., S.T. Pantelides, and N.D. Lang, "Current-induced forces in molecular wires," *Phys. Rev. Lett.*, Vol. 88, No. 4, 2002, pp. 046801/1-4.

[15] Dholakia, G.R., et al., "Transport in self-assembled molecular wires: effect of packing and order," *Phys. Rev. B*, Vol. 69, No. 15, 2004, pp. 153402/1-4.

[16] van Ruitenbeek, J.M., et al., "Adjustable nanofabricated atomic size contacts," *Rev. Sci. Instrum.*, Vol. 67, No. 1, 1996, pp. 108-111.

[17] Dulić, D., et al., "One-way optoelectronic switching of photochromic molecules on gold," *Phys. Rev. Lett.*, Vol. 91, No. 20, 2003, pp. 207402/1-4.

[18] Reed, M.A., et al., "Conductance of a molecular junction," *Science*, Vol. 278, No. 5336, 1997, pp. 252-254.

[19] Zhitenev, N.B., H. Meng, and Z. Bao, "Conductance of small molecular junctions," *Phys. Rev. Lett.*, Vol. 88, No. 22, 2002, pp. 226801/1-4.

[20] Richter, J., "Metallization of DNA," *Physica E*, Vol. 16, No. 2, 2003, pp. 157-173.

[21] Endres, R.G., D.L. Cox, and R.R.P. Singh, "The quest for high-conductance DNA," *Rev. Mod. Phys.*, Vol. 76, No.1, 2004, pp. 195-214.

[22] Storm, A.J., et al., "Insulating behavior for DNA molecules between nanoelectrodes at the 100 nm length scale," *Appl. Phys. Lett.*, Vol. 79, No. 23, 2001, pp. 3881-3883.

[23] Porath, D., et al., "Direct measurement of electrical transport through DNA molecules," *Nature*, Vol. 403, No. 6770, 2000, pp. 635-638.

[24] Rakitin, A., et al., "Metallic conduction through engineered DNA: DNA nanoelectronic building blocks," *Phys. Rev. Lett.*, Vol. 86, No. 16, 2001, pp. 3670-3673.

[25] Lee, H.Y., et al., "Control of electrical conduction in DNA using oxygen hole doping," *Appl. Phys. Lett.*, Vol. 80, No. 9, 2002, pp. 1670-1672.

[26] Pati, R., et al., "Theoretical study of electrical transport in a fullerene-doped semiconducting carbon nanotubes," *J. Appl. Phys.*, Vol. 95, No. 2, 2004, pp. 694-697.

[27] Hla, S.-W. and K.-H. Rieder, "Engineering of single molecules with a scanning tunneling microscope tip," *Superlattices and Microstructures*, Vol. 31, No. 1, 2002, pp. 63-72.

[28] Ikai, A., et al., "Nanotechnology and protein mechanics," *J. Biological Phys.*, Vol. 28, No. 4, 2002, pp. 561-572.

[29] Yan, J., D. Skoko, and J.F. Marko, "Near-field-magnetic tweezer manipulation of single DNA molecules," *Phys. Rev. E*, Vol. 70, No. 1, 2004, pp. 011905/1-5.

[30] Chiou, C.-H. and G.-B. Lee, "A micromachined DNA manipulation platform for the stretching and rotation of a single DNA molecule," *J. Micromech. Microeng.*, Vol. 15, No. 1, 2005, pp. 109-117.

[31] Burke, P.J., "Nanodielectrophoresis: electronic nanotweezers," in *Encyclopedia of Nanoscience and Nanotechnology*, H.S. Nalwa (Ed.), American Scientific Publishers, Valencia, California, Vol. 10, 2003, pp. 1-19.

[32] Cross, R.A. and N.J. Carter, "Molecular motors," *Current Biology*, Vol. 10, No. 5, 2000, pp. R177-R179.

[33] Astumian, R.D., "Thermodynamics and kinetics of a Brownian motor," *Science*, Vol. 276, No. 5314, 1997, pp. 917-922.

[34] Derényi, I. and T. Vicsek, "Realistic models of biological motion," *Physica A*, Vol. 249, No. 1-4, 1998, pp. 397-406.

[35] Koza, Z., "Maximal force exerted by a molecular motor," *Phys. Rev. E*, Vol. 65, No. 3, 2002, pp. 031905/1-5.

[36] Derényi, I., M. Bier, and R.D. Astumian, "Generalized efficiency and its application to microscopic engines," *Phys. Rev. Lett.*, Vol. 83, No. 5, 1999, pp. 903-906.

[37] Igarashi, A., S. Tsukamoto, and H. Goko, "Transport properties and efficiency of elastically coupled Brownian motors," *Phys. Rev. E*, Vol. 64, No. 5, 2001, pp. 051908/1-5.

[38] Bader, J.S., et al., "DNA transport by a micromachined Brownian ratchet device," *Proc. Natl. Acad. Sci. USA (PNAS)*, Vol. 96, No. 23, 1999, pp. 13165-13169.

[39] Porto, M., "Molecular motor based entirely on the Coulomb interaction," *Phys. Rev. E*, Vol. 63, No. 3, 2001, pp. 030102(R)/1-4.

[40] Linke, H., et al., "A quantum dot ratchet: experiment and theory," *Europhys. Lett.*, Vol. 44, No. 3, 1998, pp. 341-347.

[41] Reimann, P., M. Grifoni, and P. Hänggi, "Quantum ratchets," *Phys. Rev. Lett.*, Vol. 79, No. 1, 1997, pp. 10-13.

[42] Linke, H., et al., "Experimental tunneling ratchets," *Science*, Vol. 286, No. 5448, 1999, pp. 2314-2317.

[43] Humphrey, T.E., H. Linke, and R. Newbury, "Pumping heat with quantum ratchets," *Physica E*, Vol. 11, No. 2-3, 2001, pp. 281-286.

[44] Derényi, I. and R.D. Astumian, "Spontaneous onset of coherence and energy storage by membrane transporters in an RLC electric circuit," *Phys. Rev. Lett.*, Vol. 80, No. 20, 1998, pp. 4602-4605.

[45] Siwy, Z. and A. Fuliński, "A nanodevice for rectification and pumping ions," *Am. J. Phys.*, Vol. 72, No. 5, 2004, pp. 567-574.

[46] Hou, S., et al., "Investigation of triptycene-based surface-mounted rotors," *Nanotechnology*, Vol. 14, No. 5, 2003, pp. 566-570.

[47] Gimzewski, J.K., et al., "Rotation of a single molecule within a supramolecular bearing," *Science*, Vol. 281, No. 5376, 1998, pp. 531-533.

[48] Badjić, J.D., et al., "A molecular elevator," *Science*, Vol. 303, No. 5665, 2004, pp. 1845-1849.

[49] Simmel, F.C. and B. Yurke, "Using DNA to construct and power a nanoactuator," *Phys. Rev. E*, Vol. 63, No. 4, 2001, pp. 041913/1-5.

[50] Simmel, F.C. and B. Yurke, "A DNA-based molecular device switchable between three distinct mechanical states," *Appl. Phys. Lett.*, Vol. 80, No. 5, 2002, pp. 883-885.

[51] Fritz, J., et al., "Translating biomolecular recognition into nanomechanics," *Science*, Vol. 288, No. 5464, 2000, pp. 316-318.

[52] Li, Z., et al., "Sequence-specific label-free DNA sensors based on silicon nanowires," *Nano Lett.*, Vol. 4, No. 2, 2004, pp. 245-247.

[53] Hahm, J.-I. and C.M. Lieber, "Direct ultrasensitive electrical detection of DNA and DNA sequence variations using nanowire nanosensors," *Nano Lett.*, Vol. 4, No. 1, 2004, pp. 51-54.

[54] Haguet, V., et al., "Combined nanogap nanoparticles nanosensor for electrical detection of biomolecular interactions between polypeptides," *Appl. Phys. Lett.*, Vol. 84, No. 7, 2004, pp. 1213-1215.

[55] Huang, T.J., et al., "A nanomechanical device based on linear molecular motors," *Appl. Phys. Lett.*, Vol. 85, No. 22, 2004, pp. 5391-5393.

[56] Kis, A., et al., "Nanomechanics of microtubules," *Phys. Rev. Lett.*, Vol. 89, No. 24, 2002, pp. 248101/1-4.

[57] Hess, H., J. Howard, and V. Vogel, "A piconewton forcemeter assembled from microtubules and kinesins," *Nano Lett.*, Vol. 2, No. 10, 2002, pp. 1113-1115.

[58] Cahen, D. and G. Hodes, "Molecules and electronic materials," *Adv. Mater.*, Vol. 14, No. 11, 2002, pp. 789-798.

[59] Aviram, A. and M.A. Ratner, "Molecular rectifiers," *Chem. Phys. Lett.*, Vol. 29, No. 2, 1974, pp. 277-283.

[60] Troisi, A. and M.A. Ratner, "Conformational molecular rectifiers," *Nano Lett.*, Vol. 4, No. 4, 2004, pp. 591-595.

[61] Bezryadin, A., C. Dekker, and G. Schmid, "Electrostatic trapping of single conducting nanoparticles between nanoelectrodes," *Appl. Phys. Lett.*, Vol. 71, No. 9, 1997, pp. 1273-1275.

[62] Maubach, G. and W. Fritzsche, "Precise positioning of individual DNA structures in electrode gaps by self-organization onto guiding microstructures," *Nano Lett.*, Vol. 4, No. 4, 2004, pp. 607-611.

[63] Wegewijs, M.R., et al., "Negative differential conductance in a benzene-molecular device," *Physica E*, Vol. 18, No. 1-3, 2003, pp. 241-242.

[64] Kratochvilova, I., et al., "Room temperature negative differential resistance in molecular nanowires," *J. Mater. Chem.*, Vol. 12, No. 10, 2002, pp. 2927-2930.

[65] Guisinger, N.P., et al., "Room temperature negative differential resistance through individual organic molecules on silicon surfaces," *Nano Lett.*, Vol. 4, No. 1, 2004, pp. 55-59.

[66] Le, J.D., et al., "Negative differential resistance in a bilayer molecular junction," *Appl. Phys. Lett.*, Vol. 83, No. 26, 2003, pp. 5518-5520.

[67] Just, E.M. and M.R. Wasielewski, "Picosecond molecular switch based on the influence of photogenerated electric fields on optical charge transfer transitions," *Superlattices and Microstructures*, Vol. 28, No. 4, 2000, pp. 317-328.

[68] Emberly, E.G. and G. Kirczenow, "The smallest molecular switch," *Phys. Rev. Lett.*, Vol. 91, No. 18, 2003, pp. 188301/1-4.

[69] Gutierrez, R., et al., "Theory of an all-carbon molecular switch," *Phys. Rev. B*, Vol. 65, No. 11, 2002, pp. 113410/1-4.

[70] Alexandrov, A.S., A.M. Bratkovsky, and R.S. Williams, "Bistable tunneling current through a molecular quantum dot," *Phys. Rev. B*, Vol. 67, No. 7, 2003, pp. 075301/1-4.

[71] Orellana, P. and F. Claro, "A terahertz molecular switch," *Phys. Rev. Lett.*, Vol. 90, No. 17, 2003, pp. 178302/1-4.

[72] So, H.-M., et al., "Molecule-based single electron transistor," *Physica E*, Vol. 18, No. 1-3, 2003, pp. 243-244.

[73] Watanabe, H., et al., "Single molecule DNA device measured with triple-probe atomic force microscope," *Appl. Phys. Lett.*, Vol. 79, No. 15, 2001, pp. 2462-2464.

[74] Park, J., et al., "Coulomb blockade and the Kondo effect in single-atom transistors," *Nature*, Vol. 417, No. 6890, 2002, pp. 722-725.

[75] Park, H., et al., "Nanomechanical oscillations in a single-C_{60} transistor," *Nature*, Vol. 407, No. 6800, 2000, pp. 57-60.

[76] Park, J., et al., "Wiring up single molecule," *Thin Solid Films*, Vol. 438-439, No. 1, 2003, pp. 457-461.

[77] Liang, W., et al., "Kondo resonance in a single-molecule transistor," *Nature*, Vol. 417, No. 6890, 2002, pp. 725-729.

[78] Diez, S., et al., "Stretching and transporting DNA molecules using motor proteins," *Nano Lett.*, Vol. 3, No. 9, 2003, pp. 1251-1254.

[79] Deng, Z. and C. Mao, "DNA-templated fabrication of 1D parallel and 2D crossed metallic nanowire arrays," *Nano Lett.*, Vol. 3, No. 11, 2003, pp. 1545-1548.

[80] Nakao, H., et al., "Highly ordered assemblies of Au nanoparticles organized on DNA," *Nano Lett.*, Vol. 3, No. 10, 2003, pp. 1391-1394.

[81] Dittmer, W.U. and F.C. Simmel, "Chains of semiconductor nanoparticles templated on DNA," *Appl. Phys. Lett.*, Vol. 85, No. 4, 2004, pp. 633-635.

[82] Park, S.H., et al., "Electronic nanostructures templated on self-assembled DNA scaffolds," *Nanotechnology*, Vol. 15, No. 10, 2004, pp. S525-S527.

[83] Dujardin, E. and S. Mann, "Bio-inspired materials chemistry," *Adv. Mater.*, Vol. 14, No. 11, 2002, pp. 775-788.

[84] Keren, K., et al., "DNA-templated carbon nanotube field-effect transistor," *Science*, Vol. 302, No. 5649, 2003, pp. 1380-1382.

[85] Dwyer, C., et al., "Design tools for a DNA-guided self-assembling carbon nanotube technology," *Nanotechnology*, Vol. 15, No. 9, 2004, pp. 1240-1245.

[86] Yan, D., Y. Zhou and J. Hou, "Supramolecular self-assembly of macroscopic tubes," *Science*, Vol. 303, No. 5654, 2004, pp. 65-67.

[87] Crittenden, S., et al., "Humidity-dependent open-circuit photovoltage from a bacteriorhodopsin/indium tin oxide bioelectronic heterostructure," *Nanotechnology*, Vol. 14, No. 5, 2003, pp. 562-565.

[88] Choi, J.-W. and M. Fujihira, "Molecular-scale biophotodiode consisting of a green fluorescent protein/cytochrome *c* self-assembled heterolayer," *Appl. Phys. Lett.*, Vol. 84, No. 12, 2004, pp. 2187-2189.

[89] Zhang, C., et al., "Coherent electron transport through an azobenzene molecule: a light-driven molecular switch," *Phys. Rev. Lett.*, Vol. 92, No. 15, 2004, pp. 158301/1-4.

[90] Wakayama, Y., et al., "Optical switching of single-electron tunneling in SiO_2/molecule/SiO_2 multilayer on Si(100)," *Appl. Phys. Lett.*, Vol. 85, No. 2, 2004, pp. 329-331.

[91] Chen, Y., et al., "Nanoscale molecular-switch crossbar circuits," *Nanotechnology*, Vol. 14, No. 4, 2003, pp. 462-468.

[92] Stewart, D.R., "Molecule-independent electrical switching in Pt/organic monolayer/Ti devices," *Nano Lett.*, Vol. 4, No. 1, 2004, pp. 133-136.

[93] Li, C., et al., "Fabrication approach for molecular memory arrays," *Appl. Phys. Lett.*, Vol. 82, No.4, 2003, pp. 645-647.

[94] Wu, H.M., et al., "Ultrahigh-density data storage on a novel organic thin film achieved using a scanning tunneling microscope," *Nanotechnology*, Vol. 13, No. 6, 2002, pp. 733-735.

[95] Stadler, R., M. Forshaw, and C. Joachim, "Modulation of electron transmission for molecular data storage," *Nanotechnology*, Vol. 14, No. 2, 2003, pp. 138-142.

[96] Stadler, R., et al., "Integrating logic functions inside a single molecule," *Nanotechnology*, Vol. 15, No. 4, 2004, pp. S115-S121.

[97] Raymo, F.M., "Digital processing and communication with molecular switches," *Adv. Mater.*, Vol. 14, No. 6, 2002, pp. 401-414.

[98] Lent, C.S. and B. Isaksen, "Clocked molecular quantum-dot cellular automata," *IEEE Trans. Electron Devices*, Vol. 50, No.9, 2003, pp. 1890-1896.

[99] Kwon, Y.-K., D. Tománek, and S. Iijima, 'Bucky shuttle' memory device: synthetic approach and molecular dynamics simulations," *Phys. Rev. Lett.*, Vol. 82, No.7, 1999, pp. 1470-1473.

[100] Lewis, A., et al., "Optical computation with negative light intensity with a plastic bacteriorhodopsin film," *Science*, Vol. 275, No. 5305, 1997, pp. 1462-1464.

[101] Ben-Jacob, E., Z. Hermon, and S. Caspi, "DNA transistor and quantum bit element: realization of nano-biomolecular logical devices," *Phys. Lett. A*, Vol. 263, No. 3, 1999, pp. 199-202.

[102] Li, C. et al., "Multilevel memory based on molecular devices," *Appl. Phys. Lett.*, Vol. 84, No. 11, 2004, pp. 1949-1951.

[103] Adleman, L.M., "Molecular computation of solutions to combinatorial problems," *Science*, Vol. 266, No. 5187, 1994, pp. 1021-1024.

[104] Lipton, R.J., "DNA solution of hard computational problems," *Science*, Vol. 268, No. 5209, 1995, pp. 542-545.

[105] Yurke, B., A.P. Mills Jr., and S.L. Cheng, "DNA implementation of addition in which the input strands are separate from the operator strands," *BioSystems*, Vol. 52, No. 1-3, 1999, pp. 165-174.

[106] Jonoska, N., S.A. Karl, and M. Saito, "Three dimensional DNA structures in computing," *BioSystems*, Vol. 52, No. 1-3, 1999, pp. 143-153.

[107] Bogunia-Kubik, K. and M. Sugisaka, "From molecular biology to nanotechnology and nanomedicine," *BioSystems*, Vol. 65, No. 2-3, 2002, pp. 123-138.

[108] Dwyer, C., et al., "DNA self-assembled parallel computer architectures," *Nanotechnology*, Vol. 15, No. 11, 2004, pp. 1688-1694.

[109] Fiurášek, J., et al., "Intramolecular Hamiltonian logic gates," *Physica E*, Vol. 24, No. 3-4, 2004, pp. 161-172.

[110] Ouyang, M. and D.D. Awschalom, "Coherent spin transfer between molecularly bridged quantum dots," *Science*, Vol. 301, No. 5636, 2003, pp. 1074-1078.

[111] Simpson, M.L. et al., "Engineering in the biological substrate: information processing in genetic circuits," *Proc. IEEE*, Vol. 92, no. 5, 2004, pp. 848-863.

[112] Ji, S., "The cell as the smallest DNA-based molecular computer," *BioSystems*, Vol. 52, No. 1-3, 1999, pp. 123-133.

[113] Duan, Z., M. Holcombe, and A. Bell, "A logic for biological systems," *BioSystems*, Vol. 55, No. 1-3, 2000, pp. 93-105.

[114] Daviss, B., "Snap, crackle and pop," *New Scientist*, Vol. 171, No. 2300, 2001, p. 34.

[115] Meyer, J.-U., "Retina implant—a bioMEMS challenge," *Sensors and Actuators A*, Vol. 97-98, No. 1, 2002, pp. 1-9.

[116] Yoo, K., et al., "Fabrication of biomimetic 3-D structured diaphragms," *Sensors and Actuators A*, Vol. 97-98, No. 1, 2002, pp. 448-456.

[117] Zuccarello, G., et al., "Materials for bio-inspired optics," *Adv. Mater.*, Vol. 14, No. 18, 2002, pp. 1261-1264.

[118] Vukusic, P. and J.R. Sambles, "Photonic structures in biology," *Nature*, Vol. 424, No. 6950, 2003, pp. 852-855.

[119] Ben-Tabou de Leon, S., et al., "Neurons culturing and biophotonic sensing using porous silicon," *Appl. Phys. Lett.*, Vol. 84, No. 22, 2004, pp. 4361-4363.

[120] McKnight, T.E., et al., "Intracellular integration of synthetic nanostructures with viable cells for controlled biochemical manipulation," *Nanotechnology*, Vol. 14, No. 5, 2003, pp. 551-556.

[121] Sakiyama-Elbert, S.E. and J.A. Hubbell, "Functional biomaterials: design of novel biomaterials," *Annu. Rev. Mater. Res.*, Vol. 31, No. 1, 2001, pp. 183-201.

[122] Tsukruk, V.V., "Molecular lubricants and glues for micro- and nanodevices," *Adv. Mater.*, Vol. 13, No. 2, 2001, pp. 95-108.

[123] Kim, K.J. and M. Shahinpoor, "A novel method of manufacturing three-dimensional ionic polymer-metal composites (IPMCs) biomimetic sensors, actuators and artificial muscles," *Polymer*, Vol. 43, No. 3, 2002, pp. 797-802.

List of Acronyms

0D	zero dimensional
1D	one dimensional
2D	two dimensional
2DEG	two-dimensional electron gas
3D	three dimensional
AFM	atomic force microscope
ATP	adenosine triphosphate
BARITT	barrier injection and transit time diode
CMOS	complementary metal oxide semiconductor
CNT	carbon nanotube
CNTFET	carbon nanotube field effect transistor
CVD	chemical vapor deposition
DBR	distributed Bragg reflector
DGFET	double-gate field effect transistor
DMS	diluted magnetic semiconductors
DNA	deoxyribonucleic acid
DOS	density of states

DRAM	dynamic random access memory
EB	electron beam lithography
FET	field emission transistor
FIB	focused ion beam lithography
FM	ferromagnetic
GMR	giant magnetoresistance
HEMT	high electron mobility transistor
HOMO	highest occupied molecular orbital
IR	infrared
LDOS	local density of states
LED	light-emitting diode
LPCVD	low-pressure chemical vapor deposition
LUMO	lowest unoccupied molecular orbital
MBE	molecular beam epitaxy
MEMS	micro-electro-mechanical systems
MFM	magnetic force microscopy
MOCVD	metal organic chemical vapor deposition
MOSFET	metal oxide semiconductor field effect transistor
MVL	multiple-valued logic
MWCNT	multiwalled carbon nanotube
NEMS	nano-electro-mechanical systems

NIR	near infrared
PADOX	pattern-dependent oxidation
PBG	photonic band gap
PECVD	plasma enhanced chemical vapor deposition
PL	photoluminescence
PMMA	polymethylmethacrylate
RIE	reactive ion etching
RTD	resonant tunneling diode
SAINT	saturable absorber incorporating nanotubes
SAM	self-assembled monolayer
SEM	scanning electron microscope
SET	single-electron transistor
SNOM	scanning near field optical microscope
SOI MOSFET	silicon-on-insulator MOSFET
SPM	scanning probe microscopy
SPP	surface plasmon-polariton
SRAM	static random access memory
STM	scanning tunneling microscope
SWCNT	single-walled carbon nanotube
TBJ	three-terminal ballistic junction
UHV	ultra-high vacuum

UV	ultraviolet
VCSEL	vertical-cavity surface-emitting lasers
VLSI	very large scale integration
WKB	Wentzel-Kramers-Brillouin

About the Authors

Mircea Dragoman was born in Bucharest, Romania in 1955. He received his M.Sc. and Ph.D. degrees from the Polytechnical Institute, Bucharest, in 1980 and 1991, respectively.

He is presently Principal Researcher I at the National Research Institute for Microtechnology, Bucharest, Romania, a position equivalent to professor. He was awarded the Alexander von Humboldt Fellowship in 1991. He was a visiting professor at the University of Duisburg (Germany), the University of "Tor Vergata," Roma (Italy), Institute of Solid-State Electronics, Roma (Italy), University of Saint-Etienne (France), University of Mannheim (Germany), University of Frankfurt (Germany), University of Darmstadt (Germany), and LAAS-CNRS, Toulouse (France). He was also a NATO senior Scientist (1993) and NATO invited lecturer (1995). He authored 125 papers in the area of microwaves, millimeter-waves, terahertz, RF-MEMS, MOEMS, NEMS, optical information, and quantum devices. He is the coauthor of the books *Advanced Optoelectronic Devices*, Springer, 1999, *Optical Characterization of Solids,* Springer, 2002, and *Quantum-Classical Analogies*, Springer, 2004. His current interests are nanodevices, MEMS techniques applied to high frequencies, and quantum devices. He was awarded the Romanian Academy Prize "Gheorghe Cartianu" in 1999. In 2002, together with a research team from five European countries, he was nominated for the Descartes Prize of the European Union for excellence in the area of European research. He is the editor of the book series *Frontiers of Physics*, which is currently published by Springer.

Daniela Dragoman was born in Raduati, Romania, in 1965, and obtained her M.Sc. degree at the University of Bucharest, Romania, in 1989, and her Ph.D. degree at Limerick University, Ireland, in 1993. She was a visiting professor at the University Saint-Etienne, France, in 1997 and 2000 and awarded with the Alexander von Humboldt Fellowship in 1998. She was also a visiting professor at the University of Mannheim during 1998–1999 and 2001–2002. Presently, she is a professor at the

Physics Faculty, University of Bucharest, where she teaches integrated optoelectronic devices and the interaction of radiation with matter.

She authored about 120 papers in the area of fiber optics, optical beam characterization, quantum devices, and quantum optics. She is coauthor of the books *Advanced Optoelecronic Devices* Springer, 1999, *Optical Characterization of Solids* Springer, 2002, and *Quantum-Classical Analogies* Springer, 2004. Her current research interests include mesoscopic devices, beam characterization in linear and nonlinear media, optical micromechanical devices, quantum and classical coherence, and interference, and the foundations of physics. She is an acting reviewer at several international journals of applied physics, optics, and photonics. She was awarded the Romanian Academy Prize "Gheorghe Cartianu" in 1999. She is editor of the book series *Frontiers of Physics*, which is currently published by Springer.

Index

Recent Related Artech House Titles

Organic and Inorganic Nanostructures, Alexi Nabok

Optics of Quantum Dots and Wires, Garnett W. Bryant and Glenn S. Soloman

Semiconductor Nanostructures for Optoelectronic Applications, Todd Steiner, editor

Advances in Silicon Carbide Processing and Applications, Stephen E. Saddow and Anant Agarwal, editors

Microfluidics for Biotechnology, Jean Berthier and Pascal Silberzan

Fundamentals and Applications of Microfluidics, Nam-Trung Nguyen and Steven T. Wereley

Mathematical Handbook for Electrical Engineers, Sergey A. Leonov and Alexander I. Leonov

Nanoelectronics: Principles and Devices, Mircea Dragoman and Daniela Dragoman

For further information on these and other Artech House titles, including previously considered out-of-print books now available through our In-Print-Forever® (IPF®) program, contact:

Artech House
685 Canton Street
Norwood, MA 02062
Phone: 781-769-9750
Fax: 781-769-6334
e-mail: artech@artechhouse.com

Artech House
46 Gillingham Street
London SW1V 1AH UK
Phone: +44 (0)20 7596-8750
Fax: +44 (0)20 7630-0166
e-mail: artech-uk@artechhouse.com

Find us on the World Wide Web at: www.artechhouse.com